The McGraw-Hill Engineering Reference Guide Series

This series makes available to professionals and students a wide variety of engineering information and data available in McGraw-Hill's library of highly acclaimed books and publications. The books in the series are drawn directly from this vast resource of titles. Each one is either a condensation of a single title or a collection of sections culled from several titles. The Project Editors responsible for the books in the series are highly respected professionals in the engineering areas covered. Each Editor selected only the most relevant and current information available in the McGraw-Hill library, adding further details and commentary where necessary.

Hicks · CIVIL ENGINEERING CALCULATIONS REFERENCE GUIDE

Hicks · MACHINE DESIGN CALCULATIONS REFERENCE GUIDE

Hicks · PLUMBING DESIGN AND INSTALLATION REFERENCE GUIDE

Hicks · POWER GENERATION CALCULATIONS REFERENCE GUIDE

Hicks · POWER PLANT EVALUATION AND DESIGN REFERENCE GUIDE

Johnson & Jasik · ANTENNA APPLICATIONS REFERENCE GUIDE

Markus and Weston · CLASSIC CIRCUITS REFERENCE GUIDE

Merritt · CIVIL ENGINEERING REFERENCE GUIDE

oodson · HUMAN FACTORS REFERENCE GUIDE FOR ELECTRONICS AND COMPUTER PROFESSIONALS

dson · HUMAN FACTORS REFERENCE GUIDE FOR PROCESS PLANTS

POWER GENERATION CALCULATIONS REFERENCE GUIDE

POWER GENERATION CALCULATIONS REFERENCE GUIDE

TYLER G. HICKS., P.E., EDITOR
International Engineering Associates

Contributors

EDGAR J. KATES, P.E.
Consulting Engineer

B.G.A. SKROTZKI, P.E.
Power Magazine

RAYMOND J. ROARK
Professor, University of Wisconsin

S.W. SPIELVOGEL
Piping Engineering Consultant

RUFUS OLDENBURGER
Professor, Purdue University

LYMAN F. SCHEEL
Consulting Engineer

GERALD M. EISENBERG
Project Engineering Administrator
American Society of Mechanical Engineers

STEPHEN M. EBER, P.E.
Ebasco Services, Inc.

JEROME F. MUELLER, P.E.
Mueller Engineering Corp.

SAMUEL C. LIND
Consultant
United States Atomic Energy Commission

McGRAW-HILL BOOK COMPANY

New York St. Louis San Francisco Auckland Bogotá Hamburg
Johannesburg London Madrid Mexico Milan Montreal New Delhi
Panama Paris São Paulo Singapore Sydney Tokyo Toronto

Library of Congress Cataloging-in-Publication Data

Power generation calculations reference guide.

(The McGraw-Hill engineering reference guide series)
"Condensation of the Standard handbook of engineering calculations, 2nd edition"—Pref.
Includes index.
1. Power (Mechanics)—Handbooks, manuals, etc.
I. Hicks, Tyler Gregory. II. Kates, Edgar J. (Edgar Jesse). III. Standard handbook of engineering calculations. IV. Series.
TJ163.9.P665 1987 621.042 86-33797
ISBN 0-07-028800-3

POWER GENERATION CALCULATIONS REFERENCE GUIDE

1234567890 DOC/DOC 893210987

ISBN 0-07-028800-3

The material in this volume has been published previously in *Standard Handbook of Engineering Calculations,* Second Edition, edited by Tyler G. Hicks.

Printed and bound by R.R. Donnelley and Sons Company

CONTENTS

PREFACE

This reference guide is a concise coverage of the key areas of power generation, fluid flow, heat transfer, refrigeration, and materials handling. The guide is a condensation of the *Standard Handbook of Engineering Calculations*, 2nd Edition.

Fully metricated, the guide contains hundreds of step-by-step calculation procedures for making quick and accurate analyses of many common and uncommon design situations. Specific topics covered include: combustion, steam charts and tables, steam processes, various steam cycles (Rankine, bleed-steam, reheat, reheat-regenerative), feedwater heating, steam boiler analysis and performance, internal-combustion engines, air and gas compressors and vacuum systems, pumps and pumping systems, materials handling, piping and fluid flow, heat transfer and heat exchangers, refrigeration, energy conservation, and nuclear power.

Most procedures given have related calculations—that is, other items which can be computed using the same general technique. This expands the coverage of the guide enormously. The result is that engineers, scientists, and designers have a powerful guide for quick and accurate solutions of hundreds of calculation procedures, along with worked-out real-life applications.

Both USCS and SI units are used throughout the guide. This permits easier and faster use of the guide in both the United States and overseas, where SI is widely used. Thus, a designer in the United States can easily do power-generation work overseas. And an overseas designer can easily do power-generation work for applications in the United States.

The guide is thoroughly up-to-date in its coverage, with calculations devoted to the design of wind energy systems, fuel savings from the use of high-temperature hot water in heating systems, heat-recovery fuel savings, cost of heat loss from uninsulated pipes, heat-rate improvement from using turbine-driven boiler fans, cogeneration cost analyses, boiler conversion from oil or gas to coal, return on investment for energy-saving projects, fuel savings using direct digital control, small hydro power system analysis, sizing flash tanks to save energy, plus other related topics. Such coverage helps users of the guide cope with new design jobs they meet in their daily work.

Users who will find the guide most helpful include electrical and mechanical engineers, nuclear engineers, scientists, designers, drafters, estimators and schedulers in any field associated with power generation. Further, the algorithms given are ideal for use in micro, mini, and mainframe computers to effect a greater time saving for users.

The editor thanks the many contributors whose work is cited in the guide. And if readers find any errors or deficiencies in the guide, the editor asks that they be pointed out to him. He will be grateful to every reader detecting such flaws who writes him in care of the publisher.

TYLER G. HICKS, P.E.

POWER GENERATION CALCULATIONS REFERENCE GUIDE

Combustion

REFERENCES: Chigier—*Energy, Combustion and Environment*, McGraw-Hill; Lewis and Von Elbe—*Combustion, Flames and Explosion of Gases*, Pergamon Press; Zung—*Evaporation-Combustion of Fuel*, American Chemical Society; Johnson and Auth—*Fuels and Combustion Handbook*, McGraw-Hill; Babcock & Wilcox Company—*Steam: Its Generation and Use*; Combustion Engineering Corporation—*Combustion Engineering*; Gaffert—*Steam Power Stations*, McGraw-Hill; Skrotzki and Vopat—*Applied Energy Conversion*, McGraw-Hill; Popovich-Hering—*Fuels and Lubricants*, Wiley; ASME—*Power Test Code for Steam Boilers*; Moore—*Coal*, Wiley; Moore—*Liquid Fuels*, The Technical Press, Ltd., London; American Gas Association—*Combustion*; Dunstan—*Science of Petroleum*, Oxford, London; Trinks—*Industrial Furnaces*, Wiley; Perry—*Chemical Engineers Handbook*, McGraw-Hill.

COMBUSTION OF COAL FUEL IN A FURNACE

A coal has the following ultimate analysis (or percent by weight): C = 0.8339; H_2 = 0.0456; O_2 = 0.0505; N_2 = 0.0103; S = 0.0064; ash = 0.0533; total = 1.000 lb (0.45 kg). This coal is burned in a steam-boiler furnace. Determine the weight of air required for theoretically perfect combustion, the weight of gas formed per pound (kilogram) of coal burned, and the volume of flue gas, at the boiler exit temperature of 600°F (316°C) per pound (kilogram) of coal burned; air required with 20 percent excess air, and the volume of gas formed with this excess; the CO_2 percentage in the flue gas on a dry and wet basis.

Calculation Procedure:

1. *Compute the weight of oxygen required per pound of coal*

To find the weight of oxygen required for theoretically perfect combustion of coal, set up the following tabulation, based on the ultimate analysis of the coal:

Element	×	Molecular-weight ratio	=	lb (kg) O_2 required
C; 0.8339	×	32/12	=	2.2237 (1.001)
H_2; 0.0456	×	16/2	=	0.3648 (0.164)
O_2; 0.0505; decreases external O_2 required			=	
N_2; 0.0103 is inert in combustion and is ignored				
S; 0.0064	×	32/32	=	0.0064 (0.003)
Ash 0.0533 is inert in combustion and is ignored				
Total 1.0000				
lb (kg) external O_2 per lb (kg) fuel			=	2.5444 (1.168)

Note that of the total oxygen needed for combustion, 0.0505 lb (0.023 kg), is furnished by the fuel itself and is assumed to reduce the total external oxygen required by the amount of oxygen present in the fuel. The molecular-weight ratio is obtained from the equation for the chemical reaction of the element with oxygen in combustion. Thus, for carbon $C + O_2 \rightarrow CO_2$, or $12 + 32 = 44$, where 12 and 32 are the molecular weights of C and O_2, respectively.

2. *Compute the weight of air required for perfect combustion*

Air at sea level is a mechanical mixture of various gases, principally 23.2 percent oxygen and 76.8 percent nitrogen by weight. The nitrogen associated with the 2.5444 lb (1.145 kg) of oxygen required per pound (kilogram) of coal burned in this furnace is the product of the ratio of the nitrogen and oxygen weights in the air and 2.5444, or $(2.5444)(0.768/0.232) = 8.4219$ lb (3.790 kg). Then the weight of air required for perfect combustion of 1 lb (2.2 kg) of coal = sum of nitrogen and oxygen required = $8.4219 + 2.5444 = 10.9663$ lb (4.935 kg) of air per pound (kilogram) of coal burned.

3. *Compute the weight of the products of combustion*

Find the products of combustion by addition:

Fuel constituents	+	Oxygen	→	Products of combustion		lb	kg
C; 0.8339	+	2.2237	→	CO_2	=	2.0576	0.926
H; 0.0456	+	0.3648	→	H_2O	=	0.4104	0.185
O_2; 0.0505; this is *not* a product of combustion							
N_2; 0.0103; inert but passes through furnace					=	0.0103	0.005
S; 0.0064	+	0.0064	→	SO_2	=	0.0128	0.006
Outside nitrogen from step 2				$= N_2$	=	8.4219	3.790
lb (kg) of flue gas per lb (kg) of coal burned					=	11.9130	4.912

4. *Convert the flue-gas weight to volume*

Use Avogadro's law, which states that under the same conditions of pressure and temperature, 1 mol (the molecular weight of a gas expressed in lb) of any gas will occupy the same volume.

At 14.7 lb/in^2 (abs) (101.3 kPa) and 32°F (0°C), 1 mol of any gas occupies 359 ft^3 (10.2 m^3). The volume per pound of any gas at these conditions can be found by dividing 359 by the molecular weight of the gas and correcting for the gas temperature by multiplying the volume by the ratio of the absolute flue-gas temperature and the atmospheric temperature. To change the weight analysis (step 3) of the products of combustion to volumetric analysis, set up the calculation thus:

Products	Weight, lb	Weight, kg	Molecular weight	Temperature correction		Volume at 600°F, ft^3	Volume at 316°C, m^3
CO_2	3.0576	1.3898	44	(359/44)(3.0576)(2.15)	=	53.8	1.523
H_2O	0.4104	0.1865	18	(359/18)(0.4104)(2.15)	=	17.6	0.498
Total N_2	8.4322	3.8328	28	(359/28)(8.4322)(2.15)	=	233.0	6.596
SO_2	0.0128	0.0058	64	(359/64)(0.0128)(2.15)	=	0.17	0.005
ft^3 (m^3) of flue gas per lb (kg) of coal burned					=	304.57	8.622

In this calculation, the temperature correction factor 2.15 = absolute flue-gas temperature, °R/absolute atmospheric temperature, °R = (600 + 460)/(32 + 460). The total weight of N_2 in the flue gas is the sum of the N_2 in the combustion air and the fuel, or 8.4219 + 0.0103 = 8.4322 lb (3.794 kg). This value is used in computing the flue-gas volume.

5. *Compute the CO_2 content of the flue gas*

The volume of CO_2 in the products of combustion at 600°F (316°C) is 53.8 ft^3 (1.523 m^3), as computed in step 4; and the total volume of the combustion products is 304.57 ft^3 (8.622 m^3). Therefore, the percent CO_2 on a wet basis (i.e., including the moisture in the combustion products) = ft^3 CO_2/total ft^3 = 53.8/304.57 = 0.1765, or 17.65 percent.

The percent CO_2 on a dry, or Orsat, basis is found in the same manner except that the weight of H_2O in the products of combustion, 17.6 lb (7.83 kg) from step 4, is subtracted from the total gas weight. Or, percent CO_2, dry, or Orsat basis = (53.8)/(304.57 − 17.6) = 0.1875, or 18.75 percent.

6. *Compute the air required with the stated excess flow*

With 20 percent excess air, the air flow required = (0.20 + 1.00)(air flow with no excess) = 1.20 (10.9663) = 13.1596 lb (5.922 kg) of air per pound (kilogram) of coal burned. The air flow with no excess is obtained from step 2.

7. *Compute the weight of the products of combustion*

The excess air passes through the furnace without taking part in the combustion and increases the weight of the products of combustion per pound (kilogram) of coal burned. Therefore, the weight of the products of combustion is the sum of the weight of the combustion products without the excess air and the product of (percent excess air)(air for perfect combustion, lb); or, given the weights from steps 3 and 2, respectively, = 11.9130 + (0.20)(10.9663) = 14.1063 lb (6.348 kg) of gas per pound (kilogram) of coal burned with 20 percent excess air.

8. *Compute the volume of the combustion products and the percent CO_2*

The volume of the excess air in the products of combustion is obtained by converting from the weight analysis to the volumetric analysis and correcting for temperature as in step 4, using the air weight from step 2 for perfect combustion and the excess-air percentage, or (10.9663)(0.20)(359/28.95)(2.15) = 58.5 ft^3 (1.656 m^3). In this calculation the value 28.95 is the molecular weight of air. The total volume of the products of combustion is the sum of the column for perfect combustion, step 4, and the excess-air volume, above, or 304.57 + 58.5 = 363.07 ft^3 (10,279 m^3).

By using the procedure in step 5, the percent CO_2, wet basis = 53.8/363.07 = 14.8 percent. The percent CO_2, dry basis = 53.8/(363.07 − 17.6) = 15.6 percent.

Related Calculations: Use the method given here when making combustion calculations for any type of coal—bituminous, semibituminous, lignite, anthracite, cannel, or coking—from any coal field in the world used in any type of furnace—boiler, heater, process, or waste-heat. When the air used for combustion contains moisture, as is usually true, this moisture is added to the combustion-formed moisture appearing in the products of combustion. Thus, for 80°F (26.7°C) air of 60 percent relative humidity, the moisture content is 0.013 lb/lb (0.006 kg/kg) of dry air. This amount appears in the products of combustion for each pound of air used and is a commonly assumed standard in combustion calculations.

COMBUSTION OF FUEL OIL IN A FURNACE

A fuel oil has the following ultimate analysis: C = 0.8543; H_2 = 0.1131; O_2 = 0.0270; N_2 = 0.0022; S = 0.0034; total = 1.0000. This fuel oil is burned in a steam-boiler furnace. Determine the weight of air required for theoretically perfect combustion, the weight of gas formed per pound (kilogram) of oil burned, and the volume of flue gas, at the boiler exit temperature of 600°F (316°C), per pound (kilogram) of oil burned; the air required with 20 percent excess air, and the volume of gas formed with this excess; the CO_2 percentage in the flue gas on a dry and wet basis.

Calculation Procedure:

1. *Compute the weight of oxygen required per pound (kilogram) of oil*

The same general steps as given in the previous calculation procedure will be followed. Consult that procedure for a complete explanation of each step.

Using the molecular weight of each element, we find

Element	×	Molecular-weight ratio	=	lb (kg) O_2 required
C; 0.8543	×	32/12	=	2.2781 (1.025)
H_2; 0.1131	×	16/2	=	0.9048 (0.407)
O_2; 0.0270; decreases external O_2 required			=	−0.0270 (−0.012)
N_2; 0.0022 is inert in combustion and is ignored				
S; 0.0034	×	32/32	=	0.0034 (0.002)
Total 1.0000				
lb (kg) external O_2 per lb (kg) fuel			=	3.1593 (1.422)

2. *Compute the weight of air required for perfect combustion*

The weight of nitrogen associated with the required oxygen = (3.1593)(0.768/0.232) = 10.458 lb (4.706 kg). The weight of air required = 10.4583 + 3.1593 = 13.6176 lb/lb (6.128 kg/kg) of oil burned.

3. *Compute the weight of the products of combustion*

As before,

Fuel constituents	+	Oxygen	=	Products of combustion
C; 0.8543 + 2.2781	=	3.1324	=	CO_2
H_2; 0.1131 + 0.9148	=	1.0179	=	H_2O
O_2; 0.270; *not* a product of combustion				
N_2; 0.0022; inert but passes through furnace	=	0.0022	=	N_2
S; 0.0034 + 0.0034	=	0.0068	=	SO_2
Outside N_2 from Step 2	=	10.458	=	N_2
lb (kg) of flue gas per lb (kg) of oil burned	=	14.6173 (6.578)		

4. *Convert the flue-gas weight to volume*

As before,

Products	Weight, lb	Weight, kg	Molecular weight	Temperature correction		Volume at 600°F, ft³	Volume at 316°C, m³
CO_2	3.1324	1.4238	44	(359/44)(3.1324)(2.15)	=	55.0	1.557
H_2O	1.0179	0.4626	18	(359/18)(1.0179)(2.15)	=	43.5	1.231
N_2 (total)	10.460	4.7545	28	(359/28)(10.460)(2.15)	=	288.5	8.167
SO_2	0.0068	0.0031	64	(359/64)(0.0068)(2.15)	=	0.82	0.023
ft³ (m³) of flue gas per lb (kg) of oil burned					=	387.82	10.978

In this calculation, the temperature correction factor 2.15 = absolute flue-gas temperature, °R/absolute atmospheric temperature, °R = (600 + 460)/(32 + 460). The total weight of N_2 in the flue gas is the sum of the N_2 in the combustion air and the fuel, or 10.4580 + 0.0022 = 10.4602 lb (4.707 kg).

5. *Compute the CO_2 content of the flue gas*

CO_2, wet basis, = 55.0/387.82 = 0.142, or 14.2 percent. CO_2, dry basis, = 55.0/(387.2 − 43.5) = 0.160, or 16.0 percent.

6. *Compute the air required with stated excess flow*

The pounds (kilograms) of air per pound (kilogram) of oil with 20 percent excess air = (1.20)(13.6176) = 16.3411 lb (7.353 kg) of air per pound (kilogram) of oil burned.

7. *Compute the weight of the products of combustion*

The weight of the products of combustion = product weight for perfect combustion, lb + (percent excess air)(air for perfect combustion, lb) = 14.6173 + (0.20)(13.6176) = 17.3408 lb (7.803 kilogram) of flue gas per pound (kilogram) of oil burned with 20 percent excess air.

8. *Compute the volume of the combustion products and the percent CO_2*

The volume of excess air in the products of combustion is found by converting from the weight to the volumetric analysis and correcting for temperature as in step 4, using the air weight from step 2 for perfect combustion and the excess-air percentage, or (13.6176)(0.20)(359/28.95)(2.15) = 72.7 ft³ (2.058 m³). Add this to the volume of the products of combustion found in step 4, or 387.82 + 72.70 = 460.52 ft³ (13.037 m³).

By using the procedure in step 5, the percent CO_2, wet basis = 55.0/460.52 = 0.1192, or 11.92 percent. The percent CO_2, dry basis = 55.0/(460.52 − 43.5) = 0.1318, or 13.18 percent.

Related Calculations: Use the method given here when making combustion calculations for any type of fuel oil—paraffin-base, asphalt-base, Bunker C, no. 2, 3, 4, or 5—from any source, domestic or foreign, in any type of furnace—boiler, heater, process, or waste-heat. When the air used for combustion contains moisture, as is usually true, this moisture is added to the combustion-formed moisture appearing in the products of combustion. Thus, for 80°F (26.7°C) air of 60 percent relative humidity, the moisture content is 0.013 lb/lb (0.006 kg/kg) of dry air. This amount appears in the products of combustion for each pound (kilogram) of air used and is a commonly assumed standard in combustion calculations.

COMBUSTION OF NATURAL GAS IN A FURNACE

A natural gas has the following volumetric analysis at 60°F (15.5°C): CO_2 = 0.004; CH_4 = 0.921; C_2H_6 = 0.041; N_2 = 0.034; total = 1.000. This natural gas is burned in a steam-boiler furnace. Determine the weight of air required for theoretically perfect combustion, the weight of gas formed per pound of natural gas burned, and the volume of the flue gas, at the boiler exit temperature of 650°F (343°C), per pound (kilogram) of natural gas burned; air required with 20 percent excess air, and the volume of gas formed with this excess; CO_2 percentage in the flue gas on a dry and wet basis.

Calculation Procedure:

1. ***Compute the weight of oxygen required per pound of gas***

The same general steps as given in the previous calculation procedures will be followed, except that they will be altered to make allowances for the differences between natural gas and coal.

The composition of the gas is given on a volumetric basis, which is the usual way of expressing a fuel-gas analysis. To use the volumetric-analysis data in combustion calculations, they must be converted to a weight basis. This is done by dividing the weight of each component by the total weight of the gas. A volume of 1 ft^3 (1 m^3) of the gas is used for this computation. Find the weight of each component and the total weight of 1 ft^3 (1 m^3) as follows, using the properties of the combustion elements and compounds given in Table 1:

Component	Percent by volume	Density		Component weight = column 1 × column 2	
		lb/ft^3	kg/m^3	lb/ft^3	kg/m^3
CO_2	0.004	0.1161	1.859	0.0004644	0.007
CH_4	0.921	0.0423	0.677	0.0389583	0.624
C_2H_6	0.041	0.0792	1.268	0.0032472	0.052
N_2	0.034	0.0739	0.094	0.0025026	0.040
Total	1.000			0.0451725	0.723

$$\text{Percent } CO_2 = 0.0004644/0.0451725 = 0.01026, \text{ or } 1.03 \text{ percent}$$
$$\text{Percent } CH_4 \text{ by weight} = 0.0389583/0.0451725 = 0.8625 \text{ or } 86.25 \text{ percent}$$
$$\text{Percent } C_2H_6 \text{ by weight} = 0.0032472/0.0451725 = 0.0718, \text{ or } 7.18 \text{ percent}$$
$$\text{Percent } N_2 \text{ by weight} = 0.0025026/0.0451725 = 0.0554, \text{ or } 5.54 \text{ percent}$$

The sum of the weight percentages = 1.03 + 86.25 + 7.18 + 5.54 = 100.00. This sum checks the accuracy of the weight calculation, because the sum of the weights of the component parts should equal 100 percent.

Next, find the oxygen required for combustion. Since both the CO_2 and N_2 are inert, they do not take part in the combustion; they pass through the furnace unchanged. Using the molecular weights of the remaining components in the gas and the weight percentages, we have

Compound	×	Molecular-weight ratio	=	lb (kg) O_2 required
CH_4; 0.8625	×	64/16	=	3.4500 (1.553)
C_2H_6; 0.0718	×	112/30	=	0.2920 (0.131)
lb (kg) external O_2 required per lb (kg) fuel			=	3.7420 (1.684)

In this calculation, the molecular-weight ratio is obtained from the equation for the combustion chemical reaction, or $CH_4 + 2O_2 = CO_2 + 2H_2O$, that is, 16 + 64 = 44 + 36, and $C_2H_6 + \frac{7}{2}O_2 = 2CO_2 + 3H_2O$, that is, 30 + 112 = 88 + 54. See Table 2 from these and other useful chemical reactions in combustion.

2. ***Compute the weight of air required for perfect combustion***

The weight of nitrogen associated with the required oxygen = (3.742)(0.768/0.232) = 12.39 lb (5.576 kg). The weight of air required = 12.39 + 3.742 = 16.132 lb/lb (7.259 kg/kg) of gas burned.

TABLE 1 Properties of Combustion Elements[*]

Element or compound	Formula	Molecular weight	At 14.7 lb/in^2 (abs) (101.3 kPa), 60°F (15.6°C)		Nature			Heat value, Btu (kJ)	
			Weight, lb/ft^3 (kg/m^3)	Volume, ft^3/lb (m^3/kg)	Gas or solid	Combustible	Per lb (kg)	Per ft^3 (m^3) at 14.7 lb/in^2 (abs) (101.3 kPa), 60°F (15.6°C)	Per mole
Carbon	C	12	. . .	. . .	S	Yes	14,540 (33,820)	. . .	174,500
Hydrogen	H_2	2.02[†]	0.0053 (0.0849)	188 (11.74)	G	Yes	61,000 (141,886)	325 (12,109)	123,100
Sulfur	S	32	. . .	. . .	S	Yes	4,050 (9,420)	. . .	129,600
Carbon monoxide	CO	28	0.0739 (1.183)	13.54 (0.85)	G	Yes	4,380 (10,187)	323 (12,035)	122,400
Methane	CH_4	16	0.0423 (0.677)	23.69 (1.48)	G	Yes	24,000 (55,824)	1,012 (37,706)	384,000
Acetylene	C_2H_2	26	0.0686 (1.098)	14.58 (0.91)	G	Yes	21,500 (50,009)	1,483 (55,255)	562,000
Ethylene	C_2H_4	28	0.0739 (1.183)	13.54 (0.85)	G	Yes	22,200 (51,637)	1,641 (61,141)	622,400
Ethane	C_2H_6	30	0.0792 (1.268)	12.63 (0.79)	G	Yes	22,300 (51,870)	1,762 (65,650)	668,300
Oxygen	O_2	32	0.0844 (1.351)	11.84 (0.74)	G				
Nitrogen	N_2	28	0.0739 (1.183)	13.52 (0.84)	G				
Air[‡]	. . .	29	0.0765 (1.225)	13.07 (0.82)	G				
Carbon dioxide	CO_2	44	0.1161 (1.859)	8.61 (0.54)	G				
Water	H_2O	18	0.0475 (0.760)	21.06 (1.31)	G				

[*] P. W. Swain and L. N. Rowley, "Library of Practical Power Engineering" (collection of articles published in *Power*).
[†] For most practical purposes, the value of 2 is sufficient.
[‡] The molecular weight of 29 is merely the weighted average of the molecular weight of the constituents.

TABLE 2 Chemical Reactions

Combustible substance	Reaction	Mols	lb (kg)*
Carbon to carbon monoxide	$C + \frac{1}{2}O_2 = CO$	$1 + \frac{1}{2} = 1$	$12 + 16 = 28$
Carbon to carbon dioxide	$C + O_2 = CO_2$	$1 + 1 = 1$	$12 + 16 = 28$
Carbon monoxide to carbon dioxide	$CO + \frac{1}{2}O_2 = CO_2$	$1 + \frac{1}{2} = 1$	$28 + 16 = 44$
Hydrogen	$H_2 + \frac{1}{2}O_2 = H_2O$	$1 + \frac{1}{2} = 1$	$2 + 16 = 18$
Sulfur to sulfur dioxide	$S + O_2 = SO_2$	$1 + 1 = 1$	$32 + 32 = 64$
Sulfur to sulfur trioxide	$S + \frac{3}{2}O_2 = SO_3$	$1 + \frac{3}{2} = 1$	$32 + 48 = 80$
Methane	$CH_4 + 2O_2 = CO_2 + 2H_2O$	$1 + 2 = 1 + 2$	$16 + 64 = 44 + 36$
Ethane	$C_2H_6 + \frac{7}{2}O_2 = 2CO_2 + 3H_2O$	$1 + \frac{7}{2} = 2 + 3$	$30 + 112 = 88 + 54$
Propane	$C_3H_8 + 5O_2 = 3CO_2 + 4H_2O$	$1 + 5 = 3 + 4$	$44 + 160 = 132 + 72$
Butane	$C_4H_{10} + \frac{13}{2}O_2 = 4CO_2 + 5H_2O$	$1 + \frac{13}{2} = 4 + 5$	$58 + 208 = 176 + 90$
Acetylene	$C_2H_2 + \frac{5}{2}O_2 = 2CO_2 + H_2O$	$1 + \frac{5}{2} = 2 + 2$	$26 + 80 = 88 + 18$
Ethylene	$C_2H_4 + 3O_2 = 2CO_2 + 2H_2O$	$1 + 3 = 2 + 2$	$28 + 96 = 88 + 36$

*Substitute the molecular weights in the reaction equation to secure lb (kg). The lb (kg) on each side of the equation must balance.

3. *Compute the weight of the products of combustion*

Fuel constituents	+	Oxygen	=	Products of combustion, lb	Products of combustion, kg
CO_2; 0.0103; inert but passes through the furnace			=	0.010300	0.005
CH_4; 0.8625	+	3.45	=	4.312500	1.941
C_2H_6; 0.003247	+	0.2920	=	0.032447	0.015
N_2; 0.0554; inert but passes through the furnace			=	0.055400	0.025
Outside N_2 from step 2			=	12.390000	5.576
lb (kg) of flue gas per lb (kg) of natural gas burned			=	16.800347	7.562

4. *Convert the flue-gas weight to volume*

The products of complete combustion of any fuel that does not contain sulfur are CO_2, H_2O, and N_2. Using the combustion equation in step 1, compute the products of combustion thus: $CH_4 + 2O_2 = CO_2 + H_2O$; $16 + 64 = 44 + 36$; or the CH_4 burns to CO_2 in the ratio of 1 part CH_4 to 44/16 parts CO_2. Since, from step 1, there is 0.03896 lb CH_4 per ft^3 (0.624 kg/m^3) of natural gas, this forms $(0.03896)(44/16) = 0.1069$ lb (0.048 kg) of CO_2. Likewise, for C_2H_6, $(0.003247)(88/30) = 0.00952$ lb (0.004 kg). The total CO_2 in the combustion products $= 0.00464 + 0.1069 + 0.00952 = 0.11688$ lb (0.053 kg), where the first quantity is the CO_2 in the fuel.

Using a similar procedure for the H_2O formed in the products of combustion by CH_4, we find $(0.03896)(36/16) = 0.0875$ lb (0.039 kg). For C_2H_6, $(0.003247)(54/30) = 0.005816$ lb (0.003 kg). The total H_2O in the combustion products $= 0.0875 + 0.005816 = 0.093316$ lb (0.042 kg).

Step 2 shows that 12.39 lb (5.58 kg) of N_2 is required per lb (kg) of fuel. Since 1 ft^3 (0.028 m^3) of the fuel weighs 0.04517 lb (0.02 kg), the volume of gas which weighs 1 lb (2.2 kg) is $1/0.04517 = 22.1$ ft^3 (0.626 m^3). Therefore, the weight of N_2 per ft^3 of fuel burned $= 12.39/22.1 = 0.560$ lb (0.252 kg). This, plus the weight of N_2 in the fuel, step 1, is $0.560 + 0.0025 = 0.5625$ lb (0.253 kg) of N_2 in the products of combustion.

Next, find the total weight of the products of combustion by taking the sum of the CO_2, H_2O, and N_2 weights, or $0.11688 + 0.09332 + 0.5625 = 0.7727$ lb (0.35 kg). Now convert each weight to ft^3 at 650°F (343°C), the temperature of the combustion products, or:

Products	Weight, lb	Weight, kg	Molecular weight	Temperature correction		Volume at 650°F, ft^3	Volume at 343°C, m^3
CO_2	0.11688	0.05302	44	(379/44)(0.11688)(2.255)	=	2.265	0.0641
H_2O	0.09332	0.04233	18	(379/18)(0.09332)(2.255)	=	4.425	0.1252
N_2 (total)	0.5625	0.25515	28	(379/28)(0.5625)(2.255)	=	17.190	0.4866
ft^3 (m^3) of flue gas per ft^3 (m^3) of natural-gas fuel					=	23.880	0.6759

In this calculation, the value of 379 is used in the molecular-weight ratio because at 60°F (15.6°C) and 14.7 lb/in^2 (abs) (101.3 kPa), the volume of 1 lb (0.45 kg) of any gas = 379/gas molecular weight. The fuel gas used is initially at 60°F (15.6°C) and 14.7 lb/in^2 (abs) (101.3 kPa). The ratio $2.255 = (650 + 460)/(32 + 460)$.

5. *Compute the CO_2 content of the flue gas*

CO_2, wet basis $= 2.265/23.88 = 0.947$, or 9.47 percent. CO_2, dry basis $= 2.265/(23.88 - 4.425) = 0.1164$, or 11.64 percent.

6. *Compute the air required with the stated excess flow*

With 20 percent excess air, (1.20)(16.132) = 19.3584 lb of air per lb (8.71 kg/kg) of natural gas, or 19.3584/22.1 = 0.875 lb of air per ft^3 (13.9 kg/m^3) of natural gas. See step 4 for an explanation of the value 22.1.

7. *Compute the weight of the products of combustion*

Weight of the products of combustion = product weight for perfect combustion, lb + (percent excess air) (air for perfect combustion, lb) = 16.80 + (0.20)(16.132) = 20.03 lb (9.01 kg).

8. *Compute the volume of the combustion products and the percent CO_2*

The volume of excess air in the products of combustion is found by converting from the weight to the volumetric analysis and correcting for temperature as in step 4, using the air weight from step 2 for perfect combustion and the excess-air percentage, or (16.132/22.1)(0.20)(379/28.95)(2.255) = 4.31 ft^3 (0.122 m^3). Add this to the volume of the products of combustion found in step 4, or 23.88 + 4.31 = 28.19 ft^3 (0.798 m^3).

By the procedure in step 5, the percent CO_2, wet basis = 2.265/28.19 = 0.0804, or 8.04 percent. The percent CO_2, dry basis = 2.265/(28.19 − 4.425) = 0.0953, or 9.53 percent.

Related Calculations: Use the method given here when making combustion calculations for any type of gas used as a fuel—natural gas, blast-furnace gas, coke-oven gas, producer gas, water gas, sewer gas—from any source, domestic or foreign, in any type of furnace—boiler, heater, process, or waste-heat. When the air used for combustion contains moisture, as is usually true, this moisture is added to the combustion-formed moisture appearing in the products of combustion. Thus, for 80°F (26.7°C) air of 60 percent relative humidity, the moisture content is 0.013 lb/lb (0.006 kg/kg) of dry air. This amount appears in the products of combustion for each pound of air used and is a commonly assumed standard in combustion calculations.

COMBUSTION OF WOOD FUEL IN A FURNACE

The weight analysis of a yellow-pine wood fuel is: C = 0.490; H_2 = 0.074; O_2 = 0.406; N_2 = 0.030. Determine the weight of oxygen and air required with perfect combustion and with 20 percent excess air. Find the weight and volume of the products of combustion under the same conditions, and the wet and dry CO_2. The flue-gas temperature is 600°F (316°C). The air supplied for combustion has a moisture content of 0.013 lb/lb (0.006 kg/kg) of dry air.

Calculation Procedure:

1. *Compute the weight of oxygen required per pound of wood*

The same general steps as given in earlier calculation procedures will be followed; consult them for a complete explanation of each step. Using the molecular weight of each element, we have

Element	×	Molecular-weight ratio	=	lb (kg) O_2 required
C; 0.490	×	32/12	=	1.307 (0.588)
H_2; 0.074	×	16/2	=	0.592 (0.266)
O_2; 0.406; decreases external O_2 required			=	−0.406 (−0.183)
N_2; 0.030 inert in combustion				
Total 1.000				
lb (kg) external O_2 per lb (kg) fuel			=	1.493 (0.671)

2. *Compute the weight of air required for complete combustion*

The weight of nitrogen associated with the required oxygen = (1.493)(0.768/0.232) = 4.95 lb (2.228 kg). The weight of air required = 4.95 + 1.493 = 6.443 lb/lb (2.899 kg/kg) of wood burned, if the air is dry. But the air contains 0.013 lb of moisture per lb (0.006 kg/kg) of air. Hence, the total weight of the air = 6.443 + (0.013)(6.443) = 6.527 lb (2.937 kg).

3. *Compute the weight of the products of combustion*

Use the following relation:

Fuel constituents	+	Oxygen	=	Products of combustion, lb (kg)
C; 0.490	+	1.307	=	1.797 (0.809) = CO_2
H_2; 0.074	+	0.592	=	0.666 (0.300) = H_2O
O_2; not a product of combustion				
N_2; inert but passes through the furnace			=	0.030 (0.014) = N_2
Outside N_2 from step 2			=	4.950 (2.228) = N_2
Outside moisture from step 2			=	0.237 (0.107)
lb (kg) of flue gas per lb (kg) of wood burned			=	7.680 (3.458)

4. *Convert the flue-gas weight to volume*

Use, as before, the following tabulation:

Products	Weight lb	Weight kg	Molecular weight	Temperature correction	Volume at 600°F, ft^3	Volume at 316°C, m^3
CO_2	1.797	0.809	44	(359/44)(1.797)(2.15) =	31.5	0.892
H_2O (fuel)	0.666	0.300	18	(359/18)(0.666)(2.15) =	28.6	0.810
N_2 (total)	4.980	2.241	28	(359/28)(4.980)(2.15) =	137.2	3.884
H_2O (outside air)	0.837	0.377	18	(359/18)(0.837)(2.15) =	35.9	10.16
Cu ft (m^3) of flue gas per lb (kg) of oil					233.2	6.602

In this calculation the temperature correction factor 2.15 = (absolute flue-gas temperature, °R)/(absolute atmospheric temperature, °R) = (600 + 460)/(32 + 460). The total weight of N_2 is the sum of the N_2 in the combustion air and the fuel.

5. *Compute the CO_2 content of the flue gas*

The CO_2, wet basis = 31.5/233.2 = 0.135, or 13.5 percent. The CO_2, dry basis = 31.5/(233.2 − 28.6 − 35.9) = 0.187, or 18.7 percent.

6. *Compute the air required with the stated excess flow*

With 20 percent excess air, (1.20)(6.527) = 7.832 lb (3.524 kg) of air per lb (kg) of wood burned.

7. *Compute the weight of the products of combustion*

The weight of the products of combustion = product weight for perfect combustion, lb + (percent excess air)(air for perfect combustion, lb) = 8.280 + (0.20)(6.527) = 9.585 lb (4.313 kg) of flue gas per lb (kg) of wood burned with 20 percent excess air.

8. *Compute the volume of the combustion products and the percent CO_2*

The volume of the excess air in the products of combustion is found by converting from the weight to the volumetric analysis and correcting for temperature as in step 4, using the air weight from step 2 for perfect combustion and the excess-air percentage, or (6.527)(0.20)(359/28.95)(2.15) = 34.8 ft^3 (0.985 m^3). Add this to the volume of the products of combustion found in step 4, or 233.2 + 34.8 = 268.0 ft^3 (7.587 m^3).

By using the procedure in step 5, the percent CO_2, wet basis = 31.5/268 = 0.1174, or 11.74 percent. The percent CO_2, dry basis = 31.5/(268 − 28.6 − 35.9 − 0.20 × 0.837) = 0.155, or 15.5 percent. In the dry-basis calculation, the factor (0.20)(0.837) is the outside moisture in the excess air.

Related Calculations: Use the method given here when making combustion calculations for any type of wood or woodlike fuel—spruce, cypress, maple, oak, sawdust, wood shavings, tanbark, bagesse, peat, charcoal, redwood, hemlock, fir, ash, birch, cottonwood, elm, hickory, walnut, chopped trimmings, hogged fuel, straw, corn, cottonseed hulls, city refuse—in any type of furnace—boiler, heating, process, or waste-heat. Most of these fuels contain a small amount of ash—usually less than 1 percent. This was ignored in this calculation procedure because it does not take part in the combustion.

MOLAL METHOD OF COMBUSTION ANALYSIS

A coal fuel has this ultimate analysis: C = 0.8339; H_2 = 0.0456; O_2 = 0.0505; N_2 = 0.0103; S = 0.0064; ash = 0.0533; total = 1.000. This coal is completely burned in a boiler furnace. Using the molal method, determine the weight of air required per lb (kg) of coal with complete combustion. How much air is needed with 25 percent excess air? What is the weight of the combustion products with 25 percent excess air? The combustion air contains 0.013 lb of moisture per lb (0.006 kg/kg) of air.

Calculation Procedure:

1. *Convert the ultimate analysis to moles*

A mole of any substance is an amount of the substance having a weight equal to the molecular weight of the substance. Thus, 1 mol of carbon is 12 lb (5.4 kg) of carbon, because the molecular weight of carbon is 12. To convert an ultimate analysis of a fuel to moles, assume that 100 lb (45 kg) of the fuel is being considered. Set up a tabulation thus:

Ultimate analysis, %	Weight		Molecular weight	Moles per 100-lb (45-kg) fuel
	lb	kg		
C = 0.8339	83.39	37.526	12	6.940
H_2 = 0.0456	4.56	2.052	2	2.280
O_2 = 0.0505	5.05	2.678	32	0.158
N_2 = 0.0103	1.03	0.464	28	0.037
S = 0.0064	0.64	0.288	32	
Ash = 0.0533	5.33	2.399	Inert	
Total	100.00	45.407	...	9.435

2. *Compute the mols of oxygen for complete combustion*

From Table 2, the burning of carbon to carbon dioxide requires 1 mol of carbon and 1 mol of oxygen, yielding 1 mol of CO_2. Using the molal equations in Table 2 for the other elements in the fuel, set up a tabulation thus, entering the product of columns 2 and 3 in column 4:

(1) Element	(2) Moles per 100-lb (45-kg) fuel	(3) Moles O_2 per 100-lb (45-kg) fuel	(4) Total moles O_2
C	6.940	1.00	6.940
H_2	2.280	0.5	1.140
O_2	0.158	Reduces O_2 required	−0.158
N_2	0.037	Inert in combustion	
S	0.020	1.00	0.020
Total moles of O_2 required	...	...	7.942

TABLE 3 Molal Conversion Factors

	Mol/mol of combustible for complete combustion; no excess air					
	For combustion			Combustion products		
Element or compound	O_2	N_2	Air	CO_2	H_2O	N_2
Carbon,* C	1.0	3.76	4.76	1.0	...	3.76
Hydrogen, H_2	0.5	0.188	2.38	...	1.0	1.88
Oxygen, O_2						
Nitrogen, N_2						
Carbon monoxide, CO	0.5	1.88	2.38	1.0	...	1.88
Carbon dioxide, CO_2						
Sulfur,* S	1.0	3.76	4.76	1.0	...	3.76
Methane, CH_4	2.0	7.53	0.53	1.0	2.0	7.53
Ethane, C_2H_6	3.5	13.18	16.68	2.0	3.0	13.18

*In molal calculations, carbon and sulfur are considered as gases.

3. *Compute the moles of air for complete combustion*

Set up a similar tabulation for air, thus:

(1) Element	(2) Moles per 100-lb (45-kg) fuel	(3) Moles air per 100-lb (45-kg) fuel	(4) Total moles air
C	6.940	4.76	33.050
H_2	2.280	2.38	5.430
O_2	0.158	Reduces O_2 required	−0.752
N_2	0.037	Inert in combustion	
S	0.020	4.76	0.095
Total moles of air required		...	37.823

In this tabulation, the factors in column 3 are constants used for computing the total moles of air required for complete combustion of each of the fuel elements listed. These factors are given in the Babcock & Wilcox Company—*Steam: Its Generation and Use* and similar treatises on fuels and their combustion. A tabulation of these factors is given in Table 3.

An alternative, and simpler, way of computing the moles of air required is to convert the required O_2 to the corresponding N_2 and find the sum of the O_2 and N_2. Or, $3.76O_2 = N_2$; $N_2 + O_2$ = moles of air required. The factor 3.76 converts the required O_2 to the corresponding N_2. These two relations were used to convert the 0.158 mol of O_2 in the above tabulation to moles of air.

Using the same relations and the moles of O_2 required from step 2, we get $(3.76)(7.942) = 29.861$ mol of N_2. Then $29.861 + 7.942 = 37.803$ mol of air, which agrees closely with the 37.823 mol computed in the tabulation. The difference of 0.02 mol is traceable to slide-rule readings.

4. *Compute the air required with the stated excess air*

With 25 percent excess, the air required for combustion $= (125/100)(37.823) = 47.24$ mol.

5. *Compute the mols of combustion products*

Using data from Table 3, and recalling that the products of combustion of a sulfur-containing fuel are CO_2, H_2O, and SO_2, and that N_2 and excess O_2 pass through the furnace, set up a tabulation thus:

(1) Moles per 100-lb (45-kg) fuel	(2) Mol/mol of combustible	(3) Moles of combustion products per 100-lb (45-kg) of fuel
CO_2; 6.940	1	6.940
H_2O; 2.280 + (47.24)(0.021 + 0.158)	. . .	3.430
SO_2; 0.020	1	0.202
N_2; (47.24)(0.79)	. . .	37.320
Excess O_2; (1.25)(7.942) − 7.942	. . .	1.986
	Total moles, wet combustion products	= 49.878
	Total moles, dry combustion products	= 49.878 −3.232 = 46.646

In this calculation, the total moles of CO_2 is obtained from step 2. The moles of H_2 in 100 lb (45 kg) of the fuel, 2.280, is assumed to form H_2O. In addition, the air from step 4, 47.24 mol, contains 0.013 lb of moisture per lb (0.006 kg/kg) of air. This moisture is converted to moles by dividing the molecular weight of air, 28.95, by the molecular weight of water, 18, and multiplying the result by the moisture content of the air, or (28.95/18)(0.013) = 0.0209, say 0.021 mol of water per mol of air. The product of this and the moles of air gives the total moles of moisture (water) in the combustion products per 100 lb (45 kg) of fuel fired. To this is added the moles of O_2, 0.158, per 100 lb (45 kg) of fuel, because this oxygen is assumed to unite with hydrogen in the air to form water. The nitrogen in the products of combustion is that portion of the moles of air required, 47.24 mol from step 4, times the proportion of N_2 in the air, or 0.79. The excess O_2 passes through the furnace and adds to the combustion products and is computed as shown in the tabulation. Subtracting the total moisture, 3430 mol, from the total (or wet) combustion products gives the moles of dry combustion products.

Related Calculations: Use this method for molal combustion calculations for all types of fuels—solid, liquid, and gaseous—burned in any type of furnace—boiler, heater, process, or waste-heat. Select the correct factors from Table 3.

STEAM BOILER HEAT BALANCE DETERMINATION

A steam generator having a maximum rated capacity of 60,000 lb/h (27,000 kg/h) is operating at 45,340 lb/h (20,403 kg/h), delivering 125-lb/in^2 (gage) 400°F (862-kPa, 204°C) steam with a feedwater temperature of 181°F (82.8°C). At this generating rate, the boiler requires 4370 lb/h (1967 kg/h) of West Virginia bituminous coal having a heating value of 13,850 Btu/lb (32,215 kJ/kg) on a dry basis. The ultimate fuel analysis is: C = 0.7757; H_2 = 0.0507; O_2 = 0.0519; N_2 = 0.0120; S = 0.0270; ash = 0.0827; total = 1.0000. The coal contains 1.61 percent moisture. The boiler-room intake air and the fuel temperature = 79°F (26.1°C) dry bulb, 71°F (21.7°C) wet bulb. The flue-gas temperature is 500°F (260°C), and the analysis of the flue gas shows these percentages: CO_2 = 12.8; CO = 0.4; O_2 = 6.1; N_2 = 80.7; total = 100.0. Measured ash and refuse = 9.42 percent of dry coal; combustible in ash and refuse = 32.3 percent. Compute a heat balance for this boiler based on these test data. The boiler has four water-cooled furnace walls.

Calculation Procedure:

1. *Determine the heat input to the boiler*

In a boiler heat balance the input is usually stated in Btu per pound of fuel as fired. Therefore, input = heating value of fuel = 13,850 Btu/lb (32,215 kJ/kg).

2. *Compute the output of the boiler*

The output of any boiler = Btu/lb (kJ/kg) of fuel + the losses. In this step the first portion of the output, Btu/lb (kJ/kg) of fuel will be computed. The losses will be computed in step 3.

First find W_s lb of steam produced per lb of fuel fired. Since 45,340 lb/h (20,403 kg/h) of steam is produced when 4370 lb/h (1967 kg/h) of fuel is fired, W_s = 45,340/4370 = 10.34 lb of steam per lb (4.65 kg/kg) of fuel.

Once W_s is known, the output h_1 Btu/lb of fuel can be found from $h_1 = W_s(h_s - h_w)$, where h_s = enthalpy of steam leaving the superheater, or boiler if a superheater is not used; h_w = enthalpy of feedwater, Btu/lb. For this boiler with steam at 125 lb/in^2 (gage) [= 139.7 lb/in^2 (abs)] and 400°F (930 kPa, 204°C), h_s = 1221.2 Btu/lb (2841 kJ/kg), and h_w = 180.92 Btu/lb (420.8 kJ/kg), from the steam tables. Then h_1 = 10.34(1221.2 − 180.92) = 10,766.5 Btu/lb (25,043 kJ/kg) of coal.

3. *Compute the dry flue-gas loss*

For any boiler, the dry flue-gas loss h_2 Btu/lb (kJ/kg) of fuel is given by $h_2 = 0.24W_g \times (T_g - T_a)$, where W_g = lb of dry flue gas per lb of fuel; T_g = flue-gas exit temperature, °F; T_a = intake-air temperature, °F.

Before W_g can be found, however, it must be determined whether any excess air is passing through the boiler. Compute the excess air, if any, from excess air, percent = 100 (O_2 − ½CO)/[0.264N_2 − (O_2 − ½CO)], where the symbols refer to the elements in the flue-gas analysis. Substituting values from the flue-gas analysis gives excess air = 100(6.1 − 0.2)/[0.264 × 80.7 − (6.1 − 0.2)] = 38.4 percent.

Using the method given in earlier calculation procedures, find the air required for complete combustion as 10.557 lb/lb (4.751 kg/kg) of coal. With 38.4 percent excess air, the additional air required = (10.557)(0.384) = 4.053 lb/lb (1.82 kg/kg) of fuel.

From the same computation in which the air required for complete combustion was determined, the lb of *dry* flue gas per lb of fuel = 11.018 (4.958 kg/kg). Then, the total flue gas at 38.4 percent excess air = 11.018 + 4.053 = 15.071 lb/lb (6.782 kg/kg) of fuel.

With a flue-gas temperature of 500°F (260°C), and an intake-air temperature of 79°F (26.1°C), h_2 = 0.24(15.071)(500 − 70) = 1524 Btu/lb (3545 kJ/kg) of fuel.

4. *Compute the loss due to evaporation of hydrogen-formed water*

Hydrogen in the fuel is burned in forming H_2O. This water is evaporated by heat in the fuel, and less heat is available for producing steam. This loss is h_3 Btu/lb of fuel = $9H(1089 - T_f + 0.46T_g)$, where H = percent H_2 in the fuel ÷ 100; T_f = temperature of fuel *before* combustion, °F; other symbols as before. For this fuel with 5.07 percent H_2, h_3 = 9(5.07/100)(1089 − 79 + 0.46 × 500) = 565.8 Btu/lb (1316 kJ/kg) of fuel.

5. *Compute the loss from evaporation of fuel moisture*

This loss is h_4 Btu/lb of fuel = $W_{mf}(1089 - T_f + 0.46T_g)$, where W_{mf} = lb of moisture per lb of fuel; other symbols as before. Since the fuel contains 1.61 percent moisture, in terms of *dry* coal this is (1.61)/(100 − 1.61) = 0.0164, or 1.64 percent. Then h_4 = (1.64/100)(1089 − 79 + 0.46 × 500) = 20.34 Btu/lb (47.3 kJ/kg) of fuel.

6. *Compute the loss from moisture in the air*

This loss is h_5 Btu/lb of fuel = $0.46W_{ma}(T_g - T_a)$, W_{ma} = (lb of water per lb of dry air)(lb air supplied per lb fuel). From a psychrometric chart, the weight of moisture per lb of air at a 79°F (26.1°C) dry-bulb and 71°F (21.7°C) wet-bulb temperature is 0.014 (0.006 kg). The combustion calculation, step 3, shows that the total air required with 38.4 percent excess air = 10.557 + 4.053 = 14.61 lb of air per lb (6.575 kg/kg) of fuel. Then, W_{ma} = (0.014)(14.61) = 0.2045 lb of moisture per lb (0.092 kg/kg) of air. And h_5 = (0.46)(0.2045)(500 − 79) = 39.6 Btu/lb (92.1 kJ/kg) of fuel.

7. *Compute the loss from incomplete combustion of C to CO_2 in the stack*

This loss is h_6 Btu/lb of fuel = [CO/(CO + CO_2)](C)(10,190), where CO and CO_2 are the percent by volume of these compounds in the flue gas by Orsat analysis; C = lb carbon per lb of coal. With the given flue-gas analysis and the coal ultimate analysis, h_6 = 0.4/(0.4 + 12.8)[(77.57)/(100)](10,190) = 239.5 Btu/lb (557 kJ/kg) of fuel.

8. *Compute the loss due to unconsumed carbon in the refuse*

This loss is h_7 Btu/lb of fuel = W_c(14,150), where W_c = lb of unconsumed carbon in refuse per lb of fuel fired. With an ash and refuse of 9.42 percent of the dry coal and combustible in the ash and refuse of 32.3 percent, h_7 = (9.42/100)(32.3/100)(14,150) = 430.2 Btu/lb (1006 kJ/kg) of fuel.

9. *Find the radiation loss in the boiler furnace*

Use the American Boiler and Affiliated Industries (ABAI) chart, or the manufacturer's engineering data to approximate the radiation loss in the boiler. Either source will show that the radiation loss is 1.09 percent of the gross heat input. Since the gross heat input is 13,850 Btu/lb (32,215 kJ/kg) of fuel, the radiation loss = (13,850)(1.09/100) = 151.0 Btu/lb (351.2 kJ/kg) of fuel.

10. *Summarize the losses; find the unaccounted-for loss*

Set up a tabulation thus, entering the various losses computed earlier:

Item	Btu/lb fuel	kJ/kg fuel	Percent
1. Input	13,850.0	32,215.4	100.0
2. Output	10,770.0	25,051	77.75
Losses:			
3. Flue gas	1,524.0	3,545	11.00
4. Hydrogen	565.8	1,315	4.09
5. Water-fuel	20.3	47.2	0.15
6. Water-air	39.6	92.1	0.29
7. CO	239.5	557	1.73
8. Carbon-ash	430.2	1,001	3.11
9. Radiation	151.0	351.2	1.09
10. Unaccounted	109.6	254.9	0.79
Total	13,850.0	32,214.4	100.00

The unaccounted-for loss is found by summing all the other losses, 3 through 9, and subtracting from 100.00.

Related Calculations: Use this method to compute the heat balance for any type of boiler—watertube or firetube—in any kind of service—power, process, or heating—using any kind of fuel—coal, oil, gas, wood, or refuse. Note that step 3 shows how to compute excess air from an Orsat flue-gas analysis.

Power Generation

REFERENCES: El-Wakil—*Powerplant Technology,* McGraw-Hill; Goss—*Factors Affecting Power Plant Waste Heat Utilization,* Pergamon Press; Polimeros—*Energy Cogeneration Handbook,* Industrial Press; Yu—*Electric Power System Dynamics,* Academic Press; Hagel—*Alternative Energy Strategies,* Praeger; Aschner—*Planning Fundamentals of Thermal Power Plants,* Israel University Press; Komanoff—*Power Plant Cost Escalation,* VNR; Seeley—*Elements of Thermal Technology,* Dekker; Hunt—*Handbook of Energy Technology,* VNR; Blair, Cassel, and Edelstein—*Geothermal Energy: Investment Decisions and Commercial Development,* Wiley; Goodman and Love—*Geothermal Energy Projects: Planning and Management,* Pergamon Press; Edgerton—*Available Energy and Environmental Economics,* Heath; Meyers—*Handbook of Energy Technology and Economics,* Wiley; Babcock & Wilcox Company—*Steam: Its Generation and Use;* Combustion Engineering Corporation—*Combustion Engineering;* Skrotzki and Vopat—*Power Station Engineering and Economy,* McGraw-Hill; Heat Enchange Institute—*Steam Surface Condenser Standards;* Gaffert—*Steam Power Stations,* McGraw-Hill; ASME—*Test Code for Steam Generating Units;* Potter—*Steam Power Plants,* Ronald; Smith and Stinson—*Fuels and Combustion,* McGraw-Hill; Zerban and Nye—*Steam Power Plants,* International Textbook; Sorenson—*Gas Turbines,* Ronald; Salisbury—*Steam Turbines and Their Cycles,* Wiley; Dusinberre—*Gas Turbine Power,* International Textbook; Zemansky—*Heat and Thermodynamics,* McGraw-Hill; Jakob—*Heat Transfer,* Wiley; McAdams—*Heat Transmission,* McGraw-Hill; Buffalo Forge Company—*Fan Engineering;* Church—*Steam Turbines,* McGraw-Hill; Bleeder Heater Manufacturers Association, Inc.—*Standards;* Tubular Exchanger Manufacturers Association—*Standards.*

STEAM MOLLIER DIAGRAM AND STEAM TABLE USE

(1) Determine from the Mollier diagram for steam (*a*) the enthalpy of 100 lb/in^2 (abs) (689.5-kPa) saturated steam, (*b*) the enthalpy of 10-lb/in^2 (abs) (68.9-kPa) steam containing 40 percent

moisture, (*c*) the enthalpy of 100-lb/in^2 (abs) (689.5-kPa) steam at 600°F (315.6°C). (2) Determine from the steam tables (*a*) the enthalpy, specific volume, and entropy of steam at 145.3 lb/in^2 (gage) (1001.8 kPa); (*b*) the enthalpy and specific volume of superheated steam at 1100 lb/in^2 (abs) (7584.2 kPa) and 600°F (315.6°C); (*c*) the enthalpy and specific volume of high-pressure steam at 7500 lb/in^2 (abs) (51,710.7 kPa) and 1200°F (648.9°C); (*d*) the enthalpy, specific volume, and entropy of 10-lb/in^2 (abs) (68.9-kPa) steam containing 40 percent moisture.

Calculation Procedure:

1. *Use the pressure and saturation (or moisture) lines to find enthalpy*

(*a*) Enter the Mollier diagram by finding the 100-lb/in^2 (abs) (689.5-kPa) pressure line, Fig. 1. In the Mollier diagram for steam, the pressure lines slope upward to the right from the lower left-hand corner. For saturated steam, the enthalpy is read at the intersection of the pressure line with the saturation curve *cef*, Fig. 1.

Thus, project along the 100-lb/in^2 (abs) (689.5-kPa) pressure curve, Fig. 1, until it intersects the saturation curve, point *g*. From here project horizontally to the left-hand scale of Fig. 1 and

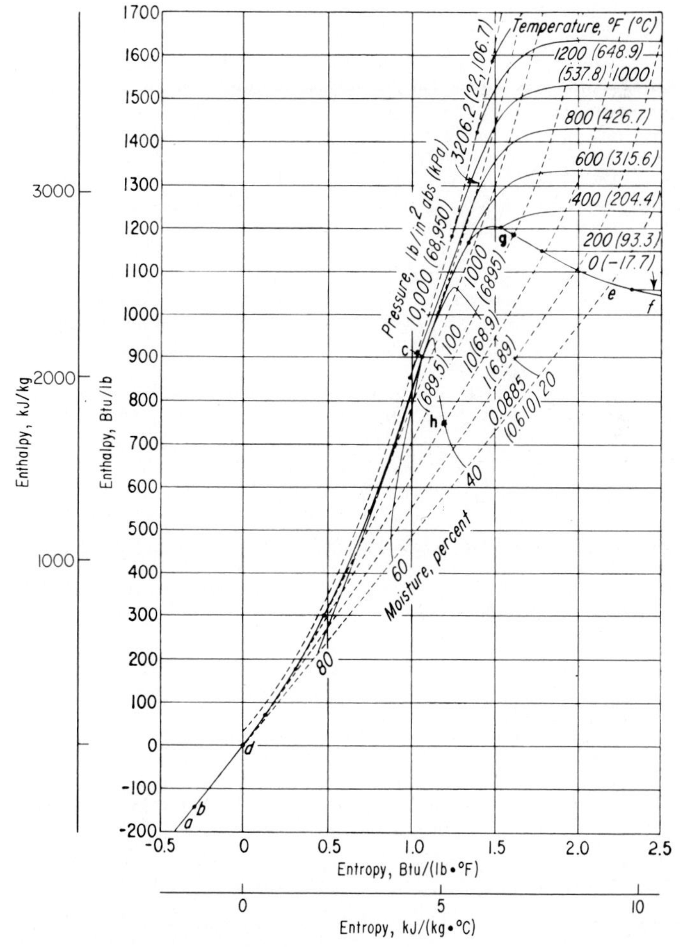

FIG. 1 Simplified Mollier diagram for steam.

read the enthalpy of 100-lb/in^2 (abs) (689.5-kPa) saturated steam as 1187 Btu/lb (2761.0 kJ/kg). (The Mollier diagram in Fig. 1 has fewer grid divisions than large-scale diagrams to permit easier location of the major elements of the diagram.)

(*b*) On a Mollier diagram, the enthalpy of wet steam is found at the intersection of the saturation pressure line with the percentage-of-moisture curve corresponding to the amount of moisture in the steam. In a Mollier diagram for steam, the moisture curves slope downward to the right from the saturated liquid line *cd*, Fig. 1.

To find the enthalpy of 10-lb/in^2 (abs) (68.9-kPa) steam containing 40 percent moisture, project along the 10-lb/in^2 (abs) (68.9-kPa) saturation pressure line until the 40 percent moisture curve is intersected, Fig. 1. From here project horizontally to the left-hand scale and read the enthalpy of 10-lb/in^2 (abs) (68.9-kPa) wet steam containing 40 percent moisture as 750 Btu/lb (1744.5 kJ/kg).

2. *Find the steam properties from the steam tables*

(*a*) Steam tables normally list absolute pressures or temperature in degrees Fahrenheit as one of their arguments. Therefore, when the steam pressure is given in terms of a gage reading, it must be converted to an absolute pressure before the table can be entered. To convert gage pressure to absolute pressure, add 14.7 to the gage pressure, or $p_a = p_g + 14.7$. In this instance, $p_a = 145.3 + 14.7 = 160.0$ lb/in^2 (abs) (1103.2 kPa). Once the absolute pressure is known, enter the saturation pressure table of the steam table at this value, and project horizontally to the desired values. For 160-lb/in^2 (abs) (1103.2-kPa) steam, using the ASME or Keenan and Keyes—*Thermodynamic Properties of Steam*, we see that the enthalpy of evaporation $h_{fg} = 859.2$ Btu/lb (1998.5 kJ/kg), and the enthalpy of saturated vapor $h_g = 1195.1$ Btu/lb (2779.8 kJ/kg), read from the respective columns of the steam tables. The specific volume v_g of the saturated vapor of 160-lb/in^2 (abs) (1103.2-kPa) steam is, from the tables, 2.834 ft^3/lb (0.18 m^3/kg), and the entropy s_g is 1.5640 Btu/(lb·°F) [6.55 kJ/(kg·°C)].

(*b*) Every steam table contains a separate tabulation of properties of superheated steam. To enter the superheated steam table, two arguments are needed—the absolute pressure and the temperature of the steam. To determine the properties of 1100-lb/in^2 (abs) (7584.5-kPa) 600°F (315.6°C) steam, enter the superheated steam table at the given absolute pressure and project horizontally from this absolute pressure [1100 lb/in^2 (abs) or 7584.5 kPa] to the column corresponding to the superheated temperature (600°F or 315.6°C) to read the enthalpy of the superheated vapor as $h = 1236.7$ Btu/lb (2876.6 kJ/kg) and the specific volume of the superheated vapor $v = 0.4532$ ft^3/lb (0.03 m^3/kg).

(*c*) For high-pressure steam use the ASME—*Steam Table*, entering it in the same manner as the superheated steam table. Thus, for 7500-lb/in^2 (abs) (51,712.5 kPa) 1200°F (648.9°C) steam, the enthalpy of the superheated vapor is 1474.9 Btu/lb (3430.6 kJ/kg), and the specific volume of the superheated vapor is 0.1060 ft^3/lb (0.0066 m^3/kg).

(*d*) To determine the enthalpy, specific volume, and the entropy of wet steam having y percent moisture by using steam tables instead of the Mollier diagram, apply these relations: $h = h_g - yh_{fg}/100$; $v = v_g - yv_{fg}/100$; $s = s_g - ys_{fg}/100$, where y = percentage of moisture expressed as a whole number. For 10-lb/in^2 (abs) (68.9-kPa) steam containing 40 percent moisture, obtain the needed values—h_g, h_{fg}, v_g, v_{fg}, s_g, and s_{fg}—from the saturation-pressure steam table and substitute in the above relations. Thus,

$$h = 1143.3 - \frac{40(982.1)}{100} = 750.5 \text{ Btu/lb } (1745.7 \text{ kJ/kg})$$

$$v = 38.42 - \frac{40(38.40)}{100} = 23.06 \text{ ft}^3\text{/lb } (1.44 \text{ m}^3\text{/kg})$$

$$s = 1.7876 - \frac{40(1.5041)}{100} = 1.1860 \text{ Btu/(lb·°F) } [4.97 \text{ kJ/(kg·°C)}]$$

Note that Keenan and Keyes, in *Thermodynamic Properties of Steam*, do not tabulate v_{fg}. Therefore, this value must be obtained by subtraction of the tabulated values, or $v_{fg} = v_g - v_f$. The value v_{fg} thus obtained is used in the relation for the volume of the wet steam. For 10-lb/in^2 (abs) (68.9-kPa) steam containing 40 percent moisture, $v_g = 38.42$ ft^3/lb (2.398 m^3/kg) and $v_f = 0.017$ ft^3/lb (0.0011 m^3/kg). Then $v_{fg} = 38.42 - 0.017 = 28.403$ ft^3/lb (1.773 m^3/kg).

In some instances, the quality of steam may be given instead of its moisture content in percentage. The quality of steam is the percentage of vapor in the mixture. In the above calculation, the quality of the steam is 60 percent because 40 percent is moisture. Thus, quality $= 1 - m$, where m = percentage of moisture, expressed as a decimal.

INTERPOLATION OF STEAM TABLE VALUES

(1) Determine the enthalpy, specific volume, entropy, and temperature of saturated steam at 151 lb/in^2 (abs) (1041.1 kPa). (2) Determine the enthalpy, specific volume, entropy, and pressure of saturated steam at 261°F (127.2°C). (3) Find the pressure of steam at 1000°F (537.8°C) if its specific volume is 2.6150 ft^3/lb (0.16 m^3/kg). (4) Calculate the enthalpy, specific volume, and entropy of 300-lb/in^2 (abs) (2068.5-kPa) steam at 567.22°F (297.3°C).

Calculation Procedure:

1. *Use the saturation-pressure table*

Study of the saturation-pressure table shows that there is no pressure value for 151 lb/in^2 (abs) (1041.1 kPa) listed. So it will be necessary to interpolate between the next higher and next lower tabulated pressure values. In this instance, these values are 152 and 150 lb/in^2 (abs) (1048.0 and 1034.3 kPa), respectively. The pressure for which properties are being found [151 lb/in^2 (abs) or 1041.1 kPa] is called the *intermediate pressure*. At 152 lb/in^2 (abs) (1048.0 kPa), h_g = 1194.3 Btu/lb (2777.5 kJ/kg); v_g = 2.977 ft^3/lb (0.19 m^3/kg); s_g = 1.5683 Btu/(lb·°F) [6.67 kJ/(kg·°C); t = 359.46°F (181.9°C). At 150 lb/in^2 (abs) (1034.3 kPa), h_g = 1194.1 Btu/lb (2777.5 kJ/kg); v_g = 3.015 ft^3/lb (0.19 m^3/kg); s_g = 1.5694 Btu/(lb·°F) [6.57 kJ/(kg·°C)]; t = 358.42°F (181.3°C).

For the enthalpy, note that as the pressure increases, so does h_g. Therefore, the enthalpy at 151 lb/in^2 (abs) (1041.1 kPa), the intermediate pressure, will equal the enthalpy at 150 lb/in^2 (abs) (1034.3 kPa) (the lower pressure used in the interpolation) plus the proportional change (difference between the intermediate pressure and the lower pressure) for a 1-lb/in^2 (abs) (6.9-kPa) pressure increase. Or, at any higher pressure, $h_{gi} = h_{gl} + [(p_i - p_l)/(p_h - p_l)](h_h - h_l)$, where h_{gi} = enthalpy at the intermediate pressure; h_{gl} = enthalpy at the lower pressure used in the interpolation; h_h = enthalpy at the higher pressure used in the interpolation; p_i = intermediate pressure; p_h and p_l = higher and lower pressures, respectively, used in the interpolation. Thus, from the enthalpy values obtained from the steam table for 150 and 152 lb/in^2 (abs) (1034.3 and 1048.0 kPa), h_{gi} = 1194.1 + [(151 − 150)/(152 − 150)](1194.3 − 1194.1) = 1194.2 Btu/lb (2777.7 kJ/kg) at 151 lb/in^2 (abs) (1041.1 kPa) saturated.

Next study the steam table to determine the direction of change of specific volume between the lower and higher pressures. This study shows that the specific volume decreases as the pressure increases. Therefore, the specific volume at 151 lb/in^2 (abs) (1041.1 kPa) (the intermediate pressure) will equal the specific volume at 150 lb/in^2 (abs) (1034.3 kPa) (the lower pressure used in the interpolation) minus the proportional change (difference between the intermediate pressure and the lower interpolating pressure) for a 1-lb/in^2 (abs) pressure increase. Or, at any pressure, $v_{gi} = v_{gl} - [(p_i - p_l)/(p_h - p_l)](v_l - v_h)$, where the subscripts are the same as above and v = specific volume at the respective pressure. With the volume values obtained from steam tables for 150 and 152 lb/in^2 (abs) (1034.3 and 1048.0 kPa), v_{gi} = 3.015 − [(151 − 150)/(152 − 150)](3.015 − 2.977) = 2.996 ft^3/lb (0.19 m^3/kg) and 151 lb/in^2 (abs) (1041.1 kPa) saturated.

Study of the steam table for the direction of entropy change shows that entropy, like specific volume, decreases as the pressure increases. Therefore, the entropy at 151 lb/in^2 (abs) (1041.1 kPa) (the intermediate pressure) will equal the entropy at 150 lb/in^2 (abs) (1034.3 kPa) (the lower pressure used in the interpolation) minus the proportional change (difference between the intermediate pressure and the lower interpolating pressure) for a 1-lb/in^2 (abs) (6.9-kPa) pressure increase. Or, at any higher pressure, $s_{gi} = s_{gl} - [(p_i - p_l)/(p_h - p_l)](s_l - s_h)$ = 1.5164 − [(151 − 150)/(152 − 150)](1.5694 − 1.5683) = 1.56885 Btu/(lb·°F) [6.6 kJ/(kg·°C)] at 151 lb/in^2 (abs) (1041.1 kPa) saturated.

Study of the steam table for the direction of temperature change shows that the saturation temperature, like enthalpy, increases as the pressure increases. Therefore, the temperature at 151 lb/in^2 (abs) (1041.1 kPa) (the intermediate pressure) will equal the temperature at 150 lb/in^2 (abs) (1034.3 kPa) (the lower pressure used in the interpolation) plus the proportional change (difference between the intermediate pressure and the lower interpolating pressure) for a 1-lb/in^2 (abs) (6.9-kPa) increase. Or, at any higher pressure, $t_{gi} = t_{gl} + [(p_i - p_l)/(p_h - p_l)](t_h - t_l) = 358.42 + [(151 - 150)/(152 - 150)](359.46 - 358.42) = 358.94$°F (181.6°C) at 151 lb/in^2 (abs) (1041.1 kPa) saturated.

2. *Use the saturation-temperature steam table*

Study of the saturation-temperature table shows that there is no temperature value of 261°F (127.2°C) listed. Therefore, it will be necessary to interpolate between the next higher and next lower tabulated values. In this instance these values are 262 and 260°F (127.8 and 126.7°C), respectively. The temperature for which properties are being found (261°F or 127.2°C) is called the intermediate temperature.

Temperature		h_g		v_g		s_g		p_g	
°F	°C	Btu/lb	kJ/kg	ft^3/lb	m^3/kg	Btu/(lb·°F)	kJ/(kg·°C)	lb/in^2 (abs)	kPa
262	127.8	1168.0	2716.8	11.396	0.71	1.6833	7.05	36.646	252.7
260	126.7	1167.3	2715.1	11.763	0.73	1.6860	7.06	35.429	244.3

For enthalpy, note that as the temperature increases, so does h_g. Therefore, the enthalpy at 261°F (127.2°C) (the intermediate temperature) will equal the enthalpy at 260°F (126.7°C) (the lower temperature used in the interpolation) plus the proportional change (difference between the intermediate temperature and the lower temperature) for a 1°F (0.6°C) temperature increase. Or, at any higher temperature, $h_{gi} = h_{gl} + [(t_i - t_l)/(t_h - t_l)](h_h - h_l)$, where h_{gl} = enthalpy at the lower temperature used in the interpolation; h_h = enthalpy at the higher temperature used in the interpolation; t_i = intermediate temperature; t_h and t_l = higher and lower temperatures, respectively, used in the interpolation. Thus, from the enthalpy values obtained from the steam table for 260 and 262°F (126.7 and 127.8°C), $h_{gi} = 1167.3 + [(261 - 260)/(262 - 260)](1168.0 - 1167.3) = 1167.65$ Btu/lb (2716.0 kJ/kg) at 261°F (127.2°C) saturated.

Next, study the steam table to determine the direction of change of specific volume between the lower and higher temperatures. This study shows that the specific volume decreases as the pressure increases. Therefore, the specific volume at 261°F (127.2°C) (the intermediate temperature) will equal the specific volume at 260°F (126.7°C) (the lower temperature used in the interpolation) minus the proportional change (difference between the intermediate temperature and the lower interpolating temperature) for a 1°F (0.6°C) temperature increase. Or, at any higher temperature, $v_{gi} = v_{gl} - [(t_i - t_l)/(t_h - t_l)](v_l - v_h) = 11.763 - [(261 - 260)/(262 - 260)](11.763 - 11.396) = 11.5795$ ft^3/lb (0.7 m^3/kg) at 261°F (127.2°C) saturated.

Study of the steam table for the direction of entropy change shows that entropy, like specific volume, decreases as the temperature increases. Therefore, the entropy at 261°F (127.2°C) (the intermediate temperature) will equal the entropy at 260°F (126.7°C) (the lower temperature used in the interpolation) minus the proportional change (difference between the intermediate temperature and the lower temperature) for a 1°F (0.6°C) temperature increase. Or, at any higher temperature, $s_{gi} = s_{gl} - [(t_i - t_l)/(t_h - t_l)](s_l - s_h) = 1.6860 - [(261 - 260)/(262 - 260)](1.6860 - 1.6833) = 1.68465$ Btu/(lb·°F) [7.1 kJ/(kg·°C)] at 261°F (127.2°C).

Study of the steam table for the direction of pressure change shows that the saturation pressure, like enthalpy, increases as the temperature increases. Therefore, the pressure at 261°F (127.2°C) (the intermediate temperature) will equal the pressure at 260°F (126.7°C) (the lower temperature used in the interpolation) plus the proportional change (difference between the intermediate temperature and the lower interpolating temperature) for a 1°F (0.6°C) temperature increase. Or, at any higher temperature, $p_{gi} = p_{gl} + [(t_i - t_l)/(t_h - t_l)](p_h - p_l) = 35.429 + [(261 - 260)(262 - 260)](36.646 - 35.429) = 36.0375$ lb/in^2 (abs) (248.5 kPa) at 261°F (127.2°C) saturated.

3. *Use the superheated steam table*

Choose the superheated steam table for steam at 1000°F (537.9°C) and 2.6150 ft^3/lb (0.16 m^3/kg) because the highest temperature at which saturated steam can exist is 705.4°F (374.1°C). This is also the highest temperature tabulated in some saturated-temperature tables. Therefore, the steam is superheated when at a temperature of 1000°F (537.9°C).

Look down the 1000°F (537.9°C) columns in the superheated steam table until a specific volume value of 2.6150 (0.16) is found. This occurs between 325 lb/in^2 (abs) (2240.9 kPa, v = 2.636 or 0.16) and 330 lb/in^2 (abs) (2275.4 kPa, v = 2.596 or 0.16). Since there is no volume value exactly equal to 2.6150 tabulated, it will be necessary to interpolate. List the values from the steam table thus:

p		t		v	
lb/in^2 (abs)	kPa	°F	°C	ft^3/lb	m^3/kg
325	2240.9	1000	537.9	2.636	0.16
330	2275.4	1000	537.9	2.596	0.16

Note that as the pressure rises, at constant temperature, the volume decreases. Therefore, the intermediate (or unknown) pressure is found by subtracting from the higher interpolating pressure [330 lb/in^2 (abs) or 2275.4 kPa in this instance] the product of the proportional change in the specific volume and the difference in the pressures used for the interpolation. Or, $p_{gi} = p_h - [(v_i - v_h)/(v_l - v_h)](p_h - p_l)$, where the subscripts h, l, and i refer to the high, low, and intermediate (or unknown) pressures, respectively. In this instance, p_{gi} = 330 − [(2.615 − 2.596)/(2.636 − 2.596)](330 − 325) = 327.62 lb/in^2 (abs) (2,258.9 kPa) at 1000°F (537.9 kPa) and a specific volume of 2.6150 ft^3/lb (0.16 m^3/kg).

4. *Use the superheated steam table*

When a steam pressure and temperature are given, determine, before performing any interpolation, the state of the steam. Do this by entering the saturation-pressure table at the given pressure and noting the saturation temperature. If the given temperature exceeds the saturation temperature, the steam is superheated. In this instance, the saturation-pressure table shows that at 300 lb/in^2 (abs) (2068.5 kPa) the saturation temperature is 417.33°F (214.1°C). Since the given temperature of the steam is 567.22°F (297.3°C), the steam is superheated because its actual temperature is greater than the saturation temperature.

Enter the superheated steam table at 300 lb/in^2 (abs) (2068.5 kPa), and find the next temperature lower than 567.22°F (297.3°C); this is 560°F (293.3°C). Also find the next higher temperature; this is 580°F (304.4°C). Tabulate the enthalpy, specific volume, and entropy for each temperature thus:

t		h		v		s	
°F	°C	Btu/lb	kJ/kg	ft^3/lb	m^3/kg	Btu/(lb·°F)	kJ/(kg·°C)
560	293.3	1292.5	3006.4	1.9218	0.12	1.6054	6.72
580	304.4	1303.7	3032.4	1.9594	0.12	1.6163	6.77

Use the same procedures for each property—enthalpy, specific volume, and entropy—as given in step 2 above; but change the sign between the lower volume and entropy and the proportional factor (temperature in this instance), because for superheated steam the volume and entropy increase as the steam temperature increases. Thus

$$h_{gi} = 1292.5 + \frac{567.22 - 560}{580 - 560}(1303.7 - 1292.5) = 1269.6 \text{ Btu/lb } (3015.9 \text{ kJ/kg})$$

$$v_{gi} = 1.9128 + \frac{567.22 - 560}{580 - 560}(1.9594 - 1.9128) = 1.9296 \text{ ft}^3\text{/lb } (0.12 \text{ m}^3\text{/kg})$$

$$s_{gi} = 1.6054 + \frac{567.22 - 560}{580 - 560}(1.6163 - 1.6054) = 1.6093 \text{ Btu/(lb·°F) } [6.7 \text{ kJ/(kg·°C)}]$$

Note: Also observe the direction of change of a property *before* interpolating. Use a *plus* or *minus* sign between the higher interpolating value and the proportional change depending on whether the tabulated value increases (+) or decreases (−).

CONSTANT-PRESSURE STEAM PROCESS

Three pounds of wet steam, containing 15 percent moisture and initially at a pressure of 400 lb/in^2 (abs) (2758.0 kPa), expands at constant pressure ($P = C$) to 600°F (315.6°C). Determine the initial temperature T_1, enthalpy H_1, internal energy E_1, volume V_1, entropy S_1, final entropy H_2, internal energy E_2, volume V_2, entropy S_2, heat added to the steam Q_1, work output W_2, change in initial energy ΔE, change in specific volume ΔV, change in entropy ΔS.

Calculation Procedure:

1. *Determine the initial steam temperature from the steam tables*

Enter the saturation-pressure table at 400 lb/in^2 (abs) (2758.0 kPa), and read the saturation temperature as 444.59°F (229.2°C).

2. *Correct the saturation values for the moisture of the steam in the initial state*

Sketch the process on a pressure-volume (P-V), Mollier (H-S), or temperature-entropy (T-S) diagram, Fig. 2. In state 1, y = moisture content = 15 percent. Using the appropriate values from the saturation-pressure steam table for 400 lb/in^2 (abs) (2758.0 kPa), correct them for a moisture content of 15 percent:

$$H_1 = h_g - yh_{fg} = 1204.5 - 0.15(780.5) = 1087.4 \text{ Btu/lb } (2529.3 \text{ kJ/kg})$$

$$E_1 = u_g - yu_{fg} = 1118.5 - 0.15(695.9) = 1015.1 \text{ Btu/lb } (2361.1 \text{ kJ/kg})$$

$$V_1 = v_g - yv_{fg} = 1.1613 - 0.15(1.1420) = 0.990 \text{ ft}^3\text{/lb } (0.06 \text{ m}^3\text{/kg})$$

$$S_1 = s_g - ys_{fg} = 1.4844 - 0.15(0.8630) = 1.2945 \text{ Btu/(lb}\cdot\text{°F) } [5.4 \text{ kJ/(kg}\cdot\text{°C)}]$$

3. *Determine the steam properties in the final state*

Since this is a constant-pressure process, the pressure in state 2 is 400 lb/in^2 (abs) (2758.0 kPa), the same as state 1. The final temperature is given as 600°F (315.6°C). This is greater than the saturation temperature of 444.59°F (229.2°C). Hence, the steam is superheated when in state 2. Use the superheated steam tables, entering at 400 lb/in^2 (abs) (2758.8 kPa) and 600°F (315.6°C). At this condition, H_2 = 1306.9 Btu/lb (3039.8 kJ/kg); V_2 = 1.477 ft^3/lb (0.09 m^3/kg). Then $E_2 = h_{2g} - P_2V_2/J$ = 1306.9 − 400(144)(1.477)/778 = 1197.5 Btu/lb (2785.4 kJ/kg). In this equation, the constant 144 converts pounds per square inch to pounds per square foot, absolute, and J = mechanical equivalent of heat = 778 ft·lb/Btu (1 N·m/J). From the steam tables, S_2 = 1.5894 Btu/(lb·°F) [6.7 kJ/(kg·°C)].

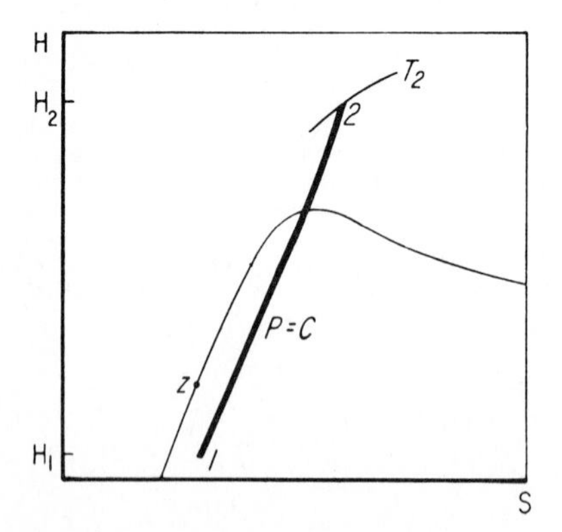

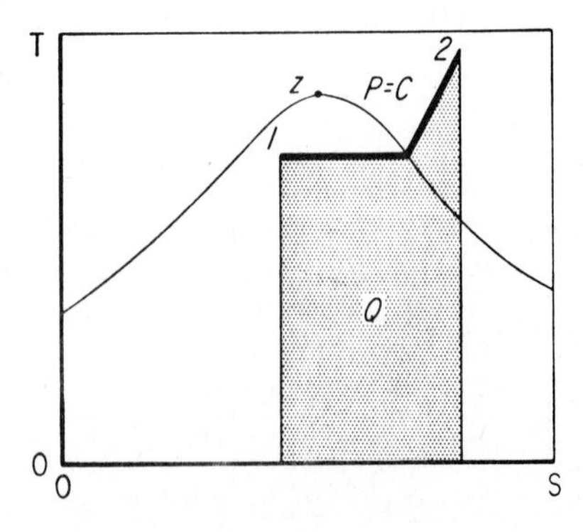

FIG. 2 Constant-pressure process.

4. *Compute the process inputs, outputs, and changes*

$W_2 = (P_1/J)(V_2 - V_1)m = [400(144)/778](1.4770 - 0.9900)(3) = 108.1$ Btu (114.1 kJ). In this equation, m = weight of steam used in the process = 3 lb (1.4 kg). Then

$$Q_1 = (H_2 - H_1)m = (1306.9 - 1087.4)(3) = 658.5 \text{ Btu } (694.4 \text{ kJ})$$

$$\Delta E = (E_2 - E_1)m = (1197.5 - 1014.1)(3) = 550.2 \text{ Btu } (580.2 \text{ kJ})$$

$$\Delta V = (V_2 - V_1)m = (1.4770 - 0.9900)(3) = 1.461 \text{ ft}^3 \ (0.041 \text{ m}^3)$$

$$\Delta S = (S_2 - S_1)m = (1.5894 - 1.2945)(3) = 0.8847 \text{ Btu/°F } (1.680 \text{ kJ/°C})$$

5. *Check the computations*

The work output W_2 should equal the change in internal energy plus the heat input, or $W_2 = E_1 - E_2 + Q_1 = -550.2 + 658.5 = 108.3$ Btu (114.3 kJ). This value very nearly equals the computed value of $W_2 = 108.1$ Btu (114.1 kJ) and is close enough for all normal engineering computations. The difference can be traced to calculator input errors. In computing the work output, the internal-energy change has a negative sign because there is a decrease in E during the process.

Related Calculations: Use this procedure for all constant-pressure steam processes.

CONSTANT-VOLUME STEAM PROCESS

Five pounds (2.3 kg) of wet steam initially at 120 lb/in^2 (abs) (827.4 kPa) with 30 percent moisture is heated at constant volume ($V = C$) to a final temperature of 1000°F (537.8°C). Determine the initial temperature T_1, enthalpy H_1, internal energy E_1, volume V_1, final pressure P_2, enthalpy H_2, internal energy E_2, volume V_2, heat added Q_1, work output W, change in internal energy ΔE, volume ΔV, and entropy ΔS.

Calculation Procedure:

1. *Determine the initial steam temperature from the steam tables*

Enter the saturation-pressure table at 120 lb/in^2 (abs) (827.4 kPa), the initial pressure, and read the saturation temperature $T_1 = 341.25$°F (171.8°C).

2. *Correct the saturation values for the moisture in the steam in the initial state*

Sketch the process on P-V, H-S, or T-S diagrams, Fig. 3. Using the appropriate values from the saturation-pressure table for 120 lb/in^2 (abs) (827.4 kPa), correct them for a moisture content of 30 percent:

$$H_1 = h_g - yh_{fg} = 1190.4 - 0.3(877.9) = 927.0 \text{ Btu/lb } (2156.2 \text{ kJ/kg})$$

$$E_1 = u_g - yu_{fg} = 1107.6 - 0.3(795.6) = 868.9 \text{ Btu/lb } (2021.1 \text{ kJ/kg})$$

$$V_1 = v_g - yv_{fg} = 3.7280 - 0.3(3.7101) = 2.6150 \text{ ft}^3\text{/lb } (0.16 \text{ m}^3\text{/kg})$$

$$S_1 = s_g - ys_{fg} = 1.5878 - 0.3(1.0962) = 1.2589 \text{ Btu/(lb}\cdot\text{°F) } [5.3 \text{ kJ/(kg}\cdot\text{°C)}]$$

3. *Determine the steam volume in the final state*

We are given $T_2 = 1000$°F (537.8°C). Since this is a constant-volume process, $V_2 = V_1 = 2.6150$ ft^3/lb (0.16 m^3/kg). The total volume of the vapor equals the product of the specific volume and the number of pounds of vapor used in the process, or total volume = 2.6150(5) = 13.075 ft^3 (0.37 m^3).

4. *Determine the final steam pressure*

The final steam temperature (1000°F or 537.8°C) and the final steam volume (2.6150 ft^3/lb or 0.16 m^3/kg) are known. To determine the final steam pressure, find in the steam tables the state corresponding to the above temperature and specific volume. Since a temperature of 1000°F (537.8°C) is higher than any saturation temperature (705.4°F or 374.1°C is the highest saturation

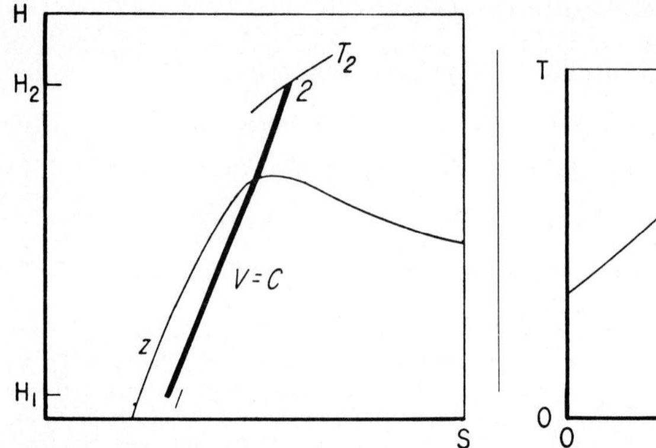

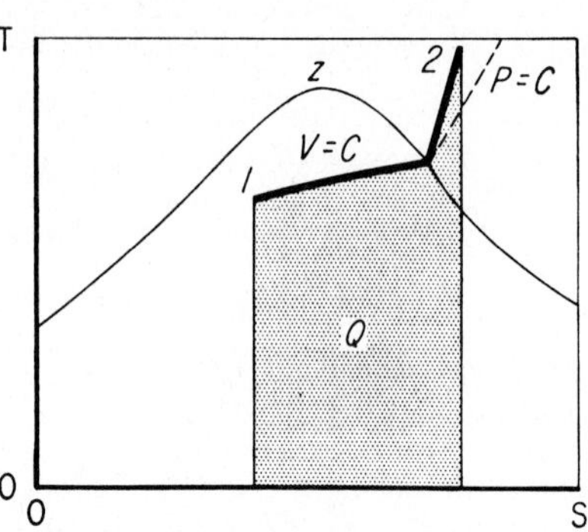

FIG. 3 Constant-volume process.

temperature for saturated steam), the steam in state 2 must be superheated. Therefore, the superheated steam tables must be used to determine P_2.

Enter the 1000°F (537.8°C) column in the steam table, and look for a superheated-vapor specific volume of 2.6150 ft³/lb (0.16 m³/kg). At a pressure of 325 lb/in² (abs) (2240.9 kPa),

$$v = 2.636 \text{ ft}^3/\text{lb } (0.16 \text{ m}^3/\text{kg})$$

$$h = 1542.5 \text{ Btu/lb } (3587.9 \text{ kJ/kg})$$

$$s = 1.7863 \text{ Btu/(lb}\cdot{}^\circ\text{F) } [7.48 \text{ kJ/(kg}\cdot{}^\circ\text{C)}]$$

and at a pressure of 330 lb/in² (abs) (2275.4 kPa)

$$v = 2.596 \text{ ft}^3/\text{lb } (0.16 \text{ m}^3/\text{kg})$$

$$h = 1524.4 \text{ Btu/lb } (3545.8 \text{ kJ/kg})$$

$$s = 1.7845 \text{ Btu/(lb}\cdot{}^\circ\text{F) } [7.47 \text{ kJ/(kg}\cdot{}^\circ\text{C)}]$$

Thus, 2.6150 lies between 325 and 330 lb/in² (abs) (2240.9 and 2275.4 kPa). To determine the pressure corresponding to the final volume, it is necessary to interpolate between the specific-volume values, or $P_2 = 330 - [(2.615 - 2.596)/(2.636 - 2.596)](330 - 325) = 327.62$ lb/in² (abs) (2258.9 kPa). In this equation, the volume values correspond to the upper [330 lb/in² (abs) or 2275.4 kPa], lower [325 lb/in² (abs) or 2240.9 kPa], and unknown pressures.

5. *Determine the final enthalpy, entropy, and internal energy*

The final enthalpy can be interpolated in the same manner, using the enthalpy at each volume instead of the pressure. Thus $H_2 = 1524.5 - [(2.615 - 2.596)/(2.636 - 2.596)](1524.5 - 1524.4) = 1524.45$ Btu/lb (3545.8 kJ/kg). Since the difference in enthalpy between the two pressures is only 0.1 Btu/lb (0.23 kJ/kg) (=1524.5 − 1524.4), the enthalpy at 327.62 lb/in² (abs) could have been assumed equal to the enthalpy at the lower pressure [325 lb/in² (abs) or 2240.9 kPa], or 1524.4 Btu/lb (3545.8 kJ/kg), and the error would have been only 0.05 Btu/lb (0.12 kJ/kg), which is negligible. However, where the enthalpy values vary by more than 1.0 Btu/lb (2.3 kJ/kg), interpolate as shown, if accurate results are desired.

Find S_2 by interpolating between pressures, or

$$S = 1.7863 - \frac{327.62 - 325}{330 - 325}(1.7863 - 1.7845) = 1.7854 \text{ Btu/(lb}\cdot{}^\circ\text{F) } [7.5 \text{ kJ/(kg}\cdot{}^\circ\text{C)}]$$

$$E_2 = H_2 - \frac{P_2V_2}{J} = 1524.4 - \frac{327.62(144)(2.615)}{778} = 1365.9 \text{ Btu/lb } (3177.1 \text{ kJ/kg})$$

6. ***Compute the changes resulting from the process***

Here $Q_1 = (E_2 - E_1)m = (1365.9 - 868.9)(5) = 2485$ Btu (2621.8 kJ); $\Delta S = (S_2 - S_1)m = (1.7854 - 1.2589)(5) = 2.6325$ Btu/°F (5.0 kJ/°C).

By definition, $W = 0$; $\Delta V = 0$; $\Delta E = Q_1$. Note that the curvatures of the constant-volume line on the *T-S* chart, Fig. 3, are different from the constant-pressure line, Fig. 2. Adding heat Q_1 to a constant-volume process affects only the internal energy. The total entropy change must take into account the total steam mass $m = 5$ lb (2.3 kg).

Related Calculations: Use this general procedure for all constant-volume steam processes.

CONSTANT-TEMPERATURE STEAM PROCESS

Six pounds (2.7 kg) of wet steam initially at 1200 lb/in^2 (abs) (8274.0 kPa) and 50 percent moisture expands at constant temperature ($T = C$) to 300 lb/in^2 (abs) (2068.5 kPa). Determine the initial temperature T_1, enthalpy H_1, internal energy E_1, specific volume V_1, entropy S_1, final temperature T_2, enthalpy H_2, internal energy E_2, volume V_2, entropy S_2, heat added Q_1, work output W_2, change in internal energy ΔE, volume ΔV, and entropy ΔS.

Calculation Procedure:

1. ***Determine the initial steam temperature from the steam tables***

Enter the saturation-pressure table at 1200 lb/in^2 (abs) (8274.0 kPa), and read the saturation temperature $T_1 = 567.22$°F (297.3°C).

2. ***Correct the saturation values for the moisture in the steam in the initial state***

Sketch the process on *P-V*, *H-S*, or *T-S* diagrams, Fig. 4. Using the appropriate values from the saturation-pressure table for 1200 lb/in^2 (abs) (8274.0 kPa), correct them for the moisture content of 50 percent:

$$H_1 = h_g - y_1 h_{fg} = 1183.4 - 0.5(611.7) = 877.5 \text{ Btu/lb } (2041.1 \text{ kJ/kg})$$

$$E_1 = u_g - y_1 u_{fg} = 1103.0 - 0.5(536.3) = 834.8 \text{ Btu/lb } (1941.7 \text{ kJ/kg})$$

$$V_1 = v_g - y_1 v_{fg} = 0.3619 - 0.5(0.3396) = 0.19 \text{ ft}^3\text{/lb } (0.012 \text{ m}^3\text{/kg})$$

$$S_1 = s_g - y_1 s_{fg} = 1.3667 - 0.5(0.5956) = 1.0689 \text{ Btu/(lb}\cdot°\text{F) } [4.5 \text{ kJ/(kg}\cdot°\text{C)}]$$

3. ***Determine the steam properties in the final state***

Since this is a constant-temperature process, $T_2 = T_1 = 567.22$°F (297.3°C); $P_2 = 300$ lb/in^2 (abs) (2068.5 kPa), given. The saturation temperature of 300 lb/in^2 (abs) (2068.5 kPa) is 417.33°F (214.1°C). Therefore, the steam is superheated in the final state because 567.22°F (297.3°C) > 417.33°F (214.1°C), the saturation temperature.

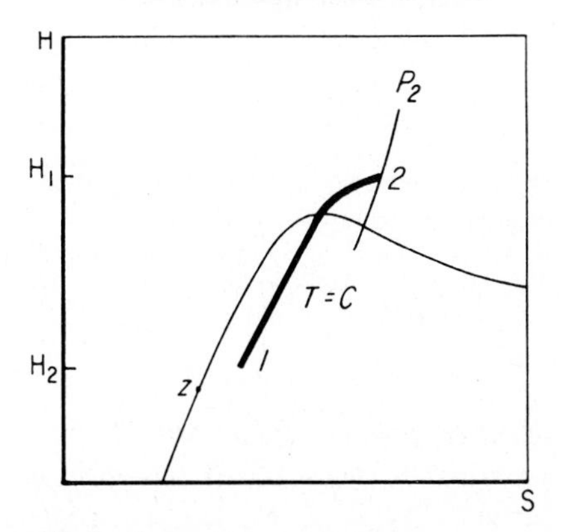

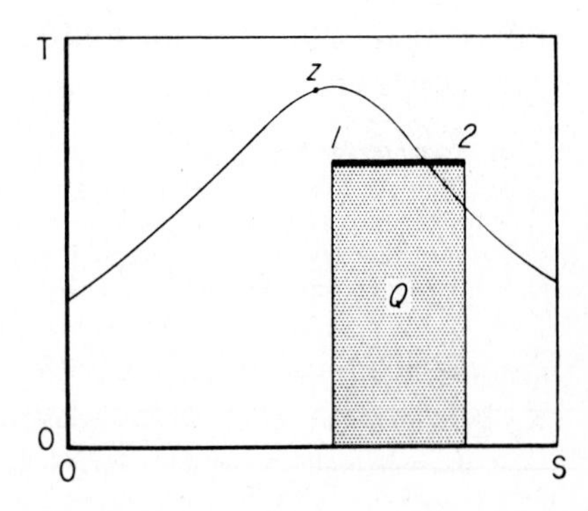

FIG. 4 Constant-temperature process.

To determine the final enthalpy, entropy, and specific volume, it is necessary to interpolate between the known final temperature and the nearest tabulated temperatures greater and less than the final temperature.

	v		h		s	
	ft³/lb	m³/kg	Btu/lb	kJ/kg	Btu/(lb·°F)	kJ/(kg·°C)
At T = 560°F (293.3°C)	1.9128	0.12	1292.5	3006.4	1.6054	6.72
At T = 580°F (304.4°C)	1.9594	0.12	1303.7	3032.4	1.6163	6.76

Then

$$H_2 = 1292.5 + \frac{567.22 - 560}{580 - 560}(1303.7 - 1292.5) = 1296.5 \text{ Btu/lb } (3015.7 \text{ kJ/kg})$$

$$S_2 = 1.6054 + \frac{567.22 - 560}{580 - 560}(1.6163 - 1.6054) = 1.6093 \text{ Btu/(lb}\cdot\text{°F) } [6.7 \text{ kJ/(kg}\cdot\text{°C)}]$$

$$V_2 = 1.9128 + \frac{567.22 - 560}{580 - 560}(1.9594 - 1.9128) = 1.9296 \text{ ft}^3\text{/lb } (0.12 \text{ m}^3\text{/kg})$$

$$E_2 = H_2 - \frac{P_2V_2}{J} = 1296.5 - \frac{300(144)(1.9296)}{778} = 1109.3 \text{ Btu/lb } (2580.2 \text{ kJ/kg})$$

4. ***Compute the process changes***

Here $Q_1 = T(S_2 - S_1)m$, where T_1 = absolute initial temperature, °R. So $Q_1 = (567.22 + 460)(1.6093 - 1.0689)(6) = 3330$ Btu (3513.3 kJ). Then

$$\Delta E = E_2 - E_1 = 1109.3 - 834.8 = 274.5 \text{ Btu/lb } (638.5 \text{ kJ/kg})$$

$$\Delta H = H_2 - H_1 = 1296.5 - 877.5 = 419.0 \text{ Btu/lb } (974.6 \text{ kJ/kg})$$

$$W_2 = (Q_1 - \Delta E)m = (555 - 274.5)(6) = 1.683 \text{ Btu } (1.8 \text{ kJ})$$

$$\Delta S = S_2 - S_1 = 1.6093 - 1.0689 = 0.5404 \text{ Btu/(lb}\cdot\text{°F) } [2.3 \text{ kJ/(kg}\cdot\text{°C)}]$$

$$\Delta V = V_2 - V_1 = 1.9296 - 0.1921 = 1.7375 \text{ ft}^3\text{/lb } (0.11 \text{ m}^3\text{/kg})$$

Related Calculations: Use this procedure for any constant-temperature steam process.

CONSTANT-ENTROPY STEAM PROCESS

Ten pounds (4.5 kg) of steam expands under two conditions—nonflow and steady flow—at constant entropy ($S = C$) from an initial pressure of 2000 lb/in² (abs) (13,790.0 kPa) and a temperature of 800°F (426.7°C) to a final pressure of 2 lb/in² (abs) (13.8 kPa). In the steady-flow process, assume that the initial kinetic energy E_{k1} = the final kinetic energy E_{k2}. Determine the initial enthalpy H_1, internal energy E_1, volume V_1, entropy S_1, final temperature T_2, percentage of moisture y, enthalpy H_2, internal energy E_2, volume V_2, entropy S_2, change in internal energy ΔE, enthalpy ΔH, entropy ΔS, volume ΔV, heat added Q_1, and work output W_2.

Calculation Procedure:

1. ***Determine the initial enthalpy, volume, and entropy from the steam tables***

Enter the superheated-vapor table at 2000 lb/in² (abs) (13,790.0 kPa) and 800°F (427.6°C), and read H_1 = 1335.5 Btu/lb (3106.4 kJ/kg); V_1 = 0.3074 ft³/lb (0.019 m³/kg); S_1 = 1.4576 Btu/(lb·°F) [6.1 kJ/(kg·°C)].

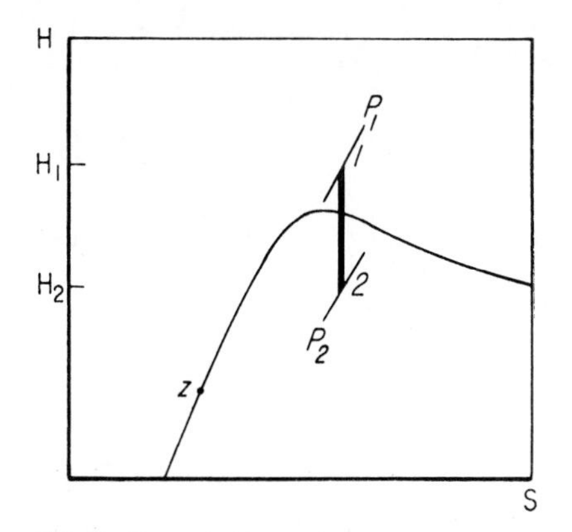

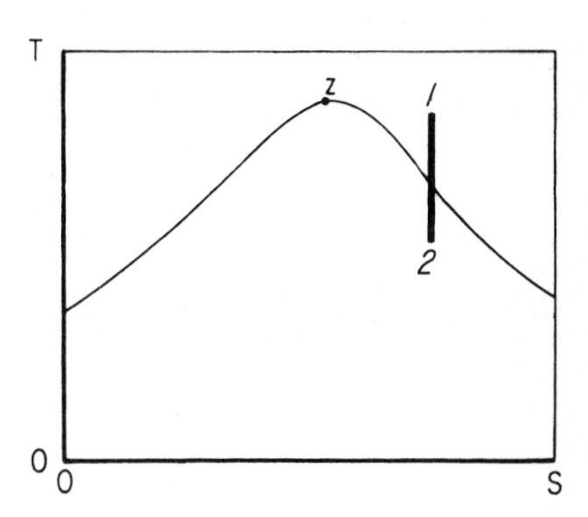

FIG. 5 Constant-entropy process.

2. *Compute the initial energy*

$$E_1 = H_1 - \frac{P_1V_1}{J} = 1335.5 - \frac{2000(144)(0.3074)}{778} = 1221.6 \text{ Btu/lb (2841.1 kJ/kg)}$$

3. *Determine the vapor properties on the final state*

Sketch the process on P-V, H-S, or T-S diagrams, Fig. 5. Note that the expanded steam is wet in the final state because the 2-lb/in² (abs) (13.8-kPa) pressure line is under the saturation curve on the H-S and T-S diagrams. Therefore, the vapor properties in the final state must be corrected for the moisture content. Read, from the saturation-pressure steam table, the liquid and vapor properties at 2 lb/in² (abs) (13.8 kPa). Tabulate these properties thus:

s_f = 0.1749 Btu/(lb·°F) [0.73 kJ/(kg·°C)] s_{fg} = 1.7451 Btu/(lb·°F) [7.31 kJ/(kg·C)]

h_f = 93.99 Btu/lb (218.6 kJ/kg) h_{fg} = 1022.2 Btu/lb (2377.6 kJ/kg)

u_f = 93.98 Btu/lb (218.6 kJ/kg) u_{fg} = 957.9 Btu/lb (2228.1 kJ/kg)

v_f = 0.016 ft³/lb (0.0010 m³/kg) v_{fg} = 173.71 ft³/lb (10.8 m³/kg)

s_g = 1.9200 Btu/(lb·°F) [8.04 kJ/(kg·C) h_g = 1116.3 Btu/lb (2596.5 kJ/kg)

u_g = 1051.9 Btu/lb (2446.7 kJ/kg) v_g = 173.73 ft³/lb (10.8 m³/kg)

Since this is a constant-entropy process, $S_2 = S_1 = s_g - y_2 s_{fg}$. Solve for y_2, the percentage of moisture in the final state. Or, $y_2 = (s_g - S_1)/s_{fg} = (1.9200 - 1.4576)/1.7451 = 0.265$, or, 26.5 percent. Then

$$H_2 = h_g - y_2 h_{fg} = 1116.2 - 0.265(1022.2) = 845.3 \text{ Btu/lb (1966.2 kJ/kg)}$$

$$E_2 = u_g - y_2 u_{fg} = 1051.9 - 0.265(957.9) = 798.0 \text{ Btu/lb (1856.1 kJ/kg)}$$

$$V_2 = v_g - y_2 v_{fg} = 173.73 - 0.265(173.71) = 127.7 \text{ ft}^3\text{/lb (8.0 m}^3\text{/kg)}$$

4. *Compute the changes resulting from the process*

The total change in properties is for 10 lb (4.5 kg) of steam, the quantity used in this process. Thus,

$$\Delta E = (E_1 - E_2)m = (1221.6 - 798.0)(10) = 4236 \text{ Btu (4469.2 kJ)}$$

$$\Delta H = (H_1 - H_2)m = (1335.5 - 845.3)(10) = 4902 \text{ Btu (5171.9 kJ)}$$

$$\Delta S = (S_1 - S_2)m = (1.4576 - 1.4576)(10) = 0 \text{ Btu/°F } (0 \text{ kJ/°C})$$

$$\Delta V = (V_1 - V_2)m = (0.3074 - 127.7)(10) = -1274 \text{ ft}^3\ (-36.1 \text{ m}^3)$$

So $Q_1 = 0$ Btu. (By definition, there is no transfer of heat in a constant-entropy process.) Nonflow $W_2 = \Delta E = 4236$ Btu (4469.2 kJ). Steady flow $W_2 = \Delta H = 4902$ Btu (5171.9 kJ).

Note: In a constant-entropy process, the nonflow work depends on the change in internal energy. The steady-flow work depends on the change in enthalpy and is larger than the nonflow work by the amount of the change in the flow work.

IRREVERSIBLE ADIABATIC EXPANSION OF STEAM

Ten pounds (4.5 kg) of steam undergoes a steady-flow expansion from an initial pressure of 2000 lb/in^2 (abs) (13,790.0 kPa) and a temperature 800°F (426.7°C) to a final pressure of 2 lb/in^2 (abs) (13.9 kPa) at an expansion efficiency of 75 percent. In this steady flow, assume $E_{k1} = E_{k2}$. Determine ΔE, ΔH, ΔS, ΔV, Q, and W_2.

Calculation Procedure:

1. *Determine the initial vapor properties from the steam tables*

Enter the superheated-vapor tables at 2000 lb/in^2 (abs) (13,790.0 kPa) and 800°F (426.7°C), and read $H_1 = 1335.5$ Btu/lb (3106.4 kJ/kg); $V_1 = 0.3074$ ft^3/lb (0.019 m^3/kg); $E_1 = 1221.6$ Btu/lb (2840.7 kJ/kg); $S_1 = 1.4576$ Btu/(lb·°F) [6.1 kJ/(kg·°C)].

2. *Determine the vapor properties in the final state*

Sketch the process on *P-V*, *H-S*, or *T-S* diagram, Fig. 6. Note that the expanded steam is wet in the final state because the 2-lb/in^2 (abs) (13.9-kPa) pressure line is under the saturation curve on the *H-S* and *T-S* diagrams. Therefore, the vapor properties in the final state must be corrected for the moisture content. However, the actual final enthalpy cannot be determined until after the expansion efficiency $[H_1 - H_2(H_1 - H_{2s})]$ is evaluated.

To determine the final enthalpy H_2, another enthalpy H_{2s} must be computed by assuming a constant-entropy expansion to 2 lb/in^2 (abs) (13.8 kPa) and a temperature of 126.08°F (52.3°C). Enthalpy H_{2s} will then correspond to a constant-entropy expansion into the wet region, and the percentage of moisture will correspond to the final state. This percentage is determined by finding the ratio of $s_g - S_1$ to s_{fg}, or $y_{2s} = s_g - S_1/s_{fg} = 1.9200 - 1.4576/1.7451 = 0.265$, where s_g and s_{fg} are entropies at 2 lb/in^2 (abs) (13.8 kPa). Then $H_{2s} = h_g - y_{2s}h_{fg} = 1116.2 - 0.265(1022.2) = 845.3$ Btu/lb (1966.2 kJ/kg). In this relation, h_g and h_{fg} are enthalpies at 2 lb/in^2 (abs) (13.8 kPa).

The expansion efficiency, given as 0.75, is $H_1 - H_2/(H_1 - H_{2s})$ = actual work/ideal work $= 0.75 = 1335.5 - H_2/(1335.5 - 845.3)$. Solve for $H_2 = 967.9$ Btu/lb (2251.3 kJ/kg).

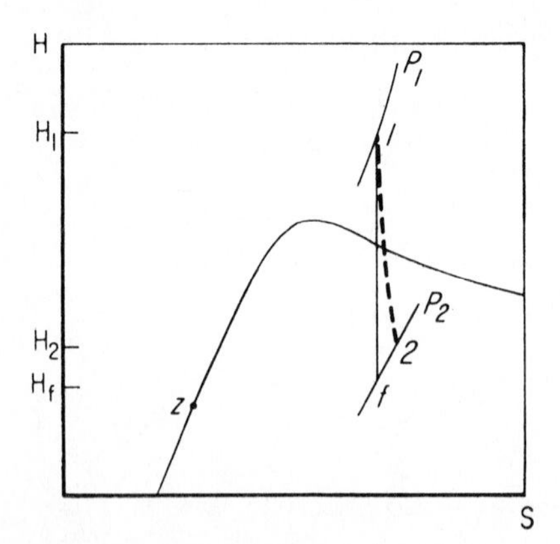

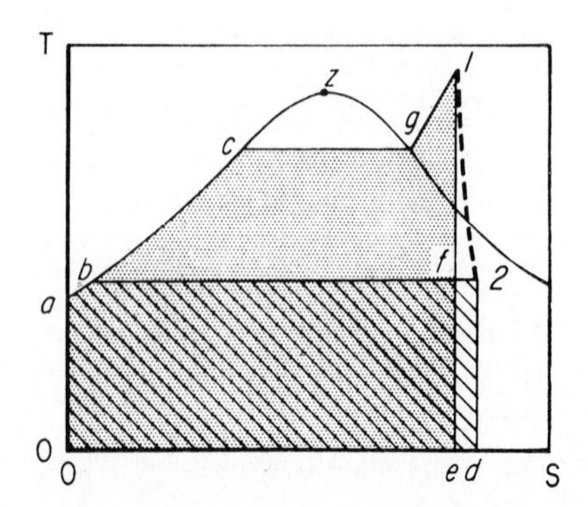

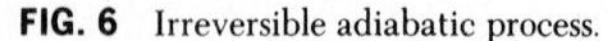
FIG. 6 Irreversible adiabatic process.

Next, read from the saturation-pressure steam table the liquid and vapor properties at 2 lb/in² (abs) (13.8 kPa). Tabulate these properties thus:

$$h_f = 93.99 \text{ Btu/lb } (218.6 \text{ kJ/kg})$$

$$h_{fg} = 1022.2 \text{ Btu/lb } (2377.6 \text{ kJ/kg})$$

$$h_g = 1116.2 \text{ Btu/lb } (2596.3 \text{ kJ/kg})$$

$$s_f = 0.1749 \text{ Btu/(lb}\cdot{}^\circ\text{F) } [0.73 \text{ kJ/(kg}\cdot{}^\circ\text{C)}]$$

$$s_{fg} = 1.7451 \text{ Btu/(lb}\cdot{}^\circ\text{F) } [7.31 \text{ kJ/(kg}\cdot{}^\circ\text{C)}]$$

$$s_g = 1.9200 \text{ Btu/(lb}\cdot{}^\circ\text{F) } [8.04 \text{ kJ/(kg}\cdot{}^\circ\text{C)}]$$

$$u_f = 93.98 \text{ Btu/lb } (218.60 \text{ kJ/kg})$$

$$u_{fg} = 957.9 \text{ Btu/lb } (2228.1 \text{ kJ/kg})$$

$$u_g = 1051.9 \text{ Btu/lb } (2446.7 \text{ kJ/kg})$$

$$v_f = 0.016 \text{ ft}^3\text{/lb } (0.0010 \text{ m}^3\text{/kg})$$

$$v_{fg} = 173.71 \text{ ft}^3\text{/lb } (10.84 \text{ m}^3\text{/kg})$$

$$v_g = 173.73 \text{ ft}^3\text{/lb } (10.85 \text{ m}^3\text{/kg})$$

Since the actual final enthalpy H_2 is different from H_{2s}, the final actual moisture y_2 must be computed by using H_2. Or, $y_2 = h_g - H_2/h_{fg} = 1116.1 - 967.9/1022.2 = 0.1451$. Then

$$E_2 = u_g - y_2 u_{fg} = 1051.9 - 0.1451(957.9) = 912.9 \text{ Btu/lb } (2123.4 \text{ kJ/kg})$$

$$V_2 = v_g - y_2 v_{fg} = 173.73 - 0.1451(173.71) = 148.5 \text{ ft}^3\text{/lb } (9.3 \text{ m}^3\text{/kg})$$

$$S_2 = s_g - y_2 s_{fg} = 1.9200 - 0.1451(1.7451) = 1.6668 \text{ Btu/(lb}\cdot{}^\circ\text{F) } [7.0 \text{ kJ/kg}\cdot{}^\circ\text{C)}]$$

3. *Compute the changes resulting from the process*

The total change in properties is for 10 lb (4.5 kg) of steam, the quantity used in this process. Thus

$$\Delta E = (E_1 - E_2)m = (1221.6 - 912.9)(10) = 3087 \text{ Btu } (3257.0 \text{ kJ})$$

$$\Delta H = (H_1 - H_2)m = (1335.5 - 967.9)(10) = 3676 \text{ Btu } (3878.4 \text{ kJ})$$

$$\Delta S = (S_2 - S_1)m = (1.6668 - 1.4576)(10) = 2.092 \text{ Btu/}^\circ\text{F } (4.0 \text{ kJ/}^\circ\text{C})$$

$$\Delta V = (V_2 - V_1)m = (148.5 - 0.3074)(10) = 1482 \text{ ft}^3 \text{ } (42.0 \text{ m}^3)$$

So $Q = 0$; by definition, $W_2 = \Delta H = 3676$ Btu (3878.4 kJ) for the steady-flow process.

IRREVERSIBLE ADIABATIC STEAM COMPRESSION

Two pounds (0.9 kg) of saturated steam at 120 lb/in² (abs) (827.4 kPa) with 80 percent quality undergoes nonflow adiabatic compression to a final pressure of 1700 lb/in² (abs) (11,721.5 kPa) at 75 percent compression efficiency. Determine the final steam temperature T_2, change in internal energy ΔE, change in entropy ΔS, work input W, and heat input Q.

Calculation Procedure:

1. *Determine the vapor properties in the initial state*

From the saturation-pressure steam tables, $T_1 = 341.25°F$ (171.8°C) at a pressure of 120 lb/in² (abs) (827.4 kPa) saturated. With $x_1 = 0.8$, $E_1 = u_f + x_1 u_{fg} = 312.05 + 0.8(795.6) = 948.5$ Btu/lb (2206.5 kJ/kg), from internal-energy values from the steam tables. The initial entropy is $S_1 = s_f + x_1 s_{fg} = 0.4916 + 0.8(1.0962) = 1.3686$ Btu/(lb·°F) [5.73 kJ/(kg·°C)].

2. *Determine the vapor properties in the final state*

Sketch a *T-S* diagram of the process, Fig. 7. Assume a constant-entropy compression from the initial to the final state. Then $S_{2s} = S_1 = 1.3686$ Btu/(lb·°F) [5.7 kJ/(kg·°C)].

The final pressure, 1700 lb/in^2 (abs) (11,721.5 kPa), is known, as is the final entropy, 1.3686 Btu/(lb·°F) [5.7 kJ/(kg·°C)] with constant-entropy expansion. The *T-S* diagram (Fig. 7) shows that the steam is superheated in the final state. Enter the superheated steam table at 1700 lb/in^2 (abs) (11,721.5 kPa), project across to an entropy of 1.3686, and read the final steam temperature as 650°F (343.3°C). (In most cases, the final entropy would not exactly equal a tabulated value, and it would be necessary to interpolate between tabulated entropy values to determine the intermediate pressure value.)

From the same table, at 1700 lb/in^2 (abs) (11.721.5 kPa) and 650°F (343.3°C), H_{2s} = 1214.4 Btu/lb (2827.4 kJ/kg); V_{2s} = 0.2755 ft^3/lb (0.017 m^3/lb). Then $E_{2s} = H_{2s} - P_2V_{2s}/J$ = 1214.4 − 1700(144)(0.2755)/788 = 1127.8 Btu/lb (2623.3 kJ/kg). Since E_1 and E_{2s} are known, the ideal work W can be computed. Or, $W = E_{2s} - E_1$ = 1127.8 − 948.5 = 179.3 Btu/lb (417.1 kJ/kg).

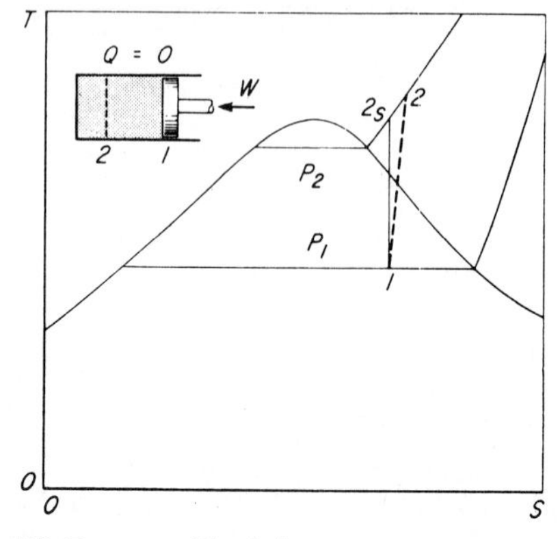

FIG. 7 Irreversible adiabatic compression process.

3. *Compute the vapor properties of the actual compression*

Since the compression efficiency is known, the actual final internal energy can be found from compression efficiency = ideal W/actual W = $E_{2s} - E_1/(E_2 - E_1)$, or 0.75 = 1127.8 − 948.5/(E_2 − 948.5); E_2 = 1187.6 Btu/lb (2762.4 kJ/kg). Then $E = (E_2 - E_1)m$ = (1187.6 − 948.5)(2) = 478.2 Btu (504.5 kJ) for 2 lb (0.9 kg) of steam. The actual work input $W = \Delta E$ = 478.2 Btu (504.5 kJ). By definition, $Q = 0$.

Last, the actual final temperature and entropy must be computed. The final actual internal energy E_2 = (1187.6 Btu/lb (2762.4 kJ/kg) is known. Also, the *T-S* diagram shows that the steam is superheated. However, the superheated steam tables do not list the internal energy of the steam. Therefore, it is necessary to assume a final temperature for the steam and then compute its internal energy. The computed value is compared with the known internal energy, and the next assumption is adjusted as necessary. Thus, assume a final temperature of 720°F (382.2°C). This assumption is higher than the ideal final temperature of 650°F (343.3°C) because the *T-S* diagram shows that the actual final temperature is higher than the ideal final temperature. Using values from the superheated steam table for 1700 lb/in^2 (abs) (11,721.5 kPa) and 720°F (382.2°C), we find

$$E = H - \frac{PV}{J} = 1288.4 - \frac{1700(144)(0.3283)}{778} = 1185.1 \text{ Btu/lb } (2756.5 \text{ kJ/kg})$$

This value is less than the actual internal energy of 1187.6 Btu/lb (2762.4 kJ/kg). Therefore, the actual temperature must be higher than 720°F (382.2°C), since the internal energy increases with temperature. To obtain a higher value for the internal energy to permit interpolation between the lower, actual, and higher values, assume a higher final temperature—in this case, the next temperature listed in the steam table, or 740°F (393.3°C). Then, for 1700 lb/in^2 (abs) (11,721.5 kPa) and 740°F (393.3°C),

$$E = 1305.8 - \frac{1700(144)(0.3410)}{778} = 1198.5 \text{ Btu/lb } (2757.7 \text{ kJ/kg})$$

This value is greater than the actual internal energy of 1187.6 Btu/lb (2762.4 kJ/kg). Therefore, the actual final temperature of the steam lies somewhere between 720 and 740°F (382.2 and 393.3°C). Interpolate between the known internal energies to determine the final steam temperature and final entropy. Or,

$$T_2 = 720 + \frac{1178.6 - 1185.1}{1198.5 - 1185.1}(740 - 720) = 723.7°\text{F } (384.3°\text{C})$$

$$S_2 = 1.4333 + \frac{1187.6 - 1185.1}{1198.5 - 1185.1}(1.4480 - 1.4333) = 1.4360 \text{ Btu/(lb}\cdot{}^\circ\text{F) [6.0 kJ/(kg}\cdot{}^\circ\text{C)]}$$

$$\Delta S = (S_2 - S_1)m = (1.4360 - 1.3686)(2) = 0.1348 \text{ Btu/}^\circ\text{F (0.26 kJ/}^\circ\text{C)}$$

Note that the final actual steam temperature is 73.7°F (40.9°C) higher than that (650°F or 343.3°C) for the ideal compression.

Related Calculations: Use this procedure for any irreversible adiabatic steam process.

THROTTLING PROCESSES FOR STEAM AND WATER

A throttling process begins at 500 lb/in^2 (abs) (3447.5 kPa) and ends at 14.7 lb/in^2 (abs) (101.4 kPa) with (1) steam at 500 lb/in^2 (abs) (3447.5 kPa) and 500°F (260.0°C); (2) steam at 500 lb/in^2 (abs) (3447.5 kPa) and 4 percent moisture; (3) steam at 500 lb/in^2 (abs) (3447.5 kPa) with 50 percent moisture; and (4) saturated water at 500 lb/in^2 (abs) (3447.5 kPa). Determine the final enthalpy H_2, temperature T_2, and moisture content y_2 for each process.

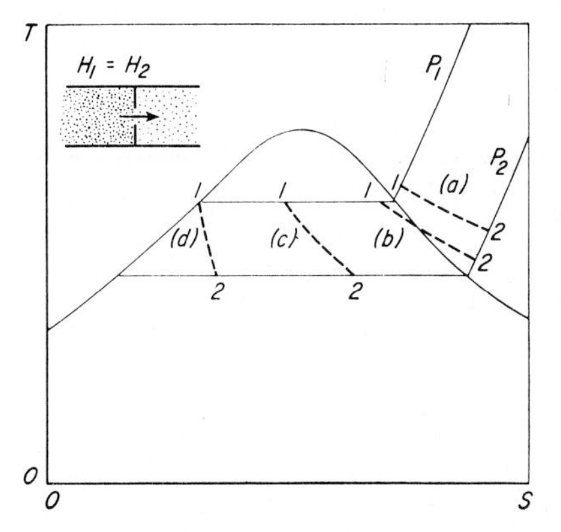

FIG. 8 Throttling process for steam.

Calculation Procedure:

1. *Compute the final-state conditions of the superheated steam*

From the superheated steam table for 500 lb/in^2 (abs) (3447.5 kPa) and 500°F (260.0°C), H_1 = 1231.3 Btu/lb (2864.0 kJ/kg). By definition of a throttling process, $H_1 = H_2 = 1231.3$ Btu/lb (2864.0 kJ/kg). Sketch the T-S diagram for a throttling process, Fig. 8*a*.

To determine the final temperature, enter the superheated steam table at 14.7 lb/in^2 (abs) (101.4 kPa), the final pressure, and project across to an enthalpy value equal to or less than the known enthalpy, 1231.3 Btu/lb (2864.0 kJ/kg). (The superheated steam table is used because the T-S diagram, Fig. 8, shows that the steam is superheated in the final state.) At 14.7 lb/in^2 (abs) (101.4 kPa) there is no tabulated enthalpy value that exactly equals 1231.3 Btu/lb (2864.0 kJ/kg). The next lower value is 1230 Btu/lb (2861.0 kJ/kg) at T = 380°F (193.3°C). The next higher value at 14.7 lb/in^2 (abs) (101.4 kPa) is 1239.9 Btu/lb (2884.0 kJ/kg) at T = 400°F (204.4°C). Interpolate between these enthalpy values to find the final steam temperature:

$$T_2 = 380 + \frac{1231.3 - 1230.5}{1239.9 - 1230.5}(400 - 380) = 381.7^\circ\text{F (194.3}^\circ\text{C)}$$

The steam does not contain any moisture in the final state because it is superheated.

2. *Compute the final-state conditions of the slightly wet steam*

Determine the enthalpy of 500-lb/in^2 (abs) (3447.5-kPa) saturated steam from the saturation-pressure steam table:

$$h_g = 1204.4 \text{ Btu/lb (2801.4 kJ/kg)} \qquad h_{fg} = 755.0 \text{ Btu/lb (1756.1 kJ/kg)}$$

Correct the enthalpy for moisture:

$$H_1 = h_g - y_1 h_{fg} = 1204.4 - 0.04(755.0) = 1174.2 \text{ Btu/lb (2731.2 kJ/kg)}$$

Then, by definition, $H_2 = H_1 = 1174.2$ Btu/lb (2731.2 kJ/kg).

Determine the final condition of the throttled steam (wet, saturated, or superheated) by studying the T-S diagram. If a diagram were not drawn, you would enter the saturation-pressure steam table at 14.7 lb/in^2 (abs) (101.4 kPa), the final pressure, and check the tabulated h_g. If the tabulated h_g were greater than H_1, the throttled steam would be superheated. If the tabulated h_g were less than H_1, the throttled steam would be saturated. Examination of the saturation-pressure steam table shows that the throttled steam is superheated because $H_1 > h_g$.

Next, enter the superheated steam table to find an enthalpy value of H_1 at 14.7 lb/in^2 (abs) (101.4 kPa). There is no value equal to 1174.2 Btu/lb (2731.2 kJ/kg). The next lower value is 1173.8 Btu/lb (2730.3 kJ/kg) at T = 260°F (126.7°C). The next higher value at 14.7 lb/in^2 (abs) (101.4 kPa) is 1183.3 Btu/lb (2752.4 kJ/kg) at T = 280°F (137.8°C). Interpolate between these enthalpy values to find the final steam temperature:

$$T_2 = 260 + \frac{1174.2 - 1173.8}{1183.3 - 1173.8}(280 - 260) = 260.8°\text{F } (127.1°\text{C})$$

This is higher than the temperature of saturated steam at 14.7 lb/in^2 (abs) (101.4 kPa)—212°F (100°C)—giving further proof that the throttled steam is superheated. The throttled steam, therefore, does not contain any moisture.

3. *Compute the final-state conditions of the very wet steam*

Determine the enthalpy of 500-lb/in^2 (abs) (3447.5-kPa) saturated steam from the saturation-pressure steam table. Or, h_g = 1204.4 Btu/lb (2801.4 kJ/kg); h_{fg} = 755.0 Btu/lb (1756.1 kJ/kg). Correct the enthalpy for moisture:

$$H_1 = H_2 = h_g - y_1 h_{fg} = 1204.4 - 0.5(755.0) = 826.9 \text{ Btu/lb } (1923.4 \text{ kJ/kg})$$

Then, by definition, $H_2 = H_1$ = 826.9 Btu/lb (1923.4 kJ/kg).

Compare the final enthalpy, H_2 = 826.9 Btu/lb (1923.4 kJ/kg), with the enthalpy of saturated steam at 14.7 lb/in^2 (abs) (101.4 kPa), or 1150.4 Btu/lb (2675.8 kJ/kg). Since the final enthalpy is less than the enthalpy of saturated steam at the same pressure, the throttled steam is wet. Since $H_1 = h_g - y_2 h_{fg}$, $y_2 = (h_g - H_1)/h_{fg}$. With a final pressure of 14.7 lb/in^2 (abs) (101.4 kPa), use h_g and h_{fg} values at this pressure. Or,

$$y_2 = \frac{1150.4 - 826.9}{970.3} = 0.3335, \text{ or } 33.35\%$$

The final temperature of the steam T_2 is the same as the saturation temperature at the final pressure of 14.7 lb/in^2 (abs) (101.4 kPa), or T_2 = 212°F (100°C).

4. *Compute the final-state conditions of saturated water*

Determine the enthalpy of 500-lb/in^2 (abs) (3447.5-kPa) saturated water from the saturation-pressure steam table at 500 lb/in^2 (abs) (3447.5 kPa); $H_1 = h_f$ = 449.4 Btu/lb (1045.3 kJ/kg) = H_2, by definition. The *T-S* diagram, Fig. 8, shows that the throttled water contains some steam vapor. Or, comparing the final enthalpy of 449.4 Btu/lb (1045.3 kJ/kg) with the enthalpy of saturated liquid at the final pressure, 14.7 lb/in^2 (abs) (101.4 kPa), 180.07 Btu/lb (418.8 kJ/kg), shows that the liquid contains some vapor in the final state because its enthalpy is greater.

Since $H_1 = H_2 = h_g - y_2 h_{fg}$, $y_2 = (h_g - H_1)/h_{fg}$. Using enthalpies at 14.7 lb/in^2 (abs) (101.4 kPa) of h_g = 1150.4 Btu/lb (2675.8 kJ/kg) and h_{fg} = 970.3 Btu/lb (2256.9 kJ/kg) from the saturation-pressure steam table, we get y_2 = 1150.4 − 449.4/970.3 = 0.723. The final temperature of the steam is the same as the saturation temperature at the final pressure of 14.7 lb/in^2 (abs) (101.4 kPa), or T_2 = 212°F (100°C).

Note: Calculation 2 shows that when you start with slightly wet steam, it can be throttled (expanded) through a large enough pressure range to produce superheated steam. This procedure is often used in a throttling calorimeter to determine the initial quality of the steam in a pipe. When very wet steam is throttled, calculation 3, the net effect may be to produce drier steam at a lower pressure. Throttling saturated water, calculation 4, can produce partial or complete flashing of the water to steam. All these processes find many applications in power-generation and process-steam plants.

REVERSIBLE HEATING PROCESS FOR STEAM

Subcooled water at 1500 lb/in^2 (abs) (10,342.5 kPa) and 140°F (60.0°C), state 1, Fig. 9, is heated at constant pressure to state 4, superheated steam at 1500 lb/in^2 (abs) (10,342.5 kPa) and 1000°F

(537.8°C). Find the heat added (1) to raise the compressed liquid to saturation temperature, (2) to vaporize the saturated liquid to saturated steam, (3) to superheat the steam to 1000°F (537.8°C), and (4) Q_1, ΔV, and ΔS from state 1 to state 4.

Calculation Procedure:

1. *Sketch the T-S diagram for this process*

Figure 9 is typical of a steam boiler and superheater. Feedwater fed to a boiler is usually subcooled liquid. If the feedwater pressure is relatively high, subcooling must be taken into account, if accurate results are desired. Some authorities recommend that at pressures below 400 lb/in² (abs) (2758.0 kPa) subcooling be ignored and values from the saturated-steam table be used. This means that the enthalpies and other properties listed in the steam table corresponding to the actual water temperature are sufficiently accurate. But above 400 lb/in² (abs) (2758.0 kPa), the compressed-liquid table should be used.

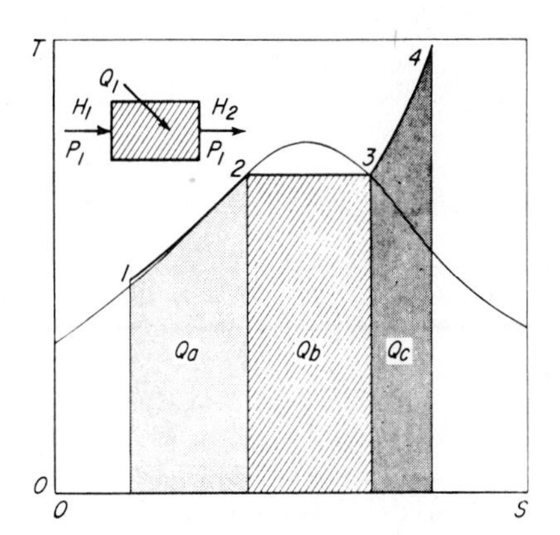

FIG. 9 Reversible heating process.

2. *Determine the initial properties of the liquid*

In the saturation-temperature steam table read, at 140°F (60.0°C), h_f = 107.89 Btu/lb (251.0 kJ/kg); p_f = 2.889 lb/in² (19.9 kPa); v_f = 0.01629 ft³/lb (0.0010 m³/kg); s_f = 0.1984 Btu/(lb·°F) [0.83 kJ/(kg·°C)].

Next, the enthalpy, volume, and entropy of the water at 1500 lb/in² (abs) (10,342.5 kPa) and 140°F (60.0°C) must be found. Since the water is at a much higher pressure than that corresponding to its temperature [1500 versus 2.889 lb/in² (abs)], the compressed-liquid portion of the steam table must be used. This table shows that three desired properties are plotted for 32, 100, and 200°F (0.0, 37.8, and 93.3°C) and higher temperatures. However, 140°F (60.0°C) is not included. Therefore, it is necessary to interpolate between 100 and 200°F (37.8 and 93.3°C). Thus, at 1500 lb/in² (abs) (10,342.5 kPa) in the compressed-liquid table:

Property	Temperature		Interpolation
	100°F (37.8°C)	200°F (93.3°C)	
$h - h_f$, Btu/lb (kJ/kg)	+3.99 (+9.28)	+3.36 (+7.82)	+3.74 (+8.70)
$(v - v_f)10^5$, ft³/lb (m³/kg)	−7.5 (−0.47)	−8.1 (−0.51)	−7.7 (−0.48)
$(s - s_f)10^3$, Btu/(lb·°F) [kJ/(kg·°C)]	−0.86 (−3.60)	−1.79 (−7.49)	−1.23 (−5.15)

Each property is interpolated in the following way:

$$h - h_f = 3.99 - \frac{3.99 - 3.36}{200 - 100}(140 - 100) = 3.99 - 0.25$$

$$= 3.74 \text{ Btu/lb } (8.70 \text{ kJ/kg})$$

$$(v - v_f)10^5 = -7.5 - \frac{8.1 - 7.5}{200 - 100}(140 - 100) = -7.5 - 0.24$$

$$= -7.74 \text{ ft}^3\text{/lb } (-0.48 \text{ m}^3\text{/kg})$$

$$(s - s_f)10^3 = -0.86 - \frac{1.79 - 0.86}{200 - 100}(140 - 100) = -0.86 - 0.37$$

$$= -1.23 \text{ Btu/(lb}\cdot{}^\circ\text{F)} [-5.15 \text{ kJ/(kg}\cdot{}^\circ\text{C)}]$$

These interpolated values must now be used to correct the saturation data at 140°F (60.0°C) to the actual subcooled state 1 properties. Thus, at 1500 lb/in² (abs) (10,342.5 kPa) and 140°F (60.0°C),

$$H_1 = h_f + \text{interpolated } h = 107.89 + 3.74 = 111.63 \text{ Btu/lb } (259.7 \text{ kJ/kg})$$

$$V_1 = v_f - \frac{\text{interpolated } v}{10^5} = 0.01629 - \frac{7.74}{10^5} = 0.01621 \text{ ft}^3\text{/lb } (0.0010 \text{ m}^3\text{/kg})$$

$$S_1 = s_f - \frac{\text{interpolated } s}{10^3} = 0.1984 - \frac{1.23}{10^3} = 0.1972 \text{ Btu/(lb}\cdot{}^\circ\text{F)} [0.83 \text{ kJ/(kg}\cdot{}^\circ\text{C)}]$$

3. ***Compute the heat added to raise the compressed liquid to the saturation temperature***

From the saturation-pressure steam table for 1500 lb/in² (abs) (10,342.5 kPa), the enthalpy of the saturated liquid H_2 = 611.6 Btu/lb (1422.6 kJ/kg). The heat added Q_a to raise the compressed liquid to the saturation temperature is $Q_a = H_2 - H_1 = 611.6 - 111.6 = 500$ Btu/lb (1163.0 kJ/kg).

4. ***Compute the heat added to vaporize the saturated liquid***

Read from the saturation-pressure steam table the enthalpy of saturated vapor at 1500 lb/in² (abs) (10,342.5 kPa), H_3 = 1167.9 Btu/lb (2716.5 kJ/kg). Then the heat added to vaporize the saturated water $Q_b = H_3 - H_2 = 1167.9 - 611.6 = 556.3$ Btu/lb (1294.0 kJ/kg).

5. ***Compute the heat added to superheat the steam***

Find in the superheated steam table for 1500 lb/in² (abs) (10,342.5 kPa) and 1000°F (537.8°C) the properties of the superheated steam: H_4 = 1490.1 Btu/lb (3466.0 kJ/kg); V_4 = 0.5390 ft³/lb (0.034 m³/kg); S_4 = 1.6001 Btu/(lb·°F) [6.7 kJ/(kg·°C)]. Then the heat added to superheat the saturated steam $Q_4 = H_4 - H_3 = 1490.1 - 1167.9 = 322.2$ Btu/lb (749.4 kJ/kg).

6. ***Determine the property changes during the process***

$$Q_1 = Q_a + Q_b + Q_c = H_4 - H_1 = 1490.1 - 111.6 = 1378.5 \text{ Btu/lb } (3206.4 \text{ kJ/kg})$$

$$\Delta V = V_4 - V_1 = 0.5390 - 0.01621 = 0.5228 \text{ ft}^3\text{/lb } (0.033 \text{ m}^3\text{/kg})$$

$$\Delta S = S_4 - S_1 = 1.6001 - 0.1972 = 1.4029 \text{ Btu/(lb}\cdot{}^\circ\text{F)} [5.9 \text{ kJ/(kg}\cdot{}^\circ\text{C)}]$$

BLEED-STEAM REGENERATIVE CYCLE LAYOUT AND *T-S* PLOT

Sketch the cycle layout, *T-S* diagram, and energy-flow chart for a regenerative bleed-steam turbine plant having three feedwater heaters and four feed pumps. Write the equations for the work-output available energy and the energy rejected to the condenser.

Calculation Procedure:

1. ***Sketch the cycle layout***

Figure 10 shows a typical practical regenerative cycle having three feedwater heaters and four feedwater pumps. Number each point where steam enters and leaves the turbine and where steam enters or leaves the condenser and boiler. Also number the points in the feedwater cycle where feedwater enters and leaves a heater. Indicate the heater steam flow by m with a subscript corresponding to the heater number. Use W_p and a suitable subscript to indicate the pump work for each feed pump, except the last, which is labeled W_{pF}. The heat input to the steam generator is Q_a; the work output of the steam turbine is W_e; the heat rejected by the condenser is Q_r.

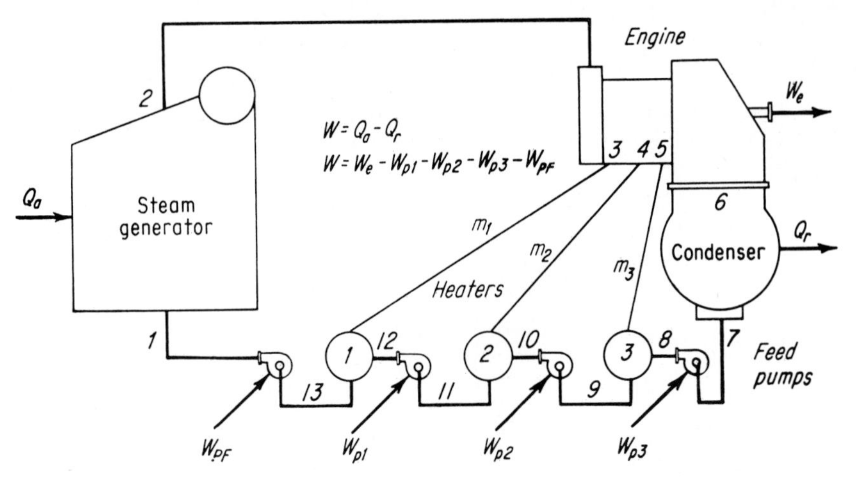

FIG. 10 Regenerative steam cycle uses bleed steam.

2. *Sketch the T-S diagram for the cycle*

To analyze any steam cycle, trace the flow of 1 lb (0.5 kg) of steam through the system. Thus, in this cycle, 1 lb (0.5 kg) of steam leaves the steam generator at point 2 and flows to the turbine. From state 2 to 3, 1 lb (0.5 kg) of steam expands at constant entropy (assumed) through the turbine, producing work output $W_1 = H_2 - H_3$, represented by area 1-*a*-2-3 on the *T-S* diagram, Fig. 11*a*. At point 3, some steam is bled from the turbine to heat the feedwater passing through heater 1. The quantity of steam bled, m_1 lb is less than the 1 lb (0.5 kg) flowing between points 2 and 3. Plot states 2 and 3 on the *T-S* diagram, Fig. 11*a*.

From point 3 to 4, the quantity of steam flowing through the turbine is $1 - m_1$ lb. This steam produces work output $W_2 = H_3 - H_4$. Plot point 4 on the *T-S* diagram. Then, area 1-3-4-12 represents the work output W_2, Fig. 11*a*.

At point 4, steam is bled to heater 2. The weight of this steam is m_2 lb. From point 4, the steam continues to flow through the turbine to point 5, Fig. 11*a*. The weight of the steam flowing between points 4 and 5 is $1 - m_1 - m_2$ lb. Plot point 5 on the *T-S* diagram, Fig. 11*a*. The work output between points 4 and 5, $W_3 = H_4 - H_5$, is represented by area 4-5-10-11 on the *T-S* diagram.

At point 5, steam is bled to heater 3. The weight of this bleed steam is m_3 lb. From point 5, steam continues to flow through the turbine to exhaust at point 6, Fig. 11*a*. The weight of steam flowing between points 5 and 6 is $1 - m_1 - m_2 - m_3$ lb. Plot point 6 on the *T-S* diagram, Fig. 11*a*.

The work output between points 5 and 6 is $W_4 = H_5 - H_6$, represented by area 5-6-7-9 on the *T-S* diagram, Fig. 11*a*. Area Q_r represents the heat given up by 1 lb (0.5 kg) of exhaust steam. Similarly, the area marked Q_a represents the heat absorbed by 1 lb (0.5 kg) of water in the steam generator.

3. *Alter the T-S diagram to show actual cycle conditions*

As plotted in Fig. 11*a*, Q_a is true for this cycle since 1 lb (0.5 kg) of water flows through the steam generator and the first section of the turbine. But Q_r is much too large; only $1 - m_1 - m_2 - m_3$ lb of steam flows through the condenser. Likewise, the net areas for W_2, W_3, and W_4, Fig. 11*a*, are all too large, because less than 1 lb (0.5 kg) of steam flows through the respective turbine sections. The area for W_1, however, is true.

A true *proportionate-area* diagram can be plotted by applying the factors for actual flow, as in Fig. 11*b*. Here W_2, outlined by the heavy lines, equals the similarly labeled area in Fig. 11*a*, multiplied by $1 - m_1$. The states marked 11′ and 12′, Fig. 11*b*, are not true state points because of the ratioing factor applied to the area for W_2. The true state points 11 and 12 of the liquid before and after heater pump 3 stay as shown in Fig. 11*a*.

Apply $1 - m_1 - m_2$ to W_3 of Fig. 11*a* to obtain the proportionate area of Fig. 11*b*; to obtain W_4, multiply by $1 - m_1 - m_2 - m_3$. Multiplying by this factor also gives Q_r. Then all the areas in Fig. 11*b* will be in proper proportion for 1 lb (0.5 kg) of steam entering the turbine throttle but less in other parts of the cycle.

In Fig. 11*b*, the work can be measured by the difference of the area Q_a and the area Q_r. There is no simple net area left, because the areas coincide on only two sides. But the area enclosed by

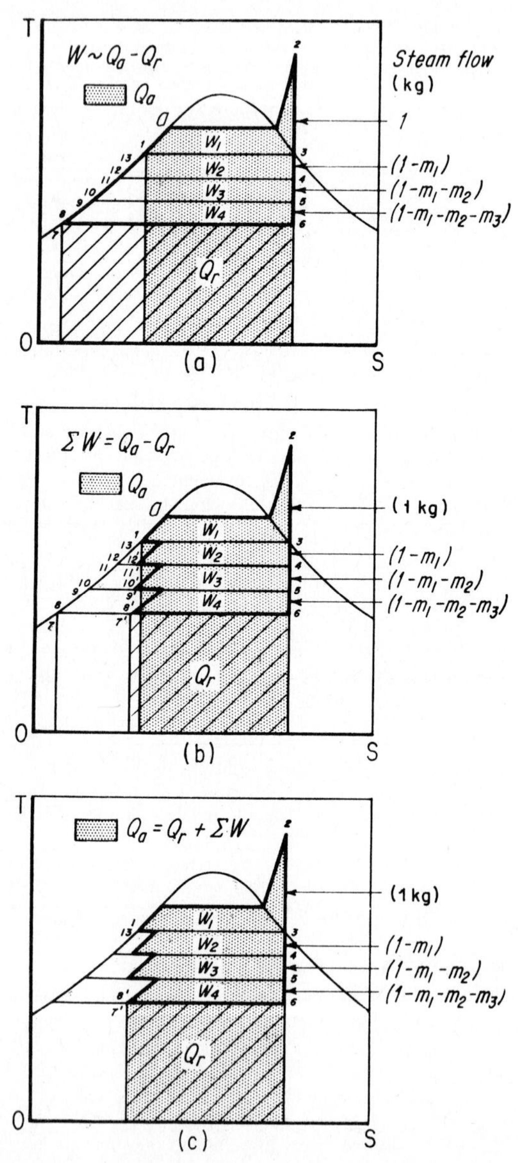

FIG. 11 (*a*) *T-S* chart for the bleed-steam regenerative cycle in Fig. 10; (*b*) actual fluid flow in the cycle; (*c*) alternative plot of (*b*).

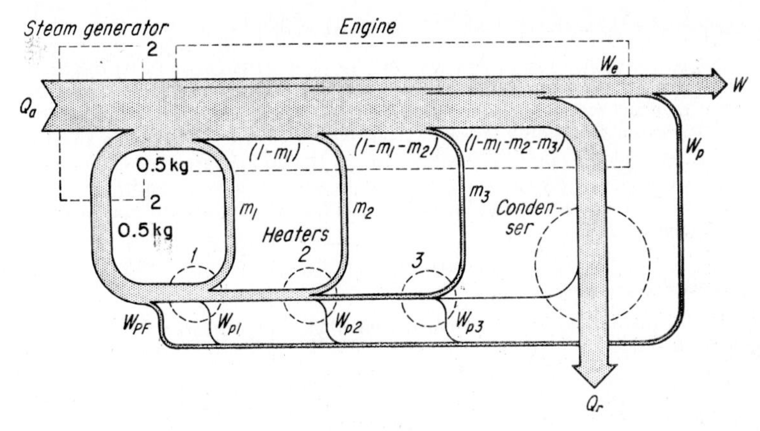

FIG. 12 Energy-flow chart of cycle in Fig. 10.

the heavy lines *is* the total net work W for the cycle, equal to the sum of the work produced in the various sections of the turbine, Fig. 11*b*. Then Q_a is the alternate area $Q_r + W_1 + W_2 + W_3 + W_4$, as shaded in Fig. 11*c*.

The sawtooth appearance of the liquid-heating line shows that as the number of heaters in the cycle increases, the heating line approaches a line of constant entropy. The best number of heaters for a given cycle depends on the steam state of the turbine inlet. Many medium-pressure and medium-temperature cycles use five to six heaters. High-pressure and high-temperature cycles use as many as nine heaters.

4. *Draw the energy-flow chart*

Choose a suitable scale for the heat content of 1 lb (0.5 kg) of steam leaving the steam generator. A typical scale is 0.375 in per 1000 Btu/lb (0.41 cm per 1000 kJ/kg). Plot the heat content of 1 lb (0.5 kg) of steam vertically on line 2-2, Fig. 12. Using the same scale, plot the heat content in energy streams m_1, m_2, m_3, W_e, W, W_p, W_{pF}, and so forth. In some cases, as W_{p1}, W_{p2}, and so forth, the energy stream may be so small that it is impossible to plot it to scale. In these instances, a single thin line is used. The completed diagram, Fig. 12, provides a useful concept of the distribution of the energy in the cycle.

Related Calculations: The procedure given here can be used for all regenerative cycles, provided that the equations are altered to allow for more, or fewer, heaters and pumps. The following calculation procedure shows the application of this method to an actual regenerative cycle.

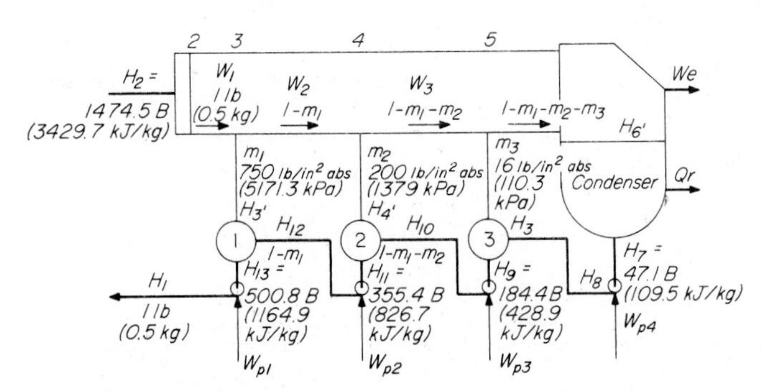

FIG. 13 Bleed-regenerative steam cycle.

BLEED REGENERATIVE STEAM CYCLE ANALYSIS

Analyze the bleed regenerative cycle shown in Fig. 13, determining the heat balance for each heater, plant thermal efficiency, turbine or engine thermal efficiency, plant heat rate, turbine or engine heat rate, and turbine or engine steam rate. Throttle steam pressure is 2000 lb/in^2 (abs) (13,790.0 kPa) at 1000°F (537.8°C); steam-generator efficiency = 0.88; station auxiliary steam consumption (excluding pump work) = 6 percent of the turbine or engine output; engine efficiency of each turbine or engine section = 0.80; turbine or engine cycle has three feedwater heaters and bleed-steam pressures as shown in Fig. 13; exhaust pressure to condenser is 1 inHg (3.4 kPa) absolute.

Calculation Procedure:

1. *Determine the enthalpy of the steam at the inlet of each heater and the condenser*

From a superheated-steam table, find the throttle enthalpy H_2 = 1474.5 Btu/lb (3429.7 kJ/kg) at 2000 lb/in^2 (abs) (13,790.0 kPa) and 1000°F (537.8°C). Next find the throttle entropy S_2 = 1.5603 Btu/(lb·°F) [6.5 kJ/(kg·°C)], at the same conditions in the superheated-steam table.

Plot the throttle steam conditions on a Mollier chart, Fig. 14. Assume that the steam expands from the throttle conditions at constant entropy = constant S to the inlet of the first feedwater heater, 1, Fig. 13. Plot this constant S expansion by drawing the straight vertical line 2-3 on the Mollier chart, Fig. 14, between the throttle condition and the heater inlet pressure of 750 lb/in^2 (abs) (5171.3 kPa).

Read on the Mollier chart H_3 = 1346.7 Btu/lb (3132.4 kJ/kg). Since the engine or turbine efficiency $e_e = H_2 - H_3/(H_2 - H_3) = 0.8 = 1474.5 - H_3/(1474.5 - 1346.7)$; H_3 = actual enthalpy of the steam at the inlet to heater 1 = 1474.5 − 0.8(1474.5 − 1346.7) = 1372.2 Btu/lb (3191.7 kJ/kg). Plot this enthalpy point on the 750-lb/in^2 (abs) (5171.3-kPa) pressure line of the Mollier chart, Fig. 14. Read the entropy at the heater inlet from the Mollier chart as $S_{3'}$ =

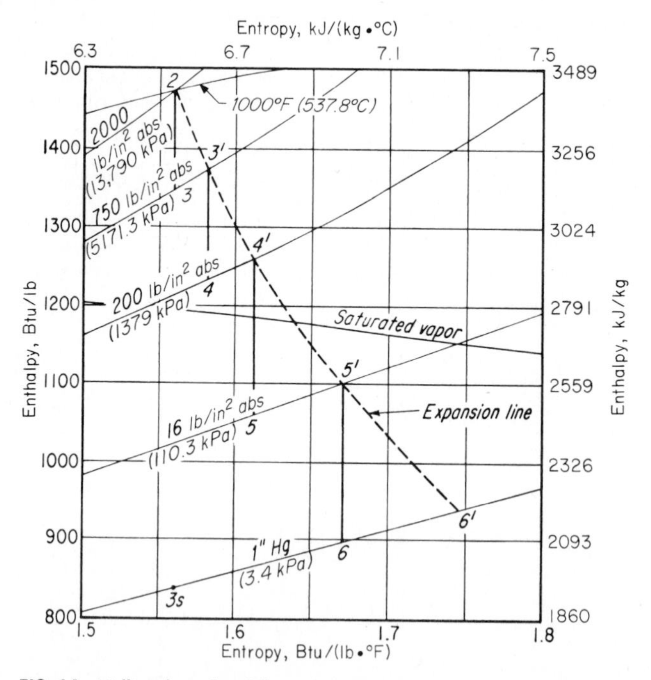

FIG. 14 Mollier-chart plot of the cycle in Fig. 13.

1.5819 Btu/(lb·°F) [6.6 kJ/(kg·°C)] at 750 lb/in^2 (abs) (5171.3 kPa) and 1372.2 Btu/lb (3191.7 kJ/kg).

Assume constant-S expansion from $H_{3'}$ to H_4 at 200 lb/in^2 (abs) (1379.0 kPa), the inlet pressure for feedwater heater 2. Draw the vertical straight line 3′-4 on the Mollier chart, Fig. 14. By using a procedure similar to that for heater 1, $H_{4'} = H_{3'} - e_e(H_{3'} - H_4) = 1372.2 - 0.8(1372.2 - 1230.0) = 1258.4$ Btu/lb (2927.0 kJ/kg). This is the actual enthalpy of the steam at the inlet to heater 2. Plot this enthalpy on the 200-lb/in^2 (abs) (1379.0-kPa) pressure line of the Mollier chart, and find $S_{4'} = 1.613$ Btu/(lb·°F) [6.8 kJ/(kg·°C)], Fig. 14.

Using the same procedure with constant-S expansion from $H_{4'}$, we find $H_5 = 1059.5$ Btu/lb (2464.4 kJ/kg) at 16 lb/in^2 (abs) (110.3 kPa), the inlet pressure to heater 3. Next find $H_{5'} = H_{4'} - e_e(H_{4'} - H_5) = 1258.4 - 0.8(1258.4 - 1059.5) = 1099.2$ Btu/lb (2556.7 kJ/kg). From the Mollier chart find $S_{5'} = 1.671$ Btu/(lb·°F) [7.0 kJ/(kg·°C)], Fig. 14.

Using the same procedure with constant-S expansion from $H_{5'}$ to H_6, find $H_6 = 898.2$ Btu/lb (2089.2 kJ/kg) at 1 inHg absolute (3.4 kPa), the condenser inlet pressure. Then $H_{6'} = H_{5'} - e_e(H_{5'} - H_6) = 1099.2 - 0.8(1099.2 - 898.2) = 938.4$ Btu/lb (2182.7 kJ/kg), the actual enthalpy of the steam at the condenser inlet. Find, on the Mollier chart, the moisture in the turbine exhaust = 15.1 percent.

2. *Determine the overall engine efficiency*

Overall engine efficiency e_e is higher than the engine-section efficiency because there is partial available-energy recovery between sections. Constant-S expansion from the throttle to the 1-inHg absolute (3.4-kPa) exhaust gives H_{3S}, Fig. 14, as 838.3 Btu/lb (1949.4 kJ/kg), assuming that all the steam flows to the condenser. Then, overall $e_e = H_2 - H_{6'}/(H_2 - H_{3S}) = 1474.5 - 938.4/1474.5 - 838.3 = 0.8425$, or 84.25 percent, compared with 0.8 or 80 percent, for individual engine sections.

3. *Compute the bleed-steam flow to each feedwater heater*

For each heater, energy in = energy out. Also, the heated condensate leaving each heater is a saturated liquid at the heater bleed-steam pressure. To simplify this calculation, assume negligible steam pressure drop between the turbine bleed point and the heater inlet. This assumption is permissible when the distance between the heater and bleed point is small. Determine the pump work by using the chart accompanying the compressed-liquid table in Keenan and Keyes—*Thermodynamic Properties of Steam*, or the ASME—*Steam Tables*.

For heater 1, energy in = energy out, or $H_{3'}m_1 + H_{12}(1 - m_1) = H_{13}$, where m = bleed-stream flow to the feedwater heater, lb/lb of throttle steam flow. (The subscript refers to the heater under consideration.) Then, $H_{3'}m_1 + (H_{11} + W_{p2})(1 - m_1) = H_{13}$, where W_{p2} = work done by pump 2, Fig. 13, in Btu/lb per pound of throttle flow. Then $1372.2m_1 + (355.4 + 1.7)(1 - m_1) = 500.8$; $m_1 = 0.1416$ lb/lb (0.064 kg/kg) throttle flow; $H_1 = H_{13} + W_{p1} = 500.8 + 4.7 = 505.5$ Btu/lb (1175.8 kJ/kg), where W_{p1} = work done by pump 1, Fig. 13. For each pump, find the work from the chart accompanying the compressed-liquid table in Keenan and Keyes—*Steam Tables* by entering the chart at the heater inlet pressure and projecting vertically at constant entropy to the heater outlet pressure, which equals the next heater inlet pressure. Read the enthalpy values at the respective pressures, and subtract the smaller from the larger to obtain the pump work during passage of the feedwater through the pump from the lower to the higher pressure. Thus, $W_{p2} = 1.7 - 0.0 = 1.7$ Btu/lb (4.0 kJ/kg), from enthalpy values for 200 lb/in^2 (abs) (1379.0 kPa) and 750 lb/in^2 (abs) (5171.3 kPa), the heater inlet and discharge pressures, respectively.

For heater 2, energy in = energy out, or $H_{4'}m_2 + H_{10}(1 - m_1 - m_2) = H_{11}(1 - m_1)H_{4'}m_2 + (H_9 + W_{p3})(1 - m_1 - m_2) = H_{11}(1 - m_1)1258.4m_2 + (184.4 + 0.5)(0.8584 - m_2) = 355.4(0.8584)m_2 = 0.1365$ lb/lb (0.0619 kg/kg) throttle flow.

For heater 3, energy in = energy out, or $H_{5'}m_3 + H_8(1 - m_1 - m_2 - m_3) = H_9(1 - m_1 - m_2)H_{5'}m_3 + (H_7 + W_{p4})(1 - m_1 - m_2 - m_3) = H_9(1 - m_1 - m_2)1099.2m_3 + (47.1 + 0.1)(0.7210 - m_3) = 184.4(0.7219)m_3 = 0.0942$ lb/lb (0.0427 kg/kg) throttle flow.

4. *Compute the turbine work output*

The work output per section W Btu is $W_1 = H_2 - H_{3'} = 1474.5 - 1372.1 = 102.3$ Btu (107.9 kJ), from the previously computed enthalpy values. Also $W_2 = (H_{3'} - H_{4'})(1 - m_1) = (1372.2 - 1258.4)(1 - 0.1416) = 97.7$ Btu (103.1 kJ); $W_3 = (H_{4'} - H_{5'})(1 - m_1 - m_2) = (1258.4 - 1099.2)(1 - 0.1416 - 0.1365) = 115.0$ Btu (121.3 kJ); $W_4 = (H_{5'} - H_{6'})(1 - m_1 - m_2 - m_3)$

= (1099.2 − 938.4)(1 − 0.1416 − 0.1365 − 0.0942) = 100.9 Btu (106.5 kJ). The total work output of the turbine = $W_e = \Sigma W$ = 102.3 + 97.7 + 115.0 + 100.9 = 415.9 Btu (438.8 kJ). The total $W_p = \Sigma W_p = W_{p1} + W_{p2} + W_{p3} + W_{p4}$ = 4.7 + 1.7 + 0.5 + 0.1 = 7.0 Btu (7.4 kJ).

Since the station auxiliaries consume 6 percent of W_e, the auxiliary consumption = 0.6(415.9) = 25.0 Btu (26.4 kJ). Then, net station work w = 415.9 − 7.0 − 25.0 = 383.9 Btu (405.0 kJ).

5. *Check the turbine work output*

The heat added to the cycle Q_a Btu/lb = $H_2 - H_1$ = 1474.5 − 505.5 = 969.0 Btu (1022.3 kJ). The heat rejected from the cycle Q_r Btu/lb = $(H_{6'} - H_7)(1 - m_1 - m_2 - m_3)$ = (938.4 − 47.1)(0.6277) = 559.5 Btu (590.3 kJ). Then $W_e - W_p = Q_a - Q_r$ = 969.0 − 559.5 = 409.5 Btu (432.0 kJ).

Compare this with $W_e - W_p$ computed earlier, or 415.9 − 7.0 = 408.9 Btu (431.4 kJ), or a difference of 409.5 − 408.9 = 0.6 Btu (0.63 kJ). This is an accurate check; the difference of 0.6 Btu (0.63 kJ) comes from errors in Mollier chart and calculator readings. Assume 408.9 Btu (431.4 kJ) is correct because it is the lower of the two values.

6. *Compute the plant and turbine efficiencies*

Plant energy input = Q_a/e_b, where e_b = boiler efficiency. Then plant energy input = 969.0/0.88 = 1101.0 Btu (1161.6 kJ). Plant thermal efficiency = $W/(Q_a/e_b)$ = 383.9 = 1101.0 = 0.3486. Turbine thermal efficiency = W_e/Q_a = 415.9/969.0 = 0.4292. Plant heat rate = 3413/0.3486 = 9970 Btu/kWh (10,329.0 kJ/kWh), where 3413 = Btu/kWh. Turbine heat rate = 3413/0.4292 = 7950 Btu/kWh (8387.7 kJ/kWh). Turbine throttle steam rate = (turbine heat rate)/$(H_2 - H_1)$ = 7950/(1474.5 − 505.5) = 8.21 lb/kWh (3.7 kg/kWh).

Related Calculations: By using the procedures given, the following values can be computed for any actual steam cycle: engine or turbine efficiency e_e; steam enthalpy at the main-condenser inlet; bleed-steam flow to a feedwater heater; turbine or engine work output per section; total turbine or engine work output; station auxiliary power consumption; net station work output; plant energy input; plant thermal efficiency; turbine or engine thermal efficiency; plant heat rate; turbine or engine heat rate; turbine throttle heat rate. To compute any of these values, use the equations given and insert the applicable variables.

REHEAT-STEAM CYCLE PERFORMANCE

A reheat-steam cycle has a 2000 lb/in² (abs) (13,790-kPa) throttle pressure at the turbine inlet and a 400-lb/in² (abs) (2758-kPa) reheat pressure. The throttle and reheat temperature of the steam is 1000°F (537.8°C); condenser pressure is 1 inHg absolute (3.4 kPa); engine efficiency of the high-pressure and low-pressure turbines is 80 percent. Find the cycle thermal efficiency.

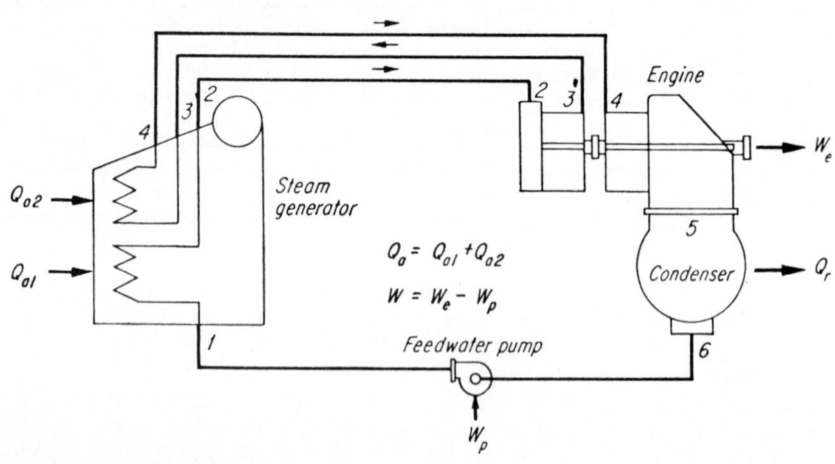

FIG. 15 Typical steam reheat cycle.

Calculation Procedure:

1. *Sketch the cycle layout and cycle T-S diagram*

Figures 15 and 16 show the cycle layout and T-S diagram with each important point numbered. Use a cycle layout and T-S diagram for every calculation of this type because it reduces the possibility of errors.

2. *Determine the throttle-steam properties from the steam tables*

Use the superheated steam tables, entering at 2000 lb/in^2 (abs) (13,790 kPa) and 1000°F (537.8°C) to find throttle-steam properties. Applying the symbols of the T-S diagram in Fig. 16, we get H_2 = 1474.5 Btu/lb (3429.7 kJ/kg); S_2 = 1.5603 Btu/(lb·°F) [6.5 kJ/(kg·°C)].

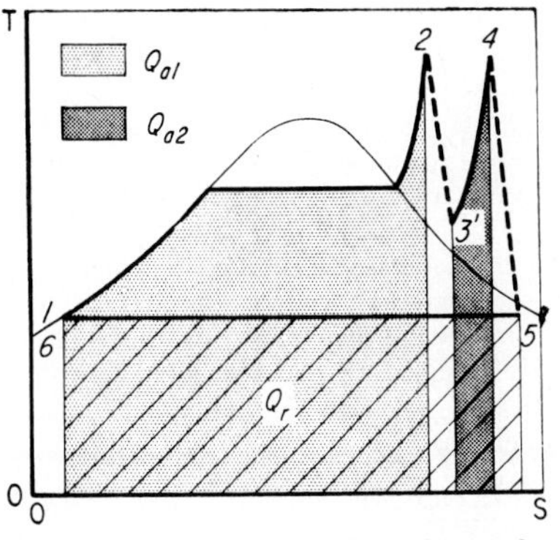

FIG. 16 Irreversible expansion in reheat cycle.

3. *Find the reheat-steam enthalpy*

Assume a constant-entropy expansion of the steam from 2000 to 400 lb/in^2 (13,790 to 2758 kPa). Trace this expansion on a Mollier (H-S) chart, Fig. 1, where a constant-entropy process is a vertical line between the initial [2000 lb/in^2 (abs) or 13,790 kPa] and reheat [400 lb/in^2 (abs) or 2758 kPa] pressures. Read on the Mollier chart H_3 = 1276.8 Btu/lb (2969.8 kJ/kg) at 400 lb/in^2 (abs) (2758 kPa).

4. *Compute the actual reheat properties*

The ideal enthalpy drop, throttle to reheat = $H_2 - H_3$ = 1474.5 − 1276.8 = 197.7 Btu/lb (459.9 kJ/kg). The actual enthalpy drop = (ideal drop)(turbine efficiency) = $H_2 - H_{3'}$ = 197.5(0.8) = 158.2 Btu/lb (368.0 kJ/kg) = W_{e1} = work output in the high-pressure section of the turbine.

Once W_{e1} is known, $H_{3'}$ can be computed from $H_{3'} = H_2 - W_{e1}$ = 1474.5 − 158.2 = 1316.3 Btu/lb (3061.7 kJ/kg).

The steam now returns to the boiler and leaves at condition 4, where P_4 = 400 lb/in^2 (abs) (2758 kPa); T_4 = 1000°F (537.8°C); S_4 = 1.7623 Btu/(lb·°F) [7.4 kJ/(kg·°C)]; H_4 = 1522.4 Btu/lb (3541.1 kJ/kg) from the superheated-steam table.

5. *Compute the exhaust-steam properties*

Use the Mollier chart and an assumed constant-entropy expansion to 1 inHg (3.4 kPa) absolute to determine the ideal exhaust enthalpy, or H_5 = 947.4 Btu/lb (2203.7 kJ/kg). The ideal work of the low-pressure section of the turbine is then $H_4 - H_5$ = 1522.4 − 947.4 = 575.0 Btu/lb (1338 kJ/kg). The actual work output of the low-pressure section of the turbine is $W_{e2} = H_4 - H_{5'}$ = 575.0(0.8) = 460.8 Btu/lb (1071.1 kJ/kg).

Once W_{e2} is known, $H_{5'}$ can be computed from $H_{5'} = H_4 - W_{e2}$ = 1522.4 − 460.0 = 1062.4 Btu/lb (2471.1 kJ/kg).

The enthalpy of the saturated liquid at the condenser pressure is found in the saturation-pressure steam table at 1 inHg absolute (3.4 kPa) = H_6 = 47.1 Btu/lb (109.5 kJ/kg).

The pump work W_p from the compressed-liquid table diagram in the stream tables is W_p = 5.5 Btu/lb (12.8 kJ/kg). Then the enthalpy of the water entering the boiler $H_1 = H_6 + W_p$ = 47.1 + 5.5 = 52.6 Btu/lb (122.3 kJ/kg).

6. *Compute the cycle thermal efficiency*

For any reheat cycle,

e = cycle thermal efficiency

$$= \frac{(H_2 - H_{3'}) + (H_4 - H_{5'}) - W_p}{(H_2 - H_1) + (H_4 - H_{3'})} = \frac{(1474.5 - 1316.3) + (1522.4 - 1062.4) - 5.5}{(1474.5 - 52.6) + (1522.4 - 1316.3)}$$

= 0.3766, or 37.66 percent

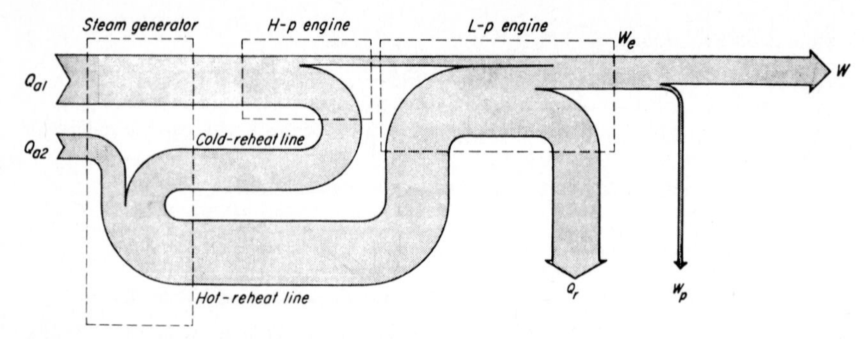

FIG. 17 Energy-flow diagram for reheat cycle in Fig. 15.

Figure 17 is an energy-flow diagram for the reheat cycle analyzed here. This diagram shows that the fuel burned in the steam generator to produce energy flow Q_{a1} is the largest part of the total energy input. The cold-reheat line carries the major share of energy leaving the high-pressure turbine.

Related Calculations: Reheat-regenerative cycles are used in some large power plants. Figure 18 shows a typical layout for such a cycle having three stages of feedwater heating and one stage of reheating. The heat balance for this cycle is computed as shown above, with the bleed-flow terms m computed by setting up an energy balance around each heater, as in earlier calculation procedures.

By using a T-S diagram, Fig. 19, the cycle thermal efficiency is

$$e = \frac{W}{Q_a} = \frac{Q_a - Q_r}{Q_a} = 1 - \frac{Q_r}{Q_{a1} + Q_{a2}}$$

Based on 1 lb (0.5 kg) of working fluid entering the steam generator and turbine throttle,

$$Q_r = (1 - m_1 - m_2 - m_3)(H_7 - H_8)$$

$$Q_{a1} = (H_2 - H_1)$$

$$Q_{a2} = (1 - m_1)(H_4 - H_3)$$

Figure 20 shows the energy-flow chart for this cycle.

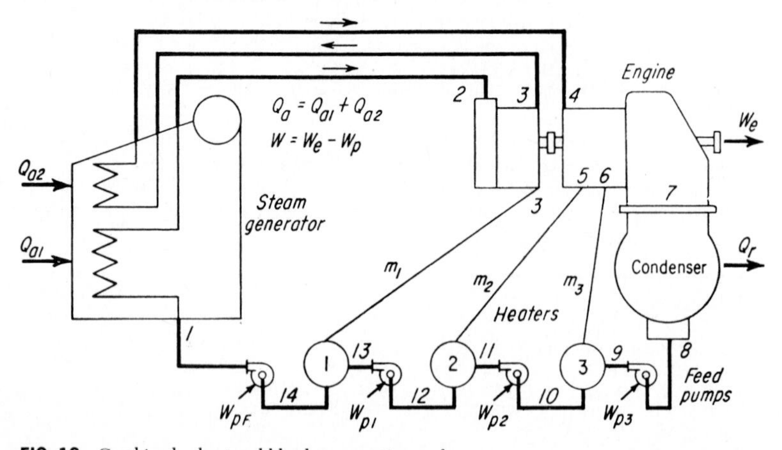

FIG. 18 Combined reheat and bleed-regenerative cycle.

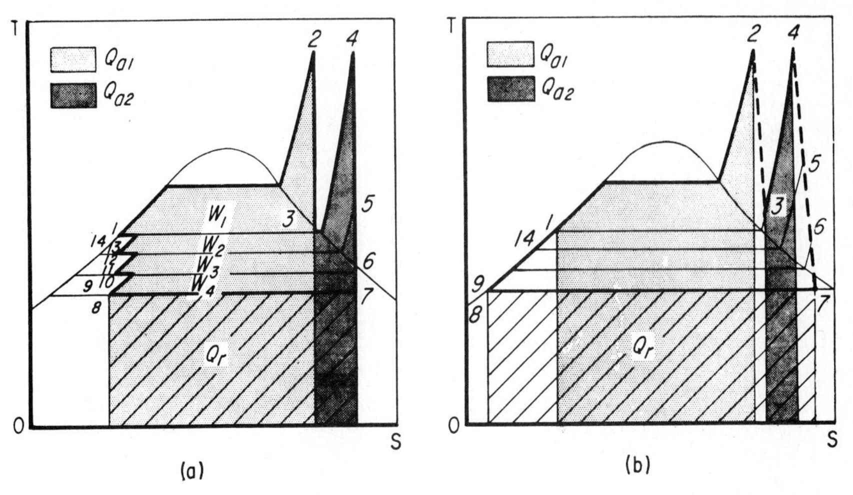

FIG. 19 (*a*) *T-S* diagram for ideal reheat-regenerative-bleed cycle; (*b*) *T-S* diagram for actual cycle.

Some high-pressure plants use two stages of reheating, Fig. 21, to raise the cycle efficiency. With two stages of reheating, the maximum number generally used, and values from Fig. 21,

$$e = \frac{(H_2 - H_3) + (H_4 - H_5) + (H_6 - H_7) - W_p}{(H_2 - H_1) + (H_4 - H_3) + (H_6 - H_5)}$$

MECHANICAL-DRIVE STEAM-TURBINE POWER-OUTPUT ANALYSIS

Show the effect of turbine engine efficiency on the condition lines of a turbine having engine efficiencies of 100 (isentropic expansion), 75, 50, 25, and 0 percent. How much of the available energy is converted to useful work for each engine efficiency? Sketch the effect of different steam inlet pressures on the condition line of a single-nozzle turbine at various loads. What is the available energy, Btu/lb of steam, in a noncondensing steam turbine having an inlet pressure of 1000 lb/in^2 (abs) (6895 kPa) and an exhaust pressure of 100 lb/in^2 (gage) (689.5 kPa)? How much work

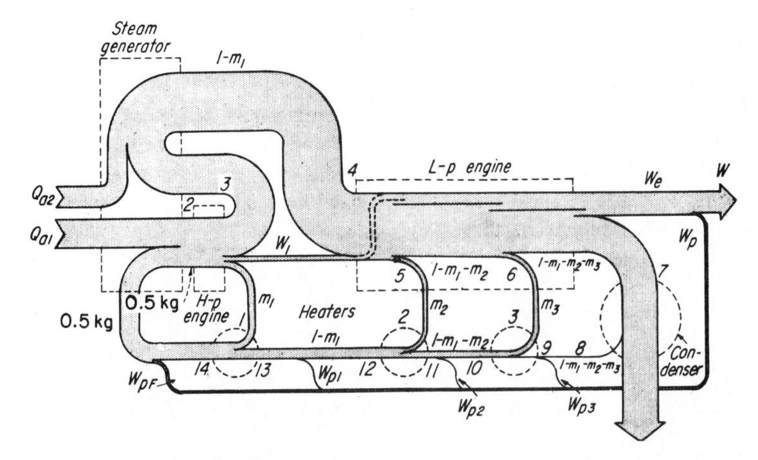

FIG. 20 Energy flow of cycle in Fig. 18.

will this turbine perform if the steam flow rate to it is 1000 lb/s (453.6 kg/s) and the engine efficiency is 40 percent?

Calculation Procedure:

1. *Sketch the condition lines on the Mollier chart*

Draw on the Mollier chart for steam initial- and exhaust-pressure lines, Fig. 22, and the initial-temperature line. For an isentropic expansion, the entropy is constant during the expansion, and the engine efficiency = 100 percent. The expansion or condition line is a vertical trace from h_1 on the initial-pressure line to h_{2s} on the exhaust-pressure line. Draw this line as shown in Fig. 22.

For zero percent engine efficiency, the other extreme in the efficiency range, $h_1 = h_2$ and the condition line is a horizontal line. Draw this line as shown in Fig. 22.

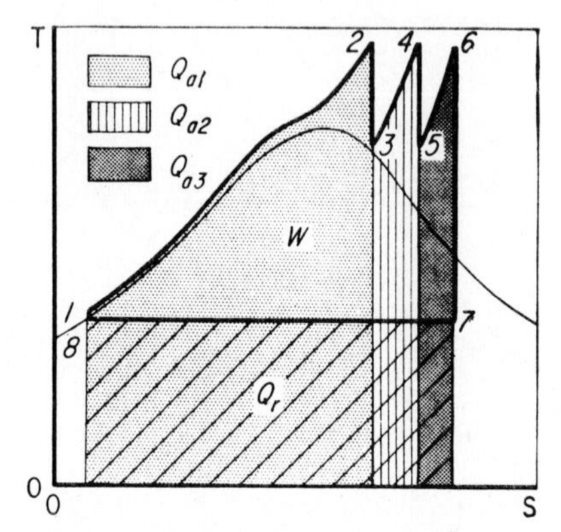

FIG. 21 *T-S* diagram for multiple reheat stages.

Between 0 and 100 percent engine efficiency, the condition lines become more nearly vertical as the engine efficiency approaches 100 percent, or an isentropic expansion. Draw the condition lines for 25, 50, and 75 percent efficiency, as shown in Fig. 22.

For the isentropic expansion, the available energy = $h_1 - h_{2s}$, Btu/lb of steam. This is the energy that an ideal turbine would make available.

For actual turbines, the enthalpy at the exhaust pressure $h_2 = h_1 -$ (available energy)(engine efficiency)/100, where available energy = $h_1 - h_{2s}$ for an ideal turbine working between the same initial and exhaust pressures. Thus, the available energy converted to useful work for any engine efficiency = (ideal available energy, Btu/lb)(engine efficiency, percent)/100. Using this relation, the available energy at each of the given engine efficiencies is found by substituting the ideal available energy and the actual engine efficiency.

2. *Sketch the condition lines for various throttle pressures*

Draw the throttle- and exhaust-pressure lines on the Mollier chart, Fig. 23. Since the inlet control valve throttles the steam flow as the load on the turbine decreases, the pressure of the steam

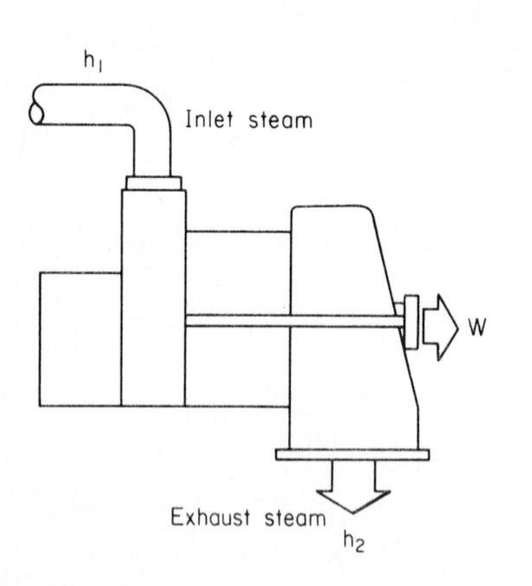

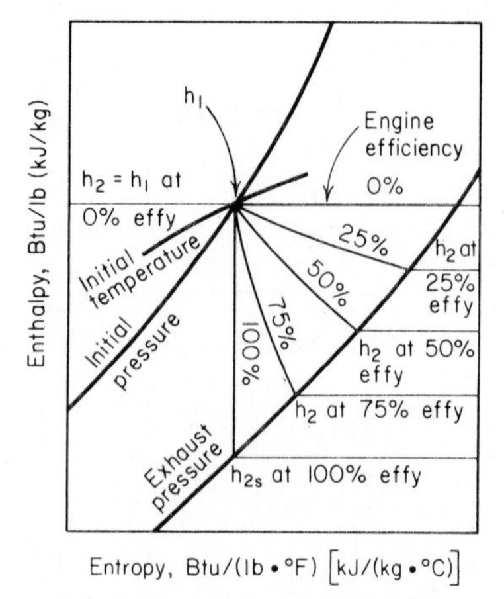

FIG. 22 Mollier chart of turbine condition lines.

entering the turbine nozzle is lower at reduced loads. Show this throttling effect by indicating the lower inlet pressure lines, Fig. 23, for the reduced loads. Note that the lowest inlet pressure occurs at the minimum plotted load—25 percent of full load—and the maximum inlet pressure at 125 percent of full load. As the turbine inlet steam pressure decreases, so does the available energy, because the exhaust enthalpy rises with decreasing load.

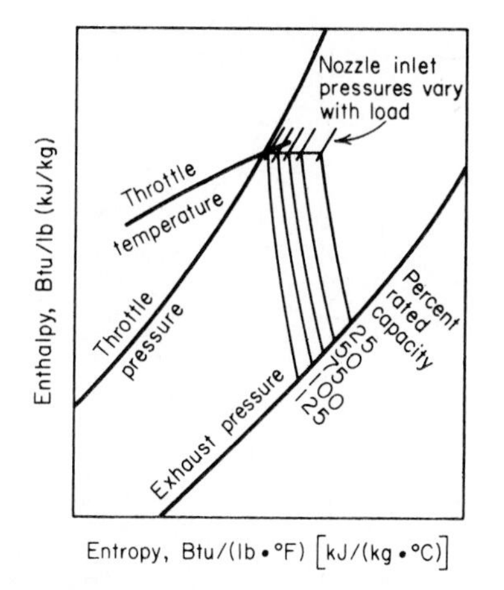

FIG. 23 Turbine condition line shifts as the inlet steam pressure varies.

3. *Compute the turbine available energy and power output*

Use a noncondensing-turbine performance chart, Fig. 24, to determine the available energy. Enter the bottom of the chart at 1000 lb/in^2 (abs) (6895 kPa) and project vertically upward until the 100-lb/in^2 (gage) (689.5-kPa) exhaust-pressure curve is intersected. At the left, read the available energy as 205 Btu/lb (476.8 kJ/kg) of steam.

With the available energy, flow rate, and engine efficiency known, the work output = (available energy, Btu/lb)(flow rate, lb/s)(engine efficiency/100)/[550 ft·lb/(s·hp)]. [*Note:* 550 ft·lb/(s·hp) = 1 N·m/(W·s).] For this turbine, work output = (205 Btu/lb)(1000 lb/s)(40/100)/550 = 149 hp (111.1 kW).

Related Calculations: Use the steps given here to analyze single-stage noncondensing mechanical-drive turbines for stationary, portable, or marine applications. Performance curves such as Fig. 24 are available from turbine manufacturers. Single-stage noncondensing turbines are used for feed-pump, draft-fan, and auxiliary-generator drive.

CONDENSING STEAM-TURBINE POWER-OUTPUT ANALYSIS

What is the available energy in steam supplied to a 5000-kW turbine if the inlet steam conditions are 1000 lb/in^2 (abs) (6895 kPa) and 800°F (426.7°C) and the turbine exhausts at 1 inHg absolute (3.4 kPa)? Determine the theoretical and actual heat rate of this turbine if its engine efficiency is 74 percent. What are the full-load output and steam rate of the turbine?

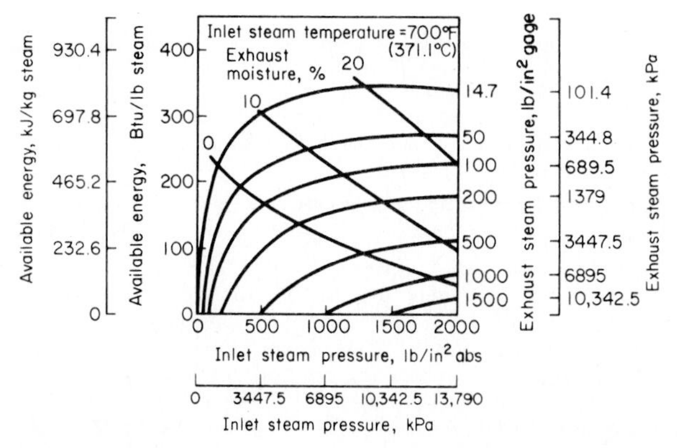

FIG. 24 Available energy in turbine depends on the initial steam state and the exhaust pressure.

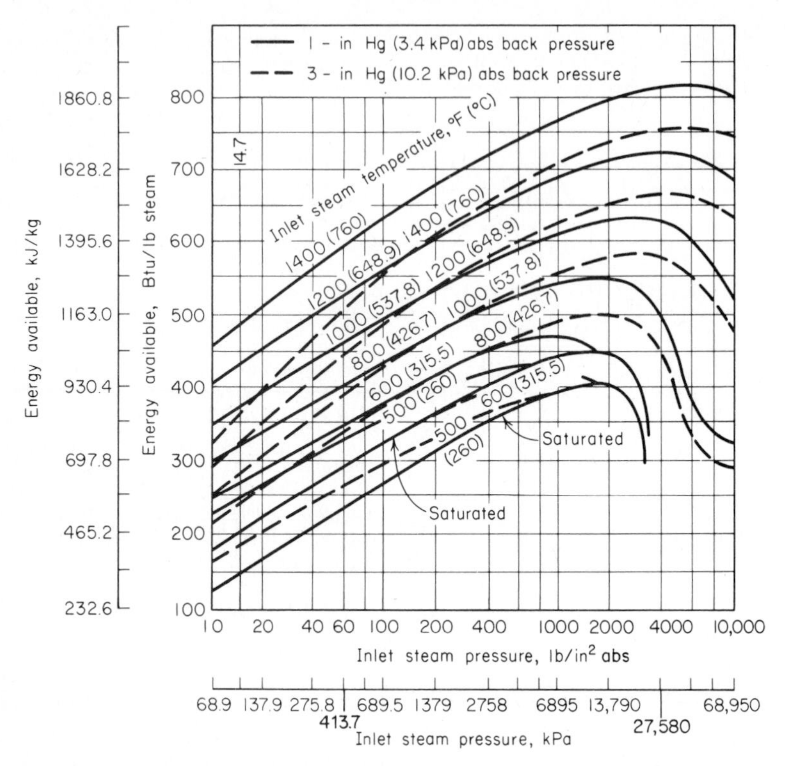

FIG. 25 Available energy for typical condensing turbines.

1. *Determine the available energy in the steam*

Enter Fig. 25 at the bottom at 1000-lb/in^2 (abs) (6895.0-kPa) inlet pressure, and project vertically upward to the 800°F (426.7°C) 1-in (3.4-kPa) exhaust-pressure curve. At the left, read the available energy as 545 Btu/lb (1267.7 kJ/kg) of steam.

2. *Determine the heat rate of the turbine*

Enter Fig. 26 at an initial steam temperature of 800°F (426.7°C), and project vertically upward to the 1000-lb/in^2 (abs) (6895.0-kPa) 1-in (3.4-kPa) curve. At the left, read the theoretical heat rate as 8400 Btu/kWh (8862.5 kJ/kWh).

When the theoretical heat rate is known, the actual heat rate is found from: actual heat rate HR, Btu/kWh = (theoretical heat rate, Btu/kWh)/(engine efficiency). Or, actual HR = 8400/0.74 = 11,350 Btu/kWh (11,974.9 kJ/kWh).

3. *Compute the full-load output and steam rate*

The energy converted to work, Btu/lb of steam = (available energy, Btu/lb of steam)(engine efficiency) = (545)(0.74) = 403 Btu/lb of steam (937.4 kJ/kg).

For any prime mover driving a generator, the full-load output, Btu = (generator kW rating)(3413 Btu/kWh) = (5000)(3413) = 17,060,000 Btu/h (4999.8 kJ/s).

The steam flow = (full-load output, Btu/h)/(work output, Btu/lb) = 17,060,000/403 = 42,300 lb/h (19,035 kg/h) of steam. Then the full-load steam rate of the turbine, lb/kWh = (steam flow, lb/h)/(kW output at full load) = 42,300/5000 = 8.46 lb/kWh (3.8 kg/kWh).

Related Calculations: Use this general procedure to determine the available energy, theoretical and actual heat rates, and full-load output and steam rate for any stationary, marine, or portable condensing steam turbine operating within the ranges of Figs. 25 and 26. If the actual

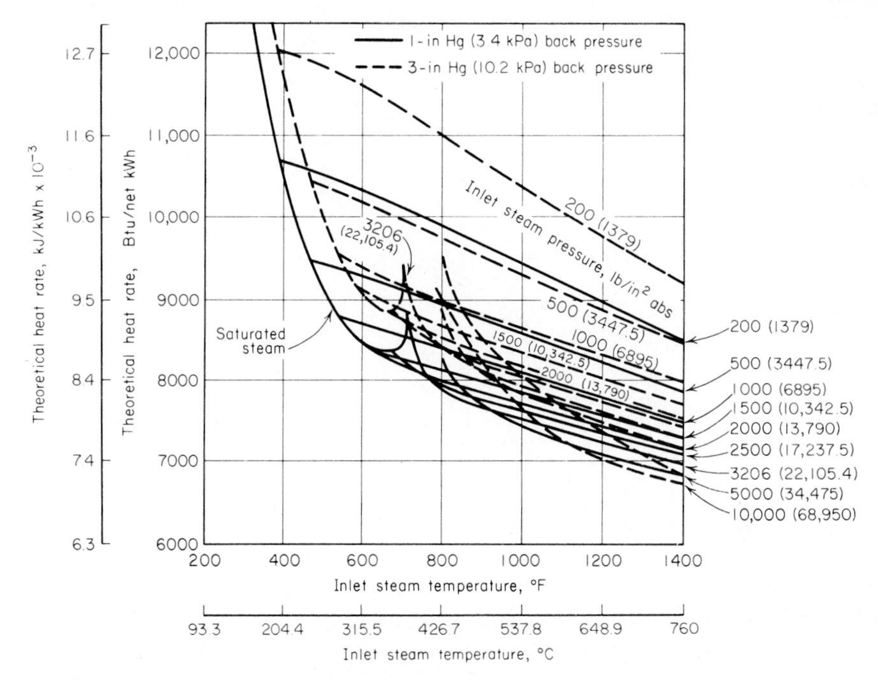

FIG. 26 Theoretical heat rate for condensing turbines.

performance curves are available, use them instead of Figs. 25 and 26. The curves given here are suitable for all preliminary estimates for condensing turbines operating with exhaust pressures of 1 or 3 inHg absolute (3.4 or 10.2 kPa). Many modern turbines operate under these conditions.

STEAM-TURBINE REGENERATIVE-CYCLE PERFORMANCE

When throttle steam is at 1000 lb/in^2 (abs) (6895 kPa) and 800°F (426.7°C) and the exhaust pressure is 1 inHg (3.4 kPa) absolute, a 5000-kW condensing turbine has an actual heat rate of 11,350 Btu/kWh (11,974.9 kJ/kWh). Three feedwater heaters are added to the cycle, Fig. 27, to heat the feedwater to 70 percent of the maximum possible enthalpy rise. What is the actual heat rate of the turbine? If 10 heaters instead of 3 were used and the water enthalpy were raised to 90 percent of the maximum possible rise in these 10 heaters, would the reduction in the actual heat rate be appreciable?

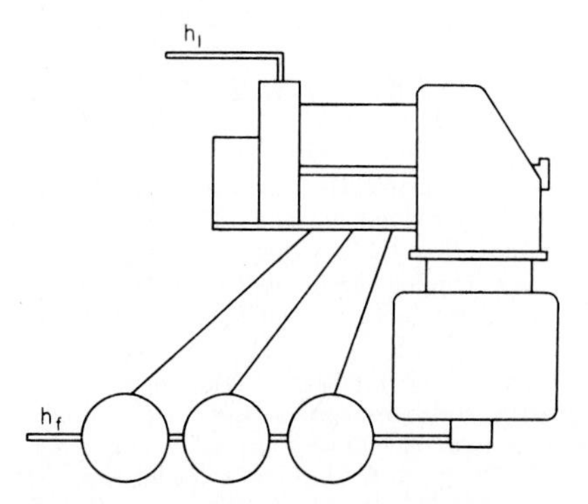

FIG. 27 Regenerative feedwater heating.

Calculation Procedure:

1. *Determine the actual enthalpy rise of the feedwater*

Enter Fig. 28 at the throttle pressure of 1000 lb/in^2 (abs) (6895 kPa), and project vertically upward to the 1-inHg (3.4-kPa) absolute back-pressure curve. At the left, read the maximum possible feedwater enthalpy rise as 495 Btu/lb (1151.4 kJ/kg). Since the actual rise is limited to 70 percent of the maximum possible rise by the conditions of the design, the actual enthalpy rise = (495)(0.70) = 346.5 Btu/lb (805.9 kJ/kg).

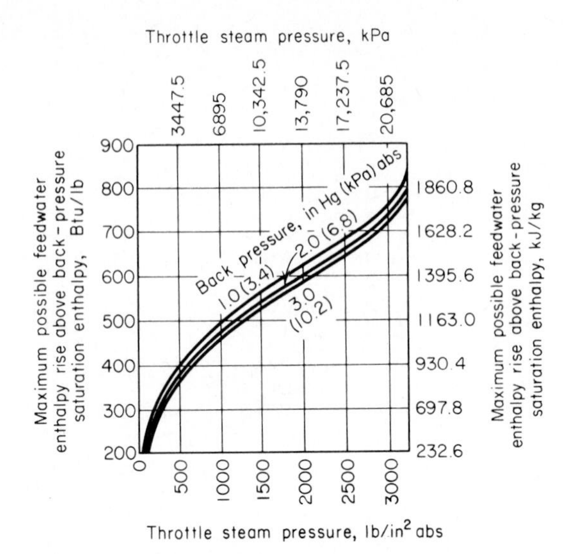

FIG. 28 Feedwater enthalpy rise.

2. ***Determine the heat-rate and heater-number correction factors***

Find the theoretical reduction in straight-condensing (no regenerative heaters) heat rates from Fig. 29. Enter the bottom of Fig. 29 at the inlet steam temperature, 800°F (426.7°C), and project vertically upward to the 1000-lb/in^2 (abs) (6895-kPa) 1-inHg (3.4-kPa) back-pressure curve. At the left, read the reduction in straight-condensing heat rate as 14.8 percent.

Next, enter Fig. 30 at the bottom at 70 percent of maximum possible rise in feedwater enthalpy, and project vertically to the three-heater curve. At the left, read the reduction in straight-condensing heat rate for the number of heaters and actual enthalpy rise as 0.71.

3. ***Apply the heat-rate and heater-number correction factors***

Full-load regenerative-cycle heat rate, Btu/kWh = (straight-condensing heat rate, Btu/kWh) [1 − (heat-rate correction factor)(heater-number correction factor)] = (13,350)[1 − (0.148)(0.71)] = 10,160 Btu/kWh (10,719.4 kJ/kWh).

4. ***Find and apply the correction factors for the larger number of heaters***

Enter Fig. 30 at 90 percent of the maximum possible enthalpy rise, and project vertically to the 10-heater curve. At the left, read the heat-rate reduction for the number of heaters and actual enthalpy rise as 0.89.

Using the heat-rate correction factor from step 2 and 0.89, found above, we see that the full-load 10-heater regenerative-cycle heat rate = (11,350)[1 − (0.148)(0.89)] = 9850 Btu/kWh (10,392.3 kJ/kWh), by using the same procedure as in step 3. Thus, adding 10 − 3 = 7 heaters reduces the heat rate by 10,160 − 9850 = 310 Btu/kWh (327.1 kJ/kWh). This is a reduction of 3.05 percent.

To determine whether this reduction in heat rate is appreciable, the carrying charges on the extra heaters, piping, and pumps must be compared with the reduction in annual fuel costs resulting from the lower heat rate. If the fuel saving is greater than the carrying charges, the larger number of heaters can usually be justified. In this case, tripling the number of heaters would probably increase the carrying charges to a level exceeding the fuel savings. Therefore, the reduction in heat rate is probably not appreciable.

Related Calculations: Use the procedure given here to compute the actual heat rate of steam-turbine regenerative cycles for stationary, marine, and portable installations. Where necessary, use the steps of the previous procedure to compute the actual heat rate of a straight-condensing cycle before applying the present procedure. The performance curves given here are suitable for first approximations in situations where actual performance curves are unavailable.

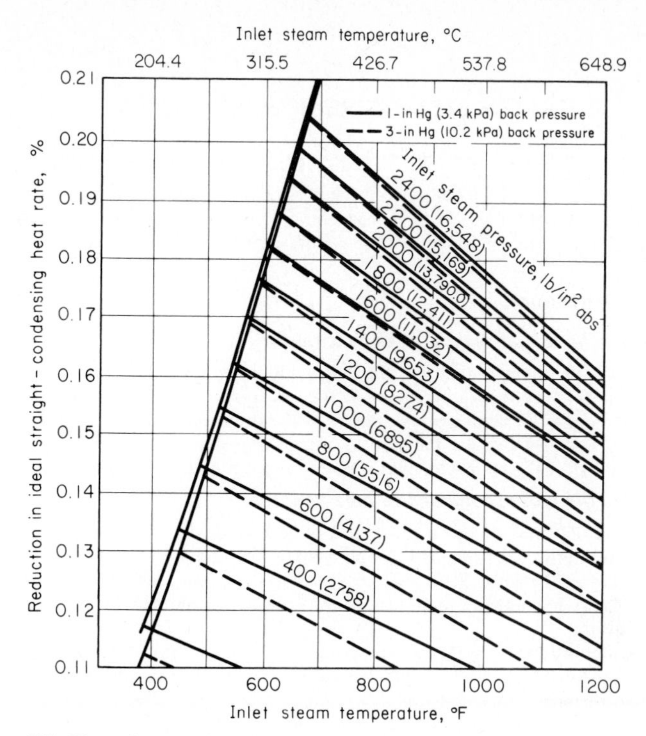

FIG. 29 Reduction in straight-condensing heat rate obtained by regenerative heating.

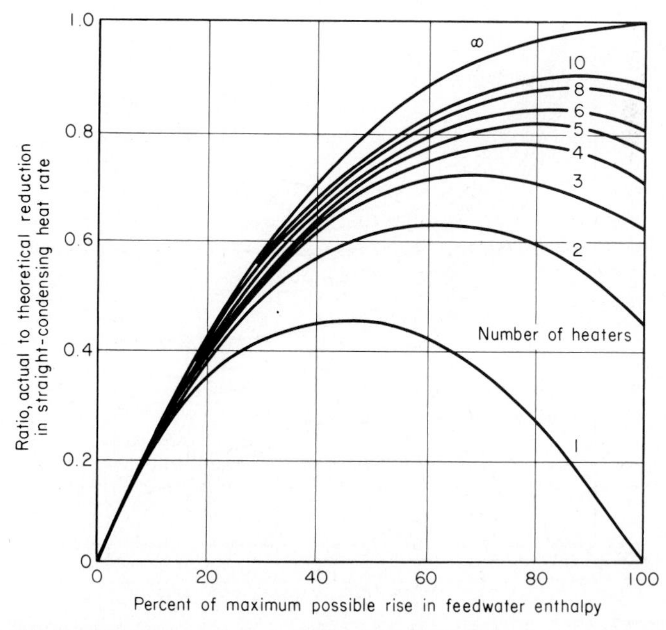

FIG. 30 Maximum possible rise in feedwater enthalpy varies with the number of heaters used.

REHEAT-REGENERATIVE STEAM-TURBINE HEAT RATES

What are the net and gross heat rates of a 300-kW reheat turbine having an initial steam pressure of 3500 lb/in^2 (gage) (24,132.5 kPa) with initial and reheat steam temperatures of 1000°F (537.8°C) with 1.5 inHg (5.1 kPa) absolute back pressure and six stages of regenerative feedwater heating? Compare this heat rate with that of 3500 lb/in^2 (gage) (24,132.5 kPa) 600-mW cross-compound four-flow turbine with 3600/1800 r/min shafts at a 300-mW load.

Calculation Procedure:

1. *Determine the reheat-regenerative heat rate*

Enter Fig. 31 at 3500-lb/in^2 (gage) (24,132.5-kPa) initial steam pressure, and project vertically to the 300-mW capacity net-heat-rate curve. At the left, read the net heat rate as 7680 Btu/kWh (8102.6 kJ/kWh). On the same vertical line, read the gross heat rate as 7350 Btu/kWh (7754.7 kJ/kWh). The gross heat rate is computed by using the generator-terminal output; the net heat rate is computed after the feedwater-pump energy input is deducted from the generator output.

2. *Determine the cross-compound turbine heat rate*

Enter Fig. 32 at 350 mW at the bottom, and project vertically upward to 1.5-inHg (5.1-kPa) exhaust pressure midway between the 1- and 2-inHg (3.4-and 6.8-kPa) curves. At the left, read the net heat rate as 7880 Btu/kWh (8313.8 kJ/kWh). Thus, the reheat-regenerative unit has a lower net heat rate. Even at full rated load of the cross-compound turbine, its heat rate is higher than the reheat unit.

Related Calculations: Use this general procedure for comparing stationary and marine high-pressure steam turbines. The curves given here are typical of those supplied by turbine manufacturers for their turbines.

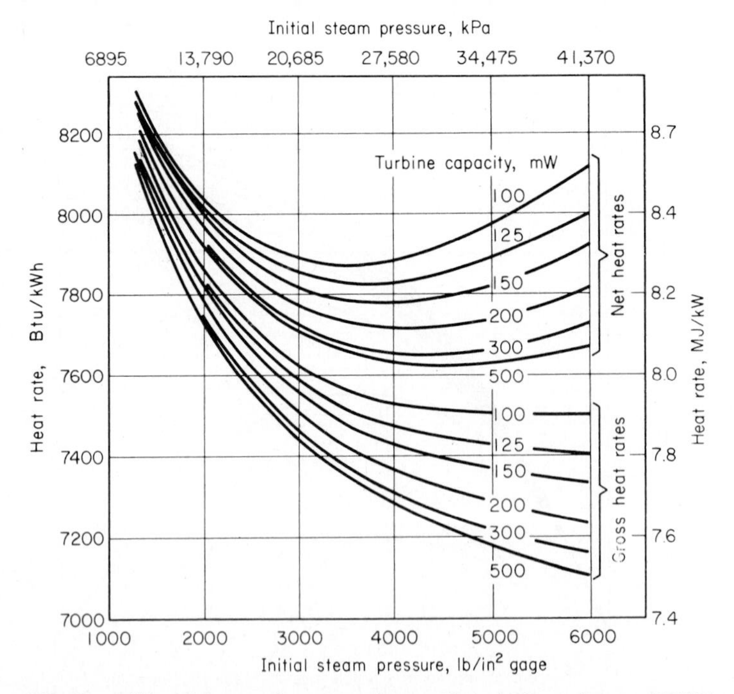

FIG. 31 Full-load heat rates for steam turbines with six feedwater heaters, 1000°F/1000°F (538°C/538°C) steam, 1.5-in (38.1-mm) Hg (abs) exhaust pressure.

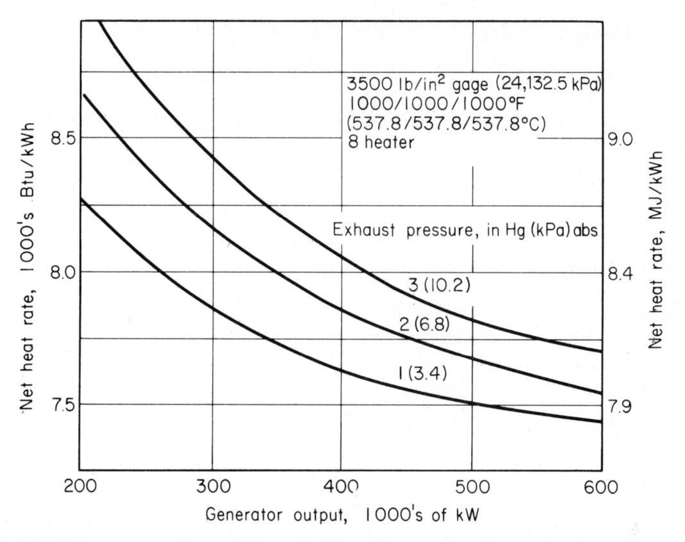

FIG. 32 Heat rate of a cross-compound four-flow steam turbine with 3600/1800-r/min shafts.

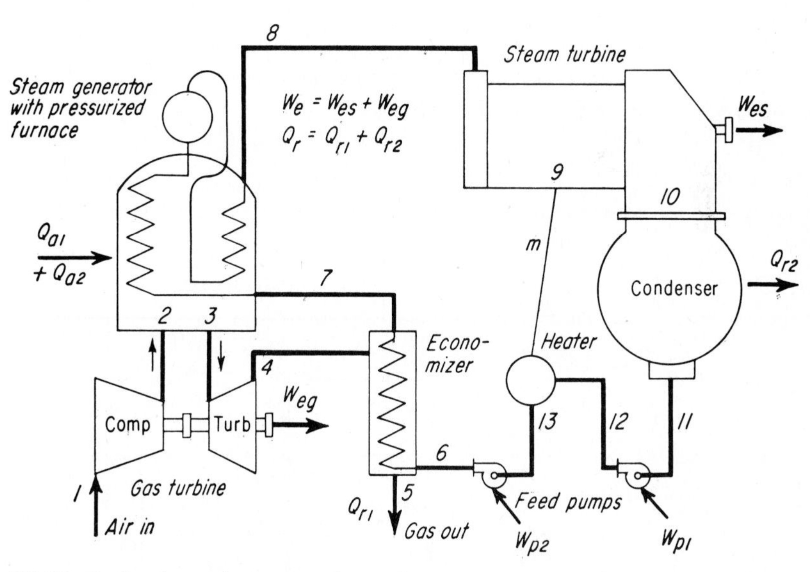

FIG. 33 Combined gas turbine–steam turbine cycle.

STEAM TURBINE–GAS TURBINE CYCLE ANALYSIS

Sketch the cycle layout, *T-S* diagram, and energy-flow chart for a combined steam turbine–gas turbine cycle having one stage of regenerative feedwater heating and one stage of economizer feedwater heating. Compute the thermal efficiency and heat rate of the combined cycle.

Calculation Procedure:

1. *Sketch the cycle layout*

Figure 33 shows the cycle. Since the gas-turbine exhaust-gas temperature is usually higher than the bleed-steam temperature, the economizer is placed after the regenerative feedwater heater. The feedwater will be progressively heated to a higher temperature during passage through the regenerative heater and the gas-turbine economizer. The cycle shown here is only one of many possible combinations of a steam plant and a gas turbine.

2. *Sketch the T-S diagram*

Figure 34 shows the *T-S* diagram for the combined gas turbine–steam turbine cycle. There is irreversible heat transfer Q_T from the gas-turbine exhaust to the feedwater in the economizer, which helps reduce the required energy input Q_{a2}.

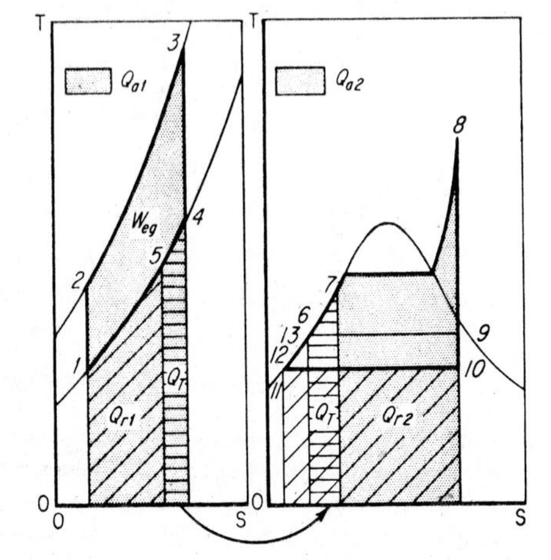

FIG. 34 *T-S* charts for combined gas turbine–steam turbine cycle have irreversible heat transfer *Q* from gas-turbine exhaust to the feedwater.

3. *Sketch the energy-flow chart*

Choose a suitable scale for the energy input, and proportion the energy flow to each of the other portions of the cycle. Use a single line when the flow is too small to plot to scale. Figure 35 shows the energy-flow chart.

4. *Determine the thermal efficiency of the cycle*

Since $e = W/Q_a$, $e = Q_a - Q_r/Q_a = 1 - [Q_{r1} + Q_{r2}/(Q_{a1} + Q_{a2})]$, given the notation in Figs. 33, 34, and 35.

The relative weight of the gas w_g to 1 lb (0.5 kg) of water must be computed by taking an energy balance about the economizer. Or, $H_7 - H_6 = w_g(H_4 - H_5)$. Using the actual values for the enthalpies, solve this equation for w_g.

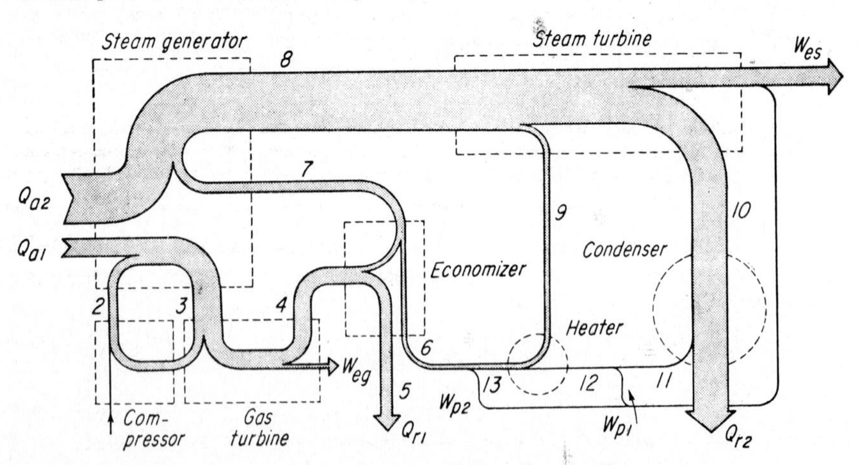

FIG. 35 Energy-flow chart of the gas turbine–steam turbine cycle in Fig. 33.

With w_g known, the other factors in the efficiency computation are

$$Q_{r1} = w_g(H_5 - H_1)$$

$$Q_{r2} = (1 - m)(H_{10} - H_{11})$$

$$Q_{a1} = w_g(H_3 - H_2)$$

$$Q_{a2} = H_8 - H_7$$

The bleed-steam flow m is calculated from an energy balance about the feedwater heater. Note that the units for the above equations can be any of those normally used in steam- and gas-turbine analyses.

STEAM-CONDENSER PERFORMANCE ANALYSIS

(*a*) Find the required tube surface area for a shell-and-tube type of condenser serving a steam turbine when the quantity of steam condensed S is 25,000 lb/h (3.1 kg/s); condenser back pressure = 2 inHg absolute (6.8 kPa); steam temperature t_s = 101.1°F (38.4°C); inlet water temperature t_1 = 80°F (26.7°C); tube length per pass L = 14 ft (4.3 m); water velocity V = 6.5 ft/s (2.0 m/s); number of passes = 2; tube size and gage: ¾ in (1.9 cm), no. 18 BWG; cleanliness factor = 0.80. (*b*) Compute the required area and cooling-water flow rate for the same conditions as (*a*) except that cooling water enters at 85°F (29.4°C). (*c*) If the steam flow through the condenser in (*a*) decreases to 15,000 lb/h (1.9 kg/s), what will be the absolute steam pressure in the condenser shell?

Calculation Procedure:

1. *Sketch the condenser, showing flow conditions*

(*a*) Figure 36 shows the condenser and the flow conditions prevailing.

2. *Determine the condenser heat-transfer coefficient*

Use standard condenser-tube engineering data available from the manufacturer or Heat Exchange Institute. Table 1 and Fig. 37 show typical condenser-tube data used in condenser selection. These data are based on a minimum water velocity of 3 ft/s (0.9 m/s) through the condenser tubes, a minimum absolute pressure of 0.7 inHg (2.4 kPa) in the condenser shell, and a minimum Δt terminal temperature difference $t_s - t_2$ of 5°F (2.8°C). These conditions are typical for power-plant surface condensers.

Enter Fig. 37 at the bottom at the given water velocity, 6.5 ft/s (2.0 m/s), and project vertically upward until the ¾-in (1.9-cm) OD tube curve is intersected. From this point, project horizontally to the left to read the heat-transfer coefficient U = 690 Btu/(ft²·°F) [14,104.8 kJ/(m²·°C)] LMTD (log mean temperature difference). Also read from Fig. 37 the temperature correction factor for an inlet-water temperature of 80°F (26.7°C) by entering at the bottom at 80°F (26.7°C) and projecting vertically upward to the temperature-correction curve. From the intersection with this curve, project to the right to read the correction as 1.04. Correct U for temperature and cleanliness by multiplying the value obtained from the chart by the correction factors, or U = 690(1.04)(0.80) = 574 Btu/(ft²·h·°F) [11,733.6 kJ/(m²·h·°C)] LMTD.

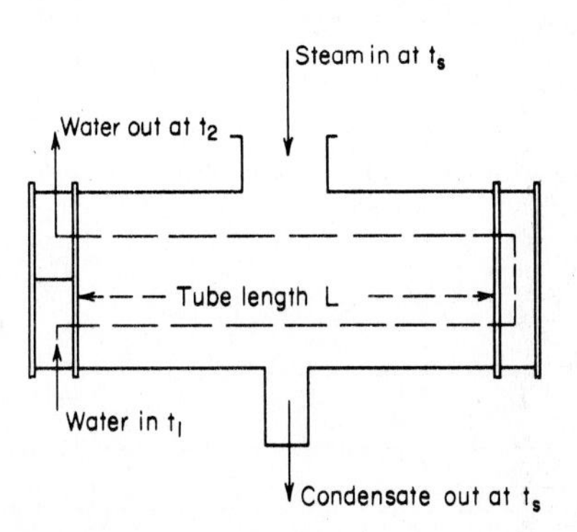

FIG. 36 Temperatures governing condenser performance.

3. *Compute the tube constant*

Read from Table 1, for two passes through ¾-in (1.9-cm) OD 18 BWG tubes, k = a constant = 0.377. Then kL/V = 0.377(14)/6.5 = 0.812.

TABLE 1 Standard Condenser Tube Data

Tube OD, in (cm)	Tube gage BWG	Tube ID, in (cm)	Surface area, ft²/ft (m²/m)		Velocity, ft/s for 1 gal/min (m/s for 1 L/min)	Value of k for number of tube passes		
			Outside	Inside		One	Two	Three
¾ (1.9)	18	0.652 (1.656)	0.1963 (0.0598)	0.1706 (0.0520)	0.9611 (0.0774)	0.188	0.377	0.565
	16	0.620 (1.575)	0.1963 (0.0598)	0.1613 (0.0492)	1.063 (0.0856)	0.208	0.417	0.625
	14	0.584 (1.483)	0.1963 (0.0598)	0.1528 (0.0466)	1.198 (0.0965)	0.235	0.470	0.705

4. *Compute the outlet-water temperature*

The equation for outlet-water temperature is $t_2 = t_s - (t_s - t_1/e^x)$, where $x = (kL/V)(U/500)$, or $x = 0.812(574/500) = 0.932$. Then $e^x = 2.7183^{0.932} = 2.54$. With this value known, $t_2 = 101.1 - (101.1 - 80/2.54) = 92.8$°F (33.8°C). Check to see that $\Delta t(t_s - t_2)$ is less than the minimum 5°F (2.8°C) terminal difference. Or, $101.1 - 92.8 = 8.3$°F (4.6°C), which is greater than 5°F (2.8°C).

5. *Compute the required tube surface area*

The required cooling-water flow, gal/min $= 950S/[500(t_2 - t_1)] = 950(25{,}000)/[500(92.8 - 80)] = 3700$ gal/min (233.4 L/s). This equation assumes that 950 Btu is to be removed from each

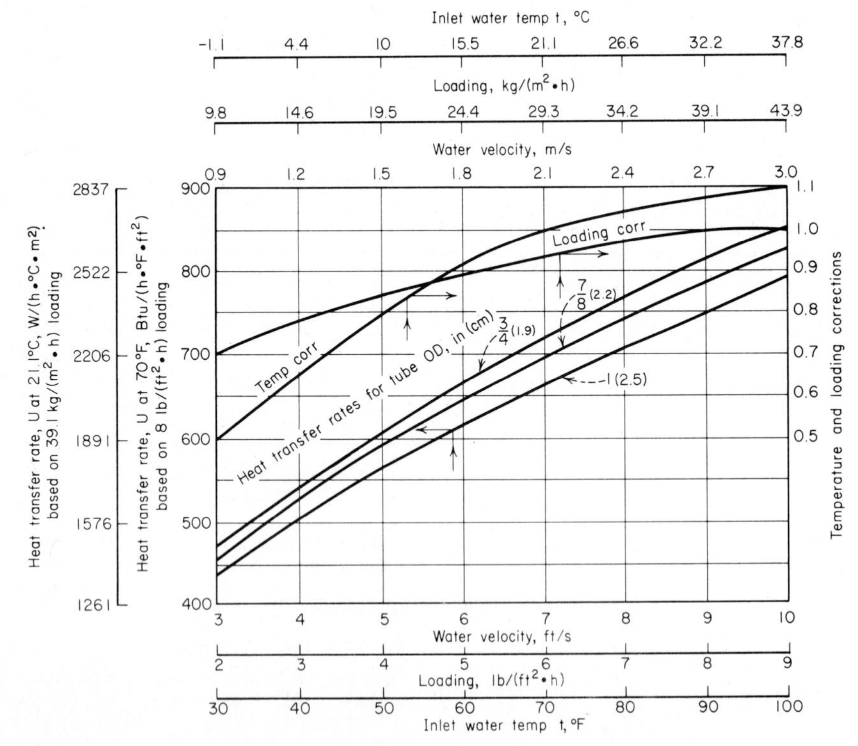

FIG. 37 Heat-transfer and correction curves for calculating surface-condenser performances.

pound (2209.7 kJ/kg) of steam condensed. When a different quantity of heat must be removed, use the actual quantity in place of the 950 in this equation.

With the tube constant kL/V and cooling-water flow rate known, the required area is computed from $A = (kL/V)\,(gpm) = (0.812)(3700) = 3000$ ft^2 (278.7 m^2).

Since the value of U was not corrected for condenser loading, it is necessary to check whether such a correction is needed. Condenser loading $= S/A = 25{,}000/3000 = 8.33$ lb/ft^2 (40.7 kg/m^2). Figure 37 shows that no correction (correction factor = 1.0) is necessary for loadings greater than 8.0 lb/ft^2 (39.1 kg/m^2). Therefore, the loading for this condenser is satisfactory without correction.

This step concludes the general calculation procedure for a surface condenser serving any steam turbine. The next procedure shows the method to follow when a higher cooling-water inlet temperature prevails.

6. *Compute the cooling-water outlet temperature*

(*b*) Higher cooling water temperature. From Fig. 37 for 85°F (29.4°C) cooling-water inlet temperature and a 0.80 cleanliness factor, $U = 690(1.06)(0.80) = 585$ Btu/(ft$^2\cdot$h$\cdot$°F) [3.3 kJ/(m$^2\cdot$°C$\cdot$s)] LMTD.

Given data from Table 1, the tube constant $kL/V = 0.377(14)/6.5 = 0.812$. Then $x = (kL/V)(U/500) = 0.812(585/500) = 0.950$. Using this exponent, we get $e^x = 2.8183^{0.950} = 2.586$. The cooling-water outlet temperature is then $t_2 = t_s - (t_s - t_1/e^x) = 101.1 - (101.1 - 85)/2.586 = 94.9$°F (34.9°C). Check to see that $\Delta t(t_s - t_2)$ is greater than the minimum 5°F (2.8°C) terminal temperature difference. Or, $101.1 - 94.9 = 6.5$°F (3.6°C), which is greater than 5°F (2.8°C).

7. *Compute the water flow rate, required area, and loading*

The required cooling-water flow, gal/min $= 950S/[500(t_2 - t_1)] = 950(25{,}000)/[500(94.9 - 85)] = 4800$ gal/min (302.8 L/s).

With the tube constant kL/V and cooling-water flow rate known, the required area is computed from $A = (kL/V)(gpm) = 0.812(4800) = 3900$ ft^2 (362.3 m^2). Then loading $= S/A = 25{,}000/3900 = 6.4$ lb/ft^2 (31.2 kg/m^2).

Since the loading is less than 8 lb/ft^2 (39.1 kg/m^2), refer to Fig. 37 to obtain the loading correction factor. Enter at the bottom at 6.4 lb/ft^2 (31.2 kg/m^2), and project vertically to the loading curve. At the right, read the loading correction factor as 0.95. Now the value of U already computed must be corrected, and all dependent quantities recalculated.

8. *Recalculate the condenser proportions*

First, correct U for loading. Or, $U = 585(0.95) = 555$. Then $x = 0.812(555/500) = 0.90$; $e^x = 2.7183^{0.90} = 2.46$; $t_2 = 101.1 - (101.1 - 85/2.46) = 94.6$°F (34.8°C). Check $\Delta t = t_s - t_2 = 101.1 - 94.6 = 6.5$°F (3.6°C), which is greater than 5°F (2.8°C). The cooling-water flow rate, gal/min $= 950\,(25{,}000)/[500(94.6 - 85)] = 4950$ gal/min (312.3 L/s). Then $A = 0.812(4950) = 4020$ ft^2 (373.5 m^2), and loading $= 25{,}000/4020 = 6.23$ lb/ft^2 (30.4 kg/m^2).

Check the correction factor for this loading in Fig. 37. The correction factor is 0.94, compared with 0.95 for the first calculation. Since the value of U would be changed only about 1 percent by using the lower factor, the calculations need not be revised further. Where U would change by a larger amount—say 5 percent or more—it would be necessary to repeat the procedure just detailed, applying the new correction factor.

Note that the 5°F (2.8°C) increase in cooling-water temperature (from 80 to 85°F or 26.7 to 29.4°C) requires an additional 1020 ft^2 (94.8 m^2) of condenser surface and 125 gal/min (7.9 L/s) of cooling-water flow to maintain the same back pressure. These increments will vary, depending on the temperature level at which the increase occurs. The effect of reduced steam flow on the steam pressure in the condenser shell will not be computed because the recalculation above is the last step in part (*b*) of this procedure.

(*c*) Reduced steam flow to condenser.

9. *Determine the condenser loading*

From procedure (*a*) above, the cooling-water flow = 3700 gal/min (233.4 L/s); condenser surface $A = 3000$ ft^2 (278.7 m^2). Then, with a 15,000-lb/h (1.9-kg/s) steam flow, loading $= S/A = 15{,}000/3000 = 5$ lb/ft^2 (24.4 kg/m^2).

10. *Compute the heat-transfer coefficient*

Correct the previous heat-transfer rate U = 690 Btu/(ft^2·h·°F) [3.9 kJ/(m^2·°C·s)] LMTD for temperature, cleanliness, and loading. Or, U = 690(1.04)(0.80)(0.89) = 511 Btu/(ft^2·h·°F) [2.9 kJ/(m^2·°C·s)] LMTD, given the correction factors from Fig. 37.

11. *Compute the final steam temperature*

As before, $x = (kL/V)(U/500) = (0.377)(14/6.5)(511/500) = 0.830$. Then $\Delta t = t_2 - t_1 = 950S/(500gpm) = 950(15{,}000)/[500(3700)] = 7.7$°F (4.3°C). With t_1 = 80°F (26.7°C), $t_2 = \Delta t + t_1 = 7.7 + 80 = 87.7$°F (30.9°C). Since $t_2 = t_s - (t_s - t_1)/e^x$, $e^x = t_s - t_1/(t_s - t_2)$, or $2.7183^{0.830} = t_s - 80/(t_s - 87.7)$. Solve for t_s; or, $t_s = 201.1 - 80/1.294 = 93.6$°F (34.2°C).

At a saturation temperature of 93.6°F (34.2°C), the steam table (saturation temperature) shows that the steam pressure in the condenser shell is 1.59 inHg (5.4 kPa).

Check the Δt terminal temperature difference. Or, $\Delta t = t_s - t_2 = 93.6 - 87.7 = 5.9$°F (3.3°C). Since the terminal temperature difference is greater than 5°F (2.8°C), the calculated performance can be realized.

Related Calculations: The procedures and data given here can be used to compute the required cooling-water flow, cooling-water temperature rise, quantity of steam condensed by a given cooling-water flow rate and temperature rise, required condenser surface area, tube length per pass, water velocity, steam temperature in condenser, cleanliness factor, and heat-transfer rate. Whereas Fig. 37 is suitable for all usual condenser calculations for the ranges given, check the Heat Exchange Institute for any new curves that might have been made available before you make the final selection of very large condensers (more than 100,000 lb/h or 12.6 kg/s of steam flow).

Note: The design water temperature used for condensers is either the average summer water temperature or the average annual water temperature, depending on which is higher. The design steam load is the maximum steam flow expected at the full-load rating of the turbine or engine. Usual shell-and-tube condensers have tubes that vary in length from about 8 ft (2.4 m) in the smallest sizes to about 40 ft (12.2 m) or more in the largest sizes. Each square foot of tube surface will condense 7 to 20 lb/h (0.88 to 2.5 g/s) of steam with a cooling-water circulating rate of 0.1 to 0.25 gal/(lb·min) [0.014 to 0.035 L/(kg·s)] of steam condensed. The method presented here is the work of Glenn C. Boyer.

STEAM-CONDENSER SELECTION

Select a condenser for a steam turbine exhausting 150,000 lb/h (18.9 kg/s) of steam at 2 inHg absolute (6.8 kPa) with a cooling-water inlet temperature of 75°F (23.9°C). Assume a 0.85 condition factor, ⅞-in (2.2-cm) no. 18 BWG tubes, and an 8-ft/s (2.4-m/s) water velocity. The water supply is restricted. Obtain condenser constants from the Heat Exchange Institute, *Steam Surface Condenser Standards*.

Calculation Procedure:

1. *Select the $t_s - t_1$ temperature difference*

Table 2 shows customary design conditions for steam condensers. With an inlet-water temperature of 75°F (23.9°C) and an exhaust steam pressure of 2.0 inHg absolute (6.8 kPa), the customary temperature difference $t_s - t_1$ = 26.1°F (14.5°C). With a sufficient water supply and a siphonic circuitry, $(t_2 - t_1)/(t_s - t_1)$ is usually between 0.5 and 0.55. For a restricted water supply or high frictional resistance and static head, the value of this factor ranges from 0.55 to 0.75.

2. *Compute the LMTD across the condenser*

With 75°F (23.9°C) inlet water, $t_s - t_1 = 101.14 - 75 = 26.14$°F (14.5°C), given the steam temperature in the saturation-pressure table. Once $t_s - t_1$ is known, it is necessary to assume a value for the ratio $(t_2 - t_1)/(t_s - t_1)$. As a trial, assume 0.60, since the water supply is restricted. Then $(t_2 - t_1)/(t_s - t_1) = 0.60 = (t_2 - t_1)/26.14$. Solving, we get $t_2 - t_1 = 15.68$°F (8.7°C). The difference between the steam temperature t_s and the outlet temperature t_2 is then $t_s - t_2 = 26.14 - 15.68 = 10.46$°F (5.8°C). Checking, we find $t_2 = t_1 + (t_2 - t_1) = 75 + 15.68 =$

TABLE 2 Typical Design Conditions for Steam Condensers

Cooling water temperature		Steam pressure		Temperature difference $t_s - t_1$	
°F	°C	inHg	kPa	°F	°C
70	21.1	1.5–2.0	5.1–6.8	21.7–31.1	12.1–17.3
75	23.9	2.0–2.5	6.8–8.5	26.1–33.7	14.5–18.7
80	26.7	2.0–4.0	6.8–13.5	21.1–45.4	11.7–25.2

90.68°F (50.38°C); $t_s - t_2$ = 101.14 − 90.68 = 10.46°F (5.8°C). This value is greater than the required minimum value of 5°F (2.8°C) for $t_s - t_2$. The assumed ratio 0.60 is therefore satisfactory.

Were $t_s - t_2$ less than 5°F (2.8°C), another ratio value would be assumed and the difference computed again. You would continue doing this until a value of $t_s - t_2$ greater than 5°F (2.8°C) were obtained. Then LMTD = $(t_2 - t_1)/\ln[(t_s - t_1)/(t_s - t_2)]$; LMTD = 15.68/ln (26.1/10.46) = 17.18°F (9.5°C).

3. *Determine the heat-transfer coefficient*

From the Heat Exchange Institute or manufacturer's data U is 740 Btu/(ft²·h·°F) [4.2 kJ/(m²·°C·s)] LMTD for a water velocity of 8 ft/s (2.4 m/s). If these data are not available, Fig. 37 can be used with complete safety for all preliminary selections.

Now U must be corrected for the inlet-water temperature, 75°F (23.9°C), and the condition factor, 0.85, which is a term used in place of the correction factor by some authorities. From Fig. 37, the correction for 75°F (23.9°C) inlet water = 1.04. Then actual U = 740(1.04)(0.85) = 655 Btu/(ft²·h·°F) [3.7 kJ/(m²·°C·s)] LMTD.

4. *Compute the steam condensation rate*

The heat-transfer rate per square foot of condenser surface with a 17.18°F (9.5°C) LMTD is U(LMTD) = 655(17.18) = 11,252.9 Btu/(ft²·h) [35.5 kJ/(m²·s)].

Condensers serving steam turbines are assumed, for design purposes, to remove 950 Btu/lb (2209.7 kJ/kg) of steam condensed. Therefore, the steam condensation rate for any condenser is [Btu/(ft²·h)]/950, or 1252.9/950 = 11.25 lb/(ft²·h) [15.3 g/(m²·s)].

5. *Compute the required surface area and water flow*

The required surface area = steam flow (lb/h)/[condensation rate, lb/(ft²·h)], or with a 150,000-lb/h (18.9-kg/s) flow, 150,000/11.25 = 13,320 ft² (1237.4 m²).

The water flow rate, gal/min = $950S/[500(t_2 - t_1)]$ = 950(150,000)/[500(15.68)] = 18,200 gal/min (1148.1 L/s).

Related Calculations: See the previous calculation procedure for steps in determining the water-pressure loss through a surface condenser.

To choose a surface condenser for a steam engine, use the same procedures as given above, except that the heat removed from the exhaust steam is 1000 Btu/lb (2326.9 kJ/kg). Use a condition (cleanliness) factor of 0.65 for steam engines because the oil in the exhaust steam fouls the condenser tubes, reducing the rate of heat transfer. The condition (cleanliness) factor for steam turbines is usually assumed to be 0.8 to 0.9 for relatively clean, oil-free cooling water.

At loads greater than 50 percent of the design load, $t_s - t_1$ follows a straight-line relationship. Thus, in the above condenser, $t_s - t_1$ = 26.14°F (14.5°C) at the full load of 150,000 lb/h (18.9 kg/s). If the load falls to 60 percent (90,000 lb/h or 11.3 kg/s), then $t_s - t_1$ = 26.14(0.60) = 15.7°F (8.7°C). At 120 percent load (180,000 lb/h or 22.7 kg/s), $t_s - t_l$ = 26.14(1.20) = 31.4°F (17.4°C). This straight-line law is valid with constant inlet-water temperature and cooling-water flow rate. It is useful in analyzing condenser operating conditions at other than full load.

Single- or multiple-pass surface condensers may be used in power services. When a liberal supply of water is available, the single-pass condenser is often chosen. With a limited water supply, a two-pass condenser is often chosen.

AIR-EJECTOR ANALYSIS AND SELECTION

Choose a steam-jet air ejector for a condenser serving a 250,000-lb/h (31.5-kg/s) steam turbine exhausting at 2 inHg absolute (6.8 kPa). Determine the number of stages to use, the approximate steam consumption, and the quantity of air and vapor mixture the ejector will handle.

Calculation Procedure:

1. *Select the number of stages for the ejector*

Use Fig. 38 as a preliminary guide to the number of stages required in the ejector. Enter at 2-inHg absolute (6.8-kPa) condenser pressure, and project horizontally to the stage area. This shows that a two-stage ejector will probably be satisfactory.

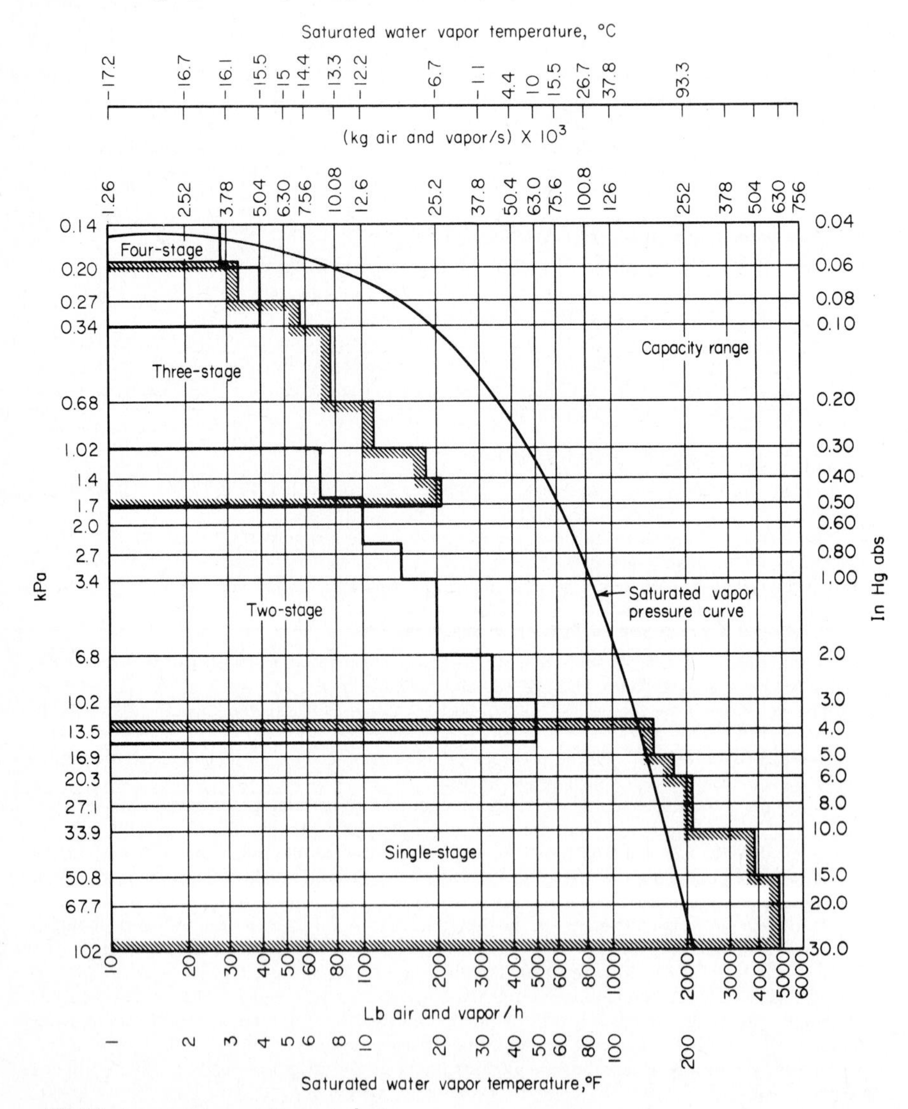

FIG. 38 Steam-ejector capacity-range chart.

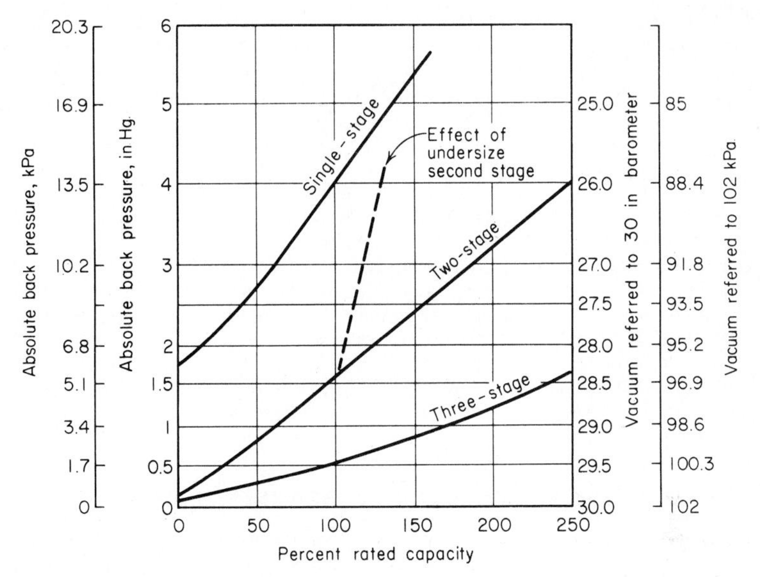

FIG. 39 Steam-jet ejector characteristics.

Check the number of stages above against the probable overload range of the prime mover by using Fig. 39. Enter at 2-inHg absolute (6.8-kPa) condenser pressure, and project to the two-stage curve. This curve shows that a two-stage ejector can readily handle a 25-percent overload of the prime mover. Also, the two-stage curve shows that this ejector could handle up to 50 percent overload with an increase in the condenser absolute pressure of only 0.4 inHg (1.4 kPa). This is shown by the pressure, 2.4 inHg absolute (8.1 kPa), at which the two-stage curve crosses the 150 percent overload ordinate (Fig. 39).

2. *Determine the ejector operating conditions*

Use the Heat Exchange Institute or manufacturer's data. Table 3 excerpts data from the Heat Exchange Institute for condensers in the range considered in this procedure.

Study of Table 3 shows that a two-stage condensing ejector unit serving a 250,000-lb/h (31.5-kg/s) steam turbine will require 450 lb/h (56.7 g/s) of 300-lb/in^2 (gage) (2068.5-kPa) steam. Also, the ejector will handle 7.5 ft^3/min (0.2 m^3/min) of free, dry air, or 33.75 lb/h (4.5 g/s) of air. It will remove up to 112.5 lb/h (14.2 g/s) of an air-vapor mixture.

The actual air leakage into a condenser varies with the absolute pressure in the condenser, the

TABLE 3 Air-Ejector Capacities for Surface Condensers for Steam Turbines*

Steam load, lb/h (kg/s)	Free, dry air at 70°F (21.1°C), ft^3/min (cm^3/s)	Air, lb/h (g/s)	Air-vapor mixture at 30 percent dry air, lb/h (g/s)	Steam consumption at 300 lb/in^2 (gage) (2068.5 kPa), lb/h (g/s)
100,001–250,000	7.5	33.75	112.5	450
(12.6–31.5)	(3539.6)	(4.3)	(14.2)	(56.7)

*Two-stage condensing ejector unit.

tightness of the joints, and the conditions of the tubes. Some authorities cite a maximum leakage of about 250-lb/h (31.5-g/s) steam flow. At 400,000 lb/h (50.4 kg/s), the leakage is 160 lb/h (20.2 g/s); at 250,000 lb/h (31.5 kg/s), it is 130 lb/h (16.4 g/s) of air-vapor mixture. A condenser in good condition will usually have less leakage.

For an installation in which the manufacturer supplies data on the probable air leakage, use a psychrometric chart to determine the weight of water vapor contained in the air. Thus, at 2 inHg absolute (6.8 kPa) and 80°F (26.7°C), each pound of air will carry with it 0.68 lb (0.68 kg/kg) of water vapor. In a surface condenser into which 20 lb (9.1 kg) of air leaks, the ejector must handle 20 + 20(0.68) = 33.6 lb/h (4.2 g/s) of air-vapor mixture. Table 3 shows that this ejector can readily handle this quantity of air-vapor mixture.

Related Calculations: When you choose an air ejector for steam-engine service, double the Heat Exchange Institute steam-consumption estimates. For most low-pressure power-plant service, a two-stage ejector with inter- and aftercondensers is satisfactory, although some steam engines operating at higher absolute exhaust pressures require only a single-stage ejector. Twin-element ejectors have two sets of stages; one set serves as a spare and may also be used for capacity regulation in stationary and marine service. The capacity of an ejector is constant for a given steam pressure and suction pressure. Raising the steam pressure will not increase the ejector capacity.

SURFACE-CONDENSER CIRCULATING-WATER PRESSURE LOSS

Determine the circulating-water pressure loss in a two-pass condenser having 12,000 ft^2 (1114.8 m^2) of condensing surface, a circulating-water flow rate of 10,000 gal/min (630.8 L/s), ¾-in (1.9-cm) no. 16 BWG tubes, a water flow rate of 7 ft/s (2.1 m/s), external friction of 20 ft of water (59.8 kPa), and a 10-ft-of-water (29.9-kPa) siphonic effect on the circulating-water discharge.

Calculation Procedure:

1. *Determine the water flow rate per tube*

Use a tabulation of condenser-tube engineering data available from the manufacturer or the Heat Exchange Institute, or complete the water flow rate from the physical dimensions of the tube thus: ¾ in (1.9 cm) no. 16 BWG tube ID = 0.620 in (1.6 cm) from a tabulation of condenser-tube data, such as Table 1. Assume a water velocity of 1 ft/s (0.3 m/s). Then a 1-ft (0.3-m) length of the tube will contain $(12)(0.620)^2\pi/4 = 3.62$ in^3 (59.3 cm^3) of water. This quantity of water will flow through the tube for each foot of length per second of water velocity [194.6 cm^3/(m·s)]. The flow per minute will be 3.62 (60 s/min) = 217.2 in^3/min (3559.3 cm^3/min). Since 1 U.S. gal = 231 in^3 (3.8 L), the gal/min flow at a 1 ft/s (0.3 m/s) velocity = 217.2/231 = 0.94 gal/min (0.059 L/s).

With an actual velocity of 7 ft/s (2.1 m/s), the water flow rate per tube is 7(0.94) = 6.58 gal/min (0.42 L/s).

2. *Determine the number of tubes and length of water travel*

Since the water flow rate through the condenser is 10,000 gal/min (630.8 L/s) and each tube conveys 6.58 gal/min (0.42 L/s), the number of tubes = 10,000/6.58 = 1520 tubes per pass.

Next, the total length of water travel for a condenser having A ft^2 of condensing surface is computed from A(number of tubes)(outside area per linear foot, ft^2). The outside area of each tube can be obtained from a table of tube properties, such as Table 1, or computed from (OD, in)(π)(12)/144, or (0.75)(π)(12)/144 = 0.196 ft^2/lin ft (0.06 m^2/m). Then, total length of travel = 12,000/[(1520)(0.196)] = 40.2 ft (12.3 m). Since the condenser has two passes, the length of tube per pass = 40.2/2 = 20.1 ft (6.1 m). Since each pass has an equal number of tubes and there are two passes, the total number of tubes in the condenser = 2 passes (1520 tubes per pass) = 3040 tubes.

3. *Compute the friction loss in the system*

Use the Heat Exchange Institute or manufacturer's curves to find the friction loss per foot of condenser tube. At 7 ft/s (2.1 m/s), the Heat Exchange Institute curve shows the head loss is 0.4 ft of head per foot (3.9 kPa/m) of travel for ¾-in (1.9-cm) no. 16 BWG tubes. With a total length of 40.2 ft (12.3 m), the tube head loss is 0.4(40.2) = 16.1 ft (48.1 kPa).

Use the Heat Exchange Institute or manufacturer's curves to find the head loss through the condenser water boxes. From the first reference, for a velocity of 7 ft/s (2.1 m/s), head loss = 1.4 ft (4.2 kPa) of water for a single-pass condenser. Since this is a two-pass condenser, the total water-box head loss = 2(1.4) = 2.8 ft (8.4 kPa).

The total condenser friction loss is then the sum of the tube and water-box losses, or 16.1 + 2.8 = 18.9 ft (56.5 kPa) of water. With an external friction loss of 20 ft (59.8 kPa) in the circulating-water piping, the total loss in the system, without siphonic assistance, is 18.9 + 20 = 38.9 ft (116.3 kPa). Since there is 10 ft (29.9 kPa) of siphonic assistance, the total friction loss in the system with siphonic assistance is 38.9 − 10 = 28.9 ft (86.3 kPa). In choosing a pump to serve this system, the frictional resistance of 28.9 ft (86.3 kPa) would be rounded to 30 ft (89.7 kPa), and any factor of safety added to this value of head loss.

Note: The most economical cooling-water velocity in condenser tubes is 6 to 7 ft/s (1.8 to 2.1 m/s); a velocity greater than 8 ft/s (2.4 m/s) should not be used, unless warranted by special conditions.

SURFACE-CONDENSER WEIGHT ANALYSIS

A turbine exhaust nozzle can support a weight of 100,000 lb (444,822.2 N). Determine what portion of the total weight of a surface condenser must be supported by the foundation if the weight of the condenser is 275,000 lb (1,223,261.1 N), the tubes and water boxes have a capacity of 8000 gal (30,280.0 L), and the steam space has a capacity of 30,000 gal (113,550.0 L) of water.

Calculation Procedure:

1. *Compute the maximum weight of the condenser*

The maximum weight on a condenser foundation occurs when the shell, tubes, and water boxes are full of water. This condition could prevail during accidental flooding of the steam space or during tests for tube leaks when the steam space is purposefully flooded. In either circumstance, the condenser foundation and spring supports, if used, must be able to carry the load imposed on them. To compute this load, find the sum of the individual weights:

Condenser weight, dry	275,000 lb (1,223,261.1 N)
Water in tubes and boxes = (8.33 lb/gal)(8000 gal)	66,640 lb (296,429.5 N)
Water in steam space = (8.33)(30,000)	249,900 lb (1,111,610.7 N)
Maximum weight when full of water	592,540 lb (2,631,301.2 N)

2. *Compute the foundation load*

The turbine nozzle can support 100,000 lb (444,822.2 N). Therefore, the foundation must support 591,540 − 100,000 = 491,540 lb (2,186,479.0 N). For foundation design purposes this would be rounded to 495,000 lb (2,201,869.9 N).

Related Calculations: When you design a condenser foundation, do the following: (1) Leave enough room at one end to permit withdrawal of faulty tubes and insertion of new tubes. Since some tubes may exceed 40 ft (12.3 m) in length, careful planning is needed to provide sufficient installation space. During the design of a power plant, a template representing the tube length is useful for checking the tube clearance on a scale plan and side view of the condenser installation. When there is insufficient room for tube removal with one shape of condenser, try another shape with shorter tubes.

(2) Provide enough headroom under the condenser to produce the required submergence on the condensate-pump impeller. Most condensate pumps require at least 3-ft (0.9-m) submergence. If necessary, the condensate pump can be installed in a pit under the condenser, but this should be avoided if possible.

BAROMETRIC-CONDENSER ANALYSIS AND SELECTION

Select a countercurrent barometric condenser to serve a steam turbine exhausting 25,000 lb/h (3.1 kg/s) of steam at 5 inHg absolute (16.9 kPa). Determine the quantity of cooling water required if the water inlet temperature is 50°F (10.0°C). What is the required dry-air capacity of the ejector? What is the required pump head if the static head is 40 ft (119.6 kPa) and the pipe friction is 15 ft of water (44.8 kPa)?

Calculation Procedure:

1. *Find the steam properties from the steam tables*

At 5 inHg absolute (16.9 kPa), h_g = 1119.4 Btu/lb (2603.7 kJ/kg), from the saturation-pressure table. If the condensing water were to condense the steam without subcooling the condensate, the final temperature of the condensate, from the steam tables, would be 133.76°F (56.5°C), corresponding to the saturation temperature. However, subcooling almost always occurs, and the usual practice in selecting a countercurrent barometric condenser is to assume the final condensate temperature t_c will be 5°F (2.8°C) below the saturation temperature corresponding to the absolute pressure in the condenser. Given a 5°F (2.8°C) difference, t_c = 133.76 − 5 = 128.76°F (53.7°C). Interpolating in the saturation-temperature steam table, we find the enthalpy of the condensate h_f at 128.76°F (53.7°C) is 96.6 Btu/lb (224.8 kJ/kg).

2. *Compute the quantity of condensing water required*

In any countercurrent barometric condenser, the quantity of cooling water Q lb/h required is $Q = W(h_g - h_t)/(t_c - t_1)$, where W = weight of steam condensed, lb/h; t_1 = cooling-water inlet temperature, °F. Then Q = 25,000(1119.4 − 96.66)/(128.76 − 50) = 325,000 lb/h (40.9 kg/s). By converting to gallons per minute, Q = 325,000/500 = 650 gal/min (41.0 L/s).

3. *Determine the required ejector dry-air capacity*

Use the Heat Exchanger Institute or a manufacturer's tabulation of free, dry-air leakage and the allowance for air in the cooling water to determine the required dry-air capacity. Thus, from Table 4, the free, dry-air leakage for a barometric condenser serving a turbine is 3.0 ft^3/min (0.08 m^3/min) of air and vapor. The allowance for air in the 50°F (10.0°C) cooling water is 3.3 ft^3/min (0.09 m^3/min) of air at 70°F (21.1°C) per 1000 gal/min (63.1 L/s) of cooling water, Fig. 40. The total dry-air leakage is the sum, or 3.0 + 3.3 = 6.3 ft^3/min (0.18 m^3/min). Thus, the ejector must be capable of handling at least 6.3 ft^3/min (0.18 m^3/min) of dry air to serve this barometric condenser at its rated load of 25,000 lb/h (3.1 kg/s) of steam.

Where the condenser will operate at a lower vacuum (i.e., a higher absolute pressure), overloads up to 50 percent may be met. To provide adequate dry-air handling capacity at this overload with the same cooling-water inlet temperature, find the free, dry-air leakage at the higher condensing rate from Table 4 and add this to the previously found allowance for air in the cooling

TABLE 4 Free, Dry-Air Leakage

[*ft^3/min (m^3/s) at 70°F or 21.1°C air and vapor mixture, 7½° below vacuum temperature or 4.2° for Celsius*]

Barometric and low-level jet condensers									
Maximum steam condensation		Serving turbines				Serving engines			
lb/h	kg/s	ft^3/min	m^3/s	lb/h	g/s	ft^3/min	m^3/s	lb/h	g/s
75,000–150,000	9.4–18.9	6.5	0.0031	97.5	12.3	13.0	0.0061	195.0	24.6
150,001–250,000	18.9–31.5	8.5	0.0040	127.5	16.1				
250,001–350,000	31.5–44.1	10.0	0.0047	150.0	18.9				

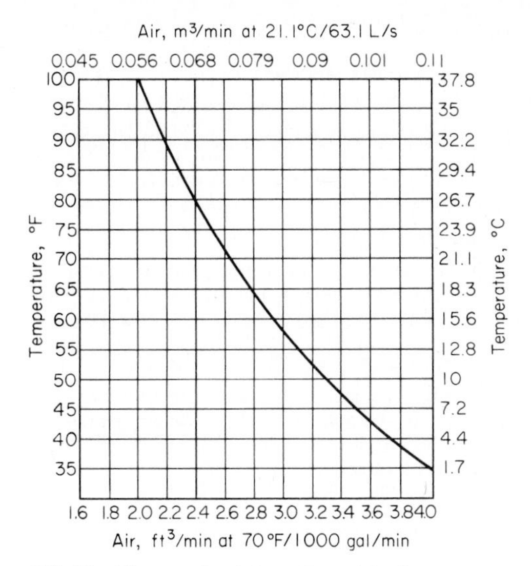

FIG. 40 Allowance for air in condenser injection water.

water. Or, 4.5 + 3.3 = 7.8 ft^3/min (0.22 m^3/min). An ejector capable of handling up to 10 ft^3/min (0.32 m^3/min) would be a wise choice for this countercurrent barometric condenser.

4. ***Determine the pump head required***

Since a countercurrent barometric condenser operates at pressures below atmospheric, it assists the cooling-water pump by "sucking" the water into the condenser. The maximum assist that can be assumed is $0.75V$, where V = design vacuum, inHg.

In this condenser with a 26-in (88.0-kPa) vacuum, the maximum assist is 0.75(26) = 19.5 inHg (66.0 kPa). Converting to feet of water, using 1.0 inHg = 1.134 ft (3.4 kPa) of water, we find 19.5(1.134) = 22.1 ft (66.1 kPa) of water. The total head on the pump is then the sum of the static and friction heads less $0.75V$, expressed in feet of water. Or, the total head on the pump = 40 + 15 − 22.1 = 32.9 ft (98.4 kPa). A pump with a total head of at least 35 ft (104.6 kPa) of water would be chosen for this condenser. Where corrosion or partial clogging of the piping is expected, a pump with a total head of 50 ft (149.4 kPa) would probably be chosen to ensure sufficient head even though the piping is partially clogged.

Related Calculations: (1) When a condenser serving a steam engine is being chosen, use the appropriate dry-air leakage value from Table 4. (2) For ejector-jet barometric condensers, assume the final condensate temperature t_c as 10 to 20°F (5.6 to 11.1°C) below the saturation temperature corresponding to the absolute pressure in the condenser. This type of condenser does not use an ejector, but it requires 25 to 50 percent more cooling water than the countercurrent barometric condenser for the same vacuum. (3) The total pump head for an ejector-jet barometric condenser is the sum of the static and friction heads plus 10 ft (29.9 kPa). The additional positive head is required to overcome the pressure loss in spray nozzles.

COOLING-POND SIZE FOR A KNOWN HEAT LOAD

How many spray nozzles and what surface area are needed to cool 10,000 gal/min (630.8 L/s) of water from 120 to 90°F (48.9 to 32.2°C) in a spray-type cooling pond if the average wet-bulb temperature is 650°F (15.6°C)? What would the approximate dimensions of the cooling pond be? Determine the total pumping head if the static head is 10 ft (29.9 kPa), the pipe friction is 35 ft of water (104.6 kPa), and the nozzle pressure is 8 lb/in^2 (55.2 kPa).

Calculation Procedure:

1. *Compute the number of nozzles required*

Assume a water flow of 50 gal/min (3.2 L/s) per nozzle; this is a typical flow rate for usual cooling-pond nozzles. Then the number of nozzles required = (10,000 gal/min)/(50 gal/min per nozzle) = 200 nozzles. If 6 nozzles are used in each spray group, a series of crossed arms, with each arm containing one or more nozzles, then 200 nozzles/6 nozzles per spray group = 33⅓ spray groups will be needed. Since a partial spray group is seldom used, 34 spray groups would be chosen.

2. *Determine the surface area required*

Usual design practice is to provide 1 ft^2 (0.09 m^2) of pond area per 250 lb (113.4 kg) of water cooled for water quantities exceeding 1000 gal/min (63.1 L/s). Thus, in this pond, the weight of water cooled = (10,000 gal/min)(8.33 lb/gal)(60 min/h) = 4,998,000, say 5,000,000 lb/h (630.0 kg/s). Then, the area required, given 1 ft^2 of pond area per 250 lb of water (0.82 m^2 per 1000 kg) cooled = 5,000,000/250 = 20,000 ft^2 (1858.0 m^2).

As a cross-check, use another commonly accepted area value: 125 Btu/($ft^2 \cdot$°F) [2555.2 kJ/($m^2 \cdot$°C)] is the difference between the air wet-bulb temperature and the warm entering-water temperature. This is the equivalent of (120 − 60)(125) = 7500 Btu/ft^2 (85,174 kJ/m^2) in this spray pond, because the air wet-bulb temperature is 60°F (15.6°C) and the warm-water temperature is 120°F (48.9°C). The heat removed from the water is (lb/h of water)(temperature decrease, °F)(specific heat of water) = (5,000,000)(120 − 90)(1.0) = 150,000,000 Btu/h (43,960.7 kW). Then, area required = (heat removed, Btu/h)/(heat removal, Btu/ft^2) = 150,000,000/7500 = 20,000 ft^2 (1858.0 m^2). This checks the previously obtained area value.

3. *Determine the spray-pond dimensions*

Spray groups on the same header or pipe main are usually arranged on about 12-ft (3.7-m) centers with the headers or pipe mains spaced on about 25-ft (7.6-m) centers, Fig. 41. Assume that 34 spray groups are used, instead of the required 33⅓, to provide an equal number of groups in two headers and a small extra capacity.

Sketch the spray pond and headers, Fig. 41. This shows that the length of each header will be about 204 ft (62.2 m) because there are seventeen 12-ft (3.7-m) spaces between spray groups in each header. Allowing 3 ft (0.9 m) at each end of a header for fittings and clean-outs gives an overall header length of 210 ft (64.0 m). The distance between headers is 25 ft (7.6 m). Allow 25 ft (7.6 m) between the outer sprays and the edge of the pond. This gives an overall width of 85 ft (25.9 m) for the pond, if we assume the width of each arm in a spray group is 10 ft (3.0 m). The overall length will then be 210 + 25 + 25 = 260 ft (79.2 m). A cold well for the pump

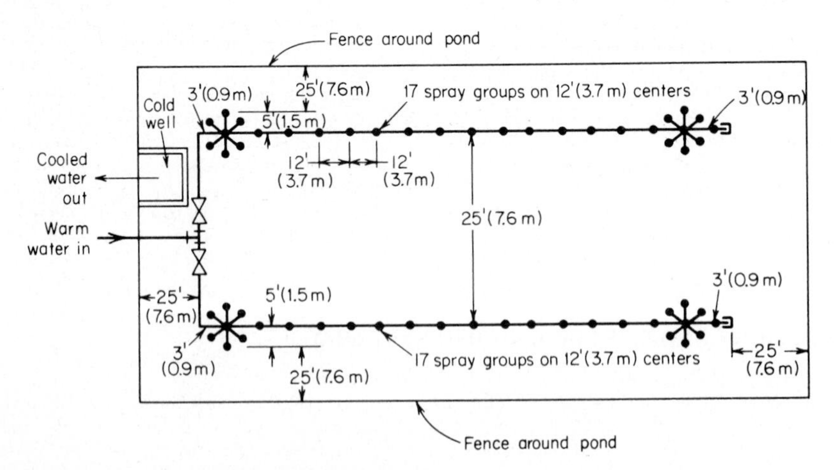

FIG. 41 Spray-pond nozzle and piping layout.

suction and suitable valving for control of the incoming water must be provided, as shown in Fig. 41. The water depth in the pond should be 2 to 3 ft (0.6 to 0.9 m).

4. *Compute the total pumping head*

The total head, ft of water = static head + friction head + required nozzle head = 10 + 35 + 80(0.434) = 48.5 ft (145.0 kPa) of water. A pump having a total head of at least 50 ft (15.2 m) of water would be chosen for this spray pond. If future expansion of the pond is anticipated, compute the probable total head required at a future date, and choose a pump to deliver that head. Until the pond is expanded, the pump would operate with a throttled discharge. Normal nozzle inlet pressures range from about 6 to 10 lb/in^2 (41.4 to 69.0 kPa). Higher pressures should not be used, because there will be excessive spray loss and rapid wear of the nozzles.

Related Calculations: Unsprayed cooling ponds cool 4 to 6 lb (1.8 to 2.7 kg) of water from 100 to 70°F/ft^2 (598.0 to 418.6°C/m^2) of water surface. An alternative design rule is to assume that the pond will dissipate 3.5 Btu/(ft^2·h) (11.0 W/m^2) water surface per degree difference between the wet-bulb temperature of the air and the entering warm water.

DIRECT-CONTACT FEEDWATER HEATER ANALYSIS

Determine the outlet temperature of water leaving a direct-contact open-type feedwater heater if 250,000 lb/h (31.5 kg/s) of water enters the heater at 100°F (37.8°C). Exhaust steam at 10.3 lb/in^2 (gage) (71.0 kPa) saturated flows to the heater at the rate of 25,000 lb/h (31.5 kg/s). What saving is obtained by using this heater if the boiler pressure is 250 lb/in^2 (abs) (1723.8 kPa)? Determine the approximate volume of the heater if a 2-min storage capacity is provided in it.

Calculation Procedure:

1. *Compute the water outlet temperature*

Assume the heater is 90 percent efficient. Then $t_o = t_i w_w + 0.9 w_s h_g/(w_w + 0.9 w_s)$, where t_o = outlet water temperature, °F; t_i = inlet water temperature, °F; w_w = weight of water flowing through heater, lb/h; 0.9 = heater efficiency, expressed as a decimal; w_s = weight of steam flowing to the heater, lb/h; h_g = enthalpy of the steam flowing to the heater, Btu/lb.

For saturated steam at 10.3 lb/in^2 (gage) (71.0 kPa), or 10.3 + 14.7 = 25 lb/in^2 (abs) (172.4 kPa), h_g = 1160.6 Btu/lb (2599.6 kJ/kg), from the saturation pressure steam tables. Then

$$t_o = \frac{100(250{,}000) + 0.9(25{,}000)(1160.6)}{250{,}000 + 0.9(25{,}000)} = 187.5°\text{F } (86.4°\text{C})$$

2. *Compute the savings obtained by feed heating*

The percentage of saving, expressed as a decimal, obtained by heating feedwater is $(h_o - h_i)/(h_b - h_i)$ where h_o and h_i = enthalpy of the water leaving and entering the heater, respectively, Btu/lb; h_b = enthalpy of the steam at the boiler operating pressure, Btu/lb. For this plant, from the steam tables, $h_o - h_i/(h_b - h_i)$ = 155.44 – 67.97/(1201.1 – 67.97) = 0.077, or 7.7 percent.

A popular rule of thumb states that for every 11°F (6.1°C) rise in feedwater temperature in a heater, there is approximately a 1 percent saving in the fuel that would otherwise be used to heat the feedwater. Checking the above calculation with this rule of thumb shows reasonably good agreement.

3. *Determine the heater volume*

With a capacity of W lb/h of water, the volume of a direct-contact or open-type heater can be approximated from $v = W/10{,}000$, where v = heater internal volume, ft^3. For this heater, v = 250,000/10,000 = 25 ft^3 (0.71 m^3).

Related Calculations: Most direct-contact or open feedwater heaters store in 2-min supply of feedwater when the boiler load is constant, and the feedwater supply is all makeup. With little or no makeup, the heater volume is chosen so that there is enough capacity to store 5 to 30 min feedwater for the boiler.

CLOSED FEEDWATER HEATER ANALYSIS AND SELECTION

Analyze and select a closed feedwater heater for the third stage of a regenerative steam-turbine cycle in which the feedwater flow rate is 37,640 lb/h (4.7 kg/s), the desired temperature rise of the water during flow through the heater is 80°F (44.4°C) (from 238 to 318°F or, 114.4 to 158.9°C), bleed heating steam is at 100 lb/in² (abs) (689.5 kPa) and 460°F (237.8°C), drains leave the heater at the saturation temperature corresponding to the heating steam pressure [100 lb/in² (abs) or 689.5 kPa], and ⅝-in (1.6-cm) OD admiralty metal tubes with a maximum length of 6 ft (1.8 m) are used. Use the *Standards of the Bleeder Heater Manufacturers Association, Inc.*, when analyzing the heater.

Calculation Procedure:

1. *Determine the LMTD across heater*

When heat-transfer rates in feedwater heaters are computed, the average film temperature of the feedwater is used. In computing this, the *Standards of the Bleeder Heater Manufacturers Association* specify that the *saturation temperature* of the heating steam be used. At 100 lb/in² (abs) (689.5 kPa), t_s = 327.81°F (164.3°C). Then

$$\text{LMTD} = t_m = \frac{(t_s - t_i) - (t_s - t_o)}{\ln [t_s - t_i/(t_s - t_o)]}$$

where the symbols are as defined in the previous calculation procedure. Thus,

$$t_m = \frac{(327.81 - 238) - (327.81 - 318)}{\ln [327.81 - 238/(327.81 - 318)]}$$

$$= 36.5°\text{F } (20.3°\text{C})$$

The average film temperature t_f for any closed heater is then

$$t_f = t_s - 0.8t_m$$

$$= 327.81 - 29.2 = 298.6°\text{F } (148.1°\text{C})$$

2. *Determine the overall heat-transfer rate*

Assume a feedwater velocity of 8 ft/s (2.4 m/s) for this heater. This velocity value is typical for smaller heaters handling less than 100,000-lb/h (12.6-kg/s) feedwater flow. Enter Fig. 42 at 8 ft/s (2.4 m/s) on the lower horizontal scale, and project vertically upward to the 250°F (121.1°C) average film temperature curve. This curve is used even though t_f = 298.6°F (148.1°C), because the standards recommend that heat-transfer rates higher than those for a 250°F (121.1°C) film temperature not be used. So, from the 8-ft/s (2.4-m/s) intersection with the 250°F (121.1°C) curve in Fig. 42, project to the left to read U = the overall heat-transfer rate = 910 Btu/(ft²·°F·h) [5.2 kJ/(m²·°C·s)].

Next, check Table 5 for the correction factor for U. Assume that no. 18 BWG ⅝-in (1.6-cm) OD arsenical copper tubes are used in this exchanger. Then the correction factor from Table 5 is 1.00, and U_{corr} = 910(1.00) = 910. If no. 9 BWG tubes are chosen, U_{corr} = 910(0.85) = 773.5 Btu/(ft²·°F·h) [4.4 kJ/(m²·°C·s)], given the correction factor from Table 5 for arsenical copper tubes.

3. *Compute the amount of heat transferred by the heater*

The enthalpy of the entering feedwater at 238°F (114.4°C) is, from the saturation-temperature steam table, h_{fi} = 206.32 Btu/lb (479.9 kJ/kg). The enthalpy of the leaving feedwater at 318°F (158.9°C) is, from the same table, h_{fo} = 288.20 Btu/lb (670.4 kJ/kg). Then the heater transferred H_t Btu/h is $H_t = w_w(h_{fo} - h_{fi})$, where w_w = feedwater flow rate, lb/h. Or, H_t = 37,640(288.20 − 206.32) = 3,080,000 Btu/h (902.7 kW).

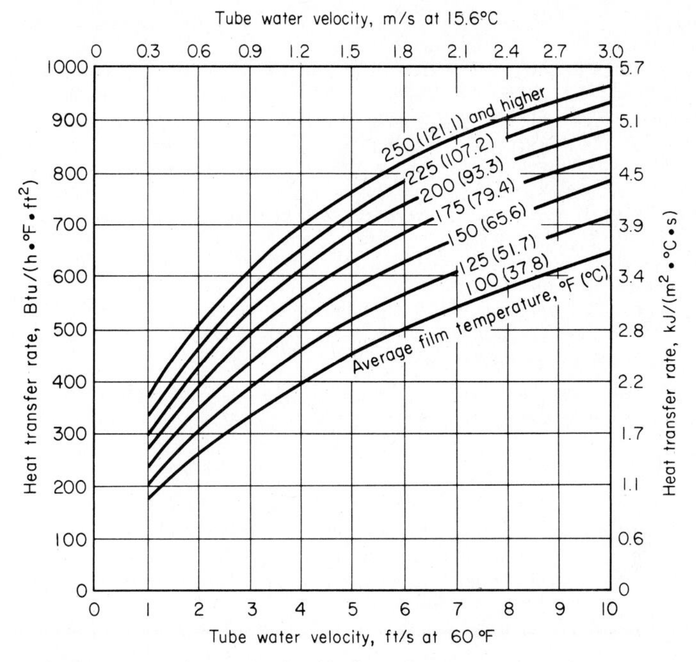

FIG. 42 Heat-transfer rates for closed feedwater heaters. *(Standards of Bleeder Heater Manufacturers Association, Inc.)*

4. *Compute the surface area required in the exchanger*

The surface area required A ft^2 = H_t/Ut_m. Then A = 3,080,000/[(910)(36.5)] = 92.7 ft^2 (8.6 m^2).

5. *Determine the number of tubes per pass*

Assume the heater has only one pass, and compute the number of tubes required. Once the number of tubes is known, a decision can be made about the number of passes required. In a closed

TABLE 5 Multipliers for Base Heat-Transfer Rates

[*For tube OD ⅝ to 1 in (1.6 to 2.5 cm) inclusive*]

	Tube material					
BWG	As-Cu	Adm	90/10 Cu-Ni	80/20 Cu-Ni	70/30 Cu-Ni	Monel
18	1.00	1.00	0.97	0.95	0.92	0.89
17	1.00	1.00	0.94	0.91	0.87	0.85
16	1.00	1.00	0.91	0.88	0.84	0.82
15	1.00	0.99	0.89	0.86	0.82	0.79
14	1.00	0.96	0.85	0.82	0.77	0.75
13	0.98	0.93	0.81	0.78	0.73	0.70
12	0.95	0.90	0.77	0.73	0.68	0.65
11	0.92	0.87	0.74	0.70	0.65	0.62
10	0.89	0.83	0.69	0.66	0.60	0.58
9	0.85	0.80	0.65	0.62	0.56	0.54

heater, number of tubes = w_w (passes) (ft^3/s per tube)/[v(ft^2 per tube open area)], where w_w = lb/h of feedwater passing through heater; v = feedwater velocity in tubes, ft/s.

Since the feedwater enters the heater at 238°F (114.4°C) and leaves at 318°F (158.9°C), its specific volume at 278°F (136.7°C), midway between t_i and t_o, can be considered the average specific volume of the feedwater in the heater. From the saturation-pressure steam table, v_f = 0.01691 ft^3/lb (0.0011 m^3/kg) at 278°F (136.7°C). Convert this to cubic feet per second per tube by dividing this specific volume by 3600 (number of seconds in 1 h) and multiplying by the pounds per hour of feedwater per tube. Or, ft^3/s per tube = (0.01691/3600)(lb/h per tube).

Since no. 18 BWG ⅝-in (1.6-cm) OD tubes are being used, ID = 0.625 − 2(thickness) = 0.625 − 2(0.049) = 0.527 in (1.3 cm). Then, open area per tube, ft^2 = $(\pi d^2/4)/144$ = 0.7854(0.527)^2/144 = 0.001525 ft^2 (0.00014 m^2) per tube. Alternatively, this area could be obtained from a table of tube properties.

With these data, compute the total number of tubes from number of tubes = [(37,640)(1)(0.01681/3600)]/[(8)(0.001525)] = 14.49 tubes.

6. *Compute the required tube length*

Assume that 14 tubes are used, since the number required is less than 14.5. Then, tube length l, ft = A/(number of tubes per pass)(passes)(area per ft of tube). Or, tube length for 1 pass = 92.7/[(14)(1)(0.1636)] = 40.6 ft (12.4 m). The area per ft of tube length is obtained from a table of tube properties or computed from 12π(OD)/144 = 12π(0.625)/155 = 0.1636 ft^2 (0.015 m^2).

7. *Compute the actual number of passes and the actual tube length*

Since the tubes in this heater cannot exceed 6 ft (1.8 m) in length, the number of passes required = (length for one pass, ft)/(maximum allowable tube length, ft) = 40.6/6 = 6.77 passes. Since a fractional number of passes cannot be used and an even number of passes permit a more convenient layout of the heater, choose eight passes.

From the same equation for tube length as in step 6, l = tube length = 92.7/[(14)(8)(0.1636)] = 5.06 ft (1.5 m).

8. *Determine the feedwater pressure drop through heater*

In any closed feedwater heater, the pressure loss Δp lb/in^2 is $\Delta p = F_1F_2(L + 5.5D)N/D^{1.24}$, where Δp = pressure drop in the feedwater passing through the heater, lb/in^2; F_1 and F_2 = correction factors from Fig. 43; L = total lin ft of tubing divided by the number of tube holes in

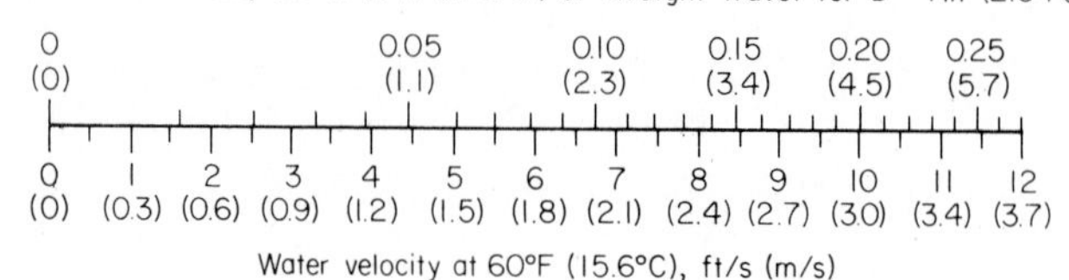

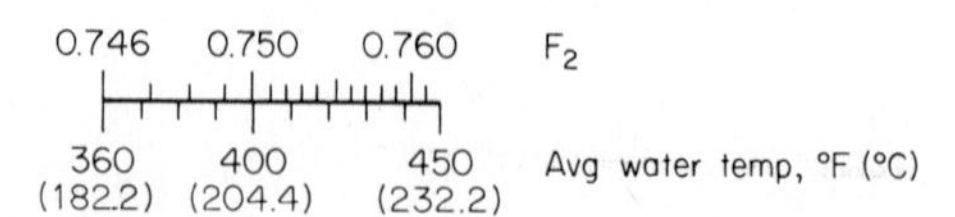

FIG. 43 Correction factors for closed feedwater heaters. *(Standards of Bleeder Heater Manufactuers Association, Inc.)*

one tube sheet; D = tube ID; N = number of passes. In finding F_2, the average water temperature is taken as $t_s - t_m$.

For this heater, using correction factors from Fig. 43,

$$\Delta p = (0.136)(0.761)\left[\frac{5.06(8)(14)}{(8)(14)} + 5.5(0.527)\right]\frac{8}{0.527^{1.24}}$$

$$= 14.6 \text{ lb/in}^2 \text{ (100.7 kPa)}$$

9. *Find the heater shell outside diameter*

The total number of tubes in the heater = (number of passes)(tubes per pass) = 8(14) = 112 tubes. Assume that there is 3/8-in (1.0-cm) clearance between each tube and the tube alongside, above, or below it. Then the pitch or center-to-center distance between tubes = pitch + tube OD = 3/8 + 5/8 = 1 in (2.5 cm).

The number of tubes per ft² of tube sheet = 166/(pitch)², or 166/1² = 166 tubes per ft² (1786.8 per m²). Since the heater has 112 tubes, the area of the tube sheet = 112/166 = 0.675 ft², or 97 in² (625.8 cm²).

The inside diameter of the heater shell = (tube sheet area, in²/0.7854)$^{0.5}$ = (97/0.7854)$^{0.5}$ = 11.1 in (28.2 cm). With a 0.25-in (0.6-cm) thick shell, the heater shell OD = 11.1 + 2(0.25) = 11.6 in (29.5 cm).

10. *Compute the quantity of heating steam required*

Steam enters the heater at 100 lb/in² (abs) (689.5 kPa) and 460°F (237.8°C). The enthalpy at this pressure and temperature is, from the superheated steam table, h_g = 1258.8 Btu/lb (2928.0 kJ/kg). The steam condenses in the heater, leaving as condensate at the saturation temperature corresponding to 100 lb/in² (abs) (689.5 kPa), or 327.81°F (164.3°C). The enthalpy of the saturated liquid at this temperature is, from the steam tables, h_f = 298.4 Btu/lb (694.1 kJ/kg).

The heater steam consumption for any closed-type feedwater heater is W, lb/h = $w_w(\Delta t)(h_g - h_f)$, where Δt = temperature rise of feedwater in heater, °F; c = specific heat of feedwater, Btu/(lb·°F). Assume c = 1.00 for the temperature range in this heater, and W = (37,640)(318 − 238)(1.00)/(1258.8 − 298.4) = 3140 lb/h (0.40 kg/s).

Related Calculations: The procedure used here can be applied to closed feedwater heaters in stationary and marine service. A similar procedure is used for selecting hot-water heaters for buildings, marine, and portable service. Various authorities recommend the following terminal difference (heater condensate temperature minus the outlet feedwater temperature) for closed feedwater heaters:

Feedwater outlet temperature		Terminal difference	
°F	°C	°F	°C
86 to 230	30.0 to 110.0	5	2.8
230 to 300	110.0 to 148.9	10	5.6
300 to 400	148.9 to 204.4	15	8.3
400 to 525	204.4 to 273.9	20	11.1

POWER-PLANT HEATER EXTRACTION-CYCLE ANALYSIS

A steam power plant operates at a boiler-drum pressure of 460 lb/in² (abs) (3171.7 kPa), a turbine throttle pressure of 415 lb/in² (abs) (2861.4 kPa) and 725°F (385.0°C), and a turbine capacity of 10,000 kW (or 13,410 hp). The Rankine-cycle efficiency ratio (including generator losses) is: full load, 75.3 percent; three-quarters load, 74.75 percent; half load, 71.75 percent. The turbine exhaust pressure is 1 inHg absolute (3.4 kPa); steam flow to the steam-jet air ejector is 1000 lb/h (0.13 kg/s). Analyze this cycle to determine the possible gains from two stages of extraction for feedwater heating, with the first stage a closed heater and the second stage a direct-contact or mixing heater. Use engineering-office methods in analyzing the cycle.

Calculation Procedure:

1. *Sketch the power-plant cycle*

Figure 44*a* shows the plant with one closed heater and one direct-contact heater. Values marked on Fig. 44*a* will be computed as part of this calculation procedure. Enter each value on the diagram as soon as it is computed.

2. *Compute the throttle flow without feedwater heating extraction*

Use the superheated steam tables to find the throttle enthalpy h_f = 1375.5 Btu/lb (3199.4 kJ/kg) at 415 lb/in^2 (abs) (2861.4 kPa) and 725°F (385.0°C).

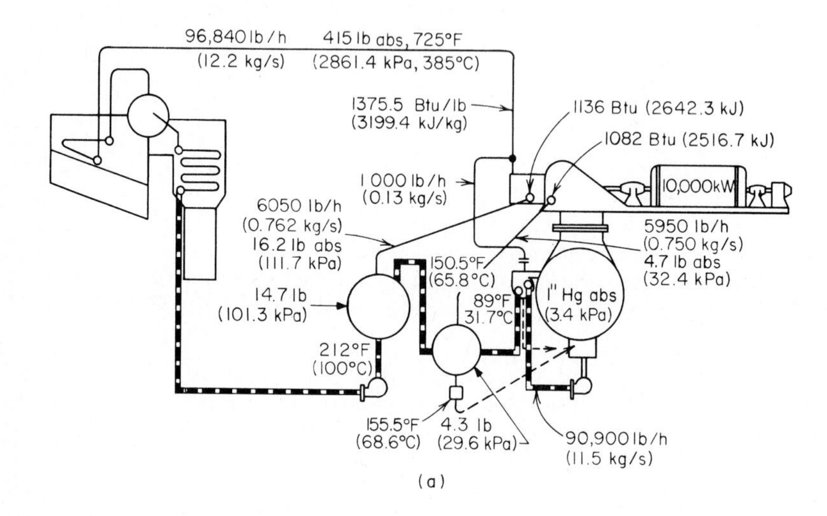

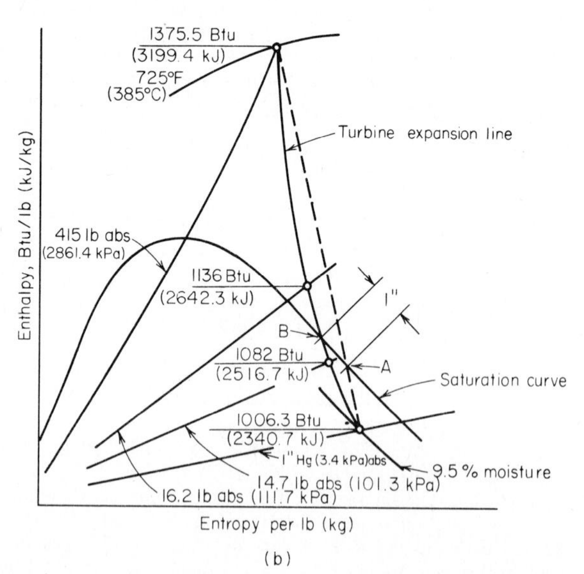

FIG. 44 (*a*) Two stages of feedwater heating in a steam plant; (*b*) Mollier chart of the cycle in (*a*).

Assume an irreversible adiabatic expansion between throttle conditions and the exhaust pressure of 1 inHg (3.4 kPa). Compute the final enthalpy H_{2s} by the same method used in earlier calculation procedures by finding y_{2s}, the percentage of moisture at the exhaust conditions with 1-inHg absolute (3.4-kPa) exhaust pressure. Do this by setting up the ratio $y_{2s} = (s_y - S_1)/s_{fg}$, where s_g and s_{fg} are entropies at the exhaust pressure; S_1 is entropy at throttle conditions. From the steam tables, $y_{2s} = 2.0387 - 1.6468/1.9473 = 0.201$. Then $H_{2s} = h_g - y_{2s}h_{fg}$, where h_g and h_{fg} are enthalpies at 1 inHg absolute (3.4 kPa). Substitute values from the steam table for 1 inHg absolute (3.4 kPa); or, $H_{2s} = 1096.3 - 0.201(1049.2) = 885.3$ Btu/lb (2059.2 kJ/kg).

The available energy in this irreversible adiabatic expansion is the difference between the throttle and exhaust conditions, or 1375.5 − 885.3 = 490.2 Btu/lb (1140.2 kJ/kg). The work at full load on the turbine is: (Rankine-cycle efficiency)(adiabatic available energy) = (0.753)(490.2) = 369.1 Btu/lb (858.5 kJ/kg). Enthalpy at the exhaust of the actual turbine = throttle enthalpy minus full-load actual work, or 1375.5 − 369.1 = 1006.4 Btu/lb (2340.9 kJ/kg). Use the Mollier chart to find, at 1.0 inHg absolute (3.4 kPa) and 1006.4 Btu/lb (2340.9 kJ/kg), that the exhaust steam contains 9.5 percent moisture.

Now the turbine steam rate SR = 3413(actual work output, Btu). Or, SR = 3413/369.1 = 9.25 lb/kWh (4.2 kg/kWh). With the steam rate known, the nonextraction throttle flow is (SR)(kW output) = 9.25(10,000) = 92,500 lb/h (11.7 kg/s).

3. *Determine the heater extraction pressures*

With steam extraction from the turbine for feedwater heating, the steam flow to the main condenser will be reduced, even with added throttle flow to compensate for extraction.

Assume that the final feedwater temperature will be 212°F (100.0°C) and that the heating range for each heater is equal. Both assumptions represent typical practice for a moderate-pressure cycle of the type being considered.

Feedwater leaving the condenser hotwell at 1 inHg absolute (3.4 kPa) is at 79.03°F (26.1°C). This feedwater is pumped through the air-ejector intercondensers and aftercondensers, where the condensate temperature will usually rise 5 to 15°F (2.8 to 8.3°C), depending on the turbine load. Assume that there is a 10°F (5.6°C) rise in condensate temperature from 79 to 89°F (26.1 to 31.7°C). Then the temperature range for the two heaters is 212 − 89 = 123°F (68.3°C). The temperature rise per heater is 123/2 = 61.5°F (34.2°C), since there are two heaters and each will have the same temperature rise. Since water enters the first-stage closed heater at 89°F (31.7°C), the exit temperature from this heater is 89 + 61.5 = 150.5°F (65.8°C).

The second-stage heater is a direct-contact unit operating at 14.7 lb/in^2 (abs) (101.4 kPa), because this is the saturation pressure at an outlet temperature of 212°F (100.0°C). Assume a 10 percent pressure drop between the turbine and heater steam inlet. This is a typical pressure loss for an extraction heater. Extraction pressure for the second-stage heater is then 1.1(14.7) = 16.2 lb/in^2 (abs) (111.7 kPa).

Assume a 5°F (2.8°C) terminal difference for the first-stage heater. This is a typical terminal difference, as explained in an earlier calculation procedure. The saturated steam temperature in the heater equals the condensate temperature = 150.5°F (65.8°C) exit temperature + 5°F (2.8°C) terminal difference = 155.5°F (68.6°C). From the saturation-temperature steam table, the pressure at 155.5°F (68.6°C) is 4.3 lb/in^2 (abs) (29.6 kPa). With a 10 percent pressure loss, the extraction pressure = 1.1(4.3) = 4.73 lb/in^2 (abs) (32.6 kPa).

4. *Determine the extraction enthalpies*

To establish the enthalpy of the extracted steam at each stage, the actual turbine-expansion line must be plotted. Two points—the throttle inlet conditions and the exhaust conditions—are known. Plot these on a Mollier chart, Fig. 44*b*. Connect these two points by a dashed straight line, Fig. 44*b*.

Next, measure along the saturation curve 1 in (2.5 cm) from the intersection point *A* back toward the enthalpy coordinate, and locate point *B*. Now draw a gradually sloping line from the throttle conditions to point *B*; from *B* increase the slope to the exhaust conditions. The enthalpy of the steam at each extraction point is read where the lines of constant pressure cross the expansion line. Thus, for the second-stage direct-contact heater where $p = 16.2$ lb/in^2 (abs) (111.7 kPa), $h_g = 1136$ Btu/lb (2642.3 kJ/kg). For the first-stage closed heater where $p = 4.7$ lb/in^2 (abs) (32.4 kPa), $h_g = 1082$ Btu/lb (2516.7 kJ/kg).

When the actual expansion curve is plotted, a steeper slope is used between the throttle superheat conditions and the saturation curve of the Mollier chart, because the turbine stages using superheated steam (stages above the saturation curve) are more efficient than stages using wet steam (stages below the saturation curve).

5. *Compute the extraction steam flow*

To determine the extraction flow rates, two assumptions must be made—condenser steam flow rate and first-stage closed-heater extraction flow rate. The complete cycle will be analyzed, and the assumption checked. If the assumptions are incorrect, new values will be assumed, and the cycle analyzed again.

Assume that the condenser steam flow from the turbine is 84,000 lb/h (10.6 kg/s) when it is operating with extraction. Note that this value is less than the nonextraction flow of 92,500 lb/h (11.7 kg/s). The reason is that extraction of steam will reduce flow to the condenser because the steam is bled from the turbine after passage through the throttle but before the condenser inlet. Then, for the first-stage closed heater, condensate flow is as follows:

From condenser	84,000 lb/h (10.6 kg/s) assumed
From steam-jet ejector	1,000 lb/h (0.13 kg/s)
From first-stage heater	5,900 lb/h (0.74 kg/s) assumed
Total	90,900 lb/h (11.5 kg/s)

The value of 5900 lb/h (0.74 kg/s) of condensate from the first-stage heater is the second assumption made. Since it will be checked later, an error in the assumption can be detected.

Assume a 2 percent heat radiation loss between the turbine and heater. This is a typical loss. Then

Steam enthalpy at heater = 1082(0.98)	= 1060.4 Btu/lb (2465.5 kJ/kg)
Enthalpy of condensate at 155.5°F (68.6°C)	= −123.4 Btu/lb (−287.0 kJ/kg)
Heat given up per lb (kg) of steam condensed	= 937.0 Btu/lb (2179.5 kJ/kg)
Enthalpy of feedwater at 150.5°F (65.8°C)	= 118.3 Btu/lb (275.2 kJ/kg)
Enthalpy of feedwater to heater at 89°F (31.7°C)	= −57.0 Btu/lb (−132.6 kJ/kg)
Heat absorbed by feedwater	= 61.3 Btu/lb (142.6 kJ/kg)

Required extraction = (total condensate flow, lb/h) [(heat absorbed by feedwater, Btu/lb)/(heat given up per lb of steam condensed, Btu/lb)], or required extraction = (90,900)(61.3/937) = 5950 lb/h (0.75 kg/s)

Compare the required extraction, 5950 lb/h (0.75 kg/s), with the assumed extraction, 5900 lb/h (0.74 kg/s). The difference is only 50 lb/h (0.006 kg/s), which is less than 1 percent. Therefore, the assumed flow rate is satisfactory, because estimates within 1 percent are considered sufficiently accurate for all routine analyses.

For the second-stage direct-contact heater, condensate flow, lb/h, is as follows:

From the first-stage heater	90,900 lb/h (11.5 kg/s)
Steam enthalpy at heater = 1135(0.98)	= 1112.3 Btu/lb (2587.2 kJ/kg)
Enthalpy of condensate at 212°F (100.0°C)	= −180.0 Btu/lb (−418.7 kJ/kg)
Heat given up per lb of steam condensed	= 932.3 Btu/lb (2168.5 kJ/kg)
Enthalpy of feedwater at 212°F (100.0°C)	= 180.0 Btu/lb (418.7 kJ/kg)
Enthalpy of feedwater at 150.5°F (65.8°C)	= 118.3 Btu/lb (275.2 kJ/kg)
Heat absorbed by feedwater	= 61.7 Btu/lb (143.5 kJ/kg)

The required extraction, calculated in the same way as for the first-stage heater, is (90,900)(61.7/932.2) = 6050 lb/h (0.8 kg/s).

The computed extraction flow for the second-stage heater is not compared with an assumed value because an assumption was not necessary.

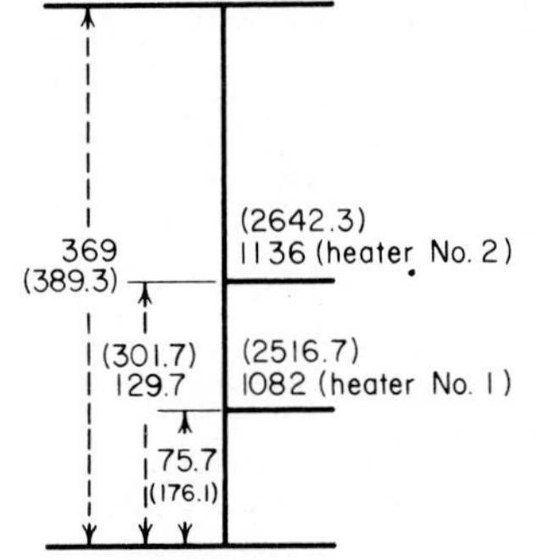

Enthalpy at exhaust 1006.3 Btu/lb (2340.7 kJ/kg)

FIG. 45 Diagram of turbine-expansion line.

6. *Compute the actual condenser steam flow*

Sketch a vertical line diagram, Fig. 45, showing the enthalpies at the throttle, heaters, and exhaust. From this diagram, the work lost by the extracted steam can be computed. As Fig. 45 shows, the total enthalpy drop from the throttle to the exhaust is 369 Btu/lb (389.3 kJ/kg). Each pound of extracted steam from the first- and second-stage bleed points causes a work loss of 75.7 Btu/lb (176.1 kJ/kg) and 129.7 Btu/lb (301.7 kJ/kg), respectively. To carry the same load, 10,000 kW, with extraction, it will be necessary to supply the following additional compensation steam to the turbine throttle: (heater flow, lb/h)(work loss, Btu/h)/(total work, Btu/h). Then

	lb/h	kg/s
First-stage closed heater:		
(5950)(75.7/369)	1220	0.15
Second-stage direct-contact heater:		
(6050)(129.7/369)	2120	0.27
Total additional throttle flow to compensate for extraction	3340	0.42

Check the assumed condenser flow using nonextraction throttle flow + additional throttle flow − heater extraction = condenser flow. Set up a tabulation of the flows as follows:

Flow	lb/h	kg/s
Throttle; nonextraction	92,500	11.65
Added flow (compensation)	3,340	0.42
Throttle; extraction	95,840	12.07
Extraction (5950 + 6050)	−12,000	−1.51
Condenser flow	83,840	10.56

Compare this actual flow, 83,840 lb/h (10.6 kg/s), with the assumed flow, 84,000 lb/h (10.6 kg/s). The difference, 160 lb/h (0.02 kg/s), is less than 1 percent. Since an accuracy within 1 percent is sufficient for all normal power-plant calculations, it is not necessary to recompute the cycle. Had the difference been greater than 1 percent, a new condenser flow would be assumed and the cycle recomputed. Follow this procedure until a difference of less than 1 percent is obtained.

7. *Determine the economy of the extraction cycle*

For a nonextraction cycle operating in the same pressure range,

	Btu/lb	kJ/kg
Enthalpy of throttle steam	1375.3	3198.9
Enthalpy of condensate at 79°F (26.1°C)	−47.0	−109.3
Heat supplied by boiler	1328.3	3089.6

Heat chargeable to turbine = (throttle flow + air-ejector flow)(heat supplied by boiler)/(kW output of turbine) = (92,500 + 1000)(1328.3)/10,000 = 12,410 Btu/kWh (13,093.2 kJ/kWh), which is the actual heat rate HR of the nonextraction cycle.

For the extraction cycle using two heaters,

	Btu/lb	kJ/kg
Enthalpy of throttle steam	1375.3	3198.9
Enthalpy of feedwater leaving second heater	−180.0	−418.7
Heat supplied by boiler	1195.3	2780.3

As before, heat chargeable to turbine = (95,840 + 1000)(1195.3)/10,000 = 11,580 Btu/kWh (12,217.5 kJ/kWh). Therefore, the improvement = (nonextraction HR − extraction HR)/nonextraction HR = (12,410 − 11,580)/12,410 = 0.0662, or 6.62 percent.

Related Calculations: (1) To determine the percent improvement in a steam cycle resulting from additional feedwater heaters in the cycle, use the same procedure as given above for three, four, five, six, or more heaters. Plot the percent improvement vs. number of stages of extraction, Fig. 46, to observe the effect of additional heaters. A plot of this type shows the decreasing gains made by additional heaters. Eventually the gains become so small that the added expenditure for an additional heater cannot be justified.

(2) Many simple marine steam plants use only two stages of feedwater heating. To analyze such a cycle, use the procedure given, substituting the hp output for the kW output of the turbine.

(3) Where a marine plant has more than two stages of feedwater heating, follow the procedure given in (1) above.

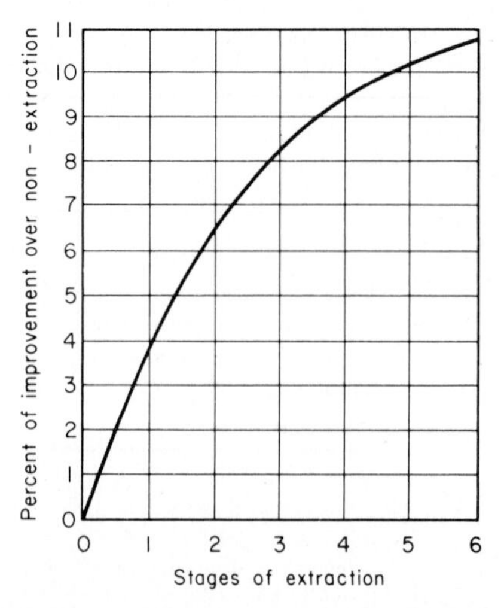

FIG. 46 Percentage of improvement in turbine heat rate vs. stages of extraction.

STEAM BOILER, ECONOMIZER, AND AIR-HEATER EFFICIENCY

Determine the overall efficiency of a steam boiler generating 56,000 lb/h (7.1 kg/s) of 600 lb/in^2 (abs) (4137.0 kPa) 800°F (426.7°C) steam. The boiler is continuously blown down at the rate of 2500 lb/h (0.31 kg/s). Feedwater enters the economizer at 300°F (148.9°C). The furnace burns 5958 lb/h (0.75 kg/s) of 13,100-Btu/lb (30,470.6-kJ/kg) HHV (higher heating value) coal having an ultimate analysis of 68.5 percent C, 5 percent H_2, 8.9 percent O_2, 1.2 percent N_2, 3.2 percent S, 8.7 percent ash, and

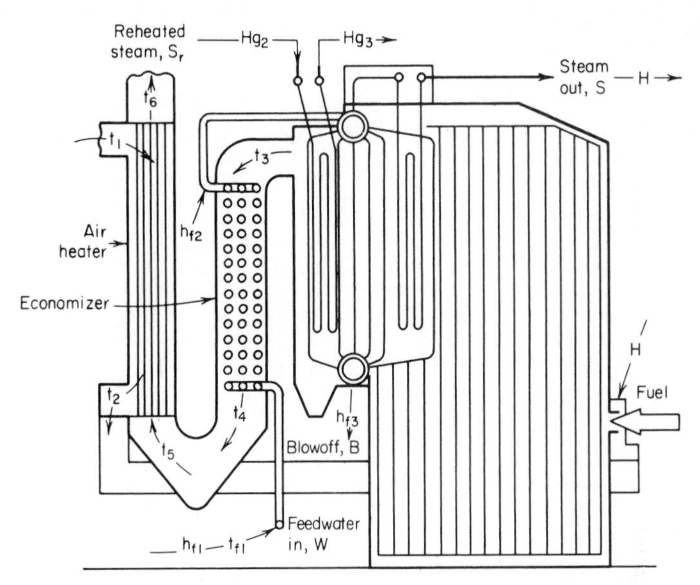

FIG. 47 Points in a steam generator where temperatures and enthalpies are measured in determining the boiler efficiency.

4.5 percent moisture. Air enters the boiler at 63°F (17.2°C) dry-bulb and 56°F (13.3°C) wet-bulb temperature, with 56 gr of vapor per lb (123.5 gr/kg) of dry air. Carbon in the fuel refuse is 7 percent, refuse is 0.093 lb/lb (0.2 kg/kg) of fuel. Feedwater leaves the economizer at 370°F (187.8°C). Flue gas enters the economizer at 850°F (454.4°C) and has an analysis of 15.8 percent CO_2, 2.8 percent O_2, and 81.4 percent N_2. Air enters the air heater at 63°F (17.2°C) with 56 gr/lb (123.5 gr/kg) of dry air; air leaves the heater at 480°F (248.9°C). Gas enters the air heater at 570°F (298.9°C), and 14 percent of the air to the furnace comes from the mill fan. Determine the steam generator overall efficiency, economizer efficiency, and air-heater efficiency. Figure 47 shows the steam generator and the flow factors that must be considered.

Calculation Procedure:

1. *Determine the boiler output*

The boiler output $= S(h_g - h_{f1}) + S_r(h_{g3} - h_{g2}) + B(h_{f3} - h_{f1})$, where S = steam generated, lb/h; h_g = enthalpy of the generated steam, Btu/lb; h_{f1} = enthalpy of inlet feedwater; S_r = reheated steam flow, lb/h (if any); h_{g3} = outlet enthalpy of reheated steam; h_{g2} = inlet enthalpy of reheated steam; B = blowoff, lb/h; h_{f3} = blowoff enthalpy, where all enthalpies are in Btu/lb. Using the appropriate steam table and deleting the reheat factor because there is no reheat, we get boiler output = 56,000(1407.7 − 269.6) + 2500(471.6 − 269.6) = 64,238,600 Btu/h (18,826.5 kW).

2. *Compute the heat input to the boiler*

The boiler input $= FH$, where F = fuel input, lb/h (as fired); H = higher heating value, Btu/lb (as fired). Or, boiler input = 5958(13,100) = 78,049,800 Btu/h (22,874.1 kW).

3. *Compute the boiler efficiency*

The boiler efficiency = (output, Btu/h)/(input, Btu/h) = 64,238,600/78,049,800 = 0.822, or 82.2 percent.

4. *Determine the heat absorbed by the economizer*

The heat absorbed by the economizer, Btu/h = $w_w(h_{f2} - h_{f1})$, where w_w = feedwater flow, lb/h; h_{f2} and h_{f1} = enthalpies of feedwater leaving and entering the economizer, respectively, Btu/lb. For this economizer, with the feedwater leaving the economizer at 370°F (187.8°C) and entering at 300°F (148.9°C), heat absorbed = (56,000 + 2500)(342.79 − 269.59) = 4,283,000 Btu/h (1255.2 kW). Note that the total feedwater flow w_w is the sum of the steam generated and the continuous blowdown rate.

5. *Compute the heat available to the economizer*

The heat available to the economizer, Btu/h = H_gF, where H_g = heat available in flue gas, Btu/lb of fuel = heat available in dry gas + heat available in flue-gas vapor, Btu/lb of fuel = $(t_3 - t_{f1})(0.24G) + (t_3 - t_{f1})(0.46)\{M_f + 8.94H_2 + M_a[G - C_b - N_2 - 7.94(H_2 - O_2/8)]\}$, where $G = \{[11CO_2 + 8O_2 + 7(N_2 + CO)]/[3(CO_2 + CO)]\}(C_b + S/2.67) + S/1.60$; M_f = lb of moisture per lb fuel burned; M_a = lb of moisture per lb of dry air to furnace; C_b = lb of carbon burned per lb of fuel burned = $C - RC_r$; C_r = lb of combustible per lb of refuse; R = lb of refuse per lb of fuel; H_2, N_2, C, O_2, S = lb of each element per lb of fuel, as fired; CO_2, CO, O_2, N_2 = percentage parts of volumetric analysis of dry combustion gas entering the economizer. Substituting gives C_b = 0.685 − (0.093)(0.07) = 0.678 lb/lb (0.678 kg/kg) fuel; G = [11(0.158) + 8(0.028) + 7(0.814)]/[3(0.158)] × (0.678 + 0.032/2.67) + 0.032/1.60; G = 11.18 lb/lb (11.18 kg/kg) fuel. H_g = (800 − 300)(0.24) × (11.18) + (800 − 300)(0.46){0.045 + (8.9)(0.05) + 56/7000[11.18 − 0.678 − 0.012 − 7.94 × (0.05 − 0.089/8)]}; H_g = 1473 Btu/lb (3426.2 kJ/kg) fuel. Heat available = H_gF = (1473)(5958) = 8,770,000 Btu/h (2570.2 kW).

6. *Compute the economizer efficiency*

The economizer efficiency = (heat absorbed, Btu/h)/(heat available, Btu/h) = 4,283,000/8,770,000 = 0.488, or 48.8 percent.

7. *Compute the heat absorbed by air heater*

The heat absorbed by the air heater, Btu/lb of fuel, = $A_h(t_2 - t_1)(0.24 + 0.46M_a)$, where A_h = air flow through heater, lb/lb fuel = $A - A_m$; A = total air to furnace, lb/lb fuel = $G - C_b - N_2 - 7.94(H_2 - O_2/8)$; G = similar to economizer but based on gas at the furnace exit; A_m = external air supplied by the mill fan or other source, lb/lb of fuel. Substituting shows G = [11(0.16) + 8(0.26) + 7(0.184)]/[3(0.16)](0.678 + 0.032/2.67) + 0.032/1.60; G = 11.03 lb/lb (11.03 kg/kg) fuel; A = 11.03 − 0.69 − 0.012 − 7.94(0.05 − 0.089/8); A = 10.02 lb/lb (10.02 kg/kg) fuel. Heat absorbed = (1 − 0.15)(10.02)(480 − 63)(0.24 + 56/7000) = 865.5 Btu/lb (2013.2 kJ/kg) fuel.

8. *Compute the heat available to the air heater*

The heat available to the air heater, Btu/h = $(t_5 - t_1)0.24G + (t_5 - t_1)0.46(M_f + 8.94H_2 + M_aA)$. In this relation, all symbols are the same as for the economizer except that G and A are based on the gas entering the heater. Substituting gives G = [11(0.15) + 8(0.036) + 7(0.814)]/[3(0.15)](0.678 + 0.032/2.67) + 0.032/1.60; G = 11.72 lb/lb (11.72 kg/kg) fuel. And A = 11.72 − 0.69 − 0.012 − 7.94(0.05 − 0.089/8) = 10.71 lb/lb (10.71 kg/kg) fuel. Heat available = (570 − 3)(0.24)(11.72) + (570 − 63)(0.46)[0.045 + 8.94(0.05) + 56/7000(10.71)] = 1561 Btu/lb (3630.9 kJ/kg).

9. *Compute the air-heater efficiency*

The air-heater efficiency = (heat absorbed, Btu/lb fuel)/(heat available, Btu/lb fuel) = 865.5/1561 = 0.554, or 55.4 percent.

Related Calculations: The above procedure is valid for all types of steam generators, regardless of the kind of fuel used. Where oil or gas is the fuel, alter the combustion calculations to reflect the differences between the fuels. Further, this procedure is also valid for marine and portable boilers.

FIRE-TUBE BOILER ANALYSIS AND SELECTION

Determine the heating surface in an 84-in (213.4-cm) diameter fire-tube boiler 18 ft (5.5 m) long having 84 tubes of 4-in (10.2-cm) ID if 25 percent of the upper shell ends are heat-insulated. How much steam is generated if the boiler evaporates 34.5 lb/h of water per 12 ft^2 [3.9 g/(m^2·s)] of heating surface? How much heat is added by the boiler if it operates at 200 lb/in^2 (abs) (1379.0 kPa) with 200°F (93.3°C) feedwater? What is the factor of evaporation for this boiler? How much hp is developed by the boiler if 7,000,000 Btu/h (2051.4 kW) is delivered to the water?

Calculation Procedure:

1. *Compute the shell area exposed to furnace gas*

Shell area = $\pi DL(1 - 0.25)$, where D = boiler diameter, ft; L = shell length, ft; $1 - 0.25$ is the portion of the shell in contact with the furnace gas. Then shell area = $\pi(84/12)(18)(0.75)$ = 297 ft^2 (27.0 m^2).

2. *Compute the tube area exposed to furnace gas*

Tube area = πdLN, where d = tube ID, ft; L = tube length, ft; N = number of tubes in boiler. Substituting gives tube area = $\pi(4/12)(18)(84)$ = 1583 ft^2 (147.1 m^2).

3. *Compute the head area exposed to furnace gas*

The area exposed to furnace gas is twice (since there are *two* heads) the exposed head area minus twice the area occupied by the tubes. The exposed head area is (total area)(1 − portion covered by insulation, expressed as a decimal). Substituting, we get $2\pi D^2/4 - (2)(84)\pi d^2/4 = 2\pi/4(84/12)^2(0.75) - (2)(84)\pi(4/12)^2/4$ = head area = 43.1 ft^2 (4.0 m^2).

4. *Find the total heating surface*

The total heating surface of any fire-tube boiler is the sum of the shell, tube, and head areas, or 297.0 + 1583 + 43.1 = 1923 ft^2 (178.7 m^2), total heating surface.

5. *Compute the quantity of steam generated*

Since the boiler evaporates 34.5 lb/h of water per 12 ft^2 [3.9 g/(m^2·s)] of heating surface, the quantity of steam generated = 34.5 (total heating surface, ft^2)/12 = 34.5(1923.1)/12 = 5200 lb/h (0.66 kg/s).

Note: Evaporation of 34.5 lb/h (0.0043 kg/s) from and at 212°F (100.0°C) is the definition of the now-discarded term *boiler horsepower*. However, this term is still met in some engineering examinations and is used by some manufacturers when comparing the performance of boilers. A term used in lieu of boiler horsepower, with the same definition, is *equivalent evaporation*. Both terms are falling into disuse, but they are included here because they still find some use today.

6. *Determine the heat added by the boiler*

Heat added, Btu/lb of steam = $h_g - h_{f1}$; from steam table values 1198.4 − 167.99 = 1030.41 Btu/lb (2396.7 kJ/kg). An alternative way of computing heat added is h_g − (feedwater temperature, °F, − 32), where 32 is the freezing temperature of water on the Fahrenheit scale. By this method, heat added = 1198.4 − (200 − 32) = 1030.4 Btu/lb (2396.7 kJ/kg). Thus, both methods give the same results in this case. In general, however, use of steam table values is preferred.

7. *Compute the factor of evaporation*

The factor of evaporation is used to convert from the actual to the equivalent evaporation, defined earlier. Or, factor of evaporation = (heat added by boiler, Btu/lb)/970.3, where 970.3 Btu/lb (2256.9 kJ/kg) is the heat added to develop 1 boiler hp (bhp) (0.75 kW). Thus, the factor of evaporation for this boiler = 1030.4/970.3 = 1.066.

8. *Compute the boiler hp output*

Boiler hp = (actual evaporation, lb/h) (factor of evaporation)/34.5. In this relation, the actual evaporation must be computed first. Since the furnace delivers 7,000,000 Btu/h (2051.5 kW) to the boiler water and the water absorbs 1030.4 Btu/lb (2396.7 kJ/kg) to produce 200-lb/in^2 (abs) (1379.0-kPa) steam with 200°F (93.3°C) feedwater, the steam generated, lb/h = (total heat delivered, Btu/h)/(heat absorbed, Btu/lb) = 7,000,000/1030.4 = 6670 lb/h (0.85 kg/s). Then boiler hp = (6760)(1.066)/34.5 = 209 hp (155.9 kW).

The rated hp output of horizontal fire-tube boilers with separate supporting walls is based on 12 ft^2 (1.1 m^2) of heating surface per boiler hp. Thus, the rated hp of this boiler = 1923.1/12 = 160 hp (119.3 kW). When producing 209 hp (155.9 kW), the boiler is operating at 209/160, or 1.305 times its normal rating, or (100)(1.305) = 130.5 percent of normal rating.

Note: Today most boiler manufacturers rate their boilers in terms of pounds per hour of steam generated at a stated pressure. Use this measure of boiler output whenever possible. Inclusion of the term *boiler hp* in this handbook does not indicate that the editor favors or recommends its use. Instead, the term was included to make the handbook as helpful as possible to users who might encounter the term in their work.

SAFETY-VALVE STEAM-FLOW CAPACITY

How much saturated steam at 150 lb/in^2 (abs) (1034.3 kPa) can a 2.5-in (6.4-cm) diameter safety valve having a 0.25-in (0.6-cm) lift pass if the discharge coefficient of the valve c_d is 0.75? What is the capacity of the same valve if the steam is superheated 100°F (55.6°C) above its saturation temperature?

Calculation Procedure:

1. *Determine the area of the valve annulus*

Annulus area, in^2 = $A = \pi DL$, where D = valve diameter, in; L = valve lift, in. Annulus area = $\pi(2.5)(0.25)$ = 1.966 in^2 (12.7 cm^2).

2. *Compute the ideal flow for this safety valve*

Ideal flow F_i lb/s for any safety valve handling saturated steam is $F_i = p_s^{0.97}A/60$, where p_s = saturated-steam pressure, lb/in^2 (abs). For this valve, $F_i = (150)^{0.97}(1.966)/60$ = 4.24 lb/s (1.9 kg/s).

3. *Compute the actual flow through the valve*

Actual flow $F_a = F_i c_d = (4.24)(0.75)$ = 3.18 lb/s (1.4 kg/s) = (3.18)(3600 s/h) = 11,448 lb/h (1.44 kg/s).

4. *Determine the superheated-steam flow rate*

The ideal superheated-steam flow F_{is} lb/s is $F_{is} = p_s^{0.97}A/[60(1 + 0.0065t_s)]$, where t_s = superheated temperature, above saturation temperature, °F. Then $F_{is} = (150)^{0.97}(1.966)/[60(1 + 0.0065 \times 100)]$ = 3.96 lb/s (1.8 kg/s). The actual flow is $F_{as} = F_{is}c_d = (3.96)(0.75)$ = 2.97 lb/s (1.4 kg/s) = (2.97)(3600) = 10,700 lb/h (1.4 kg/s).

Related Calculations: Use this procedure for safety valves serving any type of stationary or marine boiler.

SAFETY-VALVE SELECTION FOR A WATERTUBE STEAM BOILER

Select a safety valve for a watertube steam boiler having a maximum rating of 100,000 lb/h (12.6 kg/s) at 800 lb/in^2 (abs) (5516.0 kPa) and 900°F (482.2°C). Determine the valve diameter, size of boiler connection for the valve, opening pressure, closing pressure, type of connection, and valve material. The boiler is oil-fired and has a total heating surface of 9200 ft^2 (854.7 m^2) of which 1000 ft^2 (92.9 m^2) is in waterwall surface. Use the ASME *Boiler and Pressure Vessel Code* rules when selecting the valve. Sketch the escape-pipe arrangement for the safety valve.

Calculation Procedure:

1. *Determine the minimum valve relieving capacity*

Refer to the latest edition of the *Code* for the relieving-capacity rules. Recent editions of the *Code* require that the safety valve have a *minimum* relieving capacity based on the pounds of steam generated per hour per square foot of boiler heating surface and waterwall heating surface. In the edition of the *Code* used in preparing this handbook, the relieving requirement for oil-fired boilers was 10 lb/(ft^2·h) of steam [13.6 g/(m^2·s)] of boiler heating surface, and 16 lb/(ft^2·h) of steam [21.9 g/(m^2·s)] of waterwall surface. Thus, the minimum safety-valve relieving capacity for this boiler, based on total heating surface, would be (8200)(10) + (1000)(16) = 92,000 lb/h (11.6 kg/s). In this equation, 1000 ft^2 (92.9 m^2) of waterwall surface is deducted from the total heating surface of 9200 ft^2 (854.7 m^2) to obtain the boiler heating surface of 8200 ft^2 (761.8 m^2).

The minimum relieving capacity based on total heating surface is 92,000 lb/h (11.6 kg/s); the maximum rated capacity of the boiler is 100,000 lb/h (12.6 kg/s). Since the *Code* also requires that "the safety valve or valves will discharge all the steam that can be generated by the boiler," the minimum relieving capacity must be 100,000 lb/h (12.6 kg/s), because this is the maximum capacity of the boiler and it exceeds the valve capacity based on the heating-surface calculation. If the valve capacity based on the heating-surface steam generation were larger than the stated maximum capacity of the boiler, the *Code* heating-surface valve capacity would be used in safety-valve selection.

2. *Determine the number of safety valves needed*

Study the latest edition of the *Code* to determine the requirements for the number of safety valves. The edition of the *Code* used here requires that "each boiler shall have at least one safety valve and if it [the boiler] has more than 500 ft^2 (46.5 m^2) of water heating surface, it shall have two or more safety valves." Thus, at least two safety valves are needed for this boiler. The *Code* further specifies, in the edition used, that "when two or more safety valves are used on a boiler, they may be mounted either separately or as twin valves made by placing individual valves on Y bases, or duplex valves having two valves in the same body casing. Twin valves made by placing individual valves on Y bases, or duplex valves having two valves in the same body, shall be of equal sizes." Also, "when not more than two valves of different sizes are mounted singly, the relieving capacity of the smaller valve shall not be less than 50 percent of that of the larger valve."

Assume that two equal-size valves mounted on a Y base will be used on the steam drum of this boiler. Two or more equal-size valves are usually chosen for the steam drum of a watertube boiler.

Since this boiler handles superheated steam, check the *Code* requirements regarding superheaters. The *Code* states that "every attached superheater shall have one or more safety valves near the outlet." Also, "the discharge capacity of the safety valve, or valves, on an attached superheater may be included in determining the number and size of the safety valves for the boiler, provided there are no intervening valves between the superheater safety valve and the boiler, and provided the discharge capacity of the safety valve, or valves, on the boiler, as distinct from the superheater, is at least 75 percent of the aggregate valve capacity required."

Since the safety valves used must handle 100,000 lb/h (12.6 kg/s), and one or more superheater safety valves are required by the *Code*, assume that the two steam-drum valves will handle, in accordance with the above requirement, 80,000 lb/h (10.1 kg/s). Assume that one superheater safety valve will be used. Its capacity must then be at least 100,000 − 80,000 = 20,000 lb/h (2.5 kg/s). (Use as few superheater safety valves as possible, because this simplifies the installation and reduces cost.) With this arrangement, each steam-drum valve must handle 80,000/2 = 40,000 lb/h (5.0 kg/s) of steam, since there are two safety valves on the steam drum.

3. *Determine the valve pressure settings*

Consult the *Code*. It requires that "one or more safety valves on the boiler proper shall be set at or below the maximum allowable working pressure." For modern boilers, the maximum allowable working pressure is usually 1.5, or more, times the rated operating pressure in the lower [under 1000 lb/in^2 (abs) or 6895.0 kPa] pressure ranges. To prevent unnecessary operation of the safety valve and to reduce steam losses, the lowest safety-valve setting is usually about 5 percent higher than the boiler operating pressure. For this boiler, the lowest pressure setting would be 800 + 800(0.05) = 840 lb/in^2 (abs) (5791.8 kPa). Round this to 850 lb/in^2 (abs) (5860.8 kPa, or 6.25 percent) for ease of selection from the usual safety-valve rating tables. The usual safety-valve pressure setting is between 5 and 10 percent higher than the rated operating pressure of the boiler.

Boilers fitted with superheaters usually have the superheater safety valve set at a lower pressure than the steam-drum safety valve. This arrangement ensures that the superheater safety valve opens first when overpressure occurs. This provides steam flow through the superheater tubes at all times, preventing tube burnout. Therefore, the superheater safety valve in this boiler will be set to open at 850 lb/in^2 (abs) (5860.8 kPa), the lowest opening pressure for the safety valves chosen. The steam-drum safety valves will be set to open at a higher pressure. As decided earlier, the superheater safety valve will have a capacity of 20,000 lb/h (2.5 kg/s).

Between the steam drum and the superheater safety valve, there is a pressure loss that varies from one boiler to another. The boiler manufacturer supplies a performance chart showing the drum outlet pressure for various percentages of the maximum continuous steaming capacity of the boiler. This chart also shows the superheater outlet pressure for the same capacities. The difference between the drum and superheater outlet pressure for any given load is the superheater pressure loss. Obtain this pressure loss from the performance chart.

Assume, for this boiler, that the superheater pressure loss, plus any pressure losses in the nonreturn valve and dry pipe, at maximum rating, is 60 lb/in^2 (abs) (413.7 kPa). The steam-drum operating pressure will then be superheater outlet pressure + superheater pressure loss = 800 + 60 = 860 lb/in^2 (abs) (5929.7 kPa). As with the superheater safety valve, the steam-drum safety valve is usually set to open at about 5 percent above the drum operating pressure at maximum steam output. For this boiler then, the drum safety-valve set pressure = 860 + 860(0.05) = 903 lb/in^2 (abs) (6226.2 kPa). Round this to 900 lb/in^2 (abs) (6205.5 kPa) to simplify valve selection.

Some designers add the drum safety-valve blowdown or blowback pressure (difference between the valve opening and closing pressures, lb/in^2) to the total obtained above to find the drum operating pressure. However, the 5 percent allowance used above is sufficient to allow for the blowdown in boilers operating at less than 1000 lb/in^2 (abs) (6895.0 kPa). At pressures of 1000 lb/in^2 (abs) (6895.0 kPa) and higher, add the drum safety-valve blowdown *and* the 5 percent allowance to the superheater outlet pressure and pressure loss to find the drum pressure.

4. *Determine the required valve orifice discharge area*

Refer to a safety-valve manufacturer's engineering data listing valve capacities at various working pressures. For the two steam-drum valves, enter the table at 900 lb/in^2 (abs) (6205.5 kPa), and project horizontally until a capacity of 40,000 lb/h (5.0 kg/s), or more, is intersected. Here is an excerpt from a typical manufacturer's capacity table for safety valves handling *saturated steam:*

Set pressure		Orifice area					
lb/in^2 (abs)	kPa	0.994 in^2	6.41 cm^2	1.431 in^2	9.23 cm^2	2.545 in^2	16.42 cm^2
890	6,136.6	41,750	5.26	60,000	7.56	107,200	13.5
900	6,205.5	42,200	5.32	60,900	7.67	108,000	13.6
910	6,274.5	42,700	5.38	61,600	7.76	109,300	13.8

Thus, at 900 lb/in^2 (abs) (6205.5 kPa) a valve with an orifice area of 0.994 in^2 (6.4 cm^2) will have a capacity of 42,200 lb/h (5.3 kg/s) of saturated steam. This is 5.5 percent greater than the required capacity of 40,000 lb/h (5.0 kg/s) for each steam-drum valve. However, the usual selection cannot be made at exactly the desired capacity. Provided that the valve chosen has a greater steam relieving capacity than required, there is no danger of overpressure in the steam drum. Be careful to note that safety valves for saturated steam are chosen for the steam drum because superheating of the steam does not occur in the steam drum.

The superheater safety valve must handle 20,000 lb/h (2.5 kg/s) of 850 lb/in^2 (abs) (5860.8-kPa) steam at 900°F (482.2°C). Safety valves handling superheated steam have a smaller capacity than when handling saturated steam. To obtain the capacity of a safety valve handling superheated steam, the saturated steam capacity is multiplied by a correction factor that is less than 1.00. An alternative procedure is to divide the required superheated-steam capacity by the same correction factor to obtain the saturated-steam capacity of the valve. The latter procedure will be used here because it is more direct.

Obtain the correction factor from the safety-valve manufacturer's engineering data by entering at the steam pressure and projecting to the steam temperature, as shown below.

Set pressure		Steam temperature	
lb/in^2 (abs)	kPa	880°F (471.1°C)	900°F (482.2°C)
800	5516.0	0.80	0.80
850	6205.5	0.81	0.80
900	5860.8	0.81	0.80

Thus, at 850 lb/in^2 (abs) (5860.8 kPa) and 900°F (482.2°C), the correction factor is 0.80. The required saturated steam capacity then is 20,000/0.80 = 25,000 lb/h (3.1 kg/s).

Refer to the manufacturer's saturated-steam capacity table as before, and at 850 lb/in^2 (abs) (5860.8 kPa) find the closest capacity as 31,500 lb/h (4.0 kg/s) for a 0.785-in^2 (5.1-cm^2) orifice. As with the steam-drum valves, the actual capacity of the safety valve is somewhat greater than the required capacity. In general, it is difficult to find a valve with exactly the required steam relieving capacity.

5. *Determine the valve nominal size and construction details*

Turn to the data section of the safety-valve engineering manual to find the valve construction features. For the steam-drum valves having 0.994-in^2 (6.4-cm^2) orifice areas, the engineering data show, for 900-lb/in^2 (abs) (6205.5-kPa) service, each valve is 1½-in (3.8-cm) unit rated for temperatures up to 1050°F (565.6°C). The inlet is a 900-lb/in^2 (6205.5-kPa) 1½-in (3.8-cm) flanged connection, and the outlet is a 150-lb/in^2 (1034.3-kPa) 3-in (7.6-cm) flanged connection. Materials used in the valve include: body, cast carbon steel; disk seat, stainless steel AISI 321. The overall height is 27⅞ in (70.8 cm); dismantled height is 32¾ in (83.2 cm).

Similar data for the superheated steam valve show, for a maximum pressure of 900 lb/in^2 (abs) (6205.5 kPa), that it is a 1½-in (3.8-cm) unit rated for temperatures up to 1000°F (537.8°C). The inlet is a 900-lb/in^2 (6205.5-kPa) 1½-in (3.8-cm) flanged connection, and the outlet is a 150-lb/in^2 (1034.3-kPa) 3-in (7.6-cm) flanged connection. Materials used in the valve include: body, cast alloy steel, ASTM 217-WC6; spindle, stainless steel; spring, alloy steel; disk seat, stainless steel. Overall height is 21⅜ in (54.3 cm); dismantled height is 25¼ in (64.1 cm). Checking the *Code* shows that "every safety valve used on a superheater discharging superheated steam at a temperature over 450°F (232.2°C) shall have a casing, including the base, body, bonnet and spindle, of steel, steel alloy, or equivalent heat-resisting material. The valve shall have a flanged inlet connection."

Thus, the superheater valve selected is satisfactory.

6. *Compute the steam-drum connection size*

The *Code* requires that "when a boiler is fitted with two or more safety valves on one connection, this connection to the boiler shall have a cross-sectional area not less than the combined areas of inlet connections of all safety valves with which it connects."

The inlet area for each valve = $\pi D^2/4 = \pi(1.5)^2/4$ = 1.77 in^2 (11.4 cm^2). For two valves, the total inlet area = 2(1.77) = 3.54 in^2 (22.8 cm^2). The required minimum diameter of the boiler connection is $d = 2(A/\pi)^{0.5}$, where A = inlet area. Or, $d = 2(3.54/\pi)^{0.5}$ = 2.12 in (5.4 cm). Select a 2½ × 1½ × 1½ in (6.4 × 3.8 × 3.8 cm). Y for the two steam-drum valves and a 2½-in (6.4-cm) steam-drum outlet connection.

7. *Compute the safety-valve closing pressure*

The *Code* requires safety valves "to close after blowing down not more than 4 percent of the set pressure." For the steam-drum valves the closing pressure will be 900 − (900)(0.04) = 865 lb/in^2 (abs) (5964.2 kPa). The superheater safety valve will close at 850 − (850)(0.04) = 816 lb/in^2 (abs) (5626.3 kPa).

8. *Sketch the discharge elbow and drip pan*

Figure 48 shows a typical discharge elbow and drip-pan connection. Fit all boiler safety valves with escape pipes to carry the steam out of the building and away from personnel. Extend the escape pipe to at least 6 ft (1.8 m) above the roof of the building. Use an escape pipe having a diameter equal to the valve outlet size. When the escape pipe is more than 12 ft (3.7 m) long, some authorities recommend increasing the escape-pipe diameter by ½ in (1.3 cm) for each additional 12-ft (3.7-m) length. Excessive escape-pipe length without an increase in diameter can cause a backpressure on the safety valve because of flow friction. The safety valve may then chatter excessively.

Support the escape pipe independently of the safety valve. Fit a drain to the valve body and drip pan as shown in Fig. 48. This prevents freezing of the condensate and also eliminates the possibility of condensate in the escape pipe raising the valve opening pressure. When a muffler is fitted to the escape pipe, the inlet diameter of the muffler should be the same as, or larger than, the escape-pipe diameter. The outlet area should be greater than the inlet area of the muffler.

Related Calculations: Compute the safety-valve size for fire-tube boilers in the same way as described above, except that the *Code* gives a tabulation of the required area for safety-valve boiler connections based on boiler operating pressure and heating surface. Thus, with an operating

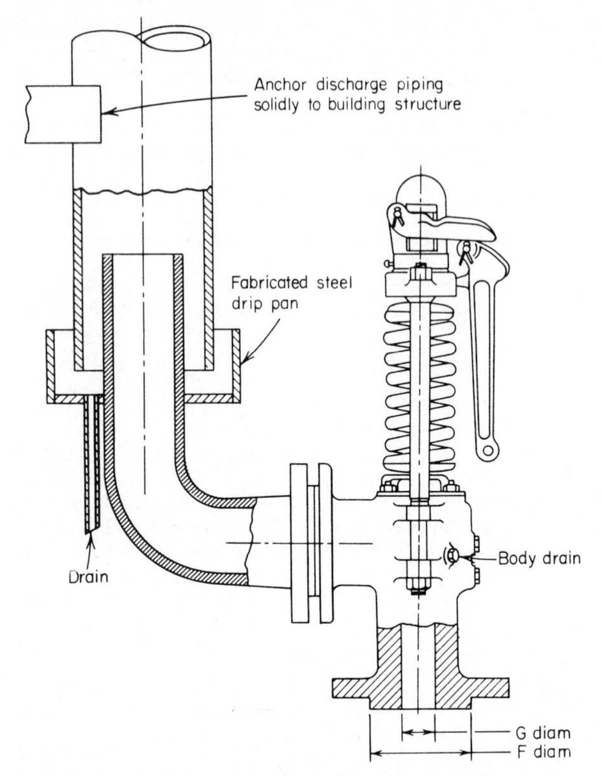

FIG. 48 Typical boiler safety-valve discharge elbow and drip-pan connection. *(Industrial Valve and Instrument Division of Dresser Industries Inc.)*

pressure of 200 lb/in^2 (gage) (1379.0 kPa) and 1800 ft^2 (167.2 m^2) of heating surface, the *Code* table shows that the safety-valve connection should have an area of at least 9.148 in^2 (59.0 cm^2). A 3½-in (8.9-cm) connection would provide this area; or two smaller connections could be used provided that the sum of their areas exceeded 9.148 in^2 (59.0 cm^2).

Note: Be sure to select safety valves approved for use under the *Code* or local law governing boilers in the area in which the boiler will be used. Choice of an unapproved valve can lead to its rejection by the bureau or other agency controlling boiler installation and operation.

STEAM-QUALITY DETERMINATION WITH A THROTTLING CALORIMETER

Steam leaves an industrial boiler at 120 lb/in^2 (abs) (827.4 kPa) and 341.25°F (171.8°C). A portion of the steam is passed through a throttling calorimeter and is exhausted to the atmosphere when the barometric pressure is 14.7 lb/in^2 (abs) (101.4 kPa). How much moisture does the steam leaving the boiler contain if the temperature of the steam at the calorimeter is 240°F (115.6°C)?

Calculation Procedure:

1. *Plot the throttling process on the Mollier diagram*

Begin with the endpoint, 14.7 lb/in^2 (abs) (101.4 kPa) and 240°F (115.6°C). Plot this point on the Mollier diagram as point A, Fig. 49. Note that this point is in the superheat region of the Mollier diagram, because steam at 14.7 lb/in^2 (abs) (101.4 kPa) has a temperature of 212°F (100.0°C), whereas the steam in this calorimeter has a temperature of 240°F (115.6°C). The enthalpy of the calorimeter steam is, from the Mollier diagram, 1164 Btu/lb (2707.5 kJ/kg).

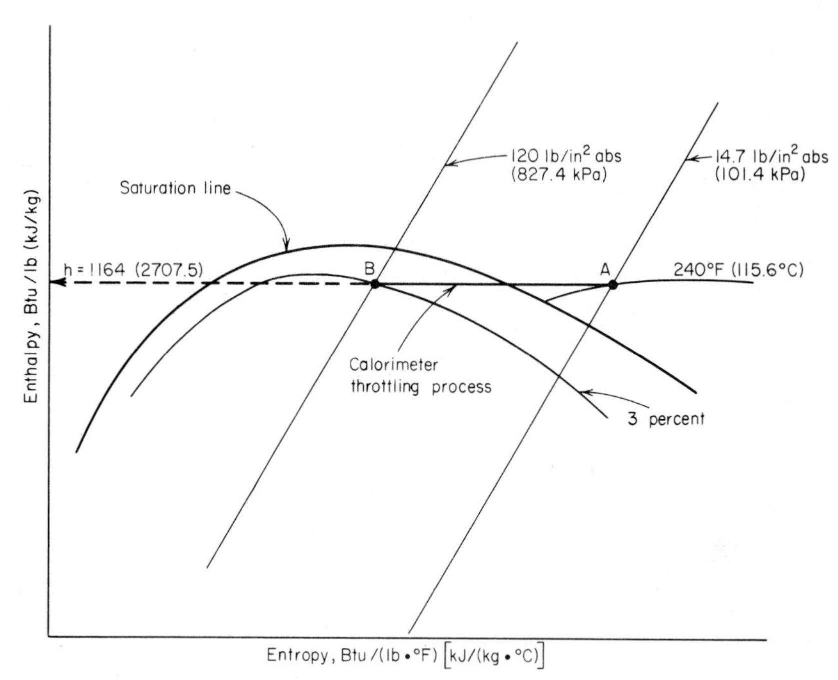

FIG. 49 Mollier-diagram plot of a throttling-calorimeter process.

2. *Trace the throttling process on the Mollier diagram*

In a throttling process, the steam expands at constant enthalpy. Draw a straight, horizontal line from point *A* to the left on the Mollier diagram until the 120-lb/in² (abs) (827.4-kPa) pressure curve is intersected, point *B*, Fig. 49. Read the moisture content of the steam as 3 percent where the 1164-Btu/lb (2707.5-kJ/kg) horizontal trace *AB*, the 120-lb/in² (abs) (827.4-kPa) pressure line, and the 3 percent moisture line intersect.

Related Calculations: A throttling calorimeter *must* produce superheated steam at the existing atmospheric pressure if the moisture content of the supply steam is to be found. Where the throttling calorimeter cannot produce superheated steam at atmospheric pressure, connect the calorimeter outlet to an area at a pressure less than atmospheric. Expand the steam from the source, and read the temperature at the calorimeter. If the steam temperature is greater than that corresponding to the absolute pressure of the vacuum area—for example, a temperature greater than 133.76°F (56.5°C) in an area of 5 inHg (16.9 kPa) absolute pressure—follow the same procedure as given above. Point *A* would then be in the below-atmospheric area of the Mollier diagram. Trace to the left to the origin pressure, and read the moisture content as before.

STEAM PRESSURE DROP IN A BOILER SUPERHEATER

What is the pressure loss in a boiler superheater handling w_s = 200,000 lb/h (25.2 kg/s) of saturated steam at 500 lb/in² (abs) (3447.5 kPa) if the desired outlet temperature is 750°F (398.9°C)? The steam free-flow area through the superheater tubes A_s ft² is 0.500, friction factor f is 0.025, tube ID is 2.125 in (5.4 cm), developed length l of a tube in one circuit is 150 in (381.0 cm), and the tube bend factor B_f is 12.0.

Calculation Procedure:

1. *Determine the initial conditions of the steam*

To compute the pressure loss in a superheater, the initial specific volume of the steam v_g and the mass-flow ratio w_s/A_s must be known. From the steam table, v_g = 0.9278 ft³/lb (0.058 m³/kg) at 500 lb/in² (abs) (3447.5 kPa) saturated. The mass-flow ratio w_s/A_s = 200,000/0.500 = 400,000.

2. *Compute the superheater entrance and exit pressure loss*

Entrance and exit pressure loss p_E lb/in² = $v_f/8(0.00001w_s/A_s)$ = 0.9278/8[(0.00001) × (400,000)]² = 1.856 lb/in² (12.8 kPa).

3. *Compute the pressure loss in the straight tubes*

Straight-tube pressure loss p_s lb/in² = $v_f lf/\text{ID}(0.00001w_s/A_s)^2$ = 0.9278(150)(0.025)/2.125[(0.00001)(400,000)]² = 26.2 lb/in² (abs) (180.6 kPa).

4. *Compute the pressure loss in the superheater bends*

Bend pressure loss p_b = $0.0833B_f(0.00001w_s/A_s)^2$ = 0.0833(12.0)[(0.00001)(400,000)]² = 16.0 lb/in² (110.3 kPa).

5. *Compute the total pressure loss*

The total pressure loss in any superheater is the sum of the entrance, straight-tube, bend, and exit-pressure losses. These losses were computed in steps 2, 3, and 4 above. Therefore, total pressure loss p_t = 1.856 + 26.2 + 16.0 = 44.056 lb/in² (303.8 kPa).

Note: Data for superheater pressure-loss calculations are best obtained from the boiler manufacturer. Several manufacturers have useful publications discussing superheater pressure losses. These are listed in the references at the beginning of this section.

SELECTION OF A STEAM BOILER FOR A GIVEN LOAD

Choose a steam boiler, or boilers, to deliver up to 250,000 lb/h (31.5 kg/s) of superheated steam at 800 lb/in² (abs) (5516 kPa) and 900°F (482.2°C). Determine the type or types of boilers to use, the capacity, type of firing, feedwater-quality requirements, and best fuel if coal, oil, and gas are all available. The normal continuous steam requirement is 200,000 lb/h (25.2 kg/s).

Calculation Procedure:

1. *Select type of steam generator*

Use Fig. 50 as a guide to the usual types of steam generators chosen for various capacities and different pressure and temperature conditions. Enter Fig. 50 at the left at 800 lb/in² (abs) (5516 kPa), and project horizontally to the right, along *AB*, until the 250,000-lb/h (31.5-kg/s) capacity ordinate *BC* is intersected. At *B*, the operating point of this boiler, Fig. 50 shows that a watertube boiler should be used.

Boiler units presently available can deliver steam at the desired temperature of 900°F (482.2°C). The required capacity of 250,000 lb/h (31.5 kg/s) is beyond the range of *packaged watertube boilers*—defined by the American Boiler Manufacturer Association as "a boiler equipped and shipped complete with fuel-burning equipment, mechanical-draft equipment, automatic controls, and accessories."

Shop-assembled boilers are larger units, where all assembly is handled in the builder's plant but with some leeway in the selection of controls and auxiliaries. The current maximum capacity of shop-assembled boilers is about 100,000 lb/h (12.6 kg/s). Thus, a standard-design, larger-capacity boiler is required.

Study manufacturers' engineering data to determine which types of watertube boilers are available for the required capacity, pressure, and temperature. This study reveals that, for this installation, a standard, field-assembled, welded-steel-cased, bent-tube, single-steam-drum boiler with a completely water-cooled furnace would be suitable. This type of boiler is usually fitted with an air heater, and an economizer might also be used. The induced- and forced-draft fans are not integral with the boiler. Capacities of this type of boiler usually available range from 50,000 to 350,000 lb/h (6.3 to 44.1 kg/s); pressure from 160 to 1050 lb/in² (1103.2 to 7239.8 kPa); steam temperature from saturation to 950°F (510.0°C); fuels—pulverized coal, oil, gas, or a combination; controls—manual to completely automatic; efficiency—to 90 percent.

2. *Determine the number of boilers required*

The normal continuous steam requirement is 200,000 lb/h (25.2 kg/s). If a 250,000-lb/h (31.5-kg/s) boiler were chosen to meet the maximum required output, the boiler would normally oper-

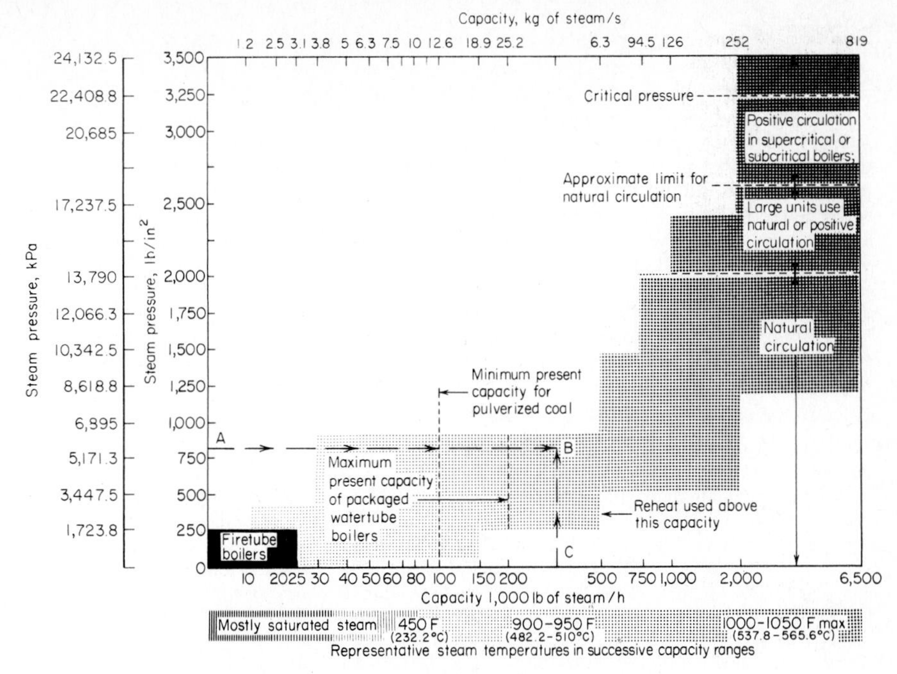

FIG. 50 Typical pressure and capacity relationships for steam generators. *(Power.)*

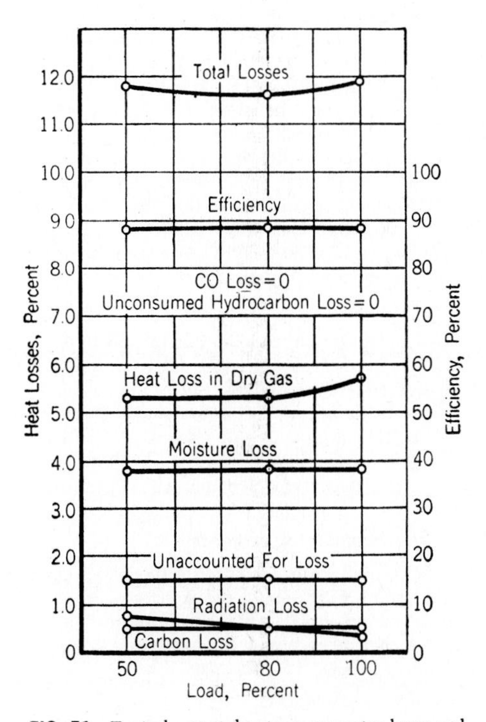

FIG. 51 Typical watertube steam-generator losses and efficiency.

ate at 200,000/250,000, or 80 percent capacity. Obtain the performance chart, Fig. 51, from the manufacturer and study it. This chart shows that at 80 percent load, the boiler efficiency is about equal to that at 100 percent load. Thus, there will not be any significant efficiency loss when the unit is operated at its normal continuous output. The total losses in the boiler are lower at 80 percent load than at full (100 percent) load.

Since there is not a large efficiency decrease at the normal continuous load, and since there are not other factors that require or make more than one boiler desirable, a single boiler unit would be most suitable for this installation. One boiler is more desirable than two or more because installation of a single unit is simpler and maintenance costs are lower. However, where the load fluctuates widely and two or more boilers could best serve the steam demand, the savings in installation and maintenance costs would be insignificant compared with the extra cost of operating a relatively large boiler installed in place of two or more smaller boilers. Therefore, each installation must be carefully analyzed and a decision made on the basis of the existing conditions.

3. *Determine the required boiler capacity*

The stated steam load is 250,000 lb/h (31.5 kg/s) at maximum demand. Study the installation to determine whether the steam demand will increase in the future. Try to determine the rate of increase in the steam demand; for example, installation of several steam-using process units each year during the next few years will increase the steam demand by a predictable amount every year. By using these data, the rate of growth and total steam demand can be estimated for each year. Where the growth will exceed the allowable overload capacity of the boiler—which can vary from 0 to 50 percent of the full-load rating, depending on the type of unit chosen—consider installing a larger-capacity boiler now to meet future load growth. Where the future load is unpredictable, or where no load growth is anticipated, a unit sized to meet today's load would be satisfactory. If this situation existed in this plant, a 250,000-lb/h unit (31.5-kg/s) would be chosen for the load. Any small temporary overloads could be handled by operating the boiler at a higher output for short periods.

Alternatively, assume that a load of 25,000 lb/h (3.1 kg/s) will be added to the maximum demand on this boiler each year for the next 5 years. This means that in 5 years the maximum demand will be $250{,}000 + 25{,}000(5) = 375{,}000$ lb/h (47.2 kg/s). This is an overload of $(375{,}000 - 250{,}000)/250{,}000 = 0.50$, or 50 percent. It is unlikely that the boiler could carry a continuous overload of 50 percent. Therefore it might be wise to install a 375,000-lb/h (47.2-kg/s) boiler to meet present and future demands. Base this decision on the accuracy of the future-demand prediction and the economic advantages or disadvantages of investing more money now for a demand that will not occur until some future date. Refer to the section on engineering economics for procedures to follow in economics calculations of this type.

Thus, with no increase in the future load, a 250,000-lb/h (31.5-kg/s) unit would be chosen. With the load increase specified, a 375,000-lb/h (47.2-kg/s) unit would be the choice, if there were no major economic disadvantages.

4. *Choose the type of fuel to use*

Watertube boilers of the type being considered will economically burn the three fuels available—coal, oil, or gas—either singly or in combination. In the design considered here, the furnace water-cooled surfaces and boiler surfaces are integral parts of each other. For this reason the boiler is well suited for pulverized-coal firing in the 50,000- to 300,000-lb/h (6.3- to 37.8-kg/s) capacity range.Thus, if a 250,000-lb/h (31.5-kg/s) unit were chosen, it could be fired by pulverized coal. With a larger unit of 375,000 lb/h (47.2 kg/s), pulverized-coal, oil, or gas firing might be used. Use an economic comparison to determine which fuel would give the lowest overall operating cost for the life of the boiler.

5. *Determine the feedwater-quality requirements*

Watertube boilers of all types require careful control of feedwater quality to prevent scale and sludge deposits in tubes and drums. Corrosion of the interior boiler surfaces must be controlled. Where all condensate is returned to the boiler, the makeup water must be treated to prevent the conditions just cited. Therefore, a comprehensive water-treating system must be planned for, particularly if the raw-water supply is poor.

6. *Estimate the boiler space requirements*

The space occupied by steam-generating units is an important consideration in plants in municipal areas and where power-plant buildings are presently crowded by existing equipment. The manufacturer's engineering data for this boiler show that for pulverized-coal firing, the hopper-type furnace bottom is best. The data also show that the smallest boiler with a hopper bottom occupies a space 21 ft (6.4 m) wide, 31 ft (9.4 m) high, and 14 ft (4.3 m) front to rear. The largest boiler occupies a space 21 ft (6.4 m) wide, 55 ft (16.8 m) high, and 36 ft (11.0 m) front to rear. Check these dimensions against the available space to determine whether the chosen boiler can be installed without major structural changes. The steel walls permit outdoor or indoor installation with top or bottom support of the boiler optional in either method of installation.

Related Calculations: Use this general procedure to select boilers for industrial, central-station, process, and marine applications.

SELECTING BOILER FORCED- AND INDUCED-DRAFT FANS

Combustion calculations show that an oil-fired watertube boiler requires 200,000 lb/h (25.2 kg/s) of air for combustion at maximum load. Select forced- and induced-draft fans for this boiler if the average temperature of the inlet air is 75°F (23.9°C) and the average temperature of the combustion gas leaving the air heater is 350°F (176.7°C) with an ambient barometric pressure of 29.9 inHg (101.0 kPa). Pressure losses on the air-inlet side are as follows, in inH_2O: air heater, 1.5 (0.37 kPa); air-supply ducts, 0.75 (0.19 kPa); boiler windbox, 1.75 (0.44 kPa); burners, 1.25 (0.31 kPa). Draft losses in the boiler and related equipment are as follows, in inH_2O: furnace pressure, 0.20 (0.05 kPa); boiler, 3.0 (0.75 kPa); superheater, 1.0 (0.25 kPa); economizer, 1.50 (0.37 kPa); air heater, 2.00 (0.50 kPa); uptake ducts and dampers, 1.25 (0.31 kPa). Determine the fan discharge pressure and horsepower input. The boiler burns 18,000 lb/h (2.3 kg/s) of oil at full load.

Calculation Procedure:

1. *Compute the quantity of air required for combustion*

The combustion calculations show that 200,000 lb/h (25.2 kg/s) of air is theoretically required for combustion in this boiler. To this theoretical requirement must be added allowances for excess air at the burner and leakage out of the air heater and furnace. Allow 25 percent excess air for this boiler. The exact allowance for a given installation depends on the type of fuel burned. However, a 25 percent excess-air allowance is an average used by power-plant designers for coal, oil, and gas firing. With this allowance, the required excess air = 200,000(0.25) = 50,000 lb/h (6.3 kg/s).

Air-heater air leakage varies from about 1 to 2 percent of the theoretically required airflow. Using 2 percent, we see the air-heater leakage allowance = 200,000(0.02) = 4000 lb/h (0.5 kg/s).

Furnace air leakage ranges from 5 to 10 percent of the theoretically required airflow. With 7.5 percent, the furnace leakage allowance = 200,000(0.075) = 15,000 lb/h (1.9 kg/s).

The total airflow required is the sum of the theoretical requirement, excess air, and leakage. Or, 200,000 + 50,000 + 4000 + 15,000 = 269,000 lb/h (33.9 kg/s). The forced-draft fan must supply at least this quantity of air to the boiler. Usual practice is to allow a 10 to 20 percent safety factor for fan capacity to ensure an adequate air supply at all operating conditions. This factor of safety is applied to the total airflow required. Using a 10 percent factor of safety, we see that fan capacity = 269,000 + 269,000(0.1) = 295,900 lb/h (37.3 kg/s). Round this to 296,000-lb/h (37.3-kg/s) fan capacity.

2. *Express the required airflow in cubic feet per minute*

Convert the required flow in pounds per hour to cubic feet per minute. To do this, apply a factor of safety to the ambient air temperature to ensure an adequate air supply during times of high ambient temperature. At such times, the density of the air is lower, and the fan discharges less air to the boiler. The usual practice is to apply a factor of safety of 20 to 25 percent to the known ambient air temperature. Using 20 percent, we see the ambient temperature for fan selection = 75 + 75(0.20) = 90°F (32.2°C). The density of air at 90°F (32.2°C) is 0.0717 lb/ft^3 (1.15 kg/m^3), found in Baumeister and Marks—*Standard Handbook for Mechanical Engineers*. Converting gives ft^3/min = (lb/h)/(60 lb/ft^3) = 296,000/60(0.0717) = 69,400 ft^3/min (32.8 m^3/s). This is the minimum capacity the forced-draft fan may have.

3. *Determine the forced-draft discharge pressure*

The total resistance between the forced-draft fan outlet and furnace is the sum of the losses in the air heater, air-supply ducts, boiler windbox, and burners. For this boiler, the total resistance,

inH_2O = 1.5 + 0.75 + 1.75 + 1.25 = 5.25 inH_2O (1.3 kPa). Apply a 15 to 30 percent factor of safety to the required discharge pressure to ensure adequate airflow at all times. Or, fan discharge pressure, with a 20 percent factor of safety = 5.25 + 5.25(0.20) = 6.30 inH_2O (1.6 kPa). The fan must therefore deliver at least 69,400 ft^3/min (32.8 m^3/s) at 6.30 inH_2O (1.6 kPa).

4. *Compute the power required to drive the forced-draft fan*

The air horsepower for any fan = $0.0001753H_fC$, where H_f = total head developed by fan, inH_2O; C = airflow, ft^3/min. For this fan, air hp = 0.0001753(6.3)(69,400) = 76.5 hp (57.0 kW). Assume or obtain the fan and fan-driver efficiencies at the rated capacity (69,400 ft^3/min, or 32.8 m^3/s) and pressure (6.30 inH_2O, or 1.6 kPa). With a fan efficiency of 75 percent and assuming the fan is driven by an electric motor having an efficiency of 90 percent, we find the overall efficiency of the fan-motor combination is (0.75)(0.90) = 0.675, or 67.5 percent. Then the motor horsepower required = air horsepower/overall efficiency = 76.5/0.675 = 113.2 hp (84.4 kW). A 125-hp (93.2-kW) motor would be chosen because it is the nearest, next larger unit readily available. Usual practice is to choose a *larger* driver capacity when the computed capacity is lower than a standard capacity. The next larger standard capacity is generally chosen, except for extremely large fans where a special motor may be ordered.

5. *Compute the quantity of flue gas handled*

The quantity of gas reaching the induced-draft fan is the sum of the actual air required for combustion from step 1, air leakage in the boiler and furnace, and the weight of fuel burned. With an air leakage of 10 percent in the boiler and furnace (this is a typical leakage factor applied in practice), the gas flow is as follows:

	lb/h	kg/s
Actual airflow required	296,000	37.3
Air leakage in boiler and furnace	29,600	3.7
Weight of oil burned	18,000	2.3
Total	343,600	43.3

Determine from combustion calculations for the boiler the density of the flue gas. Assume that the combustion calculations for this boiler show that the flue-gas density is 0.045 lb/ft^3 (0.72 kg/m^3) at the exit-gas temperature. To determine the exit-gas temperature, apply a 10 percent factor of safety to the given exit temperature, 350°F (176.6°C). Hence, exit-gas temperature = 350 + 350(0.10) = 385°F (196.1°C). Then flue-gas flow, ft^3/min = (flue-gas flow, lb/h)/(60)(flue-gas density, lb/ft^3) = 343,600/[(60)(0.045)] = 127,000 ft^3/min (59.9 m^3/s). Apply a 10 to 25 percent factor of safety to the flue-gas quantity to allow for increased gas flow. With a 20 percent factor of safety, the actual flue-gas flow the fan must handle = 127,000 + 127,000(0.20) = 152,400 ft^3/min (71.8 m^3/s), say 152,500 ft^3/min (71.9 m^3/s) for fan-selection purposes.

6. *Compute the induced-draft fan discharge pressure*

Find the sum of the draft losses from the burner outlet to the induced-draft fan inlet. These losses are as follows for this boiler:

	inH_2O	kPa
Furnace draft loss	0.20	0.05
Boiler draft loss	3.00	0.75
Superheater draft loss	1.00	0.25
Economizer draft loss	1.50	0.37
Air heater draft loss	2.00	0.50
Uptake ducts and damper draft loss	1.25	0.31
Total draft loss	8.95	2.23

Allow a 10 to 25 percent factor of safety to ensure adequate pressure during all boiler loads and furnace conditions. With a 20 percent factor of safety for this fan, the total actual pressure loss = 8.95 + 8.95(0.20) = 10.74 inH_2O (2.7 kPa). Round this to 11.0 inH_2O (2.7 kPa) for fan-selection purposes.

7. *Compute the power required to drive the induced-draft fan*

As with the forced-draft fan, air horsepower = $0.0001753H_fC$ = 0.0001753(11.0) × (127,000) = 245 hp (182.7 kW). If the combined efficiency of the fan and its driver, assumed to be an electric motor, is 68 percent, the motor horsepower required = 245/0.68 = 360.5 hp (268.8 kW). A 375-hp (279.6-kW) motor would be chosen for the fan driver.

8. *Choose the fans from a manufacturer's engineering data*

Use the next calculation procedure to select the fans from the engineering data of an acceptable manufacturer. For larger boiler units, the forced-draft fan is usually a backward-curved blade centrifugal-type unit. Where two fans are chosen to operate in parallel, the pressure curve of each fan should decrease at the same rate near shutoff so that the fans divide the load equally. Be certain that forced-draft fans are heavy-duty units designed for continuous operation with well-balanced rotors. Choose high-efficiency units with self-limiting power characteristics to prevent overloading the driving motor. Airflow is usually controlled by dampers on the fan discharge.

Induced-draft fans handle hot, dusty combustion products. For this reason, extreme care must be taken to choose units specifically designed for induced-draft service. The usual choice for large boilers is a centrifugal-type unit with forward- or backward-curved, or flat blades, depending on the type of gas handled. Flat blades are popular when the flue gas contains large quantities of dust. Fan bearings are generally water-cooled.

Related Calculations: Use the procedure given above for the selection of draft fans for all types of boilers—fire-tube, packaged, portable, marine, and stationary. Obtain draft losses from the boiler manufacturer. Compute duct pressure losses by using the methods given in later procedures in this handbook.

POWER-PLANT FAN SELECTION FROM CAPACITY TABLES

Choose a forced-draft fan to handle 69,400 ft^3/min (32.8 m^3/s) of 90°F (32.2°C) air at 6.30-inH_2O (1.6-kPa) static pressure and an induced-draft fan to handle 152,500 ft^3/min (72.0 m^3/s) of 385°F (196.1°C) gas at 11.0-inH_2O (2.7-kPa) static pressure. The boiler that these fans serve is installed at an elevation of 5000 ft (1524 m) above sea level. Use commercially available capacity tables for making the fan choice. The flue-gas density is 0.045 lb/ft^3 (0.72 kg/m^3) at 385°F (196.1°C).

Calculation Procedure:

1. *Compute the correction factors for the forced-draft fan*

Commercial fan-capacity tables are based on fans handling standard air at 70°F (21.1°C) at a barometric pressure of 29.92 inHg (101.0 kPa) and having a density of 0.075 lb/ft^3 (1.2 kg/m^3). Where different conditions exist, the fan flow rate must be corrected for temperature and altitude.

Obtain the engineering data for commercially available forced-draft fans, and turn to the temperature and altitude correction-factor tables. Pick the appropriate correction factors from these tables for the prevailing temperature and altitude of the installation. Thus, in Table 6, select the correction factors for 90°F (32.2°C) air and 5000-ft (1524.0-m) altitude. These correction factors are C_T = 1.018 for 90°F (32.2°C) air and C_A = 1.095 for 5000-ft (1524.0-m) altitude.

Find the composite correction factor (CCF) by taking the product of the temperature and altitude correction factors. Or, CCF = (1.018)(1.095) = 1.1147. Now divide the given cubic feet per minute (cfm) by the correction factor to find the capacity-table cfm. Or, capacity-table cfm = 69,400/1.147 = 62,250 ft^3/min (29.4 m^3/s).

2. *Choose the fan size from the capacity table*

Turn to the fan-capacity table in the engineering data, and look for a fan delivering 62,250 ft^3/min (29.4 m^3/s) at 6.3-inH_2O (1.6-kPa) static pressure. Inspection of the table shows that the capacities are tabulated for 6.0- and 6.5-inH_2O (1.5- and 1.6-kPa) static pressure. There is no tabulation for 6.3-inH_2O (1.57-kPa) static pressure.

TABLE 6 Fan Correction Factors

Temperature		Correction factor	Altitude		Correction factor
°F	°C		ft	m	
80	26.7	1.009	4500	1371.6	1.086
90	32.2	1.018	5000	1524.0	1.095
100	37.8	1.028	5500	1676.4	1.106
375	190.6	1.255			
400	204.4	1.273			
450	232.2	1.310			

Enter the table at the nearest capacity to that required, 62,250 ft^3/min (29.4 m^3/s), as shown in Table 7. This table, excerpted with permission from the American Standard Inc. engineering data, shows that the nearest capacity of this particular type of fan is 62,595 ft^3/min (29.5 m^3/s). The difference, or 62,595 − 62,250 = 345 ft^3/min (0.16 m^3/s), is only 345/62,250 = 0.0055, or 0.55 percent. This is a negligible difference, and the 62,595-ft^3/min (29.5-m^3/s) fan is well suited for its intended use. The extra static pressure of 6.5 − 6.3 = 0.2 inH_2O (0.05 kPa) is desirable in a forced-draft fan because furnace or duct resistance may increase during the life of the boiler. Also, the extra static pressure is so small that it will not markedly increase the fan power consumption.

3. *Compute the fan speed and power input*

Multiply the capacity-table rpm and brake horsepower (bhp) by the composite factor to determine the actual rpm and bhp. Thus, with data from Table 7, the actual rpm = (1096)(1.1147) = 1221.7 r/min. Actual bhp = (99.08)(1.1147) = 110.5 bhp (82.4 kW). This is the horsepower input required to drive the fan and is close to the 113.2 hp (84.4 kW) computed in the previous calculation procedure. The actual motor horsepower would be the same in each case because a standard-size motor would be chosen. The difference of 113.2 − 110.5 = 2.7 hp (2.0 kW) results from the assumed efficiencies that depart from the actual values. Also, a sea-level altitude was assumed in the previous calculation procedure. However, the two methods used show how accurately fan capacity and horsepower input can be estimated by judicious evaluation of variables.

4. *Compute the correction factors for the induced-draft fan*

The flue-gas density is 0.045 lb/ft^3 (0.72 kg/m^3) at 385°F (196.1°C). Interpolate in the temperature correction-factor table because a value of 385°F (196.1°C) is not tabulated. Find the correction factor for 385°F (196.1°C) thus: [(Actual temperature − lower temperature)/(higher temperature − lower temperature)] × (higher temperature correction factor − lower temperature correction factor) + lower temperature correction factor. Or, [(385 − 375)/(400 − 375)](1.273 − 1.255) + 1.255 = 1.262.

The altitude correction factor is 1.095 for an elevation of 5000 ft (1524.0 m), as shown in Table 6.

As for the forced-draft fan, CCF = $C_T C_A$ = (1.262)(1.095) = 1.3819. Use the CCF to find the capacity-table cfm in the same manner as for the forced-draft fan. Or, capacity-table cfm = (given cfm)/CCF = 152,500/1.3819 = 110,355 ft^3/min (52.1 m^3/s).

TABLE 7 Typical Fan Capacities

Capacity		Outlet velocity		Outlet velocity pressure		Ratings at 6.5-inH_2O (1.6-kPa) static pressure		
ft^3/min	m^3/s	ft/min	m/s	inH_2O	kPa	r/min	bhp	kW
61,204	28.9	4400	22.4	1.210	0.3011	1083	95.45	71.2
62,595	29.5	4500	22.9	1.266	0.3150	1096	99.08	73.9
63,975	30.2	4600	23.4	1.323	0.3212	1109	103.0	76.8

5. *Choose the fan size from the capacity table*

Check the capacity table to be sure that it lists fans suitable for induced-draft (elevated-temperature) service. Turn to the 11-inH_2O (2.7-kPa) static-pressure capacity table, and find a capacity equal to 110,355 ft^3/min (52.1 m^3/s). In the engineering data used for this fan, the nearest capacity at 11-inH_2O (2.7-kPa) static pressure is 110,467 ft^3/min (52.1 m^3/s), with an outlet velocity of 4400 ft/min (22.4 m/s), an outlet velocity pressure of 1.210 inH_2O (0.30 kPa), a speed of 1222 r/min, and an input horsepower of 255.5 bhp (190.5 kW). The tabulation of these quantities is of the same form as that given for the forced-draft fan, step 2. The selected capacity of 110,467 ft^3/min (52.1 m^3/s) is entirely satisfactory because it is only 110,467 − 110,355/110,355 = 0.00101, or 0.1 percent, higher than the desired capacity.

6. *Compute the fan speed and power input*

Multiply the capacity-table rpm and brake horsepower by the CCF to determine the actual rpm and brake horsepower. Thus, the actual rpm = (1222)(1.3819) = 1690 r/min. Actual brake horsepower = (255.5)(1.3819) = 353.5 bhp (263.6 kW). This is the horsepower input required to drive the fan and is close to the 360.5 hp (268.8 kW) computed in the previous calculation procedure. The actual motor horsepower would be the same in each case because a standard-size motor would be chosen. The difference in horsepower of 360.5 − 353.5 = 7.0 hp (5.2 kW) results from the same factors discussed in step 3.

Note: The static pressure is normally used in most fan-selection procedures because this pressure value is used in computing pressure and draft losses in boilers, economizers, air heaters, and ducts. In any fan system, the total air pressure = static pressure + velocity pressure. However, the velocity pressure at the fan discharge is not considered in draft calculations unless there are factors requiring its evaluation. These requirements are generally related to pressure losses in the fan-control devices.

Related Calculations: Use the fan-capacity table to obtain these additional details of the fan: outlet inside dimensions (length and width), fan-wheel diameter and circumference, fan maximum bhp, inlet area, fan-wheel peripheral velocity, NAFM fan class, and fan arrangement. Use the engineering data containing the fan-capacity table to find the fan dimensions, rotation and discharge designations, shipping weight, and, for some manufacturers, prices.

ANALYSIS OF BOILER AIR DUCTS AND GAS UPTAKES

Three oil-fired boilers are supplied air through the breeching shown in Fig. 52*a*. Each boiler will burn 13,600 lb/h (1.71 kg/s) of fuel oil at full load. The draft loss through each boiler is 8 inH_2O (2.0 kPa). Uptakes from the three boilers are connected as shown in Fig. 52*b*. Determine the draft loss through the entire system if a 50-ft (15.2-m) high metal stack is used and the gas temperature at the stack inlet is 400°F (204.4°C).

Calculation Procedure:

1. *Determine the airflow through the breeching*

Compute the airflow required, cubic feet per pound of oil burned, using the methods given in earlier calculation procedures. For this installation, assume that the combustion calculation shows that 250 ft^3/lb (15.6 m^3/kg) of oil burned is required. Then the total airflow required = (number of boilers)(lb/h oil burned per boiler)(ft^3/lb oil)/(60 min/h) = (3)(13,600)(250)/60 = 170,000 ft^3/min (80.2 m^3/s).

2. *Select the dimensions for each length of breeching duct*

With the airflow rate of 170,000 ft^3/min (80.2 m^3/s) known, the duct area can be determined by assuming an air velocity and computing the duct area A_d ft^2 from A_d = (airflow rate, ft^3/min)/(air velocity, ft/min). Once the area is known, the duct can be sized to give this area. Thus, if 9 ft^2 (0.8 m^2) is the required duct area, a duct 3 × 3 ft (0.9 × 0.9 m) or 2 × 4.5 ft (0.6 × 1.4 m) would provide the required area.

In the usual power plant, the room available for ducts limits the maximum allowable duct size. So the designer must try to fit a duct of the required area into the available space. This is done by changing the duct height and width until a duct of suitable area fitting the available space is

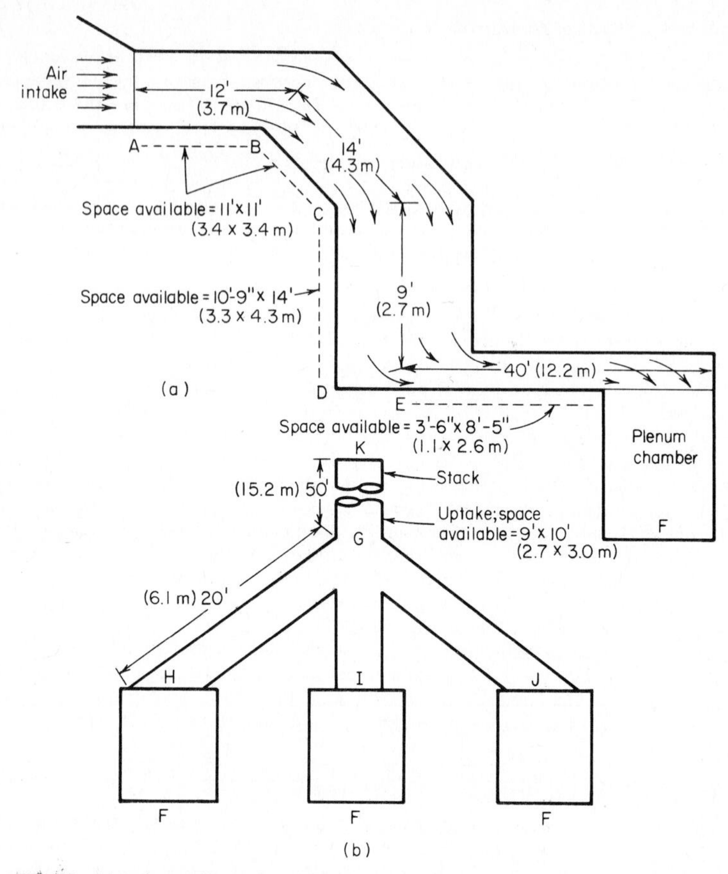

FIG. 52 (*a*) Boiler intake-air duct; (*b*) boiler uptake ducts.

found. If the duct area is reduced below that required, compute the actual air velocity to determine whether it exceeds recommended limits.

In this power plant, the space available in the open area between *A* and *C*, Fig. 52, is a square 11 × 11 ft (3.4 × 3.4 m). By allowing a 3-in (7.6-cm) clearance around the outside of the duct and using a square duct, its dimensions would be 10.5 × 10.5 ft (3.2 × 3.2 m), or a cross-sectional area of (10.5)(10.5) = 110 ft^2 (10.2 m^2), closely. With 170,000 ft^3/min (80.2 m^3/s) flowing through the duct, the air velocity v ft/min = ft^3/min/A_d = 170,000/110 = 1545 ft/min (7.8 m/s). This is a satisfactory air velocity because the usual plant air system velocity is 1200 to 3600 ft/min (6.1 to 18.3 m/s).

Between *C* and *D* the open area in this power plant is 10 ft 9 in (3.3 m) by 14 ft (4.3 m). Using the same 3-in (7.6-cm) clearance all around the duct, we find the dimensions of the vertical duct *CD* are 10.25 × 13 ft (3.1 × 4.0 m), or a cross-sectional area of 10.25 × 13 = 133 ft^2 (12.5 m^2), closely. The air velocity in this section of the duct is v = 170,000/133 = 1275 ft/min (6.5 m/s). Since it is desirable to maintain, if possible, a constant velocity in all sections of the duct where space permits, the size of this duct might be changed so it equals that of *AB*, 10.5 × 10.5 ft (3.2 × 3.2 m). However, the installation costs would probably be high because the limited space available would require alteration of the power-plant structure. Also, the velocity is section *CD* is above the usual minimum value of 1200 ft/min (6.1 m/s). For these reasons, the duct will be installed in the 10.25 × 13 ft (3.1 × 4.0 m) size.

Between E and F the vertical distance available for installation of the duct is 3.5 ft (1.1 m), and the horizontal distance is 8.5 ft (2.6 m). Using the same 3-in (7.6-cm) clearance as before gives a 3×8 ft $(0.9 \times 2.4$ m) duct size, or a cross-sectional area of $(3)(8) = 24$ ft^2 (2.2 m^2). At E the duct divides into three equal-size branches, one for each boiler, and the same area, 24 ft^2 (2.2 m^2), is available for each branch duct. The flow in any branch duct is then $170{,}000/3 = 56{,}700$ ft^3/min (26.8 m^3/s). The velocity in any of the three equal branches is $v = 56{,}700/24 = 2360$ ft/min (12.0 m/s). When a duct system has two or more equal-size branches, compute the pressure loss in one branch only because the losses in the other branches will be the same. The velocity in branch EF is acceptable because it is within the limits normally used in power-plant practice. At F the air enters a large plenum chamber, and its velocity becomes negligible because of the large flow area. The boiler forced-draft fan intakes are connected to the plenum chamber. Each of the three ducts feeds into the plenum chamber.

3. *Compute the pressure loss in each duct section*

Begin the pressure-loss calculations at the system inlet, point A, and work through each section to the stack outlet. This procedure reduces the possibility of error and permits easy review of the calculations for detection of errors. Assign letters to each point of the duct where a change in section dimensions or directions, or both, occurs. Use these letters:

Point A: Assume that 70°F (21.1°C) air having a density of 0.075 lb/ft^3 (1.2 kg/m^3) enters the system when the ambient barometric pressure is 29.92 inHg (101.3 kPa). Compute the velocity pressure at point A, in inH_2O, from $p_v = v^2/[3.06(10^4)(460 + t)]$, where t = air temperature, °F. Since the velocity of the air at A is 1545 ft/min (11.7 m/s), $p_v = (1545)^2/[3.06(10^4)(530)] = 0.147$ inH_2O (0.037 kPa) at 70°F (21.1°C).

The entrance loss at A, where there is a sharp-edged duct, is $0.5p_v$, or $0.5(0.147) = 0.0735$ inH_2O (0.018 kPa). With a rounded inlet, the loss in velocity pressure would be negligible.

Section AB: There is a pressure loss due to duct friction between A and B, and B and C. Also, there is a bend loss at points B and C. Compute the duct friction first.

For any circular duct, the static pressure loss due to friction p_s $inH_2O = (0.03L/d^{1.24})(v/1000)^{1.84}$, where L = duct length, ft; d = duct diameter, in. To convert any rectangular or square duct with sides a and b ft high and wide, respectively, to an equivalent round duct of D-ft diameter, use the relation $D = 2ab/(a + b)$. For this duct, $d = 2(10.5)(10.5)/(10.5 + 10.5) = 10.5$ ft (3.2 m) $= 126$ in (320 cm) $= d$. Since this duct is 12 ft (3.7 m) long between A and B, $p_s = [0.03(12)/126^{1.24}](1.545/1000)^{1.84} = 0.002$ inH_2O (0.50 Pa).

Point B: The 45° bend at B has, from Baumeister and Marks—*Standard Handbook for Mechanical Engineers*, a pressure drop of 60 percent of the velocity head in the duct, or $(0.60)(0.147) = 0.088$ inH_2O (20.5 Pa) loss.

Section BC: Duct friction in the 14-ft (4.3-m) long downcomer BC is $p_s = [0.03(14)/126^{1.24}](1545/1000)^{1.84} = 0.0023$ inH_2O (0.56 Pa). Point C: The 45° bend at C has a velocity head loss of 60 percent of the velocity pressure. Determine the velocity pressure in this duct in the same manner as for point A, or $p_v = (1545)^2/[3.06(10^4)(530)] = 0.147$ inH_20 (36.1 Pa), since the velocity at points B and C is the same. Then the velocity head loss $= (0.60)(0.147) = 0.088$ inH_2O (21.9 Pa).

Section CD: The equivalent round-duct diameter is $D = (2)(10.25)(13)/(10.25 + 13) = 11.45$ ft (3.5 m) $= 137.3$ in (348.7 cm). Duct friction is then $p_s = [0.03(9)/137.3^{1.24}](1275/1000)^{1.84} = 0.000934$ inH_2O (0.23 Pa). Velocity pressure in the duct is $p_v = (1275)^2/[3.06(10^4)(530)] = 0.100$ inH_2O (24.9 Pa). Since there is no room for a transition piece—that is, a duct providing a gradual change in flow area between points C and D—the decrease in velocity pressure from 0.147 to 0.100 in (36.6 to 24.9 Pa), or $0.147 - 0.10 = 0.047$ inH_2O (11.7 Pa), is not converted to static pressure and is lost.

Point E: The pressure loss in the right-angle bend at E is, from Baumeister and Marks—*Standard Handbook for Mechanical Engineers,* 1.2 times the velocity head, or $(1.2)(0.1) = 0.12$ inH_2O (29.9 Pa). Also, since this is a sharp-edged elbow, there is an additional loss of 50 percent of the velocity head, or $(0.5)(0.10) = 0.05$ inH_2O (12.4 Pa).

The velocity pressure at point E is $p_v = (2360)^2/[3.06(10^4)(530)] = 0.343$ inH_2O (85.4 Pa).

Section EF: The equivalent round-duct diameter is $D = (2)(3)(8)/(3 + 8) = 4.36$ ft (1.3 m) $= 52.4$ in (133.1 cm). Duct friction $p_s = [0.03(40)/52.4^{1.24}](2360/1000)^{1.84} = 0.0247$ inH_2O (6.2 Pa).

Air entering the large plenum chamber at F loses all its velocity. There is no static-pressure regain; therefore, the velocity-head loss $= 0.348 - 0.0 = 0.348$ inH_2O (86.6 Pa).

4. *Compute the losses in the uptake and stack*

Convert the airflow of 250 ft^3/lb (15.6 m^3/kg) of fuel oil to pounds of air per pound of fuel oil by multiplying by the density, or 250(0.075) = 18.75 lb of air per pound of oil. The flue gas will contain 18.75 lb of air + 1 lb of oil per pound of fuel burned, or (18.75 + 1)/18.75 = 1.052 times as much gas leaves the boiler as air enters; this can be termed the *flue-gas factor*.

Point *G*: The quantity of flue gas entering the stack from each boiler (corrected to a 400°F or 204.4°C outlet temperature) is, in °R, (*cfm* air to furnace)(stack, °R/air, °R)(flue-gas factor). Or stack flue-gas flow = (56,700)[(400 + 460)/(70 + 460)](1.052) = 97,000 ft^3/min (45.8 m^3/s) per boiler.

The total duct area available for the uptake leading to the stack is 9 × 10 ft (2.7 × 3.0 m) = 90 ft^2 (8.4 m^2), based on the clearance above the boilers. The flue-gas velocity for three boilers is v = (3)(97,000)/90 = 3235 ft/min (16.4 m/s). The velocity pressure in the uptake is $p = (3235)^2/[3.06(10^4)(460 + 400)] = 0.397$ inH_2O (98.8 Pa).

Point *H*: The flue-gas flow from all the boilers is divided equally between three ducts, *HG*, *IG*, *JG*, Fig. 52. It is desirable to maintain the same gas velocity in each duct and have this velocity equal to that in the uptake. The same velocity can be obtained in each duct by making each duct one-third the area of the uptake, or 90/3 = 30 ft^2 (2.8 m^2). Then v = 97,000/30 = 3235 ft/min (16.4 m/s) in each duct. Since the velocity in each duct equals the velocity in the uptake, the velocity pressure in each duct equals that in the uptake, or 0.397 inH_2O (98.8 Pa).

Ducts *HG* and *JG* have two 45° bends in them, or the equivalent of one 90° bend. The velocity-pressure loss in a 90° bend is 1.20 times the velocity head in the duct; or, for either *HG* or *JG*, (1.20)(0.397) = 0.476 inH_2O (118.5 Pa).

Section *HG*: The equivalent duct diameter for a 30-ft^2 (2.8-m^2) duct is $D = 2(30/\pi)^{0.5} = 6.19$ ft (1.9 m) = 74.2 in (188.4 cm). The duct friction in *HG*, which equals that in *JG*, is $p_s = [0.03(20)/74.2^{1.24}](530/860)(3235/1000)^{1.84} = 0.01536$ inH_2O (3.8 Pa), if we correct for the flue-gas temperature with the ratio (70 + 460)/(400 + 460) = 530/860.

Section *GK*: The stack joins the uptake at point *G*. Assume that this installation is designed for a stack-gas area of 500 lb of oil per square foot (2441.2 kg/m^2) of stack; or, for three boilers, stack area = (3)(13,600 lb/h oil)/500 = 81.5 ft^2 (7.6 m^2). The stack diameter will then be $D = 2(8.15/\pi)^{0.5} = 10.18$ ft (3.1 m) = 122 in (309.9 cm).

The gas velocity in the stack is v = (3)(97,000)/81.5 = 3570 ft/min (18.1 m/s). The friction in the stack is $p_s = [0.03(50)/122^{1.24}](3570/1000)^{1.84}(503/860) = 0.0194$ inH_2O (4.8 Pa).

5. *Compute the total losses in the system*

Tabulate the individual losses and find the sum as follows:

	inH_2O	kPa
Point *A*; entrance loss	0.0735	0.0183
Section *AB*; duct friction	0.0020	0.0005
Point *B*; bend loss	0.0880	0.0219
Section *BC*; duct friction	0.0023	0.0006
Point *C*; bend loss	0.0880	0.0219
Section *CD*; duct friction	0.0009	0.0002
Section *CD*; velocity-pressure loss	0.0470	0.0117
Point *E*; bend loss	0.1200	0.0299
Point *E*; sharp-edge loss	0.0500	0.0124
Section *EF*; duct friction	0.0247	0.0061
Section *EF*; plenum velocity-head loss	0.3480	0.0866
Boiler friction loss	8.0000	1.9907
Section *HG*; duct friction	0.0154	0.0038
Points *H* and *G* total bend loss	0.4760	0.1184
Section *GK*; stack friction	0.0194	0.0048
Total loss	9.3552	2.3279

The total loss computed here is the minimum static pressure that must be developed by the draft fans or blowers. This total static pressure can be divided between the forced- and induced-draft fans or confined solely to the forced-draft fans in plants not equipped with an induced-draft fan. If only a forced-draft fan is used, its static discharge pressure should be at least 20 percent greater than the losses, or (1.2)(9.3552) = 11.21 inH_2O (2.8 kPa) at a total airflow of 97,000 ft^3/min (45.8 m^3/s). If more than one forced-draft fan were used for each boiler, each fan would have a total static pressure of at least 11.21 inH_2O (2.8 kPa) and a capacity of less than 97,000 ft^3/min (45.8 m^3/s). In making the final selection of the fan, the static pressure would be rounded to 12 inH_2O (3.0 kPa).

Where dampers are used for combustion-air control, include the wide-open resistance of the dampers in computing the total losses in the system at full load on the boilers. Damper resistance values can be obtained from the damper manufacturer. Note that as the damper is closed to reduce the airflow at lower boiler loads, the resistance through the damper is increased. Check the fan head-capacity curve to determine whether the head developed by the fan at lower capacities is sufficient to overcome the greater damper resistance. Since the other losses in the system will decrease with smaller airflow, the fan static pressure is usually adequate.

Note: (1) Follow the notational system used here to avoid errors from plus and minus signs applied to atmospheric pressures and draft. Use of the plus and minus signs does not simplify the calculation and can be confusing.

(2) A few designers, reasoning that the pressure developed by a fan varies as the square of the air velocity, square the percentage safety-factor increase before multiplying by the static pressure. Thus, in the above forced-draft fan, the static discharge pressure with a 20 percent increase in pressure would be $(1.2)^2(9.3552)$ = 13.5 inH_2O (3.4 kPa). This procedure provides a wider margin of safety but is not widely used.

(3) Large steam-generating units, some ship propulsion plants, and some packaged boilers use only forced-draft fans. Induced-draft fans are eliminated because there is a saving in the total fan hp required, there is no air infiltration into the boiler setting, and a slightly higher boiler efficiency can be obtained.

(4) The duct system analyzed here is typical of a study-type design where no refinements are used in bends, downcomers, and other parts of the system. This type of system was chosen for the analysis because it shows more clearly the various losses met in a typical duct installation. The system could be improved by using a bellmouthed intake at *A*, dividing vanes or splitters in the elbows, a transition in the downcomer, and a transition at *F*. None of these improvements would be expensive, and they would all reduce the static pressure required at the fan discharge.

(5) Do not subtract the stack draft from the static pressure the forced- or induced-draft fan must produce. Stack draft can vary considerably, depending on ambient temperature, wind velocity, and wind direction. Therefore, the usual procedure is to ignore any stack draft in fan-selection calculations because this is the safest procedure.

Related Calculations: The procedure given here can be used for all types of boilers fitted with air-supply ducts and uptake breechings—heating, power, process, marine, portable, and packaged.

DETERMINATION OF THE MOST ECONOMICAL FAN CONTROL

Determine the most economical fan control for a forced- or induced-draft fan designed to deliver 140,000 ft^3/min (66.1 m^3/s) at 14 inH_2O (3.5 kPa) at full load. Plot the power-consumption curve for each type of control device considered.

Calculation Procedure:

1. *Determine the types of controls to consider*

There are five types of controls used for forced- and induced-draft fans: (*a*) a damper in the duct with constant-speed fan drive; (*b*) two-speed fan driver; (*c*) inlet vanes or inlet louvres with a constant-speed fan drive; (*d*) multiple-step variable-speed fan drive; and (*e*) hydraulic or electric coupling with constant-speed driver giving wide control over fan speed.

2. *Evaluate each type of fan control*

Tabulate the selection factors influencing the control decision as follows, using the control letters in step 1:

Control type	Control cost	Required power input	Advantages (A), and disadvantages (D)
a	Low	High	(A) Simplicity; (D) high power input
b	Moderate	Moderate	(A) Lower input power; (D) higher cost
c	Low	Moderate	(A) Simplicity; (D) ID fan erosion
d	Moderate	Moderate	(D) Complex; also needs dampers
e	High	Low	(A) Simple; no dampers needed

3. *Plot the control characteristics for the fans*

Draw the fan head-capacity curve for the airflow or gasflow range considered, Fig. 53. This plot shows the maximum capacity of 140,000 ft^3/min (66.1 m^3/s) and required static head of 14 inH$_2$O (3.5 kPa), point *P*.

Plot the power-input curve *ABCD* for a constant-speed motor or turbine drive with damper control—type *a*, listed above—after obtaining from the fan manufacturer, or damper builder, the input power required at various static pressures and capacities. Plotting these values gives curve *ABCD*. Fan speed is 1200 r/min.

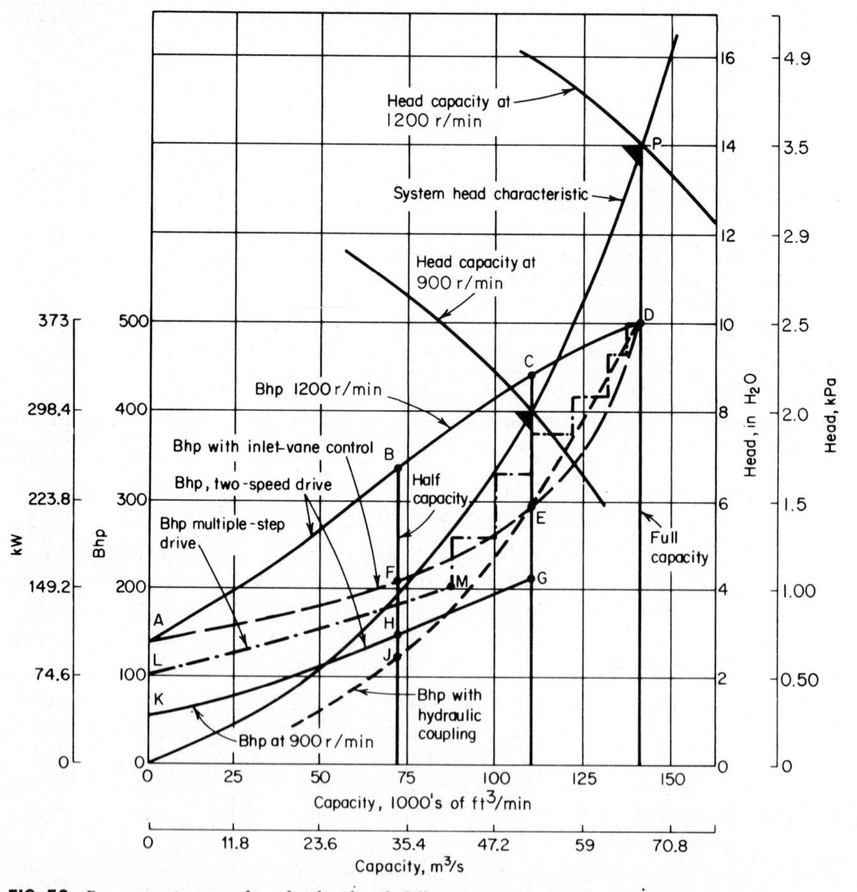

FIG. 53 Power requirements for a fan fitted with different types of controls. *(American Standard Inc.)*

TABLE 8 Fan Control Comparison

	Type of control used				
	a	*b*	*c*	*d*	*e*
Total cost, $	30,000	50,000	75,000	89,500	98,000
Extra cost, $	.....	20,000	25,000	14,500	8,500
Total power saving, $	.....	8,000	6,500	3,000	6,300
Return on extra investment, %	.....	40	26	20.7	74.2

Plot the power-input curve *GHK* for a two-speed drive, type *b*. This drive might be a motor with an additional winding, or it might be a second motor for use at reduced boiler capacities. With either arrangement, the fan speed at lower boiler capacities is 900 r/min.

Plot the power-input curve *AFED* for inlet-vane control on the forced-draft fan or inlet-louvre control on induced-draft fans. The data for plotting this curve can be obtained from the fan manufacturer.

Multiple-step variable-speed fan control, type *d*, is best applied with steam-turbine drives. In a plant with ac auxiliary motor drives, slip-ring motors with damper integration must be used between steps, making the installation expensive. Although dc motor drives would be less costly, few power plants other than marine propulsion plants have direct current available. And since marine units normally operate at full load 90 percent of the time or more, part-load operating economics are unimportant. If steam-turbine drive will be used for the fans, plot the power-input curve *LMD*, using data from the fan manufacturer.

A hydraulic coupling or electric magnetic coupling, type *e*, with a constant-speed motor drive would have the power-input curve *DEJ*.

Study of the power-input curves shows that the hydraulic and electric couplings have the smallest power input. Their first cost, however, is usually greater than any other types of power-saving devices. To determine the return on any extra investment in power-saving devices, an economic study, including a load-duration analysis of the boiler load, must be made.

4. *Compare the return on the extra investment*

Compute and tabulate the total cost of each type of control system. Then determine the extra investment for each of the more costly control systems by subtracting the cost of type *a* from the cost of each of the other types. With the extra investment known, compute the lifetime savings in power input for each of the more efficient control methods. With the extra investment and savings resulting from it known, compute the percentage return on the extra investment. Tabulate the findings as in Table 8.

In Table 8, considering control type *c*, the extra cost of type *c* over type *b* = $75,000 − 50,000 = $25,000. The total power saving of $6500 is computed on the basis of the cost of energy in the plant for the life of the control. The return on the extra investment then = $6500/$25,000 = 0.26, or 26 percent. Type *e* control provides the highest percentage return on the extra investment. It would probably be chosen if the only measure of investment desirability is the return on the extra investment. However, if other criteria are used, such as a minimum rate of return on the extra investment, one of the other control types might be chosen. This is easily determined by studying the tabulation in conjunction with the investment requirement.

Related Calculations: The procedure used here can be applied to heating, power, marine, and portable boilers of all types. Follow the same steps given above, changing the values to suit the existing conditions. Work closely with the fan and drive manufacturer when analyzing drive power input and costs.

SMOKESTACK HEIGHT AND DIAMETER DETERMINATION

Determine the required height and diameter of a smokestack to produce 1.0-inH_2O (0.25-kPa) draft at sea level if the average air temperature is 60°F (15.6°C); barometric pressure is 29.92 inHg (101.3 kPa); the boiler flue gas enters the stack at 500°F (260.0°C); the flue-gas flow rate is 100 lb/s (45.4 kg/s); the flue-gas density is 0.045 lb/ft^3 (0.72 kg/m^3); and the flue-gas velocity is

30 ft/s (9.1 m/s). What diameter and height would be required for this stack if it were located 5000 ft (1524.0 m) above sea level?

Calculation Procedure:

1. *Compute the required stack height*

The required stack height S_h ft = $d_s/0.256pK$, where d_s = stack draft, inH_2O; p = barometric pressure, inHg; $K = 1/T_a - 1/T_g$, where T_a = air temperature, °R; T_g = average temperature of stack gas, °R. In applying this equation, the temperature of the gas at the stack outlet must be known to determine the average temperature of the gas in the stack. Since the outlet temperature cannot be measured until after the stack is in use, an assumed outlet temperature must be used for design calculations. The outlet temperature depends on the inlet temperature, ambient air temperature, and materials used in the stack construction. For usual smokestacks, the gas temperature will decrease 100 to 200°F (55.6 to 111.1°C) between the stack inlet and outlet. Using a 100°F (55.6°C) gas-temperature decrease for this stack, we get $S_h = (1.0) + 0.256(29.92)(1/520 - 1/910)$ = 159 ft (48.5 m). Apply a 10 percent factor of safety. Then the stack height = (159)(1.10) = 175 ft (53.3 m).

2. *Compute the required stack diameter*

Stack diameter d_s ft is found from $d_s = 0.278(W_gT_g/Vd_gp)^{0.5}$, where W_g = flue-gas flow rate in stack, lb/s; V = flue-gas velocity in stack, ft/s; d_g = flue-gas density, lb/ft³. For this stack, $d_s = 0.278\{(100)(910)/[(30)(0.045)(29.92)]\}^{0.5}$ = 13.2 ft (4.0 m), or 13 ft 3 in (4 m 4 cm), rounding to the nearest inch diameter.

Note: Use this calculation procedure for any stack material—masonary, steel, brick, or plastic. Most boiler and stack manufacturers use charts based on the equations above to determine the economical height and diameter of a stack. Thus, the Babcock & Wilcox Company, New York, presents four charts for stack sizing, in *Steam: Its Generation and Use*. Combustion Engineering, Inc., also presents four charts for stack sizing, in *Combustion Engineering*. The equations used in the present calculation procedure are adequate for a quick, first approximation of stack height and diameter.

3. *Compute the required stack height and diameter at 5000-ft (1524.0-m) elevation*

Fuels require the same amount of oxygen for combustion regardless of the altitude at which they are burned. Therefore, this stack must provide the same draft as at sea level. But as the altitude above sea level increases, more air must be supplied to the fuel to sustain the same combustion rate, because air above sea level contains less oxygen per cubic foot than at sea level. To accommodate the larger air and flue-gas flow rate without an increase in the stack friction loss, the stack diameter must be increased.

To determine the required stack height S_e ft at an elevation above sea level, multiply the sea-level height S_h by the ratio of the sea-level and elevated-height barometric pressures, inHg. Since the barometric pressure at 5000 ft (1524.0 m) is 24.89 inHg (84.3 kPa) and the sea-level barometric pressure is 29.92 inHg (101.3 kPa), S_e = (175)(29.92/24.89) = 210.2 ft (64.1 m).

The stack diameter d_e ft at an elevation above sea level will vary as the 0.40 power of the ratio of the sea-level and altitude barometric pressures, or $d_e = d_s(p_e/p)^{0.4}$, where p_e = barometric pressure of altitude, inHg. For this stack, $d_e = (13.2)(29.92/24.89)^{0.4}$ = 14.2 ft (4.3 m), or 14 ft 3 in (4 m 34 cm).

Related Calculations: The procedure given here can be used for heating, power, marine, industrial, and residential smokestacks or chimneys, regardless of the materials used for construction. When designing smokestacks for use at altitudes above sea level, use step 3, or substitute the actual barometric pressure at the elevated location in the height and diameter equations of steps 1 and 2.

POWER-PLANT COAL-DRYER ANALYSIS

A power-plant coal dryer receives 180 tons/h (163.3 t/h) of wet coal containing 15 percent free moisture. The dryer is arranged to drain 6 percent of the moisture from the coal, and a moisture content of 1 percent is acceptable in the coal delivered to the power plant. Determine the volume and temperature of the drying gas required for the dryer, the total heat, grate area, and combustion-space volume needed. Ambient air temperature during drying is 70°F (21.1°C).

Calculation Procedure:

1. *Compute the quantity of moisture to be removed*

The total moisture in the coal = 15 percent. Of this, 6 percent is drained and 1 percent can remain in the coal. The amount of moisture to be removed is therefore 15 − 6 − 1 = 8 percent. Since 180 tons (163.3 t) of coal are received per hour, the quantity of moisture to be removed per minute is [180/(60 min/h)](2000 lb/ton)(0.08) = 480 lb/min (3.6 kg/s).

2. *Compute the airflow required through the dryer*

Air enters the dryer at 70°F (21.1°C). Assume that evaporation of the moisture on the coal takes place at 125°F (51.7°C)—this is about midway in the usual evaporation temperature range of 110 to 145°F (43.3 to 62.8°C). Determine the moisture content of saturated air at each temperature, using the psychrometric chart for air. Thus, for saturated air at 70°F (21.1°C) dry-bulb temperature, the weight of the moisture it contains is w_m lb (kg) of water per pound (kilogram) of dry air = 0.0159 (0.00721), whereas at 125°F (51.7°C), w_m = 0.09537 lb of water per pound (0.04326 kg/kg) of dry air. The weight of water removed per pound of air passing through the dryer is the difference between the moisture content at the leaving temperature, 125°F (51.7°C), and the entering temperature, 70°F (21.1°C), or 0.09537 − 0.01590 = 0.07947 lb of water per pound (0.03605 kg/kg) of dry air.

Since air at 70°F (21.1°C) has a density of 0.075 lb/ft³ (1.2 kg/m³), 1/0.075 = 13.3 ft³ (0.4 m³) of air at 70°F (21.1°C) must be supplied to absorb 0.07947 lb of water per pound (0.03605 kg/kg) of dry air. With 480 lb/min (3.6 kg/s) of water to be evaporated in the dryer, each cubic foot of air will absorb 0.07947/13.3 = 0.005945 lb (0.095 kg/m³) of moisture, and the total airflow must be (480 lb/min)/(0.005945) = 80,800 ft³/min (38.1 m³/s), given a dryer efficiency of 100 percent. However, the usual dryer efficiency is about 75 percent, not 100 percent. Therefore, the total actual airflow through the dryer should be 80,800/0.75 = 107,700 ft³/min (50.8 m³/s).

Note: If desired, a table of moist air properties can be used instead of a psychrometric chart to determine the moisture content of the air at the dryer inlet and outlet conditions. The moisture content is read in the humidity ratio W_s column. See the ASHRAE—*Guide and Data Book* for such a tabulation of moist-air properties.

3. *Compute the required air temperature*

Assume that the heating air enters at a temperature t greater than 125°F (51.7°C). Set up a heat balance such that the heat given up by the air in cooling from t to 125°F (51.7°C) = the heat required to evaporate the water on the coal + the heat required to raise the temperature of the coal and water from ambient to the evaporation temperature + radiation losses.

The heat given up by the air, Btu = (cfm)(density of air, lb/ft³)[specific heat of air, Btu/(lb·°F)](t − evaporation temperature, °F). The heat required to evaporate the water, Btu = (weight of water, lb/min)(h_{fg} at evaporation temperature). The heat required to raise the temperature of the coal and water from ambient to the evaporation temperature, Btu = (weight of coal, lb/min)(evaporation temperature − ambient temperature)[specific heat of coal, Btu/(lb·°F)] + (weight of water, lb/min)(evaporation temperature − ambient temperature)[specific heat of water, Btu/(lb·°F)]. The heat required to make up for radiation losses, Btu = {(area of dryer insulated surfaces, ft²)[heat-transfer coefficient, Btu/(ft²·°F·h)](t − ambient temperature) + (area of dryer uninsulated surfaces, ft²)[heat-transfer coefficient, Btu/(ft²·°F·h)](t − ambient temperature)}/60.

Compute the heat given up by the air, Btu, as (107,700)(0.075)(0.24)(t − 70), where 0.075 is the air density and 0.24 is the specific heat of air.

Compute the heat required to evaporate the water, Btu, as (480)(1022.9), where 1022.9 = h_{fg} at 125°F (51.7°C) from the steam tables.

Compute the heat required to raise the temperature of the coal and water from ambient to the evaporation temperature, Btu, as (6000)(t − 70)(0.30) + (480)(t − 70)(1.0), where 0.30 is the specific heat of the coal and 1.0 is the specific heat of water.

Compute the heat required to make up the radiation losses, assuming 3000 ft² (278.7 m²) of insulated and 1500 ft² (139.4 m²) of uninsulated surface in the dryer, with coefficients of heat transfer of 0.35 and 3.0 for the insulated and uninsulated surfaces, respectively. Then radiation heat loss, Btu = (3000)(0.35)(t − 70) + (1500)(3.0)(t − 70).

Set up the heat balance thus and solve for t: (107,700)(0.075)(0.24)(t − 70) = (480)(1022.9) + (6000)(125 − 70)(0.30) + (480)(125 − 70)(1.0) + [(3000)(0.35)(t − 70) + (1500)(3.0)(t −

70)]/60; so t = 406°F (207.8°C). In this heat balance, the factor 60 is divided into the radiation heat loss to convert flow in Btu/h to Btu/min because all the other expressions are in Btu/min.

4. *Determine the total heat required by the dryer*

Using the equation of step 3 with t = 406°F (207.8°C), we find the total heat = (107,770)(0.075)(0.24)(406 − 70) = 651,000 Btu/min, or 60(651,000) = 39,060,000 Btu/h (11,439.7 kW).

5. *Compute the dryer-furnace grate area*

Assume that heat for the dryer is produced from coal having a lower heating value of 13,000 Btu/lb (30,238 kJ/kg) and that 40 lb/h of coal is burned per square foot [0.05 kg/(m^2·s)] of grate area with a combustion efficiency of 70 percent.

The rate of coal firing = (Btu/min to dryer)/(coal heating value, Btu/lb)(combustion efficiency) = 651,000/(13,000)(0.70) = 71.5 lb/min, or 60(71.5) = 4990 lb/h (0.63 kg/s). Grate area = 4990/40 = 124.75 ft^2, say 125 ft^2 (11.6 m^2).

6. *Compute the dryer-furnace volume*

The usual heat-release rates for dryer furnaces are about 50,000 Btu/(h·ft^3) (517.5 kW/m^3) of furnace volume. For this furnace, which burns 4900 lb/h (0.63 kg/s) of 13,000-Btu/lb (30,238-kJ/kg) coal, the total heat released is 4990(13,000) = 64,870,000 Btu/h (18,998.8 kW). With an allowable heat release of 50,000 Btu/(h·ft^3) (517.1 kW/m^3), the required furnace volume = 64,870,000/50,000 = 1297.4 ft^3, say 1300 ft^3 (36.8 m^3).

Related Calculations: The general procedure given here can be used for any air-heated dryer used to dry moist materials. Thus, the procedure is applicable to chemical, soil, and fertilizer drying, as well as coal drying. In each case, the specific heat of the material dried must be used in place of the specific heat of coal given above.

COAL STORAGE CAPACITY OF PILES AND BUNKERS

Bituminous coal is stored in a 25-ft (7.5-m) high, 68.8-ft (21.0-m) diameter, circular-base conical pile. How many tons of coal does the pile contain if its base angle is 36°? How much bituminous coal is contained in a 25-ft (7.5-m) high rectangular pile 100 ft (30.5 m) long if the pile cross section is a triangle having a 36° base angle?

Calculation Procedure:

1. *Sketch the coal pile*

Figure 54*a* and *b* shows the two coal piles. Indicate the pertinent dimensions—height, the diameter, length, and base angle—on each sketch.

2. *Compute the volume of the coal pile*

Volume of a right circular cone, ft^3 = $\pi r^2 h/3$, where r = radius, ft; h = cone height, ft. Volume of a triangular pile = $bal/2$, where b = base length, ft; a = altitude, ft; l = length of pile, ft.

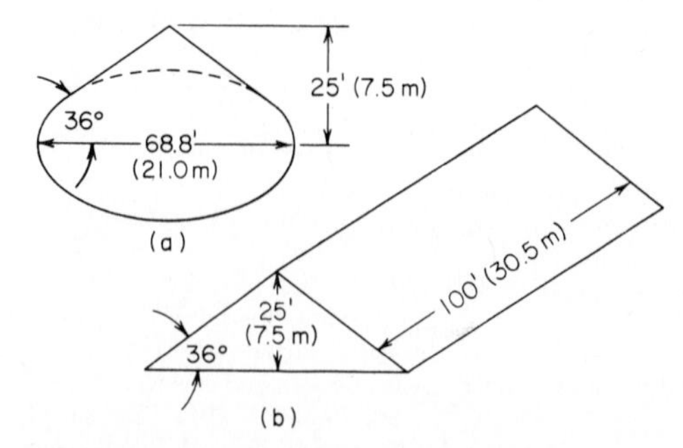

FIG. 54 (*a*) Conical coal pile; (*b*) triangular coal pile.

For this conical pile, volume = $\pi(3.4)^2(25)/3$ = 31,000 ft^3 (877.8 m^3). Since 50 lb of bituminous coal occupies about 1 ft^3 of volume (800.9 kg/m^3), the weight of coal in the conical pile = (31,000 ft^3)(50 lb/ft^3) = 1,550,000 lb, or (1,550,000 lb)/(2000 lb/ton) = 775 tons (703.1 t).

For the triangular pile, base length = $2h/\tan 36°$ = (2)(25)/0.727 = 68.8 ft (21.0 m). Then volume = (68.8)(25)(100)/2 = 86,000 ft^3 (2435.2 m^3). The weight of bituminous coal in the pile is, as for the conical pile, (86,000)(50) = 4,300,000 lb, or (4,300,000 lb)/(2000 lb/ton) = 2150 tons (1950.4 t).

Related Calculations: Use this general procedure to compute the weight of coal in piles of all shapes, and in bunkers, silos, bins, and similar storage compartments. The procedure can be used for other materials also—grain, sand, gravel, coke, etc. Be sure to use the correct density when converting the total storage volume to total weight. Refer to Baumeister and Marks—*Standard Handbook for Mechanical Engineers* for a comprehensive tabulation of the densities of various materials.

PROPERTIES OF A MIXTURE OF GASES

A 10-ft^3 (0.3-m^3) tank holds 1 lb (0.5 kg) of hydrogen (H_2), 2 lb (0.9 kg) of nitrogen (N_2), and 3 lb (1.4 kg) of carbon dioxide (CO_2) at 70°F (21.1°C). Find the specific volume, pressure, specific enthalpy, internal energy, and specific entropy of the individual gases and of the mixture and the mixture density. Use Avogadro's and Dalton's laws and Keenan and Kaye—*Thermodynamic Properties of Air, Products of Combustion and Component Gases*, Wiley, commonly termed the *Gas Tables*.

Calculation Procedure:

1. *Compute the specific volume of each gas*

Using H, N, and C as subscripts for the respective gases, we see that the specific volume of any gas v ft^3/lb = total volume of tank, ft^3/weight of gas in tank, lb. Thus, v_H = 10/1 = 10 ft^3/lb (0.6 m^3/kg); v_N = 10/2 = 5 ft^3/lb (0.3 m^3/kg); v_C = 10/3 = 3.33 ft^3/lb (0.2 m^3/kg). Then the specific volume of the mixture of gases is v_t ft^3/lb = total volume of gas in tank, ft^3/sum of weight of individual gases, lb = 10/(1 + 2 + 3) = 1.667 ft^3/lb (0.1 m^3/kg).

2. *Determine the absolute pressure of each gas*

Using $P = RTw/v_tM$, where P = absolute pressure of the gas, lb/ft^2 (abs); R = universal gas constant = 1545; T = absolute temperature of the gas, °R = °F + 459.9, usually taken as 460; w = weight of gas in the tank, lb; v_t = total volume of the gas in the tank, ft^3; M = molecular weight of the gas. Thus, P_H = (1545)(70 + 460)(1.0)/[(10)(2.0)] = 40,530 lb/ft^2 (abs) (1940.6 kPa); P_N = (1545)(70 + 460)(2.0)/[(10)(28)] = 5850 lb/ft^2 (abs) (280.1 kPa); P_C = (1545)(70 + 460)(3.0)/[(10)(44)] = 5583 lb/ft^2 (abs) (267.3 kPa); $P_t = \Sigma P_H, P_N, P_C$ = 40,530 + 5850 + 5583 = 51,963 lb/ft^2 (abs) (2488.0 kPa).

3. *Determine the specific enthalpy of each gas*

Refer to the *Gas Tables*, entering the left-hand column of the table at the absolute temperature, 530°R (294 K), for the gas being considered. Opposite the temperature, read the specific enthalpy in the h column. Thus, h_H = 1796.1 Btu/lb (4177.7 kJ/kg); h_N = 131.4 Btu/lb (305.6 kJ/kg); h_C = 90.17 Btu/lb (209.7 kJ/kg). The total enthalpy of the mixture of the gases is the sum of the products of the weight of each gas and its specific enthalpy, or (1)(1796.1) + (2)(131.4) + (3)(90.17) = 2329.4 Btu (2457.6 kJ) for the 6 lb (2.7 kg) or 10 ft^3 (0.28 m^3) of gas. The specific enthalpy of the mixture is the total enthalpy/gas weight, lb, or 2329.4/(1 + 2 + 3) = 388.2 Btu/lb (903.0 kJ/kg) of gas mixture.

4. *Determine the internal energy of each gas*

Using the *Gas Tables* as in step 3, we find E_H = 1260.0 Btu/lb (2930.8 kJ/kg); E_N = 93.8 Btu/lb (218.2 kJ/kg); E_C = 66.3 Btu/lb (154.2 kJ/kg). The total energy = (1)(1260.0) + (2)(93.8) + (3)(66.3) = 1646.5 Btu (1737.2 kJ). The specific enthalpy of the mixture = 1646.5/(1 + 2 + 3) = 274.4 Btu/lb (638.3 kJ/kg) of gas mixture.

5. *Determine the specific entropy of each gas*

Using the *Gas Tables* as in step 3, we get S_H = 15.52 Btu/(lb·°F) [65.0 kJ/(kg·°C)]; S_N = 1.558 Btu/(lb·°F) [4.7 kJ/(kg·°C)]. The entropy of the mixture = (1)(12.52) + (2)(1.558) + (3)(1.114)

= 18.978 Btu/°F (34.2 kJ/°C). The specific entropy of the mixture = 18.978/(1 + 2 + 3) = 3.163 Btu/(lb·°F) [13.2 kJ/(kg·°C)] of the gas mixture.

6. *Compute the density of the mixture*

For any gas, the total density d_t = sum of the densities of the individual gases. And since density of a gas = 1/specific volume, $d_t = 1/v_t = 1/v_H + 1/v_N + 1/v_C = 1/10 + 1/5 + 1/3.33 =$ 0.6 lb/ft³ (9.6 kg/m³) of mixture. This checks with step 1, where v_t = 1.667 ft³/lb (0.1 m³/kg), and is based on the principle that all gases occupy the same volume.

Related Calculations: Use this method for any gases stored in any type of container—steel, plastic, rubber, canvas, etc.—under any pressure from less than atmospheric to greater than atmospheric at any temperature.

REGENERATIVE-CYCLE GAS-TURBINE ANALYSIS

What is the cycle air rate, lb/kWh, for a regenerative gas turbine having a pressure ratio of 5, an air inlet temperature of 60°F (15.6°C), a compressor discharge temperature of 1500°F (815.6°C), and performance in accordance with Fig. 55? Determine the cycle thermal efficiency and work ratio. What is the power output of a regenerative gas turbine if the work input to the compressor is 4400 hp (3281.1 kW)?

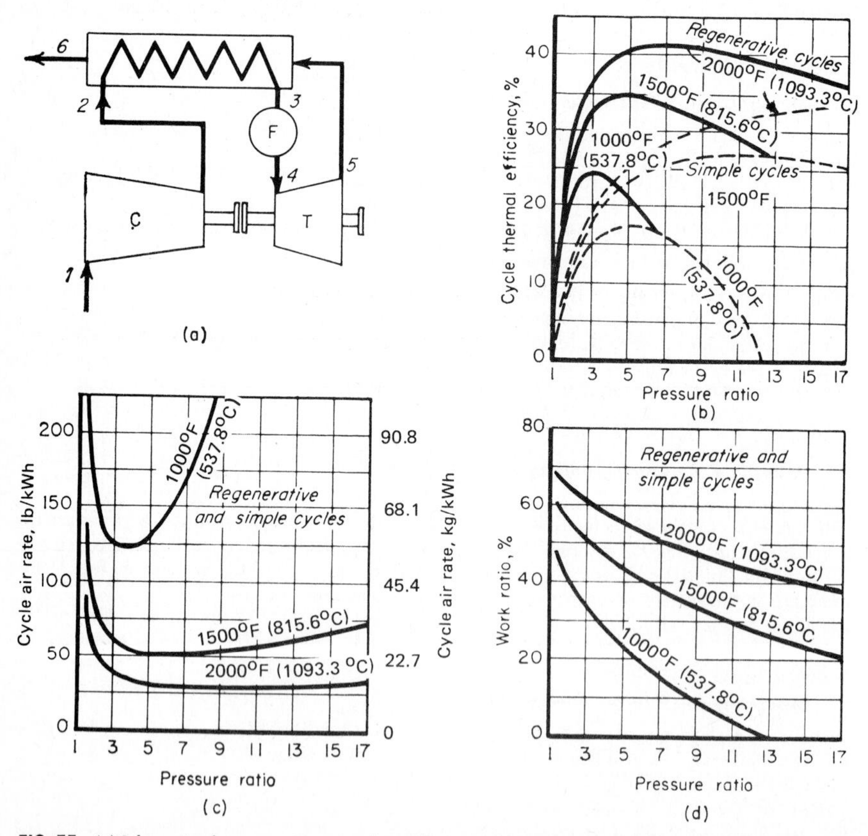

FIG. 55 (*a*) Schematic of regenerative gas turbine; (*b*), (*c*), and (*d*) gas-turbine performance based on a regenerator effectiveness of 70 percent, compressor and turbine efficiency of 85 percent; air inlet = 60°F (15.6°C); no pressure losses.

Calculation Procedure:

1. *Determine the cycle rate*

Use Fig. 55, entering at the pressure ratio of 5 in Fig. 55*c* and projecting to the 1500°F (815.6°C) curve. At the left, read the cycle air rate as 52 lb/kWh (23.6 kg/kWh).

2. *Find the cycle thermal efficiency*

Enter Fig. 55*b* at the pressure ratio of 5 and project vertically to the 1500°F (815.6°C) curve. At left, read the cycle thermal efficiency as 35 percent. Note that this point corresponds to the maximum efficiency obtainable from this cycle.

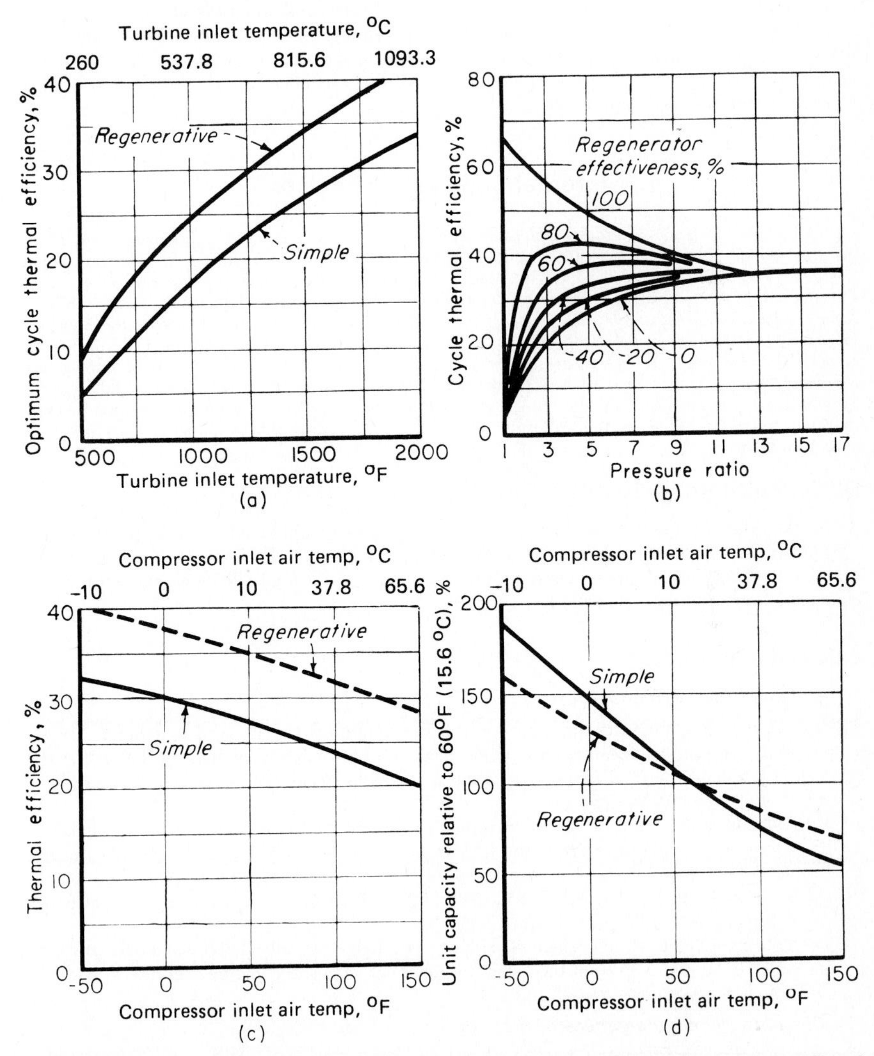

FIG. 56 (*a*) Effect of turbine-inlet on cycle performance; (*b*) effect of regenerator effectiveness; (*c*) effect of compressor inlet-air temperature; (*d*) effect of inlet-air temperature on turbine-cycle capacity. These curves are based on a turbine and compressor efficiency of 85 percent, a regenerator effectiveness of 70 percent, and a 1500°F (815.6°C) inlet-gas temperature.

3. *Find the cycle work ratio*

Enter Fig. 55*d* at the pressure ratio of 5 and project vertically to the 1500°F (815.6°C) curve. At the left, read the work ratio as 44 percent.

4. *Compute the turbine power output*

For any gas turbine, the work ratio, percent = $100w_c/w_t$, where w_c = work input to the turbine, hp; w_t = work output of the turbine, hp. Substituting gives 44 = $100(4400)/w_t$; w_t = 100(4400)/44 = 10,000 hp (7457.0 kW).

Related Calculations: Use this general procedure to analyze gas turbines for power-plant, marine, and portable applications. Where the operating conditions are different from those given here, use the manufacturer's engineering data for the turbine under consideration.

Figure 56 shows the effect of turbine-inlet temperature, regenerator effectiveness, and compressor-inlet-air temperature on the performance of a modern gas turbine. Use these curves to analyze the cycles of gas turbines being considered for a particular application if the operating conditions are close to those plotted.

Internal-Combustion Engines

REFERENCES: Benson—*Internal Combustion Engines*, Pergamon; Kates and Luck—*Diesel and High-Compression Gas Engines*, American Technical Society; Ranney—*Fuel Additives for Internal Combustion Engines*, Noyes; Blackman and Thomas—*Fuel Economy of the Gasoline Engine*, Halsted Press; Sitkei—*Heat Transfer and Thermal Loading in Internal Combustion Engines*, International Publications Services; Baxa—*Noise Control in Internal Combustion Engines*, Wiley; Diesel Engine Manufacturers Associations—*Standard Practices for Stationary Diesel Engines;* Lichty—*Internal-Combustion Engines*, McGraw-Hill; Allen—*The Modern Diesel*, Prentice-Hall; Maleev—*Internal-Combustion Engines*, McGraw-Hill; *Diesel Engineering Handbook*, Diesel Publications; Adams—*Elements of Diesel Engineering*, Henley; Severns and Degler—*Steam, Air and Gas Power*, Wiley; Ricardo—*The High-Speed Internal-Combustion Engine*, Blackie; Obert—*Internal Combustion Engines*, International Textbooks; Fors—*Practical Marine Diesel Engineering*, Simmons-Boardman.

DIESEL GENERATING UNIT EFFICIENCY

A 3000-kW diesel generating unit performs thus: fuel rate, 1.5 bbl (238.5 L) of 25° API fuel for a 900-kWh output; mechanical efficiency, 82.0 percent; generator efficiency, 92.0 percent. Compute engine fuel rate, engine-generator fuel rate, indicated thermal efficiency, overall thermal efficiency, brake thermal efficiency.

Calculation Procedure:

1. *Compute the engine fuel rate*

The fuel rate of an engine driving a generator is the weight of fuel, lb, used to generate 1 kWh at the generator input shaft. Since this engine burns 1.5 bbl (238.5 L) of fuel for 900 kW at the generator terminals, the total fuel consumption is (1.5 bbl)(42 gal/bbl) = 63 gal (238.5 L), at a generator efficiency of 92.0 percent.

To determine the weight of this oil, compute its specific gravity s from s = 141.5/(131.5 + °API), where °API = API gravity of the fuel. Hence, s = 141.5(131.5 + 25) = 0.904. Since 1 gal (3.8 L) of water weighs 8.33 lb (3.8 kg) at 60°F (15.6°C), 1 gal (3.8 L) of this oil weighs (0.904)(8.33) = 7.529 lb (3.39 kg). The total weight of fuel used when burning 63 gal is (63 gal)(7.529 lb/gal) = 474.5 lb (213.5 kg).

The generator is 92 percent efficient. Hence, the engine actually delivers enough power to generate 900/0.92 = 977 kWh at the generator terminals. Thus, the engine fuel rate = 474.5 lb fuel/977 kWh = 0.485 lb/kWh (0.218 kg/kWh).

2. *Compute the engine-generator fuel rate*

The engine-generator fuel rate takes these two units into consideration and is the weight of fuel required to generate 1 kWh at the generator terminals. Using the fuel-consumption data from step 1 and the given output of 900 kW, we see that engine-generator fuel rate = 474.5 lb fuel/900 kWh output = 0.527 lb/kWh (0.237 kg/kWh).

3. *Compute the indicated thermal efficiency*

Indicated thermal efficiency is the thermal efficiency based on the *indicated* horsepower of the engine. This is the horsepower developed in the engine cylinder. The engine fuel rate, computed in step 1, is the fuel consumed to produce the brake or shaft horsepower output, after friction losses are deducted. Since the mechanical efficiency of the engine is 82 percent, the fuel required to produce the indicated horsepower is 82 percent of that required for the brake horsepower, or (0.82)(0.485) = 0.398 lb/kWh (0.179 kg/kWh).

The indicated thermal efficiency of an internal-combustion engine driving a generator is $e_i = 3413/f_i(\text{HHV})$, where e_i = indicated thermal efficiency, expressed as a decimal; f_i = indicated fuel consumption, lb/kWh; HHV = higher heating value of the fuel, Btu/lb.

Compute the HHV for a diesel fuel from HHV = 17,680 + 60 × °API. For this fuel, HHV = 17,680 + 60(25) = 19,180 Btu/lb (44,612.7 kJ/kg).

With the HHV known, compute the indicated thermal efficiency from $e_i = 3{,}413/[(0.398)(19{,}180)] = 0.447$, or 44.7 percent.

4. *Compute the overall thermal efficiency*

The overall thermal efficiency e_o is computed from $e_o = 3413/f_o(\text{HHV})$, where f_o = overall fuel consumption, Btu/kWh; other symbols as before. Using the engine-generator fuel rate from step 2, which represents the overall fuel consumption $e_o = 3413/[(0.527)(19{,}180)] = 0.347$, or 34.7 percent.

5. *Compute the brake thermal efficiency*

The engine fuel rate, step 1, corresponds to the brake fuel rate f_b. Compute the brake thermal efficiency from $e_b = 3413/f_b(\text{HHV})$, where f_b = brake fuel rate, Btu/kWh; other symbols as before. For this engine-generator set, $e_b = 3413/[(0.485)(19{,}180)] = 0.367$, or 36.7 percent.

Related Calculations: Where the fuel consumption is given or computed in terms of lb/(hp·h), substitute the value of 2545 Btu/(hp·h) (1.0 kW/kWh) in place of the value 3413 Btu/kWh (3600.7 kJ/kWh) in the numerator of the e_i, e_o, and e_b equations. Compute the indicated, overall, and brake thermal efficiencies as before. Use the same procedure for gas and gasoline engines, except that the higher heating value of the gas or gasoline should be obtained from the supplier or by test.

ENGINE DISPLACEMENT, MEAN EFFECTIVE PRESSURE, AND EFFICIENCY

A 12 × 18 in (30.5 × 44.8 cm) four-cylinder four-stroke single-acting diesel engine is rated at 200 bhp (149.2 kW) at 260 r/min. Fuel consumption at rated load is 0.42 lb/(bhp·h) (0.25 kg/kWh). The higher heating value of the fuel is 18,920 Btu/lb (44,008 kJ/kg). What are the brake mean effective pressure, engine displacement in ft³/(min·bhp), and brake thermal efficiency?

Calculation Procedure:

1. *Compute the brake mean effective pressure*

Compute the brake mean effective pressure (bmep) for an internal-combustion engine from $bmep = 33{,}000\, bhp_n/LAn$, where $bmep$ = brake mean effective pressure, lb/in²; bhp_n = brake horsepower output delivered per cylinder, hp; L = piston stroke length, ft; a = piston area, in²; n = cycles per minute per cylinder = crankshaft rpm for a two-stroke cycle engine, and 0.5 the crankshaft rpm for a four-stroke cycle engine.

For this engine at its rated bhp, the output per cylinder is 200 bhp/4 cylinders = 50 bhp (37.3 kW). Then $bmep = 33{,}000(50)/[(18/12)(12)^2(\pi/4)(260/2)] = 74.8$ lb/in² (516.1 kPa). (The factor 12 in the denominator converts the stroke length from inches to feet.)

2. *Compute the engine displacement*

The total engine displacement V_d ft³ is given by $V_d = LAnN$, where A = piston area, ft²; N = number of cylinders in the engine; other symbols as before. For this engine, $V_d = (18/12)(12/12)^2(\pi/4)(260/2)(4) = 614$ ft³/min (17.4 m³/min). The displacement is in cubic feet per minute because the crankshaft speed is in r/min. The factor of 12 in the denominators converts the stroke and area to ft and ft², respectively. The displacement per bhp = (total displacement, ft³/min)/bhp output of engine = 614/200 = 3.07 ft³/(min·bhp) (0.12 m³/kW).

3. *Compute the brake thermal efficiency*

The brake thermal efficiency e_b of an internal-combustion engine is given by $e_b = 2545/(sfc)(\text{HHV})$, where sfc = specific fuel consumption, lb/(bhp·h); HHV = higher heating value of fuel, Btu/lb. For this engine, $e_b = 2545/[(0.42)(18{,}920)] = 0.32$, or 32.0 percent.

Related Calculations: Use the same procedure for gas and gasoline engines. Obtain the higher heating value of the fuel from the supplier, a tabulation of fuel properties, or by test.

ENGINE MEAN EFFECTIVE PRESSURE AND HORSEPOWER

A 500-hp (373-kW) internal-combustion engine has a brake mean effective pressure of 80 lb/in² (551.5 kPa) at full load. What are the indicated mean effective pressure and friction mean effective pressure if the mechanical efficiency of the engine is 85 percent? What are the indicated horsepower and friction horsepower of the engine?

Calculation Procedure:

1. *Determine the indicated mean effective pressure*

Indicated mean effective pressure *imep* lb/in² for an internal-combustion engine is found from $imep = bmep/e_m$, where $bmep$ = brake mean effective pressure, lb/in²; e_m = mechanical efficiency, percent, expressed as a decimal. For this engine, $imep = 80/0.85 = 94.1$ lb/in² (659.3 kPa).

2. *Compute the friction mean effective pressure*

For an internal-combustion engine, the friction mean effective pressure *fmep* lb/in² is found from $fmep = imep - bmep$, or $fmep = 94.1 - 80 = 14.1$ lb/in² (97.3 kPa).

3. *Compute the indicated horsepower of the engine*

For an internal-combustion engine, the mechanical efficiency $e_m = bhp/ihp$, where ihp = indicated horsepower. Thus, $ihp = bhp/e_m$, or $ihp = 500/0.85 = 588$ ihp (438.6 kW).

4. *Compute the friction hp of the engine*

For an internal-combustion engine, the friction horsepower is $fhp = ihp - bhp$. In this engine, $fhp = 588 - 500 = 88$ fhp (65.6 kW).

Related Calculations: Use a similar procedure to determine the *indicated engine efficiency* $e_{ei} = e_i/e$, where e = ideal cycle efficiency; *brake engine efficiency*, $e_{eb} = e_b\, e$; *combined engine efficiency* or *overall engine thermal efficiency* $e_{eo} = e_o/e$. Note that each of these three efficiencies is an *engine* efficiency and corresponds to an actual thermal efficiency, e_i, e_b, and e_o.

Engine efficiency $e_e = e_t/e$, where e_t = actual *engine* thermal efficiency. Where desired, the respective *actual* indicated brake, or overall, output can be susbstituted for e_i, e_b, and e_o in the numerator of the above equations if the ideal output is substituted in the denominator. The result will be the respective engine efficiency. Output can be expressed in Btu per unit time, or horsepower. Also, e_e = actual *mep*/ideal *mep*, and $e_{ei} = imep$/ideal *mep*; $e_{eb} = bmep$/ideal *mep*; e_{eo} = overall *mep*/ideal *mep*. Further, $e_b = e_m e_i$, and $bmep = e_m(imep)$. Where the actual heat supplied by the fuel, HHV Btu/lb, is known, compute $e_i e_b$ and e_o by the method given in the previous calculation procedure. The above relations apply to any reciprocating internal-combustion engine using any fuel.

SELECTION OF AN INDUSTRIAL INTERNAL-COMBUSTION ENGINE

Select an internal-combustion engine to drive a centrifugal pump handling 2000 gal/min (126.2 L/s) of water at a total head of 350 ft (106.7 m). The pump speed will be 1750 r/min, and it will run continuously. The engine and pump are located at sea level.

Calculation Procedure:

1. *Compute the power input to the pump*

The power required to pump water is $hp = 8.33GH/33{,}000e$, where G = water flow, gal/min; H = total head on the pump, ft of water; e = pump efficiency, expressed as a decimal. Typical

TABLE 1 Internal-Combustion Engine Rating Table

Diesel engines						
Continuous bhp (kW) at given rpm						
1400	1600	1750	1800	Rated bhp	No. of cylinders	Cooling*
187 (139.5)	214 (159.6)	227 (169.3)	230 (171.6)	300 (223.8)	6	*E*
230 (171.6)	256 (190.0)	275 (205.2)	280 (208.9)	438 (326.7)	12	*R*
240 (179.0)	273 (203.7)	295 (220.0)	305 (227.5)	438 (326.7)	12	*E*
Gasoline engines†						
405 (302.1)	430 (320.8)	450 (335.7)	475 (354.4)	595 (438.9)	12	*R*

*E = heat-exchanger-cooled; R = radiator-cooled.
†Use 80 percent of tabulated power if engine is to run at continuous full load.

centrifugal pumps have operating efficiencies ranging from 50 to 80 percent, depending on the pump design and condition and liquid handled. Assume that this pump has an efficiency of 70 percent. Then $hp = 8.33(2000)/(350)/[(33{,}000)(0.70)] = 252$ hp (187.9 kW). Thus, the internal-combustion engine must develop at least 252 hp (187.9 kW) to drive this pump.

2. *Select the internal-combustion engine*

Since the engine will run continuously, extreme care must be used in its selection. Refer to a tabulation of engine ratings, such as Table 1. This table shows that a diesel engine that delivers 275 continuous brake horsepower (205.2 kW) (the nearest tabulated rating equal to or greater than the required input) will be rated at 483 bhp (360.3 kW) at 1750 r/min.

The gasoline-engine rating data in Table 1 show that for continuous full load at a given speed, 80 percent of the tabulated power can be used. Thus, at 1750 r/min, the engine must be rated at 252/0.80 = 315 bhp (234.9 kW). A 450-hp (335.7-kW) unit is the only one shown in Table 1 that would meet the needs. This is too large; refer to another builder's rating table to find an engine rated at 315 to 325 bhp (234.9 to 242.5 kW) at 1750 r/min.

The unsuitable capacity range in the gasoline-engine section of Table 1 is a typical situation met in selecting equipment. More time is often spent in finding a suitable unit at an acceptable price than is spent computing the required power output.

Related Calculations: Use this procedure to select any type of reciprocating internal-combustion engine using oil, gasoline, liquefied-petroleum gas, or natural gas for fuel.

ENGINE OUTPUT AT HIGH TEMPERATURES AND HIGH ALTITUDES

An 800-hp (596.8-kW) diesel engine is operated 10,000 ft (3048 m) above sea level. What is its output at this elevation if the intake air is at 80°F (26.7°C)? What will the output at 10,000-ft (3048-m) altitude be if the intake air is at 110°F (43.4°C)? What would the output be if this engine were equipped with an exhaust turbine-driven blower?

Calculation Procedure:

1. *Compute the engine output at altitude*

Diesel engines are rated at sea level at atmospheric temperatures of not more than 90°F (32.3°C). The sea-level rating applies at altitudes up to 1500 ft (457.2 m). At higher altitudes, a correction factor for elevation must be applied. If the atmospheric temperature is higher than 90°F (32.2°C), a temperature correction must be applied.

Table 2 lists both altitude and temperature correction factors. For an 800-hp (596.8-kW) engine at 10,000 ft (3048 m) above sea level and 80°F (26.7°C) intake air, hp output = (sea-level hp) (altitude correction factor), or output = (800)(0.68) = 544 hp (405.8 kW).

TABLE 2 Correction Factors for Altitude and Temperature

Engine altitude		Engine type		Intake temperature		Correction factor
ft	m	Nonsuper-charged	Super-charged	°F	°C	
7,000	2,134	0.780	0.820	90 or less	32.3 or less	1.000
8,000	2,438	0.745	0.790	95	35	0.986
9,000	2,743	0.712	0.765	100	37.8	0.974
10,000	3,048	0.680	0.740	105	40.6	0.962
12,000	3,658	0.612	0.685	110	43.3	0.950
				115	46.1	0.937
				120	48.9	0.925
				125	51.7	0.913
				130	54.4	0.900

2. *Compute the engine output at the elevated temperature*

When the intake air is at a temperature greater than 90°F (32.3°C), a temperature correction factor must be applied. Then output = (sea-level hp)(altitude correction factor)(intake-air-temperature correction factor), or output = (800)(0.68)(0.95) = 516 hp (384.9 kW), with 110°F (43.3°C) intake air.

3. *Compute the output of a supercharged engine*

A different altitude correction factor is used for a supercharged engine, but the same temperature correction factor is applied. Table 2 lists the altitude correction factors for supercharged diesel engines. Thus, for this supercharged engine at 10,000-ft (3048-m) altitude with 80°F (26.7°C) intake air, output = (sea-level hp)(altitude correction factor) = (800)(0.74) = 592 hp (441.6 kW).

At 10,000-ft (3048-m) altitude with 110°F (43.3°C) inlet air, output = (sea-level hp)(altitude correction factor)(temperature correction factor) = (800)(0.74)(0.95) = 563 hp (420.1 kW).

Related Calculations: Use the same procedure for gasoline, gas, oil, and liquefied-petroleum gas engines. Where altitude correction factors are not available for the type of engine being used, other than a diesel, multiply the engine sea-level brake horsepower by the ratio of the altitude-level atmospheric pressure to the atmospheric pressure at sea level. Table 3 lists the atmospheric pressure at various altitudes.

An engine located below sea level can theoretically develop more power than at sea level because the intake air is denser. However, the greater potential output is generally ignored in engine-selection calculations.

TABLE 3 Atmospheric Pressure at Various Altitudes

Altitude		Pressure	
ft	m	inHg	mm
Sea Level		29.92	759.97
4,000	1,219	25.84	656.3
5,000	1,524	24.89	632.2
6,000	1,829	23.98	609.1
8,000	2,438	22.22	564.4
10,000	3,048	20.58	522.7
12,000	3,658	19.03	483.4

Note: A 500- to 1500-ft altitude is considered equivalent to sea level by the Diesel Engine Manufacturers Association if the atmospheric pressure is not less than 28.25 inHg (717.6 mmHg).

INDICATOR USE ON INTERNAL-COMBUSTION ENGINES

An indicator card taken on an internal-combustion engine cylinder has an area of 5.3 in^2 (34.2 cm^2) and a length of 4.95 in (12.7 cm). What is the indicated mean effective pressure in this cylinder? What is the indicated horsepower of this four-cycle engine if it has eight 6-in (15.6-cm) diameter cylinders, an 18-in (45.7-cm) stroke, and operates at 300 r/min? The indicator spring scale is 100 lb/in (1.77 kg/mm).

Calculation Procedure:

1. *Compute the indicated mean effective pressure*

For any indicator card, $imep$ = (card area, in^2) (indicator spring scale, lb)/(length of indicator card, in), where $imep$ = indicated mean effective pressure, lb/in^2. Thus, for this engine, $imep$ = (5.3)(100)/4.95 = 107 lb/in^2 (737.7 kPa).

2. *Compute the indicated horsepower*

For any reciprocating internal-combustion engine, $ihp = (imep)LAn/33{,}000$, where ihp = indicated horsepower per cylinder; L = piston stroke length, ft; A = piston area, in^2; n = number of cycles/min. Thus, for this four-cycle engine where n = 0.5 r/min, $ihp = (107)(18/12)(6)^2(\pi/4)(300/2)/33{,}000$ = 20.6 ihp (15.4 kW) per cylinder. Since the engine has eight cylinders, total ihp = (8 cylinders)(20.6 ihp per cylinder) = 164.8 ihp (122.9 kW).

Related Calculations: Use this procedure for any reciprocating internal-combustion engine using diesel oil, gasoline, kerosene, natural gas, liquefied-petroleum gas, or similar fuel.

ENGINE PISTON SPEED, TORQUE, DISPLACEMENT, AND COMPRESSION RATIO

What is the piston speed of an 18-in (45.7-cm) stroke 300 = r/min engine? How much torque will this engine deliver when its output is 800 hp (596.8 kW)? What are the displacement per cylinder and the total displacement if the engine has eight 12-in (30.5-cm) diameter cylinders? Determine the engine compression ratio if the volume of the combustion chamber is 9 percent of the piston displacement.

Calculation Procedure:

1. *Compute the engine piston speed*

For any reciprocating internal-combustion engine, piston speed = $fpm = 2L(rpm)$, where L = piston stroke length, ft; rpm = crankshaft rotative speed, r/min. Thus, for this engine, piston speed = 2(18/12)(300) = 9000 ft/min (2743.2 m/min).

2. *Determine the engine torque*

For any reciprocating internal-combustion engine, $T = 63{,}000(bhp)/rpm$, where T = torque developed, in·lb; bhp = engine brake horsepower output; rpm = crankshaft rotative speed, r/min. Or T = 63,000(800)/300 = 168,000 in·lb (18,981 N·m).

Where a prony brake is used to measure engine torque, apply this relation: $T = (F_b - F_o)r$, where F_b = brake scale force, lb, with engine operating; F_o = brake scale force with engine stopped and brake loose on flywheel; r = brake arm, in = distance from flywheel center to brake knife edge.

3. *Compute the displacement*

The displacement per cylinder d_c in^3 of any reciprocating internal-combustion engine is $d_c = L_iA_i$ where L_i = piston stroke, in; A = piston head area, in^2. For this engine, $d_c = (18)(12)^2(\pi/4)$ = 2035 in^3 (33,348 cm^3) per cylinder.

The total displacement of this eight-cylinder engine is therefore (8 cylinders)(2035 in^3 per cylinder) = 16,280 in^3 (266,781 cm^3).

4. *Compute the compression ratio*

For a reciprocating internal-combustion engine, the compression ratio $r_c = V_b/V_a$, where V_b = cylinder volume at the start of the compression stroke, in^3 or ft^3; V_a = combustion-space volume at the end of the compression stroke, in^3 or ft^3. When this relation is used, both volumes must be expressed in the same units.

In this engine, V_b = 2035 in^3 (33,348 cm^3); V_a = (0.09)(2035) = 183.15 in^3. Then r_c = 2035/183.15 = 11.1:1.

Related Calculations: Use these procedures for any reciprocating internal-combustion engine, regardless of the fuel burned.

INTERNAL-COMBUSTION ENGINE COOLING-WATER REQUIREMENTS

A 1000-bhp (746-kW) diesel engine has a specific fuel consumption of 0.360 lb/(bhp·h) (0.22 kg/kWh). Determine the cooling-water flow required if the higher heating value of the fuel is 10,350 Btu/lb (24,074 kJ/kg). The net heat rejection rates of various parts of the engine are, in percent: jacket water, 11.5; turbocharger, 2.0; lube oil, 3.8; aftercooling, 4.0; exhaust, 34.7; radiation, 7.5; How much 30 lb/in^2 (abs) (206.8 kPa) steam can be generated by the exhaust gas if this is a four-cycle engine? The engine operates at sea level.

Calculation Procedure:

1. *Compute the engine heat balance*

Determine the amount of heat used to generate 1 bhp·h (0.75 kWh) from: heat rate, Btu/(bhp·h) = (*sfc*)(HHV), where *sfc* = specific fuel consumption, lb/(bhp·h); HHV = higher heating value of fuel, Btu/lb. Or, heat rate = (0.36)(19,350) = 6967 Btu/(bhp·h) (2737.3 W/kWh).

Compute the heat balance of the engine by taking the product of the respective heat rejection percentages and the heat rate as follows:

			Btu/(bhp·h)	W/kWh
Jacket water	(0.115)(6967)	=	800	314.3
Turbocharger	(0.020)(6967)	=	139	54.6
Lube oil	(0.038)(6967)	=	264	103.7
Aftercooling	(0.040)(6967)	=	278	109.2
Exhaust	(0.347)(6967)	=	2420	880.1
Radiation	(0.075)(6967)	=	521	204.7
Total heat loss		=	4422	1666.6

Then the power output = 6967 − 4422 = 2545 Btu/(bhp·h) (999.9 W/kWh), or 2545/6967 = 0.365, or 36.5 percent. Note that the sum of the heat losses and power generated, expressed in percent, is 100.0.

2. *Compute the jacket cooling-water flow rate*

The jacket water cools the jackets and the turbocharger. Hence, the heat that must be absorbed by the jacket water is 800 + 139 = 939 Btu/(bhp·h) (369 W/kWh), using the heat rejection quantities computed in step 1. When the engine is developing its full rated output of 1000 bhp (746 kW), the jacket water must absorb [939 Btu/(bhp·h)](1000 bhp) = 939,000 Btu/h (275,221 W).

Apply a safety factor to allow for scaling of the heat-transfer surfaces and other unforeseen difficulties. Most designers use a 10 percent safety factor. Applying this value of the safety factor for this engine, we see the total jacket-water heat load = 939,000 + (0.10)(939,000) = 1,032,900 Btu/h (302.5 kW).

Find the required jacket-water flow from $G = H/500\Delta t$, where G = jacket-water flow, gal/min; H = heat absorbed by jacket water, Btu/h; Δt = temperature rise of the water during passage through the jackets, °F. The usual temperature rise of the jacket water during passage through a diesel engine is 10 to 20°F (5.6 to 11.1°C). Using 10°F for this engine, we find G = 1,032,900/[(500)(10)] = 206.58 gal/min (13.03 L/s), say 207 gal/min (13.06 L/s).

3. *Determine the water quantity for radiator cooling*

In the usual radiator cooling system for large engines, a portion of the cooling water is passed through a horizontal or vertical radiator. The remaining water is recirculated, after being tempered by the cooled water. Thus, the radiator must dissipate the jacket, turbocharger, and lube-oil cooler heat.

The lube oil gives off 264 Btu/(bhp·h) (103.8 W/kWh). With a 10 percent safety factor, the total heat flow is 264 + (0.10)(264) = 290.4 Btu/(bhp·h) (114.1 W/kWh). At the rated output of 1000 bhp (746 kW), the lube-oil heat load = [290.4 Btu/(bhp·h)](1000 bhp) = 290,400 Btu/h (85.1 kW). Hence, the total heat load on the radiator = jacket + lube-oil heat load = 1,032,900 + 290,400 = 1,323,300 Btu/h (387.8 kW).

Radiators (also called fan coolers) serving large internal-combustion engines are usually rated for a 35°F (19.4°C) temperature reduction of the water. To remove 1,323,300 Btu/h (387.8 kW) with a 35°F (19.4°C) temperature decrease will require a flow of $G = H/(500\Delta t) = 1{,}323{,}300/[(500)(35)] = 76.1$ gal/min (4.8 L/s).

4. *Determine the aftercooler cooling-water quantity*

The aftercooler must dissipate 278 Btu/(bhp·h) (109.2 W/kWh). At an output of 1000 bhp (746 kW), the heat load = [278 Btu/(bhp·h)](1000 bhp) = 278,000 Btu/h (81.5 kW). In general, designers do not use a factor of safety for the aftercooler because there is less chance of fouling or other difficulties.

With a 5°F (2.8°C) temperature rise of the cooling water during passage through the aftercooler, the quantity of water required $G = H/(500\Delta t) = 278{,}000/[(500)(5)] = 111$ gal/min (7.0 L/s).

5. *Compute the quantity of steam generated by the exhaust*

Find the heat available in the exhaust by using $H_e = Wc\Delta t_e$, where H_e = heat available in the exhaust, Btu/h; W = exhaust-gas flow, lb/h; c = specific heat of the exhaust gas = 0.252 Btu/(lb·°F) (2.5 kJ/kg); Δt_e = exhaust-gas temperature at the boiler inlet, °F − exhaust-gas temperature at the boiler outlet,°F.

The exhaust-gas flow from a four-cycle turbocharged diesel is about 12.5 lb/(bhp·h) (7.5 kg/kWh). At full load this engine will exhaust [12.5 lb/(bhp·h)](1000 bhp) = 12,500 lb/h (5625 kg/h).

The temperature of the exhaust gas will be about 750°F (399°C) at the boiler inlet, whereas the temperature at the boiler outlet is generally held at 75°F (41.7°C) higher than the steam temperature to prevent condensation of the exhaust gas. Steam at 30 lb/in^2 (abs) (206.8 kPa) has a temperature of 250.33°F (121.3°C). Thus, the exhaust-gas outlet temperature from the boiler will be 250.33 + 75 = 325.33°F (162.9°C), say 325°F (162.8°C). Then $H_e = (12{,}500)(0.252)(750 - 325) = 1{,}375{,}000$ Btu/h (403.0 kW).

At 30 lb/in^2 (abs) (206.8 kPa),the enthalpy of vaporization of steam is 945.3 Btu/lb (2198.9 kJ/kg), found in the steam tables. Thus, the exhaust heat can generate 1,375,000/945.3 = 1415 lb/h (636.8 kg/h) if the boiler is 100 percent efficient. With a boiler efficiency of 85 percent, the steam generated = (1415 lb/h)(0.85) = 1220 lb/h (549.0 kg/h), or (1220 lb/h)/1000 bhp = 1.22 lb/(bhp·h) (0.74 kg/kWh).

Related Calculations: Use this procedure for any reciprocating internal-combustion engine burning gasoline, kerosene, natural gas, liquified-petroleum gas, or similar fuel. Figure 1 shows typical arrangements for a number of internal-combustion engine cooling systems.

When ethylene glycol or another antifreeze solution is used in the cooling system, alter the denominator of the flow equation to reflect the change in specific gravity and specific heat of the antifreeze solution, as compared with water. Thus, with a mixture of 50 percent glycol and 50 percent water, the flow equation in step 2 becomes $G = H/(436\Delta t)$. With other solutions, the numerical factor in the denominator will change. This factor = (weight of liquid, lb/gal)(60 min/h), and the factor converts a flow rate of lb/h to gal/min when divided into the lb/h flow rate. Slant diagrams, Fig. 2, are often useful for heat-exchanger analysis.

Two-cycle engines may have a larger exhaust-gas flow than four-cycle engines because of the scavenging air. However, the exhaust temperature will usually be 50 to 100°F (27.7 to 55.6°C) lower, reducing the quantity of steam generated.

Where a dry exhaust manifold is used on an engine, the heat rejection to the cooling system is reduced by about 7.5 percent. Heat rejected to the aftercooler cooling water is about 3.5 percent

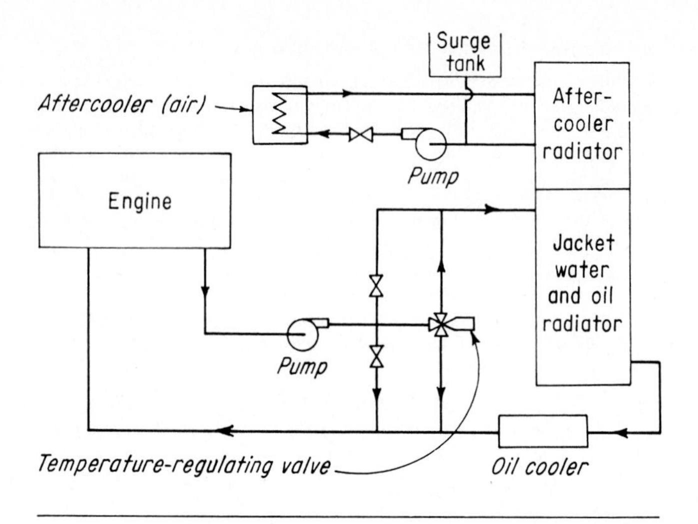

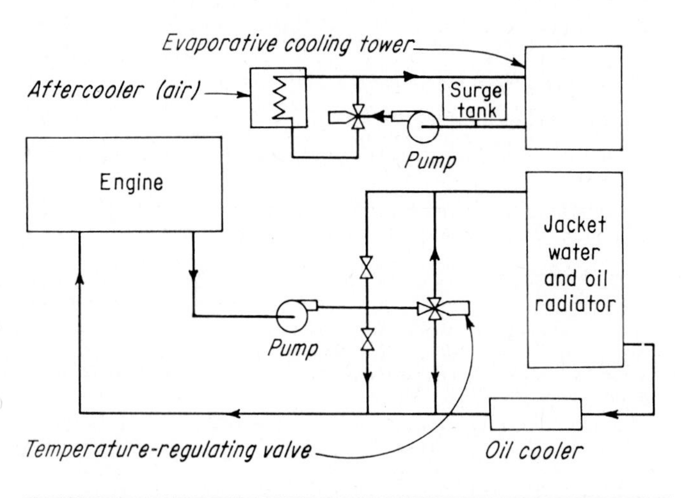

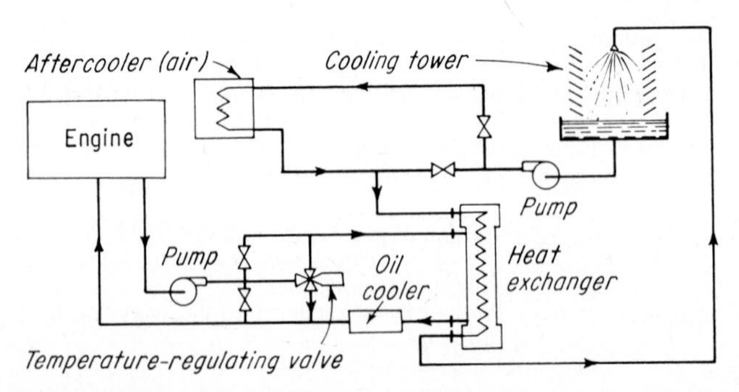

FIG. 1 Internal-combustion engine cooling systems: (*a*) radiator type; (*b*) evaporating cooling tower; (*c*) cooling tower. (*Power*.)

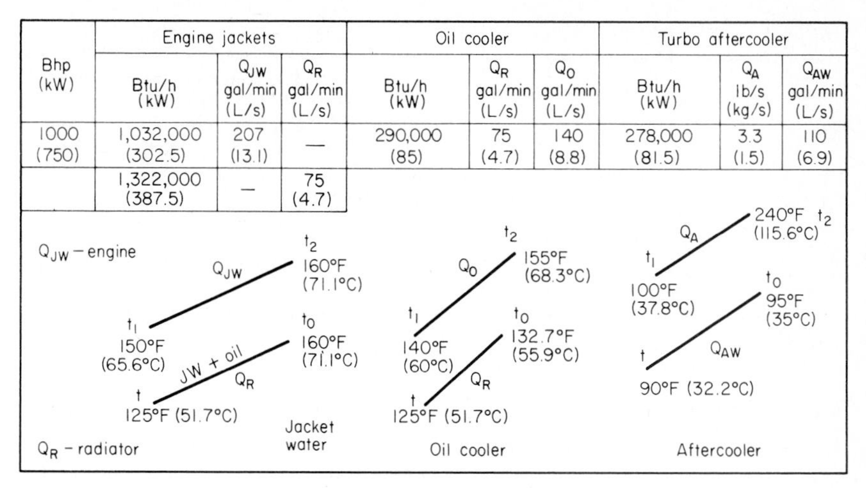

Bhp (kW)	Engine jackets			Oil cooler			Turbo aftercooler		
	Btu/h (kW)	Q_{JW} gal/min (L/s)	Q_R gal/min (L/s)	Btu/h (kW)	Q_R gal/min (L/s)	Q_O gal/min (L/s)	Btu/h (kW)	Q_A lb/s (kg/s)	Q_{AW} gal/min (L/s)
1000 (750)	1,032,000 (302.5)	207 (13.1)	—	290,000 (85)	75 (4.7)	140 (8.8)	278,000 (81.5)	3.3 (1.5)	110 (6.9)
	1,322,000 (387.5)	—	75 (4.7)						

FIG. 2 Slant diagrams for internal-combustion engine heat exhangers. *(Power.)*

of the total heat input to the engine. About 2.5 percent of the total heat input to the engine is rejected by the turbocharger jacket.

The jacket cooling water absorbs 11 to 14 percent of the total heat supplied. From 3 to 6 percent of the total heat supplied to the engine is rejected in the oil cooler.

The total heat supplied to an engine = (engine output, bhp)[heat rate, Btu/(bhp·h)]. A jacket-water flow rate of 0.25 to 0.60 gal/(min·bhp) (0.02 to 0.05 kg/kW) is usually recommended. The normal jacket-water temperature rise is 10°F (5.6°C); with a jacket-water outlet temperature of 180°F (82.2°C) or higher, the temperature rise of the jacket water is usually held to 7°F (3.9°C) or less.

To keep the cooling-water system pressure loss within reasonable limits, some designers recommend a pipe velocity equal to the nominal pipe size used in the system, or 2 ft/s for 2-in pipe (0.6 m/s for 50.8-mm); 3 ft/s for 3-in pipe (0.9 m/s for 76.2-mm); etc. The maximum recommended velocity is 10 ft/s for 10 in (3.0 m/s for 254 mm) and larger pipes. Compute the actual pipe diameter from $d = (G/2.5v)^{0.5}$, where G = cooling-water flow, gal/min; v = water velocity, ft/s.

Air needed for a four-cycle high-output turbocharged diesel engine is about 3.5 ft^3/(min·bhp) (0.13 m^3/kW); 4.5 ft^3/(min·bhp) (0.17 m^3/kW) for two-cycle engines. Exhaust-gas flow is about 8.4 ft^3/(min·bhp) (0.32 m^3/kW) for a four-cycle diesel engine; 13 ft^3/(min·bhp) (0.49 m^3/kW) for two-cycle engines. Air velocity in the turbocharger blower piping should not exceed 3300 ft/min (1006 m/min); gas velocity in the exhaust system should not exceed 6000 ft/min (1828 m/min). The exhaust-gas temperature should not be reduced below 275°F (135°C), to prevent condensation.

The method presented here is the work of W. M. Kauffman, reported in *Power*.

DESIGN OF A VENT SYSTEM FOR AN ENGINE ROOM

A radiator-cooled 60-kW internal-combustion engine generating set operates in an area where the maximum summer ambient temperature of the inlet air is 100°F (37.8°C). How much air does this engine need for combustion and for the radiator? What is the maximum permissible temperature rise of the room air? How much heat is radiated by the engine-alternator set if the exhaust pipe is 25 ft (7.6 m) long? What capacity exhaust fan is needed for this engine room if the engine room has two windows with an area of 30 ft^2 (2.8 m^2) each, and the average height between the air inlet and outlet is 5 ft (1.5 m)? Determine the rate of heat dissipation by the windows. The engine is located at sea level.

TABLE 4 Total Air Volume Needs°

Set kW	ft^3/min (m^3/min) for combustion	ft^3/min (m^3/min) for radiator	Maximum room temperature rise: Maximum ambient temperature of inlet air, °F (°C)	Maximum room temperature rise: Room air rise, °F (°C)
20	130 (3.7)	3000 (84.9)	90 (32.2)	20 (11.1)
30	195 (5.5)	5000 (141.6)	95–105 (35–40.6)	15 (8.3)
40	260 (7.4)	5500 (155.7)	110–120 (43.3–48.9)	10 (5.6)
60	390 (1.0)	6000 (169.9)		

° *Power.*

Calculation Procedure:

1. *Determine engine air-volume needs*

Table 4 shows typical air-volume needs for internal-combustion engines installed indoors. Thus, a 60-kW set requires 390 ft^3/min (11.0 m^3/min) for combustion and 6000 ft^3/min (169.9 m^3/min) for the radiator. Note that in the smaller ratings, the combustion air needed is 6.5 $ft^3/(min \cdot kW)$(0.18 m^3/kW), and the radiator air requirement is 150 $ft^3/(min \cdot kW)$(4.2 m^3/kW).

2. *Determine maximum permissible air temperature rise*

Table 4 also shows that with an ambient temperature of 95 to 105°F (35 to 40.6°C), the maximum permissible room temperature rise is 15°F (8.3°C). When you determine this value, be certain to use the highest inlet air temperature expected in the engine locality.

3. *Determine the heat radiated by the engine*

Table 5 shows the heat radiated by typical internal-combustion engine generating sets. Thus, a 60-kW radiator-and fan-cooled set radiates 2625 Btu/min (12.8 W) when the engine is fitted with a 25-ft (7.6-m) long exhaust pipe and a silencer.

4. *Compute the airflow produced by the windows*

The two windows can be used to ventilate the engine room. One window will serve as the air inlet; the other, as the air outlet. The area of the air outlet must at least equal the air-inlet area. Airflow will be produced by the stack effect resulting from the temperature difference between the inlet and outlet air.

The airflow C ft^3/min resulting from the stack effect is $C = 9.4A(h\Delta t_a)^{0.5}$, where A = free area of the air inlet, ft^2; h = height from the middle of the air-inlet opening to the middle of the

TABLE 5 Heat Radiated from Typical Internal-Combustion Units, Btu/min (W)°

	Cooling by radiator and fan		Cooling by radiator, fan, and city water	
Alternator, kW	40	60	40	60
Engine-alternator set, silencer, and 25 ft (7.6 m) of exhaust pipe, Btu/min (W)	1830 (8.94)	2625 (12.8)	1701 (8.3)	2500 (12.2)
Exhaust pipe beyond silencer:				
Length 5 ft (1.5 m)	24 (0.12)	35 (0.17)	20 (0.10)	22 (0.11)
Length 10 ft (3.0 m)	45 (0.22)	65 (0.32)	39 (0.19)	40 (0.20)
Length 15 ft (4.6 m)	65 (0.32)	89 (0.44)	57 (0.38)	55 (0.27)

° *Power.*

TABLE 6 Range of Discharge Temperature°

Room fan discharge temperature range			Wind to water gage			
			Wind velocity		Inlet pressure water gage	
°F	°C	K	mph	km/h	in	mm
80–89	26.7–31.7	57	60	96.5	1.75	44.5
90–99	32.3–37.2	58	30	48.3	0.43	10.9
100–110	37.8–43.3	59				
111–120	43.9–48.9	60				
121–130	49.4–54.4	61				

° *Power.*

air-outlet opening, ft; Δt_a = difference between the average indoor air temperature at point H and the temperature of the incoming air, °F. In this plant, the maximum permissible air temperature rise is 15°F (8.3°C), from step 2. With a 100°F (37.8°C) outdoor temperature, the maximum indoor temperature would be 100 + 15 = 115°F (46.1°C). Assume that the difference between the temperature of the incoming and outgoing air is 15°F (8.3°C). Then $C = 9.4(30)(5 \times 15)^{0.5}$ = 2445 ft^3/min (69.2 m^3/min).

5. *Compute the cooling airflow required*

This 60-kW internal-combustion engine generating set radiates 2625 Btu/min (12.8 W), step 3. Compute the cooling airflow required from $C = HK/\Delta t_a$, where C = cooling airflow required, ft^3/min; H = heat radiated by the engine, Btu/min; K = constant from Table 6; other symbols as before. Thus, for this engine with a fan discharge temperature of 111 to 120°F (43.9 to 48.9°C), Table 6, $K = 60$; Δt_a = 15°F (8.3°C) from step 4. Then $C = (2625)(60)/15 = 10{,}500$ ft^3/min (297.3 m^3/min).

The windows provide 2445 ft^3/min (69.2 m^3/min), step 4, and the engine radiator gives 6000 ft^3/min (169.9 m^3/min), step 1, or a total of 2445 + 6000 = 8445 ft^3/min (239.1 m^3/min). Thus, 10,500 − 8445 = 2055 ft^3/min (58.2 m^3/min) must be removed from the room. The usual method employed to remove the air is an exhaust fan. An exhaust fan with a capacity of 2100 ft^3/min (59.5 m^3/min) would be suitable for this engine room.

Related Calculations: Use this procedure for engines burning any type of fuel—diesel, gasoline, kerosene, or gas—in any type of enclosed room at sea level or elevations up to 1000 ft (304.8 m). Where windows or the fan outlet are fitted with louvers, screens, or intake filters, be certain to compute the net free area of the opening. When the radiator fan requires more air than is needed for cooling the room, an exhaust fan is unnecessary.

Be certain to select an exhaust fan with a sufficient discharge pressure to overcome the resistance of exhaust ducts and outlet louvers, if used. A propeller fan is usually chosen for exhaust

TABLE 7 Air Density at Various Elevations°

Elevation above sea level		Multiplying factor, A	Approximate air density percent compared with sea level for same temperature
ft	m		
4,000	1,219	1.158	86.4
5,000	1,524	1.202	83.2
6,000	1,829	1.247	80.2
7,000	2,134	1.296	77.2
10,000	3,048	1.454	68.8

° *Power.*

service. In areas having high wind velocity, an axial-flow fan may be needed to overcome the pressure produced by the wind on the fan outlet.

Table 6 shows the pressure developed by various wind velocities. When the engine is located above sea level, use the multiplying factor in Table 7 to correct the computed air quantities for the lower air density.

An engine radiates 2 to 5 percent of its total heat input. The total heat input = (engine output, bhp) [heat rate, Btu/(bhp·h)]. Provide 12 to 20 air changes per hour for the engine room. The most effective ventilators are power-driven exhaust fans or roof ventilators. Where the heat load is high, 100 air changes per hour may be provided. Auxiliary-equipment rooms require 10 air changes per hour. Windows, louvers, or power-driven fans are used. A four-cycle engine requires 3 to 3.5 ft^3/min of air per bhp (0.11 to 0.13 m^3/kW); a two-cycle engine, 4 to 5 ft^3/(min · bhp) (0.15 to 0.19 m^3/kW).

The method presented here is the work of John P. Callaghan, reported in *Power*.

DESIGN OF A BYPASS COOLING SYSTEM FOR AN ENGINE

The internal-combustion engine in Fig. 3 is rated at 402 hp (300 kW) at 514 r/min and dissipates 3500 Btu/(bhp·h) (1375 W/kW) at full load to the cooling water from the power cylinders and water-cooled exhaust manifold. Determine the required cooling-water flow rate if there is a 10°F (5.6°C) temperature rise during passage of the water through the engine. Size the piping for the cooling system, using the head-loss data in Fig. 4, and the pump characteristic curve, Fig. 5. Choose a surge tank of suitable capacity. Determine the net positive suction head requirements for this engine. The total length of straight piping in the cooling system is 45 ft (13.7 m). The engine is located 500 ft (152.4 m) above sea level.

Calculation Procedure:

1. *Compute the cooling-water quantity required*

The cooling-water quantity required is $G = H/(500\Delta t)$, where G = cooling-water flow, gal/min; H = heat absorbed by the jacket water, Btu/h = (maximum engine hp) [heat dissipated, Btu/(bhp·h)]; Δt = temperature rise of the water during passage through the engine, °F. Thus, for this engine, $G = (402)(3500)/[500(10)] = 281$ gal/min (17.7 L/s).

2. *Choose the cooling-system valve and pipe size*

Obtain the friction head-loss data for the engine, the heat exchanger, and the three-way valve from the manufacturers of the respective items. Most manufacturers have curves or tables available for easy use. Plot the head losses, as shown in Fig. 4, for the engine and heat exchanger.

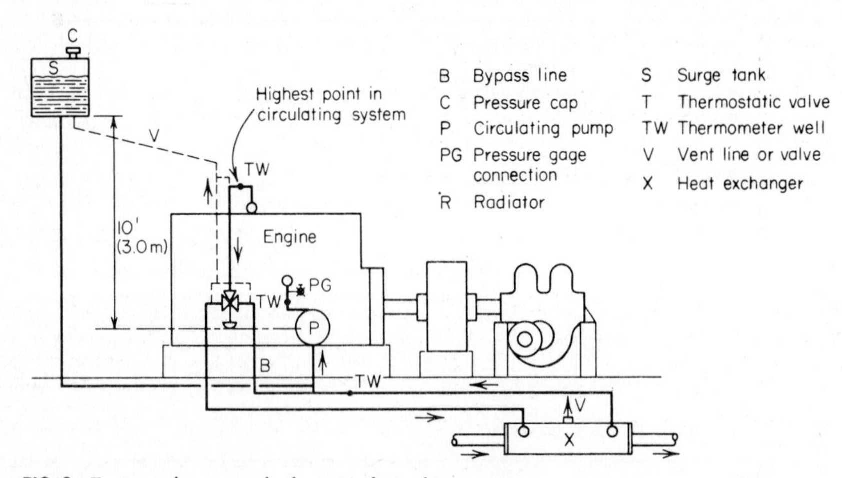

FIG. 3 Engine cooling-system hookup. *(Mechanical Engineering.)*

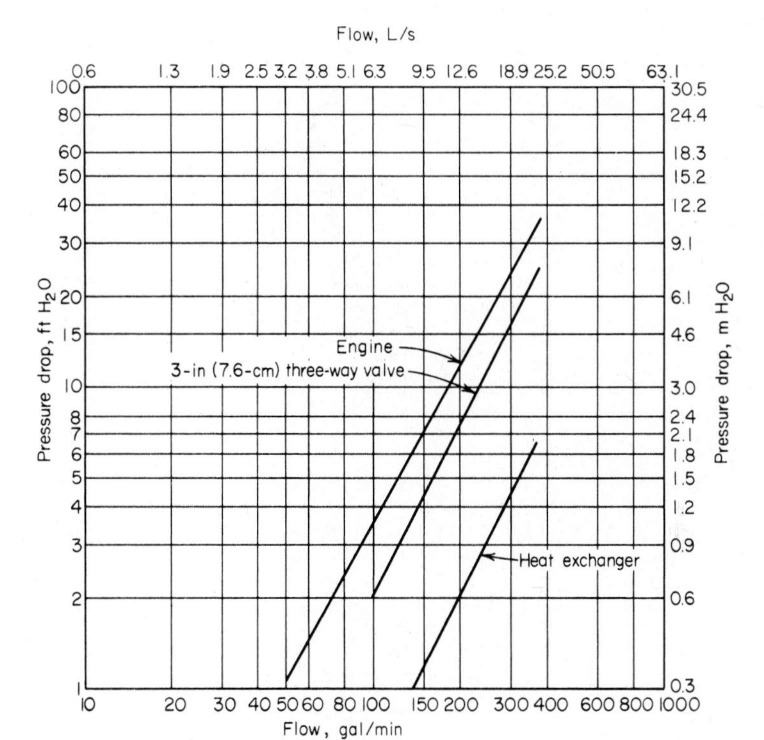

FIG. 4 Head-loss data for engine cooling-system components. *(Mechanical Engineering.)*

Before the three-way valve head loss can be plotted, a valve size must be chosen. Refer to a three-way valve capacity tabulation to determine a suitable valve size to handle a flow of 281 gal/min (17.7 L/s). One such tabulation recommends a 3-in (76.2-mm) valve for a flow of 281 gal/min (17.7 L/s). Obtain the head-loss data for the valve, and plot it as shown in Fig. 4.

Next, assume a size for the cooling-water piping. Experience shows that a water velocity of 300 to 600 ft/min (91.4 to 182.9 m/min) is satisfactory for internal-combustion engine cooling

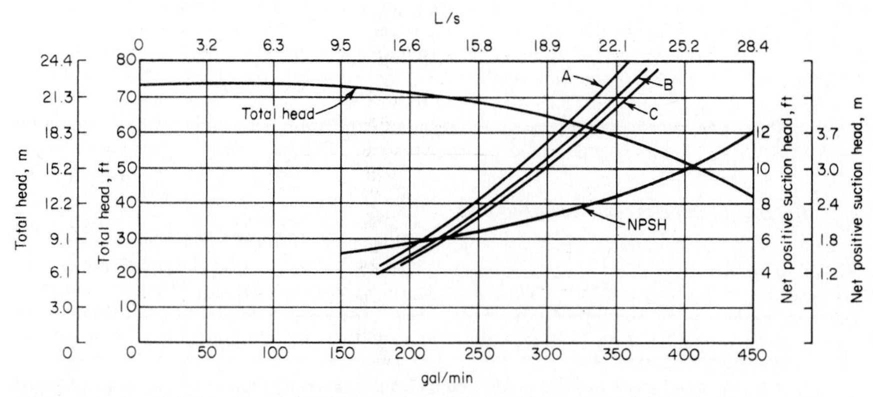

FIG. 5 Pump and system characteristics for engine cooling system. *(Mechanical Engineering.)*

TABLE 8 Sample Calculation for Full Flow through Cooling Circuit*
(Fittings and Piping in Circuit)

Fitting or pipe	Number in circuit	Equivalent length of straight pipe, ft	Equivalent length of straight pipe, m
3-in (76.2-mm) elbow	1	5.5	1.7
3 × 4 (76.2 × 101.6-mm) reducer	4	7.2	2.2
4-in (101.6-mm) elbow	7	50.4	15.4
4-in (101.6-mm) tee	1	23.0	7.0
3-in (76.2-mm) pipe	. . .	0.67	0.2
4-in (101.6-mm) pipe	. . .	45.0	13.7
Total equivalent length of pipe:			
3-in (76.2-mm) pipe, standard weight	. . .	13.37	4.1
4-in (101.6-mm) pipe, standard weight	. . .	118.4	36.1

Head loss calculation: Calculation for a flow rate of 300 gal/min (18.9 L/s) through circuit:

Using the Hazen-Williams friction-loss equation with a C factor of 130 (surface roughness constant), with 300 gal/min (18.9 L/s) flowing through the pipe, the head loss per 100 ft (30.5 m) of pipe is 21.1 ft (6.4 m) and 5.64 ft (1.1 m) for the 3-in (76.2-mm) and 4-in (101.6-mm) pipes, respectively. Thus head loss in piping is†

$$3\text{ in } \frac{21.1}{100} \times 13.37 = 2.83\text{ ft } (0.86\text{ m})$$

$$4\text{ in } \frac{5.64}{100} \times 118.4 = 6.68\text{ ft } (2.0\text{ m})$$

From Fig. 5 the head loss is:	ft	m
Through engine	26.00	7.9
Through 3-in (76.2-mm) three-way valve	17.50	5.3
Through heat exchanger	4.6	1.4
Total circuit head loss	57.61	14.6

**Mechanical Engineering.*
†Shaw and Loomis, *Cameron Hydraulic Data Book,* 12th ed., Ingersoll-Rand Company, 1951, p. 27.

systems. Using the Hydraulic Institute's *Pipe Friction Manual* or Cameron's *Hydraulic Data,* enter at 280 gal/min (17.6 L/s), the approximate flow, and choose a pipe size to give a velocity of 400 to 500 ft/min (121.9 to 152.4 m/min), i.e., midway in the recommended range.

Alternatively, compute the approximate pipe diameter from $d = 4.95\,[gpm/\text{velocity, ft/min}]^{0.5}$. With a velocity of 450 ft/min (137.2 m/min), $d = 4.95(281/450)^{0.5} = 3.92$, say 4 in (101.6 mm). The *Pipe Friction Manual* shows that the water velocity will be 7.06 ft/s (2.2 m/s), or 423.6 ft/min (129.1 m/min), in a 4-in (101.6-mm) schedule 40 pipe. This is acceptable. Using a 3½-in (88.9-mm) pipe would increase the cost because the size is not readily available from pipe suppliers. A 3-in (76.2-mm) pipe would give a velocity of 720 ft/min (219.5 m/min), which is too high.

3. *Compute the piping-system head loss*

Examine Fig. 3, which shows the cooling system piping layout. Three flow conditions are possible: (a) all the jacket water passes through the heat exchanger, (b) a portion of the jacket water passes through the heat exchanger, and (c) none of the jacket water passes through the heat exchanger—instead, all the water passes through the bypass circuit. The greatest head loss usually occurs when the largest amount of water passes through the longest circuit (or flow condition a). Compute the head loss for this situation first.

Using the method given in the piping section of this handbook, compute the equivalent length of the cooling-system fitting and piping, as shown in Table 8. Once the equivalent length of the pipe and fittings is known, compute the head loss in the piping system, using the method given in the piping section of this handbook with a Hazen-Williams constant of $C = 130$ and a rounded-off flow rate of 300 gal/min (18.9 L/s). Summarize the results as shown in Table 8.

The total head loss is produced by the water flow through the piping, fittings, engine, three-way valve, and heat exchanger. Find the head loss for the last components in Fig. 4 for a flow of 300 gal/min (18.9 L/s). List the losses in Table 8, and find the sum of all the losses. Thus, the total circuit head loss is 57.61 ft (17.6 m) of water.

Compute the head loss for 0, 0.2, 0.4, 0.6, and 0.8 load on the engine, using the same procedure as in steps 1, 2, and 3 above. Plot on the pump characteristic curve, Fig. 5, the system head loss for each load. Draw a curve *A* through the points obtained, Fig. 5.

Compute the system head loss for condition *b* with half the jacket water [150 gal/min (9.5 L/s)] passing through the heat exchanger and half [150 gal/min (9.5 L/s)] through the bypass circuit. Make the same calculation for 0, 0.2, 0.4, 0.6, and 0.8 load on the engine. Plot the result as curve *B*, Fig. 5.

Perform a similar calculation for condition *c*—full flow through the bypass circuit. Plot the results as curve *C*, Fig. 5.

4. *Compute the actual cooling-water flow rate*

Find the points of intersection of the pump total-head curve and the three system head-loss curves A, B, and C, Fig. 5. These intersections occur at 314, 325, and 330 gal/min (19.8, 20.5, and 20.8 L/s), respectively.

The initial design assumed a 10°F (5.6°C) temperature rise through the engine with a water flow rate of 281 gal/min (17.7 L/s). Rearranging the equation in step 1 gives $\Delta t = H/(400G)$. Substituting the flow rate for condition *a* gives an actual temperature rise of $\Delta t = (402)(3500)/[(500)(314)] = 8.97$°F (4.98°C). If a 180°F (82.2°C) rated thermostatic element is used in the three-way valve, holding the outlet temperature t_o to 180°F (82.2°C), the inlet temperature t_i will be $\Delta t = t_o - t_i = 8.97$; $180 - t_i = 8.97$; $t_i = 171.03$°F (77.2°C).

5. *Determine the required surge-tank capacity*

The surge tank in a cooling system provides storage space for the increase in volume of the coolant caused by thermal expansion. Compute this expansion from $E = 62.4g\Delta V$, where E = expansion, gal (L); g = number of gallons required to fill the cooling system; ΔV = specific volume, ft^3/lb (m^3/kg) of the coolant at the operating temperature − specific volume of the coolant, ft^3/lb (m^3/kg) at the filling temperature.

The cooling system for this engine must have a total capacity of 281 gal (1064 L), step 1. Round this to 300 gal (1136 L) for design purposes.The system operating temperature is 180°F (82.2°C), and the filling temperature is usually 60°F (15.6°C). Using the steam tables to find the specific volume of the water at these temperatures, we get $E = 62.4(300)(0.01651 - 0.01604) = 8.8$ gal (33.3 L).

Usual design practice is to provide two to three times the required expansion volume. Thus, a 25-gal (94.6-L) tank (nearly three times the required capacity) would be chosen. The extra volume provides for excess cooling water that might be needed to make up water lost through minor leaks in the system.

Locate the surge tank so that it is the highest point in the cooling system. Some engineers recommend that the bottom of the surge tank be at least 10 ft (3 m) above the pump centerline and connected as close as possible to the pump intake. A 1½- or 2-in (38.1- or 50.8-mm) pipe is usually large enough for connecting the surge tank to the system. The line should be sized so that the head loss of the vented fluid flowing back to the pump suction will be negligible.

6. *Determine the pump net positive suction head*

The pump characteristic curve, Fig. 5, shows the net positive suction head (NPSH) required by this pump. As the pump discharge rate increases, so does the NPSH. This is typical of a centrifugal pump.

The greatest flow, 330 gal/min (20.8 L/s), occurs in this system when all the coolant is diverted through the bypass circuit, Figs. 4 and 5. At a 330-gal/min (20.8-L/s) flow rate through the system, the required NPSH for this pump is 8 ft (2.4 m), Fig. 5. This value is found at the intersection of the 330-gal/min (20.8-L/s) ordinate and the NPSH curve.

Compute the existing NPSH, ft (m), from $\text{NPSH} = H_s - H_f + 2.31(P_s - P_v)/s$, where H_s = height of minimum surge-tank liquid level above the pump centerline, ft (m); H_f = friction loss in the suction line from the surge-tank connection to the pump inlet flange, ft (m) of liquid; P_s = pressure in surge tank, or atmospheric pressure at the elevation of the installation, lb/in^2 (abs) (kPa); P_v = vapor pressure of the coolant at the pumping temperature, lb/in^2 (abs) (kPa); s = specific gravity of the coolant at the pumping temperature.

7. *Determine the operating temperature with a closed surge tank*

A pressure cap on the surge tank, or a radiator, will permit operation at temperatures above the atmospheric boiling point of the coolant. At a 500-ft (152.4-m) elevation, water boils at 210°F (98.9°C). Thus, without a closed surge tank fitted with a pressure cap, the maximum operating temperature of a water-cooled system would be about 200°F (93.3°C).

If a 7-lb/in^2 (gage) (48.3-kPa) pressure cap were used at the 500-ft (152.4-m) elevation, then the pressure in the vapor space of the surge tank could rise to P_s = 14.4 + 7.0 = 21.4 lb/in^2 (abs) (147.5 kPa). The steam tables show that water at this pressure boils at 232°F (111.1°C). Checking the NPSH at this pressure shows that NPSH = (10 − 1.02) + 2.31(21.4 − 21.4)/0.0954 = 8.98 ft (2.7 m). This is close to the required 8-ft (2.4-m) head. However, the engine could be safely operated at a slightly lower temperature, say 225°F (107.2°C).

8. *Compute the pressure at the pump suction flange*

The pressure at the pump suction flange P lb/in^2 (gage) = $0.433s(H_s - H_f)$ = (0.433)(0.974)(10.00 − 1.02) = 3.79 lb/in^2 (gage) (26.1 kPa).

A positive pressure at the pump suction is needed to prevent the entry of air along the shaft. To further ensure against air entry, a mechanical seal can be used on the pump shaft in place of packing.

Related Calculations: Use this general procedure in designing the cooling system for any type of reciprocating internal-combustion engine—gasoline, diesel, gas, etc. Where a coolant other than water is used, follow the same procedure but change the value of the constant in the denominator of the equation of step 1. Thus, for a mixture of 50 percent glycol and 50 percent water, the constant = 436, instead of 500.

The method presented here is the work of Duane E. Marquis, reported in *Mechanical Engineering*.

HOT-WATER HEAT-RECOVERY SYSTEM ANALYSIS

An internal-combustion engine fitted with a heat-recovery silencer and a jacket-water cooler is rated at 1000 bhp (746 kW). It exhausts 13.0 lb/(bhp·h) [5.9 kg/(bhp·h)] of exhaust gas at 700°F (371.1°C). To what temperature can hot water be heated when 500 gal/min (31.5 L/s) of jacket water is circulated through the hookup in Fig. 6 and 100 gal/min (6.3 L/s) of 60°F (15.6°C) water is heated? The jacket water enters the engine at 170°F (76.7°C) and leaves at 180°F (82.2°C).

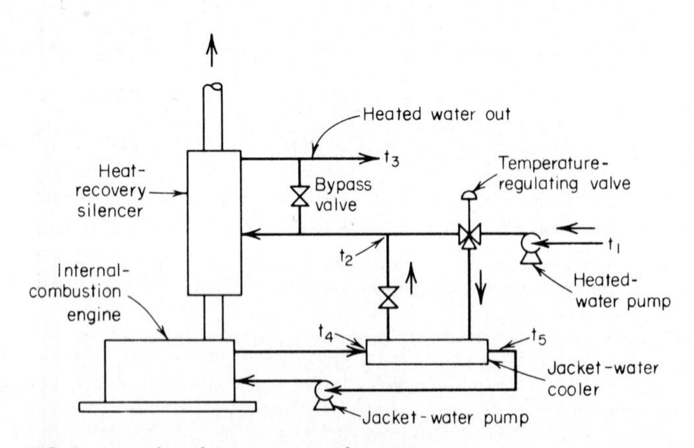

FIG. 6 Internal-combustion engine cooling system.

Calculation Procedure:

1. *Compute the exhaust heat recovered*

Find the exhaust-heat recovered from $H_e = Wc\Delta t_e$, where the symbols are the same as in the previous calculation procedures. Since the final temperature of the exhaust gas is not given, a value must be assumed. Temperatures below 275°F (135°C) are undesirable because condensation of corrosive vapors in the silencer may occur. Assume that the exhaust-gas outlet temperature from the heat-recovery silencer is 300°F (148.9°C). Then $H_e = (1000)(13)(0.252)(700 - 300) =$ 1,310,000 Btu/h (383.9 kW).

2. *Compute the heated-water outlet temperature from the cooler*

Using the temperature notation in Fig. 6, we see that the heated-water outlet temperature from the jacket-water cooler is $t_z = (w_z/w_1)(t_4 - t_5) + t_1$, where w_1 = heated-water flow, lb/h; w_z = jacket-water flow, lb/h; the other symbols are indicated in Fig. 6. To convert gal/min of water flow to lb/h, multiply by 500. Thus, w_1 = (100 gal/min)(500) = 50,000 lb/h (22,500 kg/h), and w_z = (500 gal/min)(500) = 250,000 lb/h (112,500 kg/h). Then t_z = (250,000/50,000)(180 − 170) + 60 = 110°F (43.3°C).

3. *Compute the heated-water outlet temperature from the silencer*

The silencer outlet temperature $t_3 = H_e/w_1 + t_z$, or t_3 = 1,310,000/50,000 + 110 = 136.2°F (57.9°C).

Related Calculations: Use this method for any type of engine—diesel, gasoline, or gas—burning any type of fuel. Where desired, a simple heat balance can be set up between the heat-releasing and heat-absorbing sides of the system instead of using the equations given here. However, the equations are faster and more direct.

DIESEL FUEL STORAGE CAPACITY AND COST

A diesel power plant will have six 1000-hp (746-kW) engines and three 600-hp (448-kW) engines. The annual load factor is 85 percent and is nearly uniform throughout the year. What capacity day tanks should be used for these engines? If fuel is delivered every 7 days, what storage capacity is required? Two fuel supplies are available; a 24° API fuel at $0.0825 per gallon ($0.022 per liter) and a 28° API fuel at $0.0910 per gallon ($0.024 per liter). Which is the better buy?

Calculation Procedure:

1. *Compute the engine fuel consumption*

Assume, or obtain from the engine manufacturer, the specific fuel consumption of the engine. Typical modern diesel engines have a full-load heat rate of 6900 to 7500 Btu/(bhp·h) (2711 to 3375 W/kWh), or about 0.35 lb/(bhp·h) of fuel (0.21 kg/kWh). Using this value of fuel consumption for the nine engines in this plant, we see the hourly fuel consumption at 85 percent load factor will be (6 engines)(1000 hp)(0.35)(0.85) + (3 engines)(600 hp)(0.35)(0.85) = 2320 lb/h (1044 kg/h).

Convert this consumption rate to gal/h by finding the specific gravity of the diesel oil. The specific gravity s = 141.5/(131.5 + °API). For the 24° API oil, s = 141.5/(131.5 + 24) = 0.910. Since water at 60°F (15.6°C) weighs 8.33 lb/gal (3.75 kg/L), the weight of this oil is (0.910)(8.33) = 7.578 lb/gal (3.41 kg/L). For the 28° API oil, s = 141.5/(131.5 + 28) = 0.887, and the weight of this oil is (0.887)(8.33) = 7.387 lb/gal (3.32 kg/L). Using the lighter oil, since this will give a larger gal/h consumption, we get the fuel rate = (2320 lb/h)/(7.387 lb/gal) = 315 gal/h (1192 L/h).

The daily fuel consumption is then (24 h/day)(315 gal/h) = 7550 gal/day (28,577 L/day). In 7 days the engines will use (7 days)(7550 gal/day) = 52,900, say 53,000 gal (200,605 L).

2. *Select the tank capacity*

The actual fuel consumption is 53,000 gal (200,605 L) in 7 days. If fuel is delivered exactly on time every 7 days, a fuel-tank capacity of 53,000 gal (200,605 L) would be adequate. However, bad weather, transit failures, strikes, or other unpredictable incidents may delay delivery. Therefore, added capacity must be provided to prevent engine stoppage because of an inadequate fuel supply.

Where sufficient space is available, and local regulations do not restrict the storage capacity installed, use double the required capacity. The reason is that the additional storage capacity is relatively cheap compared with the advantages gained. Where space or storage capacity is restricted, use 1½ times the required capacity.

Assuming double capacity is used in this plant, the total storage capacity will be (2)(53,000) = 106,000 gal (401,210 L). At least two tanks should be used, to permit cleaning of one without interrupting engine operation.

Consult the National Board of Fire Underwriters bulletin *Storage Tanks for Flammable Liquids* for rules governing tank materials, location, spacing, and fire-protection devices. Refer to a tank capacity table to determine the required tank diameter and length or height depending on whether the tank is horizontal or vertical. Thus, the Buffalo Tank Corporation *Handbook* shows that a 16.5-ft (5.0-m) diameter 33.5-ft (10.2-m) long horizontal tank will hold 53,600 gal (202,876 L) when full. Two tanks of this size would provide the desired capacity. Alternatively, a 35-ft (10.7-m) diameter 7.5-ft (2.3-m) high vertical tank will hold 54,000 gal (204,390 L) when full. Two tanks of this size would provide the desired capacity.

Where a tank capacity table is not available, compute the capacity of a cylindrical tank from capacity = $5.87D^2L$, where D = tank diameter, ft; L = tank length or height, ft. Consult the NBFU or the tank manufacturer for the required tank wall thickness and vent size.

3. ***Select the day-tank capacity***

Day tanks supply filtered fuel to an engine. The day tank is usually located in the engine room and holds enough fuel for a 4- to 8-h operation of an engine at full load. Local laws, insurance requirements, or the NBFU may limit the quantity of oil that can be stored in the engine room or a day tank. One day tank is usually used for each engine.

Assume that a 4-h supply will be suitable for each engine. Then the day tank capacity for a 1000-hp (746-kW) engine = (1000 hp) [0.35 lb/(bhp·h) fuel] (4 h) = 1400 lb (630 kg), or 1400/7.387 = 189.6 gal (717.6 L), given the lighter-weight fuel, step 1. Thus, one 200-gal (757-L) day tank would be suitable for each of the 1000-hp (746-kW) engines.

For the 600-hp (448-kW) engines, the day-tank capacity should be (600 hp)[0.35 lb/(bhp·h) fuel] (4 h) = 840 lb (378 kg), or 840/7.387 = 113.8 gal (430.7 L). Thus, one 125-gal (473-L) day tank would be suitable for each of the 600-hp (448-kW) engines.

4. ***Determine which is the better fuel buy***

Compute the higher heating value HHV of each fuel from HHV = 17,645 + 54(°API), or for 24° fuel, HHV = 17,645 + 54(24) = 18,941 Btu/lb (44,057 kJ/kg). For the 28° fuel, HHV = 17,645 + 54(28) = 19,157 Btu/lb (44,559 kJ/kg).

Compare the two oils on the basis of cost per 10,000 Btu (10,550 kJ), because this is the usual way of stating the cost of a fuel. The weight of each oil was computed in step 1. Thus the 24° API oil weighs 7.578 lb/gal (0.90 kg/L), while the 28° API oil weighs 7.387 lb/gal (0.878 kg/L).

Then the cost per 10,000 Btu (10,550 kJ) = (cost, $/gal)/[(HHV, Btu/lb)/10,000](oil weight, lb/gal). For the 24° API oil, cost per 10,000 Btu (10,550 kJ) = $0.0825/[(18.941/10,000)(7.578)] = $0.00574, or 0.574 cent per 10,000 Btu (10,550 kJ). For the 28° API oil, cost per 10,000 Btu = $0.0910/[(19,157/10,000)(7387)] = $0.00634, or 0.634 cent per 10,000 Btu (10,550 kJ). Thus, the 24° API is the better buy because it costs less per 10,000 Btu (10,550 kJ).

Related Calculations: Use this method for engines burning any liquid fuel. Be certain to check local laws and the latest NBFU recommendations before ordering fuel storage or day tanks.

POWER INPUT TO COOLING-WATER AND LUBE-OIL PUMPS

What is the required power input to a 200-gal/min (12.6-L/s) jacket-water pump if the total head on the pump is 75 ft (22.9 m) of water and the pump has an efficiency of 70 percent when it handles freshwater and saltwater? What capacity lube-oil pump is needed for a four-cycle 500-hp (373-kW) turbocharged diesel engine having oil-cooled pistons? What is the required power input to this pump if the discharge pressure is 80 lb/in^2 (551.5 kPa) and the efficiency of the pump is 68 percent?

Calculation Procedure:

1. ***Determine the power input to the jacket-water pump***

The power input to jacket-water and raw-water pumps serving internal-combustion engines is often computed from the relation $hp = Gh/Ce$, where hp = hp input; G = water discharged by pump, gal/min; h = total head on pump, ft of water; C = constant = 3960 for freshwater having a density of 62.4 lb/ft^3 (999.0 kg/m^3); 3855 for saltwater having a density of 64 lb/ft^3 (1024.6 kg/m^3).

For this pump handling freshwater, hp = (200)(75)/(3960)(0.70) = 5.42 hp (4.0 kW). A 7.5-hp (5.6-kW) motor would probably be selected to handle the rated capacity plus any overloads.

For this pump handling saltwater, hp = (200)(75)/[(3855)(0.70)] = 5.56 hp (4.1 kW). A 7.5-hp (5.6-kW) motor would probably be selected to handle the rated capacity plus any overloads. Thus, the same motor could drive this pump whether it handles freshwater or saltwater.

2. *Compute the lube-oil pump capacity*

The lube-oil pump capacity required for a diesel engine is found from $G = H/200\Delta t$, where G = pump capacity, gal/min; H = heat rejected to the lube oil, Btu/(bhp·h); Δt = lube-oil temperature rise during passage through the engine, °F. Usual practice is to limit the temperature rise of the oil to a range of 20 to 25°F (11.1 to 13.9°C), with a maximum operating temperature of 160°F (71.1°C). The heat rejection to the lube oil can be obtained from the engine heat balance, the engine manufacturer, or *Standard Practices for Stationary Diesel Engines*, published by the Diesel Engine Manufacturers Association. With a maximum heat rejection rate of 500 Btu/(bhp·h) (196.4 W/kWh) from *Standard Practices* and an oil-temperature rise of 20°F (11.1°C), G = [500 Btu/(bhp·h)](1000 hp)/[(200)(20)] = 125 gal/min (7.9 L/s).

By using the *lowest* temperature rise and the *highest* heat rejection rate, a safe pump capacity is obtained. Where the pump cost is a critical factor, use a higher temperature rise and a lower heat rejection rate. Thus, with a heat rejection rate of 300 Btu/(bhp·h) (117.9 W/kWh) from *Standard Practices*, the above pump would have a capacity of G = (300)(1000)/[(200)(25)] = 60 gal/min (3.8 L/s).

3. *Compute the lube-oil pump power input*

The power input to a separate oil pump serving a diesel engine is given by $hp = Gp/1720e$, where G = pump discharge rate, gal/min; p = pump discharge pressure, lb/in^2; e = pump efficiency. For this pump, hp = (125)(80)/[(1720)(0.68)] = 8.56 hp (6.4 kW). A 10-hp (7.5-kW) motor would be chosen to drive this pump.

With a capacity of 60 gal/min (3.8 L/s), the input is hp = (60)(80)/[1720)(0.68)] = 4.1 hp (3.1 kW). A 5-hp (3.7-kW) motor would be chosen to drive this pump.

Related Calculations: Use this method for any reciprocating diesel engine, two- or four-cycle. Lube-oil pump capacity is generally selected 10 to 15 percent oversize to allow for bearing wear in the engine and wear of the pump moving parts. Always check the selected capacity with the engine builder. Where a bypass-type lube-oil system is used, be sure to have a pump of sufficient capacity to handle *both* the engine and cooler oil flow.

Raw-water pumps are generally duplicates of the jacket-water pump, having the same capacity and head ratings. Then the raw-water pump can serve as a standby jacket-water pump, if necessary.

LUBE-OIL COOLER SELECTION AND OIL CONSUMPTION

A 500-hp (373-kW) internal-combustion engine rejects 300 to 600 Btu/(bhp·h) (118 to 236 W/kWh) to the lubricating oil. What capacity and type of lube-oil cooler should be used for this engine if 10 percent of the oil is bypassed? If this engine consumes 2 gal (7.6 L) of lube oil per 24 h at full load, determine its lube-oil consumption rate.

Calculation Procedure:

1. *Determine the required lube-oil cooler capacity*

Base the cooler capacity on the maximum heat rejection rate plus an allowance for overloads. The usual overload allowance is 10 percent of the full-load rating for periods of not more than 2 h in any 24 h period.

For this engine, the maximum output with a 10 percent overload is 500 + (0.10)(500) = 550 hp (410 kW). Thus, the maximum heat rejection to the lube oil would be (550 hp)[600 Btu/(bhp·h)] = 330,000 Btu/h (96.7 kW).

2. *Choose the type and capacity of lube-oil cooler*

Choose a shell-and-tube type heat exchanger to serve this engine. Long experience with many types of internal-combustion engines shows that the shell-and-tube heat exchanger is well suited for lube-oil cooling.

Select a lube-oil cooler suitable for a heat-transfer load of 330,000 Btu/h (96.7 kW) at the prevailing cooling-water temperature difference, which is usually assumed to be 10°F (5.6°C). See previous calculation procedures for the steps in selecting a liquid cooler.

3. *Determine the lube-oil consumption rate*

The lube-oil consumption rate is normally expressed in terms of bhp·h/gal. Thus, if this engine operates for 24 h and consumes 2 gal (7.6 L) of oil, its lube-oil consumption rate = (24 h)(500 bhp)/2 gal = 6000 bhp·h/gal (1183 kWh/L).

Related Calculations: Use this procedure for any type of internal-combustion engine using any fuel.

QUANTITY OF SOLIDS ENTERING AN INTERNAL-COMBUSTION ENGINE

What weight of solids annually enters the cylinders of a 1000-hp (746-kW) internal-combustion engine if the engine operates 24 h/day, 300 days/year in an area having an average dust concentration of 1.6 gr per 1000 ft^3 of air (28.3 m^3)? The engine air rate (displacement) is 3.5 ft^3/(min·bhp) (0.13 m^3/kW). What would the dust load be reduced to if an air filter fitted to the engine removed 80 percent of the dust from the air?

Calculation Procedure:

1. *Compute the quantity of air entering the engine*

Since the engine is rated at 1000 hp (746 kW) and uses 3.5 ft^3/(min·bhp) [0.133 m^3/(min·kW)], the quantity of air used by the engine each minute is (1000 hp)[3.5 ft^3/(min·hp)] = 3500 ft^3/min (99.1 m^3/min).

2. *Compute the quantity of dust entering the engine*

Each 1000 ft^3 (28.3 m^3) of air entering the engine contains 1.6 gr (103.7 mg) of dust. Thus, during every minute of engine operation, the quantity of dust entering the engine is (3500/1000)(1.6) = 5.6 gr (362.8 mg). The hourly dust intake = (60 min/h)(5.6 gr/min) = 336 gr/h (21,772 mg/h).

During the year the engine operates 24 h/day for 300 days. Hence, the annual intake of dust is (24 h/day)(300 days/year)(336 gr/h) = 2,419,200 gr (156.8 kg). Since there is 7000 gr/lb, the weight of dust entering the engine per year = 2,419,200 gr/(7000 gr/lb) = 345.6 lb/year (155.5 kg/year).

3. *Compute the filtered dust load*

With the air filter removing 80 percent of the dust, the quantity of dust reaching the engine is (1.00 − 0.80)(345.6 lb/year) = 69.12 lb/year (31.1 kg/year). This shows the effectiveness of an air filter in reducing the dust and dirt load on an engine.

Related Calculations: Use this general procedure to compute the dirt load on an engine from any external source.

INTERNAL-COMBUSTION ENGINE PERFORMANCE FACTORS

Discuss and illustrate the important factors in internal-combustion engine selection and performance. In this discussion, consider both large and small engines for a full range of usual applications.

Calculation Procedure:

1. *Plot typical engine load characteristics*

Figure 7 shows four typical load patterns for internal-combustion engines. A continuous load, Fig. 7*a*, is generally considered to be heavy-duty and is often met in engines driving pumps or electric generators.

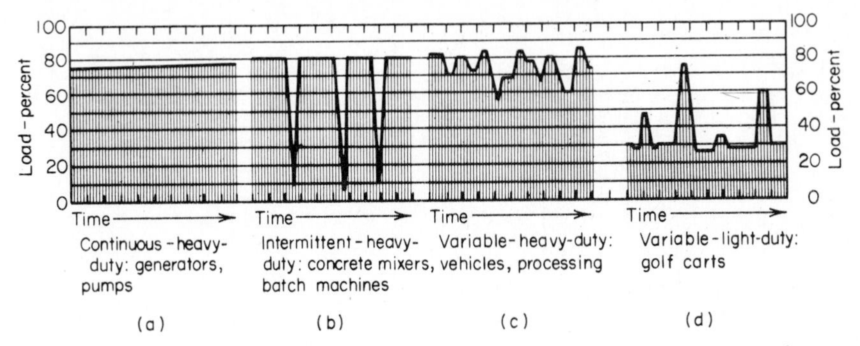

FIG. 7 Typical internal-combustion engine load cycles: (*a*) continuous, heavy-duty; (*b*) intermittent, heavy-duty; (*c*) variable, heavy-duty; (*d*) variable, light-duty. *(Product Engineering.)*

Intermittent heavy-duty loads, Fig. 7*b*, are often met in engines driving concrete mixers, batch machines, and similar loads. Variable heavy-duty loads, Fig. 7*c*, are encountered in large vehicles, process machinery, and similar applications. Variable light-duty loads, Fig. 7*d*, are met in small vehicles like golf carts, lawn mowers, chain saws, etc.

2. *Compute the engine output torque*

Use the relation $T = 5250$ bhp/(r/min) to compute the output torque of an internal-combustion engine. In this relation, bhp = engine bhp being developed at a crankshaft speed having rotating speed of *rpm*.

3. *Compute the hp output required*

Knowing the type of load on the engine (generator, pump, mixer, saw blade, etc.), compute the power output required to drive the load at a constant speed. Where a speed variation is expected, as in variable-speed drives, compute the average power needed to accelerate the load between two desired speeds in a given time.

4. *Choose the engine output speed*

Internal-combustion engines are classified in three speed categories: high (1500 r/min or more), medium (750 to 1500 r/min), and low (less than 750 r/min).

Base the speed chosen on the application of the engine. A high-speed engine can be lighter and smaller for the same hp rating, and may cost less than a medium-speed or slow-speed engine serving the same load. But medium-speed and slow-speed engines, although larger, offer a higher torque output for the equivalent hp rating. Other advantages of these two speed ranges include longer service life and, in some instances, lower maintenance costs.

Usually an application will have its own requirements, such as allowable engine weight, available space, output torque, load speed, and type of service. These requirements will often indicate that a particular speed classification must be used. Where an application has no special speed requirements, the speed selection can be made on the basis of cost (initial, installation, maintenance, and operating cost), type of parts service available, and other local conditions.

5. *Analyze the engine output torque required*

In some installations, an engine with good lugging power is necessary, especially in tractors, harvesters, and hoists, where the load frequently increases above normal. For good lugging power, the engine should have the inherent characteristic of increasing torque with drooping speed. The engine can then resist the tendency for increased load to reduce the output speed, giving the engine good lugging qualities.

One way to increase the torque delivered to the load is to use a variable-ratio hydraulic transmission. The transmission will amplify the torque so that the engine will not be forced into the lugging range.

Other types of loads, such as generators, centrifugal pumps, air conditioners, and marine drives, may not require this lugging ability. So be certain to consult the engine power curves and torque characteristic curve to determine the speed at which the maximum torque is available.

TABLE 9 Comparison of Fuels for Internal-Combustion Engines*

	Storage life (quantities)		Consistency, Btu/ft^3	Initial cost of engine, relative	Cost of fuel	Residue	Antiknock rating	Filtering necessary	Weight		Heat content			
	Small	Large							lb/gal	kg/L	Btu/vol	mJ/vol	Btu/lb	mJ/kg
Gasoline	Good	Poor (6 months)	Good	Low	High	High	Best is costly	Medium	6.000	0.714	123,039 Btu/gal	34,291 kJ/L	20,627	47.9
Diesel:														
No. 1	Good	Fair (1 year)	Good	High	Low	Low if properly filtered	. . .	High	6.850	0.815	135,800 Btu/gal	37,847 kJ/L	19,750	45.9
No. 2	Good	Fair (1 year)	Good	High	Low	Low if properly filtered	. . .	High	7.020	0.835	139,000 Btu/gal	38,739 kJ/L	19,786	46.0
Natural gas	Not necessary	Not necessary	Poor	Medium	Medium	Low	High	Very little	. . .	. . .	1,000 Btu/ft^3	37,250 kJ/m^3		
LPG:														
Propane	Good	Good	Poor	Medium	Medium	Low	Good	Very little	4.235	0.504	91,740 Btu/gal	25,568 kJ/L	21,308	49.6
Butane	Good	Good	Poor	Medium	Medium	Low	Good	Very little	4.873	0.580	103,830 Btu/gal	28,937 kJ/L	20,627	47.9

* *Product Engineering.*

TABLE 10 Performance Table for Small Internal-Combustion Engines [Less than 7 hp (5 kW)]*

	Variety of models available	Typical weight lb/hp (kg/kW)	Operating speeds		Lugging ability	Torque output	Relative life expectancy, h	Relative cost	Fuel required	Shaft direction	Noise level	Starters	Integral optional Pto's	Ignition	Cost of operation	Variety of options and accessories
			Typical maximum	Typical efficient minimum												
Lightweight: 2-stroke	Narrow	2:1 (1.2:1)	3,600 (governed) to 7,500	2,000 to 3,000	Poor to fair	Fair	500	Lowest	Gasoline oil mixed	Vertical, horizontal, or universal	High	Rope, recoil, impulse	No	Magneto	High	Standard—extremely low custom—wide
4-stroke	Wide	6:1; 10:1 (3.6:1; 6.1:1)	4,000	2,000 to 2,400	Fair to good	Good	500	1 to 2	Gasoline (LPG)	Vertical or horizontal	Moderate	Rope, recoil, impulse, electric	Several	Magneto	Moderate	Standard—wide
Heavyweight: 4-stroke	Wide	11:1; 20:1 (6.6:1; 12.1:1)	4,000	1,600 to 1,800	Good to excellent	Good	7,500	2 to 4	Gasoline (LPG)	Vertical or horizontal	Moderate	Rope, recoil, impulse, crank, electric	No	Magneto, distributor	Moderate	Standard—moderately wide
Diesel	Narrow	35:1 (21.1:1)	2,400	1,500	Excellent	Good	25,000	4	Diesel	Horizontal	Moderate to high	Electric	No	Battery, distributor, glow plugs	Low	Narrow

* *Product Engineering.*

6. *Evaluate the environmental conditions*

Internal-combustion engines are required to operate under a variety of environmental conditions. The usual environmental conditions critical in engine selection are altitude, ambient temperature, dust or dirt, and special or abnormal service. Each of these, except the last, is considered in previous calculation procedures.

Special or abnormal service includes such applications as fire fighting, emergency flood pumps and generators, and hospital standby service. In these applications, an engine must start and pick up a full load without warmup.

7. *Compare engine fuels*

Table 9 compares four types of fuels and the internal-combustion engines using them. Note that where the cost of the fuel is high, the cost of the engine is low; where the cost of the fuel is low, the cost of the engine is high. This condition prevails for both large and small engines in any service.

8. *Compare the performance of small engines*

Table 10 compares the principal characteristics of small gasoline and diesel engines rated at 7 hp (5 kW) or less. Note that engine life expectancy can vary from 500 to 25,000 h. With modern, mass-produced small engines it is often just as cheap to use short-life replaceable two-stroke gasoline engines instead of a single long-life diesel engine. Thus, the choice of a small engine is often based on other considerations, such as ease and convenience of replacement, instead of just hours of life. Chances are, however, that most long-life applications of small engines will still require a long-life engine. But the alternatives must be considered in each case.

Related Calculations: Use the general data presented here for selecting internal-combustion engines having ratings up to 200 hp (150 kW). For larger engines, other factors such as weight, specific fuel consumption, lube-oil consumption, etc., become important considerations. The method given here is the work of Paul F. Jacobi, as reported in *Product Engineering*.

Air and Gas Compressors and Vacuum Systems

REFERENCES: Hawthorne—*Aerodynamics of Turbines and Compressors*, Princeton University Press; Tramm and Dean—*Centrifugal Compressor and Pump Stability, Stall, and Surge*, ASME; *Chemical Engineering* Magazine—*Fluid Movers: Pumps, Compressors, Fans and Blowers*, McGraw-Hill; Martini—*Practical Seal Design*, Dekker; Cheremisinoff and Gupta—*Handbook of Fluids in Motion*, Butterworths; Van Atta—*Vacuum Science and Engineering*, McGraw-Hill; Dushman—*Scientific Foundations of Vacuum Technique*, Wiley; Guthrie and Wakerling—*Vacuum Equipment and Techniques*, McGraw-Hill; Yarwood—*High Vacuum Techniques*, Wiley; Lewin—*Vacuum Science and Technology*, McGraw-Hill; Pirani and Yarwood—*Principles of Vacuum Engineering*, Reinhold; Reimann—*Vacuum Technique*, Chapman and Hall; Steinherz—*Handbook of High Vacuum Engineering*, Reinhold; Compressed Air and Gas Institute—*Compressed Air and Gas Handbook*; Ingersoll-Rand Company—*Compressed Air Data*.

COMPRESSOR SELECTION FOR COMPRESSED-AIR SYSTEMS

Determine the required capacity, discharge pressure, and type of compressor for an industrial-plant compressed-air system fitted with the tools listed in Table 1. The plant is located at sea level and operates 16 h/day.

Calculation Procedure:

1. *Compute the required airflow rate*

List all the tools and devices in the compressed-air system that will consume air, Table 1. Then obtain from Table 2 the probable air consumption, ft^3/min, of each tool. Enter this value in column 1, Table 1. Next list the number of each type of tool that will be used in the system in column 2. Find the maximum probable air consumption of each tool by taking the product, line by line, of columns 1 and 2. Enter the result in column 3, Table 1, for each tool.

The air consumption values shown in column 3 represent the airflow rate required for continuous operation of each type and number of tools listed. However, few air tools operate continually. To provide for this situation, a load factor is generally used when an air compressor is selected.

TABLE 1 Typical Computation of Compressed-Air Requirements

Tool	(1) Air consumption		(2)	(3) Air required, (1) × (2)		(4)	(5) Probable air demand, (3) × (4)	
	ft³/min	m³/min	Number of tools	ft³/min	m³/min	Load factor	ft³/min	m³/min
Grinding wheel, 6 in (15.2 cm)	50	1.4	5	250	7.1	0.3	75	2.1
Rotary sander, 9-in (22.9-cm) pad	55	1.6	2	110	3.1	0.5	55	1.6
Chipping hammers, 13 lb (5.9 kg)	30	0.85	8	240	6.8	0.4	96	2.7
Nut setters, ⁵⁄₁₆ in (0.79 cm)	20	0.57	10	200	5.7	0.6	120	3.4
Paint spray	10	0.28	1	10	0.28	0.1	1	0.03
Plug drills	40	1.1	3	120	3.4	0.2	24	0.68
Riveters, 18 lb (8.1 kg)	35	0.99	5	175	4.9	0.4	70	1.9
Steel drill, ⅞ in (2.2 cm), 25 lb (11.3 kg)	80	2.3	5	400	11.3	0.4	160	4.5
Total							601*	16.9*

*To this sum must be added allowance for future needs and expected leakage loss, if any.

TABLE 2 Approximate Air Needs of Pneumatic Tools

	ft³/min	m³/min
Grinders:		
6- and 8-in (15.2- and 20.3-cm) diameter wheels	50	1.4
2- and 2½-in (5.1- and 6.4-cm) diameter wheels	14–20	0.40–0.57
File and burr machines	18	0.51
Rotary sanders, 9-in (22.9-cm) diameter pads	55	1.56
Sand rammers and tampers:		
1 × 4 in (2.5 × 10.2 cm) cylinder	25	0.71
1¼ × 5 in (3.2 × 12.7 cm) cylinder	28	0.79
1½ × 6 in (3.8 × 15.2 cm) cylinder	39	1.1
Chipping hammers:		
10 to 13 lb (4.5 to 5.9 kg)	28–30	0.79–0.85
2 to 4 lb (0.9 to 1.8 kg)	12	0.34
Nut setters:		
To ⁵⁄₁₆ in, 8 lb (0.79 cm, 3.6 kg)	20	0.57
½ to ¾ in, 18 lb (1.3 to 1.9 cm, 8.1 kg)	30	0.85
Paint spray	2–20	0.06–0.57
Plug drills	40–50	1.1–1.4
Riveters:		
³⁄₃₂- to ⅛-in (0.24- to 0.32-cm) rivets	12	0.34
Larger, weighing 18 to 22 lb (8.1 to 9.9 kg)	35	0.99
Rivet busters	35–39	0.51–0.75
Steel drills, rotary motors:		
To ¼ in (0.64 cm) weighing 1¼ to 4 lb (0.56 to 1.8 kg)	18–20	0.57–1.1
¼ to ⅜ in (0.69 to 0.95 cm) weighing 6 to 8 lb (2.7 to 3.6 kg)	20–40	1.98
½ to ¾ in (1.27 to 1.91 cm) weighing 9 to 14 lb (4.1 to 6.3 kg)	70	2.27
⅞ to 1 in (2.2 to 2.5 cm) weighing 25 lb (11.25 kg)	80	1.1
Wood borers to 1-in (2.5 cm) diameter, weighing 14 lb (6.3 kg)	40	

2. *Select the equipment load factor*

The equipment load factor = (actual air consumption of the tool or device, ft^3/min)/(full-load continuous air consumption of the tool or device, ft^3/min). Load factors for compressed-air operated devices are usually less than 1.0.

Two variables are involved in the equipment load factor. The first is the *time factor*, or the percentage of the total time the tool or device actually uses compressed air. The second is the *work factor*, or percentage of maximum possible work output done by the tool. The load factor is the product of these two variables.

Determine the load factor for a given tool or device by consulting the manufacturer's engineering data, or by estimating the factor value by using previous experience as a guide. Enter the load factor in column 4, Table 1. The values shown represent typical load factors encountered in industrial plants.

3. *Compute the actual air consumption*

Take the product, line by line, of columns 3 and 4, Table 1. Enter the result, i.e., the probable air demand, in column 5, Table 1. Find the sum of the values in column 5, or 601 ft^3/min. This is the probable air demand of the system.

4. *Apply allowances for leakage and future needs*

Most compressed-air system designs allow for 10 percent of the required air to be lost through leaks in the piping, tools, hoses, etc. Whereas some designers claim that allowing for leakage is a poor design procedure, observation of many installations indicates that air leakage is a fact of life and must be considered when an actual system is designed.

With a 10 percent leakage factor, the required air capacity = 1.1(601) = 661 ft^3/min (18.7 m^3/min).

Future requirements are best estimated by predicting what types of tools and devices will probably be used. Once this is known, prepare a tabulation similar to Table 1, listing the predicted future tools and devices and their air needs. Assume that the future air needs, column 5, are 240 ft^3/min (6.8 m^3/min). Then the total required air capacity = 661 + 240 = 901 ft^3/min (25.5 m^3/min), say 900 ft^3/min (25.47 m^3/min) = present requirements + leakage allowance + predicted future needs, all expressed in ft^3/min.

5. *Choose the compressor discharge pressure and capacity*

In selecting the type of compressor to use, two factors are of key importance: discharge pressure required and capacity required.

Most air tools and devices are designed to operate at a pressure of 90 lb/in^2 (620 kPa) at the tool inlet. Hence, usual industrial compressors are rated for a discharge pressure of 100 lb/in^2 (689 kPa), the extra lb/in^2 providing for pressure loss in the piping between the compressor and the tools. Since none of the tools used in this plant are specialty items requiring higher than the normal pressure, a 100-lb/in^2 (689-kPa) discharge pressure will be chosen.

Where the future air demands are expected to occur fairly soon—within 2 to 3 years—the general practice is to choose a compressor having the capacity to satisfy present and future needs. Hence, in this case, a 900-ft^3/min (25.5-m^3/min) compressor would be chosen.

6. *Compute the power required to compress the air*

Table 3 shows the power required to compress air to various discharge pressures at different altitudes above sea level. Study of this table shows that at sea level a single-stage compressor requires 22.1 bhp/(100 ft^3/min) (5.8 kW/m^3) when the discharge pressure is 100 lb/in^2 (689 kPa). A two-stage compressor requires 19.1 bhp (14.2 kW) under the same conditions. This is a saving of 3.0 bhp/(100 ft^3/min) (0.79 kW/m^3). Hence, a two-stage compressor would probably be a better investment because this hp will be saved for the life of the compressor. The usual life of an air compressor is 20 years. Hence, by using a two-stage compressor, the approximate required bhp = (900/100)(19.1) = 171.9 bhp (128 kW), say 175 bhp (13.1 kW).

7. *Choose the type of compressor to use*

Reciprocating compressors find the widest use for stationary plant air supply. They may be single- or two-stage, air- or water-cooled. Here is a general guide to the types of reciprocating compressors that are satisfactory for various loads and service:

Single-stage air-cooled compressor up to 3 hp (2.2 kW), pressures to 150 lb/in^2 (1034 kPa), for light and intermittent running up to 1 h/day.

TABLE 3 Air Compressor Brake Horsepower (kW) Input°

Altitude, ft (m)	Single-stage discharge pressure, lb/in^2 (gage) (kPa)			Two-stage discharge pressure, lb/in^2 (gage) (kPa)		
	60 (414)	80 (552)	100 (689)	60 (414)	80 (552)	100 (689)
0 (0)	16.3 (12.2)	19.5 (14.6)	22.1 (16.5)	14.7 (10.9)	17.1 (12.8)	19.1 (14.3)
2000 (610)	15.9 (11.9)	18.9 (14.1)	21.3 (15.9)	14.3 (10.7)	16.5 (12.3)	18.4 (13.7)
4000 (1212)	15.4 (11.5)	18.2 (13.6)	20.6 (15.4)	13.8 (10.3)	15.8 (11.8)	17.7 (13.2)
6000 (1820)	15.0 (11.2)	17.6 (13.1)	20.0 (14.9)	13.3 (9.9)	15.2 (11.3)	17.0 (12.7)

°*Courtesy Ingersoll-Rand.* Values shown are the approximate bhp input required per 100 ft^3/min (2.8 m^3/min) of free air actually delivered. The bhp input can vary considerably with the type and size of compressor.

Two-stage air-cooled compressor up to 3 hp (2.2 kW), pressures to 150 lb/in^2 (1034 kPa), for 4 to 8 h/day running time.

Single-stage air-cooled compressor up to 15 hp (11.2 kW) for pressures to 80 lb/in^2 (552 kPa); above 80 lb/in^2 (552 kPa), use two-stage air-cooled compressor.

Single-stage horizontal double-acting water-cooled compressor for pressures to 100 lb/in^2 (689 kPa) horsepowers of 10 to 100 (7.5 to 75 kW), for 24 h/day or less operating time.

Two-stage, single-acting air-cooled compressor for 10 to 100 hp (7.5 to 75 kW), 5 to 10 h/day operation.

Two-stage double-acting water-cooled compressor for 100 hp (75 kW), or more, 24 h/day, or less operating time.

Using this general guide, choose a two-stage double-acting water-cooled reciprocating compressor, because more than 100-hp (75-kW) input is required and the compressor will operate 16 h/day.

Rotary compressors are not as widely used for industrial compressed-air systems as reciprocating compressors. The reason is that usual rotary compressors discharge at pressures under 100 lb/in^2 (68.9 kPa), unless they are multistage units.

Centrifugal compressors are generally used for large airflows—several thousand ft^3/min or more. Hence, they usually find use for services requiring large air quantities, such as steel-mill blowing, copper conversion, etc. As a general rule, machines discharging at pressures of 35 lb/in^2 (241 kPa) or less are termed *blowers;* machines discharging at pressures greater than 35 lb/in^2 (241 kPa) are termed *compressors.*

Using these facts as a guide enables the designer to choose, as before, a two-stage double-acting water-cooled compressor for this application. Refer to the manufacturer's engineering data for the compressor dimensions and weight.

8. *Select the compressor drive*

Air compressors can be driven by electric motors, gasoline engines, diesel engines, gas turbines, or steam turbines. The most popular drive for reciprocating air compressors is the electric motor—either direct-connected or belt-connected. Where either dc or ac power supply is available, the usual choice is an electric-motor drive. However, special circumstances, such as the availability of low-cost fuel, may dictate another choice of drive for economic reasons. Assuming that there are no special economic reasons for choosing another type of drive, an electric motor would be chosen for this installation.

With an ac power supply, the squirrel-cage induction motor is generally chosen for belt-driven compressors. Synchronous motors are also used, particularly when power-factor correction is desired. Motor-driven air compressors generally operate at constant speed and are fitted with cylinder unloaders to vary the quantity of air delivered to the air receiver. A typical power input to a large reciprocating compressor is 22 hp (16.4 kW) per 100 ft^3/min (2.8 m^3/min) of free air compressed.

Air compressors are almost always rated in terms of *free air* capacity, i.e., air at the compressor intake location. Since the altitude, barometric pressure, and air temperature may vary at any

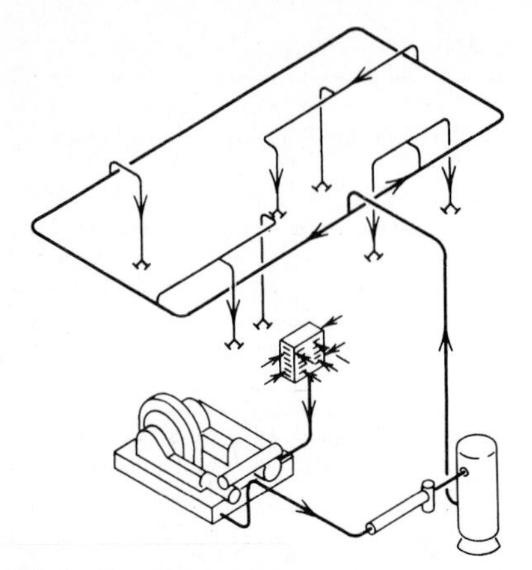

FIG. 1 Central system for compressed-air supply.

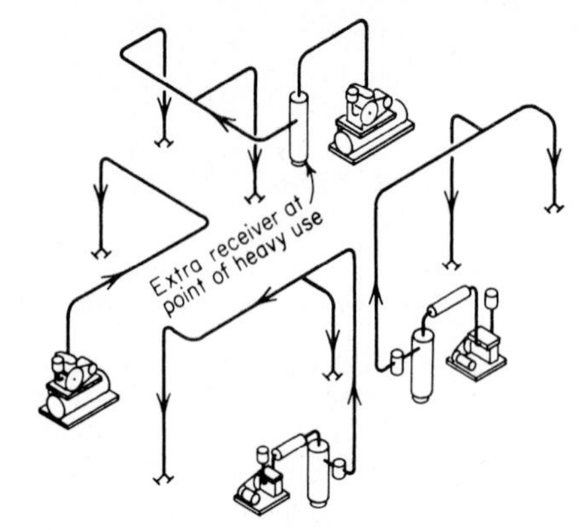

FIG. 2 Unit system for compressed-air supply.

locality, the term *free air* does not mean air under standard or uniform conditions. The displacement of an air compressor is the volume of air displaced per unit of time, usually stated in ft^3/min. In a multistage compressor, the displacement is that of the low-pressure cylinder only.

9. *Choose the type of air distribution system*

Two types of air distribution systems are in use in industrial plants: *central* and *unit*. In a central system, Fig. 1, one or more large compressors centrally located in the plant supply compressed air to the areas needing it. The supply piping often runs in the form of a loop around the areas needing air.

A unit system, Fig. 2, has smaller compressors located in the areas where air is used. In the usual plant, each compressor serves only the area in which it is located. Emergency connections between the various areas may or may not be installed.

Central systems have been used for many years in large industrial plants and give excellent service. Unit systems are used in both small and large plants but probably find more use in smaller plants today. With the large quantity of air required by this plant, a central system would probably be chosen, unless the air was needed at widely scattered locations in the plant, leading to excessive pressure losses in the distribution piping of a central system. In such a situation, a unit system with the capacity divided between compressors as necessary would be chosen.

Related Calculations: Where possible, choose a larger compressor than the calculations indicate is needed, because air use in industrial plants tends to increase. Avoid choosing a compressor having a free-air capacity less than one-third the required free-air capacity.

When choosing a water-cooled compressor instead of an air-cooled unit, remember that water cooling is more expensive than air cooling. However, the power input to water-cooled compressors is usually less than to air-cooled compressors of the same capacity. For either type of cooling, a two-stage compressor, with intercooling, is more economical when the compressor must operate 4 h or more in a 24-h period. Table 4 shows the typical cooling-water requirements of various types of water-cooled compressors.

When the inlet air temperature is above or below 60°F (15.6°C), the compressor delivery will vary. Table 5 shows the relative delivery of compressors handling air at various inlet temperatures.

SIZING COMPRESSED-AIR-SYSTEM COMPONENTS

What is the minimum capacity air receiver that should be used in a compressed-air system having a compressor displacing 800 ft^3/min (0.38 m^3/s) when the intake pressure is 14.7 lb/in^2 (abs) (101.4 kPa) and the discharge pressure is 120 lb/in^2 (abs) (827.4 kPa)? How long will it take for this compressor to pump up a 300-ft^3 (8.5-m^3) receiver from 80 to 120-lb/in^2 (551.6 to 827.4 kPa) if the average volumetric efficiency of the compressor is 68 percent? For how long can an 80-lb/in^2 (abs) (551.6-kPa) tool be operated from a 120-lb/in^2 (abs) (827.4-kPa), 300-ft^3 (8.5-m^3) receiver

TABLE 4 Cooling Water Recommended for Intercoolers, Cylinder Jackets, Aftercoolers

	Actual free air, gal/min per 100 ft^3/min (L/s per 100 m^3/s)
Intercooler separate	2.5–2.8 (334.2–374.3)
Intercooler and jackets in series	2.5–2.8 (334.2–374.3)
Aftercoolers:	
80 to 100 lb/in^2 (551.6 to 689.5 kPa), two-stage	1.25 (167.1)
80 to 100 lb/in^2 (551.6 to 689.5 kPa), single-stage	1.8 (240.6)
Two-stage jackets alone (both)	0.8 (106.9)
Single-stage jackets:	
40 lb/in^2 (275.8 kPa)	0.6 (80.2)
60 lb/in^2 (413.7 kPa)	0.8 (106.9)
80 lb/in^2 (551.6 kPa)	1.1 (147.0)
100 lb/in^2 (689.5 kPa)	1.3 (173.8)

if the tool uses 10 ft^3/min (0.005 m^3/s) of free air and the receiver pressure is allowed to fall to 85 lb/in^2 (abs) (586.1 kPa) when the atmospheric pressure is 14.7 lb/in^2 (abs) (101.4 kPa)? What diameter air piston is required to produce a 1000-lb (4448.2-N) force if the pressure of the air is 150 lb/in^2 (abs) (1034.3 kPa)?

Calculation Procedure:

1. *Compute the required volume of the air receiver*

Use the relation $V_m = dp_1/p_2$, where V_m = minimum receiver volume needed, ft^3; d = compressor displacement, ft^3/min (use only the first-stage displacement for two-stage compressors); p_1 = compressor intake pressure, lb/in^2 (abs); p_2 = compressor discharge pressure, lb/in^2 (abs). Thus, for this compressor, $V_m = 800(14.7/120) = 97$ ft^3 (2.7 m^3). To provide a reserve capacity, a receiver having a volume of 150 or 200 ft^3 (4.2 or 5.7 m^3) would probably be chosen.

2. *Compute the receiver pump-up time*

Use the relation $t = V(p_f - p_i)/(14.7de)$, where t = receiver pump-up time, min; p_f = final pressure, lb/in^2 (abs); p_i = initial receiver pressure, lb/in^2 (abs); d = compressor piston displacement, ft^3/min; e = compressor volumetric efficiency, percent. Thus, $t = 300(120 - 80)/[14.7(800)(0.68)] = 1.5$ min. When the compressor discharge capacity is given in ft^3/min of free air instead of in terms of piston displacement, drop the volumetric efficiency term from the above relation before computing the pump-up time.

TABLE 5 Effect of Initial Temperature on Delivery of Air Compressors
[Based on a nominal intake temperature of 60°F (15.6°C)]

Initial temperature				Relative delivery
°F	°R	°C	K	
40	500	4.4	277.4	1.040
50	510	10.0	283.0	1.020
60	520	15.6	288.6	1.000
70	530	21.1	294.1	0.980
80	540	26.7	299.7	0.961

TABLE 6 Air-Consumption Altitude Factors *(100-lb/in^2 or 689.5-kPa air supply)*

Altitude		Factor
ft	m	
6,000	1,828.8	1.224
8,000	2,438.3	1.310
10,000	3,048.0	1.404

Note: For pressure losses in compressed-air piping systems, see the index.

3. *Compute the air supply time*

Use the relation $t_s = V(p_{max} - p_{min})/(cp_a m)$, where t_s = time in minutes during which the receiver of volume V ft^3 will supply air from the receiver maximum pressure p_{max} lb/in^2 (abs) to the minimum pressure p_{min} lb/in^2 (abs); c = ft^3/min of free air required to operate the tool; p_a = atmospheric pressure, lb/in^2 (abs). Or, $t_s = 300(120 - 85)/[(10)(14.7)] = 7.15$ min.

Note that in this relation p_{min} is the minimum air pressure to operate the air tool. A higher minimum tank pressure was chosen here because this provides a safer estimate of the time duration for the supply of air. Had the tool operating pressure been chosen instead, the time available, by the same relation, would be $t_s = 81.5$ min.

This calculation shows that it is often wise to install an auxiliary receiver at a distance from the compressor but near the tools drawing large amounts of air. Use of such an auxiliary receiver, particularly near the end of a long distribution line, can often eliminate the need for purchasing another air compressor.

4. *Compute the required piston diameter*

Use the relation $A_p = F/p_m$, where A_p = required piston area to produce the desired force, in^2; F = force produced, lb; p_m = maximum air pressure available for the piston, lb/in^2 (abs). Or, $A_p = 1000/150 = 6.66$ in^2 (43.0 cm^2). The piston diameter d is $d = 2(A_p/\pi)^{0.5} = 2.91$ in (7.4 cm).

Related Calculations: The air consumption of power tools is normally expressed in ft^3/min of free air at sea level; the actual capacity of any type of air compressor is expressed in the same units. At locations above sea level, the quantity of free air required to operate an air tool increases because the atmospheric pressure is lower. To find the air consumption of an air tool at an altitude above sea level in terms of ft^3/min of free air at the elevation location, multiply the sea-level consumption by the appropriate factor from Table 6. Thus, a tool that consumes 10 ft^3/min (0.005 m^3/s) of free air at sea level will use 10 (1.310) = 13.1 ft^3/min (0.006 m^3/s) of 100 lb/in^2 (689.5-kPa) free air at an 8000-ft (2438.4-m) altitude.

COMPRESSED-AIR RECEIVER SIZE AND PUMP-UP TIME

What is the minimum size receiver that can be used in a compressed-air system having a compressor rated at 800 ft^3/min (0.4 m^3/s) of free air if the intake pressure is 14.7 lb/in^2 (abs) (101.4 kPa) and the discharge pressure is 120 lb/in^2 (abs) (827.4 kPa)? How long will it take the compressor to pump up the receiver from 60 lb/in^2 (abs) (413.7 kPa) to 120 lb/in^2 (abs) (827.4 kPa)? The compressor is a two-stage water-cooled unit. How much cooling water is required for the intercooler and jacket if they are piped in series and for the aftercooler?

Calculation Procedure

1. *Compute the required minimum receiver volume*

For any air compressor, the minimum receiver volume v_m ft^3 = Dp_i/p_d, where D = compressor displacement, ft^3/min free air (use only the first-stage displacement for multistage compressors); p_i = compressor inlet pressure, lb/in^2 (abs); p_d = compressor discharge pressure, lb/in^2 (abs). For this compressor, $v_m = (800)(14.7)/(120) = 98$ ft^3 (2.8 m^3). To provide a reserve supply of air, a receiver having a volume of 150 or 200 ft^3 (4.2 or 5.7 m^3) would probably be chosen. Be certain that the receiver chosen is a standard unit; otherwise, its cost may be excessive.

2. *Compute the pump-up time required*

Assume that a 150-ft^3 (4.2-m^3) receiver is chosen. Then, for any receiver, the pump-up time t min = $v_r(p_e - p_s)/De$, where v_r = receiver volume, ft^3; p_e = pressure at end of pump-up, lb/in^2 (abs); p_s = pressure at start of pump-up, lb/in^2 (abs); e = compressor volumetric efficiency, expressed as a decimal (0.50 to 0.75 for single-stage and 0.80 to 0.90 for multistage compressors). For this compressor, with a volumetric efficiency of 0.85, $t = (150)(120 - 60)/[(800)(0.85)] = 13.22$ min.

3. *Determine the quantity of cooling water required*

Use the Compressed Air and Gas Institute (CAGI) cooling-water recommendations given in the *Compressed Air and Gas Handbook*, or Baumeister and Marks—*Standard Handbook for*

Mechanical Engineers. For 80 to 125 lb/in^2 (gage) (551.6 to 861.9 kPa) discharge pressure with the intercooler and jacket in series, CAGI recommends a flow of 2.5 to 2.8 gal/min per 100 ft^3/min (334.2 to 374.3 L/s per 100 m^3/s) of free air. Using 2.5 gal/min (334.2 L/s), we see that the cooling water required for the intercooler and jackets = (2.5)(800/100) = 20.0 gal/min (2673.9 L/s). CAGI recommends 1.25 gal/min per 100 ft^3/min (167.1 L/s per 100 m^3/s) of free air for an aftercooler serving a two-stage 80 to 125 lb/in^2 (gage) (551.6- to 861.9-kPa) compressor, or (1.25)(800/100) = 10.0 gal/min (1377.3 L/s) for this compressor. Thus, the total quantity of cooling water required for this compressor is 20 + 10 = 30 gal/min (4010.9 L/s).

Related Calculations: Use this procedure for any type of air compressor serving an industrial, commercial, utility, or residential load of any capacity. Follow CAGI or the manufacturer's recommendations for cooling-water flow rate. When a compressor is located above or below sea level, multiply its rated free-air capacity by the appropriate altitude correction factor obtained from the CAGI—*Compressed Air and Gas Handbook* or Baumeister and Marks—*Standard Handbook for Mechanical Engineers*.

VACUUM-SYSTEM PUMP-DOWN TIME

An industrial vacuum system with a 200-ft^3 (5.7-m^3) receiver serving cleaning outlets is to operate to within 2.5 inHg (9.7 kPa) absolute of the barometer when the barometer is 29.8 inHg (115.1 kPa). How long will it take to evacuate the receiver to this pressure when a single-stage vacuum pump with a displacement of 60 ft^3/min (0.03 m^3/s) is used? The pump is rated to dead end at a 29.0-inHg (112.1-kPa) vacuum when the barometer is 30.0 inHg (115.9 kPa). The pump volumetric efficiency is shown in Fig. 3.

Calculation Procedure:

1. *Compute the pump operating vacuum*

The pump must operate to within 2.5 inHg (9.7 kPa) of the barometer, or a vacuum of 29.8 − 2.5 = 27.3 inHg (105.5 kPa).

2. *Compute the quantity of free air removed from the receiver*

Select a number of absolute pressures between 29.8 inHg (115.1 kPa), the actual barometric pressure, and the final receiver pressure, 2.5 inHg (9.7 kPa); and list them in the first column of a table such as Table 7. Assume equal pressure reductions—say 3 inHg (11.6 kPa)—for each step except the last few, where smaller reductions have been assumed to ensure greater accuracy.

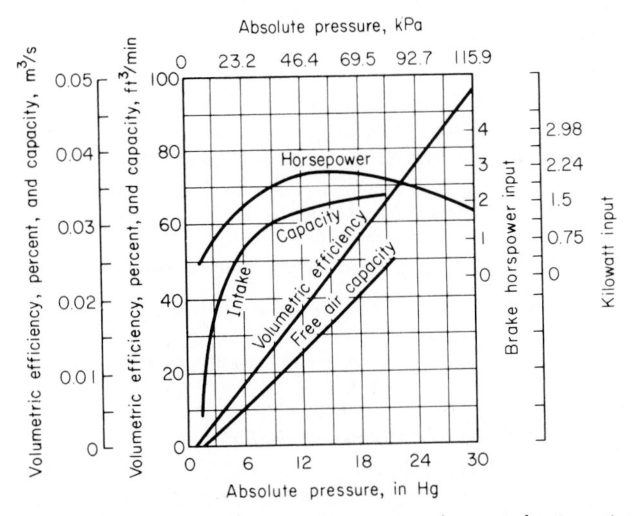

FIG. 3 Capacity, power-input, and efficiency curves for a typical reciprocating vacuum pump.

TABLE 7 Evacuation Time Calculations

Absolute pressure in receiver, inHg (kPa)	P_r/P_a	Quantity of free air, ft^3 (m^3)		Average volumetric efficiency, Fig. 2	Free-air capacity, ft^3/min (m^3/s)	Evacuation time, min
		In receiver	Removed			
29.8 (115.1)	1.000	200.0 (61.0)	0.0 (0.0)			
26.8 (103.5)	0.899	179.8 (54.8)	20.2 (6.2)	0.91	54.6 (0.026)	0.370
23.8 (92.0)	0.798	159.6 (48.6)	20.2 (6.2)	0.81	48.6 (0.023)	0.415
20.8 (80.4)	0.698	139.6 (42.6)	20.0 (6.1)	0.72	43.2 (0.020)	0.464
17.8 (68.8)	0.597	119.4 (36.4)	20.2 (6.2)	0.62	37.2 (0.018)	0.544
					Total time required	9.019

Enter in the second column of Table 7 the ratio of the absolute pressure in the receiver to the atmospheric pressure, or P_r/P_a, both expressed in inHg. Thus, for the second step, $P_r/P_a = 26.8/29.8 = 0.899$.

The amount of air remaining in the receiver, measured at atmospheric conditions, is then the product of the receiver volume, 200 ft^3 (5.7 m^3), and the ratio of the pressures. Or, for the second pressure reduction, $200(0.899) = 179.8$ ft^3 (5.1 m^3). Enter the result in the third column of Table 7. This computation is a simple application of the gas laws with the receiver temperature assumed constant. Assumption of a constant air temperature is valid because, although the air temperature varies during pumping down, the overall effect is that of a constant temperature.

Find the quantity of air removed from the receiver by successive subtraction of the values in the third column. Thus, for the second pressure step, the air removed from the receiver $= 200.0 - 179.8 = 20.2$ ft^3 (0.6 m^3) and so on for the remaining steps. Enter the result of each subtraction in the fourth column of Table 7.

3. *Compute the actual quantity of air handled by the pump*

The volumetric efficiency of a vacuum pump varies during each pressure reduction. To simplify the pump-down time calculation, an average value for the volumetric efficiency can be used for each step in the receiver pressure reduction. Find the average volumetric efficiency for this vacuum pump from Fig. 3. Thus, for the pressure reduction from 29.8 to 26.8 inHg (115.1 to 103.5 kPa), the volumetric efficiency is found at $(29.8 + 26.8)/2 = 28.3$ inHg (109.3 kPa) to be 91 percent. Enter this value in the fifth column of Table 7. Follow the same procedure to find the remaining values, and enter them as shown.

The actual quantity of free air this vacuum pump can handle is numerically equal to the product of the volumetric efficiency, column 5, Table 7, and the pump piston displacement. Or, for the above pressure reduction, free-air capacity $= 0.91(60) = 54.6$ ft^3/min (0.026 m^3/s). Enter this result in column 6, Table 7.

4. *Compute the pump-down time for each pressure reduction*

The second line of Table 7 shows, in column 4, that at an absolute pressure of 26.8 inHg (103.5 kPa), 20.2 ft^3 (0.6 m^3) of free air is removed from the receiver. However, the vacuum pump can handle 54.6 ft^3/min (0.03 m^3/s), column 6. Since the time required to remove air from the receiver is (ft^3 removed)/(cylinder capacity, ft^3/min), the time required to remove 20.2 ft^3 (0.6 m^3) is $20.2/54.6$ ft^3 $= 0.370$ min.

Compute the required time for each pressure step in the same manner. The total pump-down time is then the sum of the individual times, or 9.019 min, column 7, Table 7. This result is suitable for all usual design purposes because it closely approximates the actual time required, and the errors involved are so slight as to be negligible. Leakage into industrial-plant vacuum systems often equals the volume handled by the vacuum pump.

5. *Use the pump-down time for compressor selection*

To choose an industrial vacuum pump using the pump-down procedure described in steps 1 to 4, (*a*) obtain the characteristics curves for several makes and capacities of vacuum pumps; (*b*) compute the pump-down time for each pump, using the procedure in steps 1 to 4; (*c*) compute the air inflow to the system, based on the free-air capacity of each outlet and the number of outlets in the system; (*d*) compute how long the pump must run to handle the air inflow; and (*e*) choose the pump having the shortest running time and smallest required power input.

Thus, with 10 vacuum outlets each having a free-air flow of 50 ft^3/h (1.4 m^3/s), the total air inflow is 10(50) = 500 ft^3/h (14.2 m^3/s). This means that a 200-ft^3 (5.7-m^3) receiver would be filled 500/200 = 2.5 times per hour. Since the pump discussed in steps 1 to 4 requires approximately 9 min to reduce the receiver pressure from atmospheric to 2.5 inHg absolute (9.7 kPa), its running time to serve these outlets would be 9(2.5) = 22.5 min, approximately. The power input to this vacuum pump, Fig. 3, ranges from a minimum of about 1 hp (0.7 kW) to a maximum of about 3 hp (2.2 kW).

If another pump could evacuate this receiver in 6 min and needed only 2.5 hp (1.9 kW) as the maximum power input, it might be a better choice, provided that its first cost were not several times that of the other pump. Use the methods of engineering economics to compare the economic merits of the two pumps.

Related Calculations: Note carefully that the procedure given here applies to industrial vacuum systems used for cleaning, maintenance, and similar purposes. The procedure should not be used for high-vacuum systems applied to production processes, experimental laboratories, etc. Use instead the method given in the next calculation procedure in this section.

To be certain that the correct pump-down time is obtained, many engineers include the volume of the system piping in the computation. This is done by computing the volume of all pipes in the system and adding the result to the receiver volume. This, in effect, increases the receiver volume that must be pumped down and gives a more accurate estimate of the probable pump-down time. Some engineers also add a leakage allowance of up to 100 percent of the sum of the receiver and piping volume. Thus, if the piping volume in the above system were 50 ft^3 (1.4 m^3), the total volume to be evacuated would be 2(200 + 50) = 500 ft^3 (14.2 m^3). The factor 2 in this expression was inserted to reflect the 100 percent leakage; i.e., the pump must handle the receiver and piping volume plus the leakage, or twice the sum of the receiver and piping volume.

Some industrial vacuum pumps are standard reciprocating air compressors run in the reverse of their normal direction after slight modification. The vacuum lines are connected to the receiver, from which the compressor takes it suction. After removing air from the receiver, the compressor discharges to the atmosphere.

VACUUM-PUMP SELECTION FOR HIGH-VACUUM SYSTEMS

Choose a mechanical vacuum pump for use in a laboratory fitted with a vacuum system having a total volume, including the piping, of 12,000 ft^3 (339.8 m^3). The operating pressure of the system is 0.10 torr (0.02 kPa), and the optimum pump-down time is 150 min. (*Note:* 1 torr = 1 mmHg = 0.2 kPa.)

Calculation Procedure:

1. *Make a tentative choice of pump type*

Mechanical vacuum pumps of the reciprocating type are well suited for system pressures in the 0.0001- to 760-torr (2×10^{-5} to 115.6-kPa) range. Hence, this type of pump will be considered first to see whether it meets the desired pump-down time.

2. *Obtain the pump characteristic curves*

Many manufacturers publish pump-down factor curves such as those in Fig. 4*a* and *b*. These curves are usually published as part of the engineering data for a given line of pumps. Obtain the curves from the manufacturers whose pumps are being considered.

3. *Compute the pump-down time for the pumps being considered*

Three reciprocating pumps can serve this system: a single-stage pump, a compound or two-stage pump, or a combination of a mechanical booster and a single-stage backing or roughing-down pump. Figure 4 gives the pump-down factor for each type of pump.

To use the pump-down factor, apply this relation: $t = VF/d$, where t = pump-down time, min; V = system volume, ft^3; F = pump-down factor for the pump; d = pump displacement, ft^3/min.

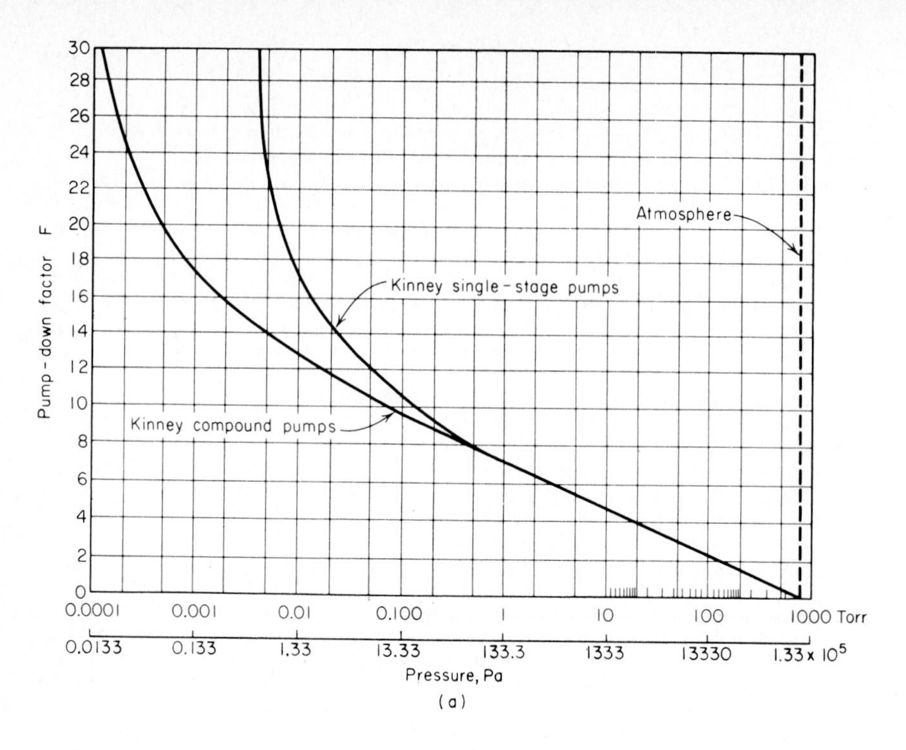

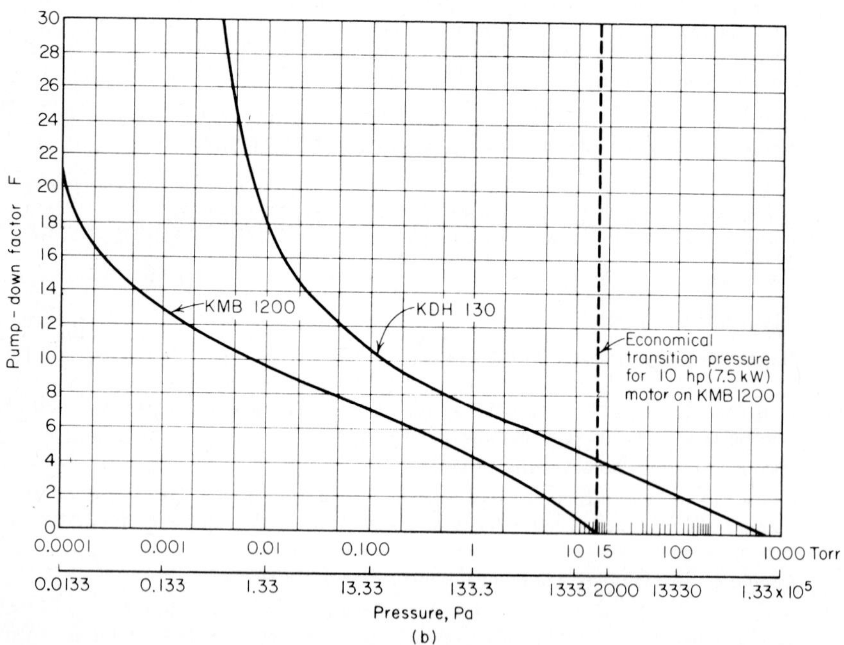

FIG. 4 (*a*) Pump-down factor for single-stage and compound vacuum pumps; (*b*) pump-down factor for mechanical booster and backing pump. *(After Kinney Vacuum Division, The New York Air Brake Company, and Van Atta.)*

Thus, for a single-stage pump, Fig. 4*a* shows that $F = 10.8$ for a pressure of 0.10 torr (1.5 kPa). Assuming a pump displacement of 1000 ft^3/min (0.5 m^3/s), $t = 12{,}000(10.8)/1000 = 129.6$ min, say 130 min.

For a compound pump, $F = 9.5$ from Fig. 4*a*. Hence, a compound pump having the same displacement, or 1000 ft^3/min (0.5 m^3/s), will require $t = 12{,}000(9.5)/1000 = 114.0$ min.

With a combination arrangement, the backing or roughing pump, a 130-ft^3/min (0.06-m^3/s) unit, reduces the system pressure from atmospheric, 760 torr (115.6 kPa), to the economical transition pressure, 15 torr (2.3 kPa), Fig. 4*b*. Then the single-stage mechanical booster pump, a 1200-ft^3/min (0.6-m^3/s) unit, takes over and in combination with the backing pump reduces the pressure to the desired level, or 0.10 torr (1.5 Pa). During this part of the cycle, the unit operates as a two-stage pump. Hence, the total pump-down time consists of the sum of the backing-pump and booster-pump times. The pump-down factors are, respectively, 4.2 for the backing pump at 15 torr (2.3 kPa) and 6.9 for the booster pump at 0.10 torr (1.5 Pa). Hence, the respective pump-down times are $t_1 = 12{,}000(4.2)/130 = 388$ min; $t_2 = 12{,}000(6.9)/1200 = 69$ min. The total time is thus $388 + 69 = 457$ min.

The pump-down time with the combination arrangment is greater than the optimum 150 min. Where a future lower operating pressure is anticipated, making the combination arrangement desirable, an additional large-capacity single-stage roughing pump can be used to assist the 130-ft^3/min (0.06-m^3/s) unit. This large-capacity unit is operated until the transition pressure is reached and roughing down is finished. The pump is then shut off, and the balance of the pumping down is carried on by the combination unit. This keeps the power consumption at a minimum.

Thus, if a 1200-ft^3/min (0.06-m^3/s) single-stage roughing pump were used to reduce the pressure to 15 torr (2.3 kPa), its pump-down time would be $t = 12{,}000(4.0)/1200 = 40$ min. The total pump-down time for the combination would then be $40 + 69 = 109$ min, using the time computed above for the two pumps in combination.

4. *Apply the respective system factors*

Studies and experience show that the calculated pump-down time for a vacuum system must be corrected by an appropriate system factor. This factor makes allowance for the normal outgassing of surfaces exposed to atmospheric air. It also provides a basis for judging whether a system is pumping down normally or whether some problem exists that must be corrected. Table 8 lists typical system factors that have proved reliable in many tests. To use the system factor for any pump, apply it this way: $t_a = tS$, where t_a = actual pump-down time, min; t = computed pump-down time from step 3, min; S = system factor for the type of pump being considered.

Thus, by using the appropriate system factor for each pump, the actual pump-down time for the single-stage mechanical pump is $t_a = 130(1.5) = 195$ min. For the compound mechanical pump, $t_a = 114(1.25) = 142.5$ min. For the combination mechanical booster pump, $t_a = 190(1.35) = 147$ min.

TABLE 8 Recommended System Factors*

Pressure range		System factors		
torr	Pa	Single-stage mechanical pump	Compound mechanical pump	Mechanical booster pump*
760–20	115.6 kPa–3000	1.0	1.0	. . .
20–1	3000–150	1.1	1.1	1.15
1–0.5	150–76	1.25	1.25	1.15
0.5–0.1	76–15	1.5	1.25	1.35
0.1–0.02	15–3	. . .	1.25	1.35
0.02–0.001	3–0.15	. . .	. . .	2.0

*Based on bypass operation until the booster pump is put into operation. Larger system factors apply if rough pumping flow must pass through the idling mechanical booster. Any time needed for operating valves and getting the mechanical booster pump up to speed must also be added.

Source: From Van Atta—*Vacuum Science and Engineering*, McGraw-Hill.

5. *Choose the pump to use*

Based on the actual pump-down time, either the compound mechanical pump or the combination mechanical booster pump can be used. In the final choice of the pump, other factors should be taken into consideration—first cost, operating cost, maintenance cost, reliability, and probable future pressure requirements in the system. Where future lower pressure requirements are not expected, the compound mechanical pump would be a good choice. However, if lower operating pressures are anticipated in the future, the combination mechanical booster pump would probably be a better choice.

Van Atta[1] gives the following typical examples of pumps chosen for vacuum systems:

Pressure range, torr	Typical pump choice
Down to 50 (7.6 kPa)	Single-stage oil-sealed rotary; large water or vapor load may require use of refrigerated traps
0.05 to 0.01 (7.6 to 1.5 Pa)	Single-stage or compound oil-sealed pump plus refrigerated traps, particularly at the lower pressure limit
0.01 to 0.005 (1.5 to 0.76 Pa)	Compound oil-sealed plus refrigerated traps, or single-stage pumps backing diffusion pumps if a continuous large evolution of gas is expected
1 to 0.0001 (152.1 to 0.015 Pa)	Mechanical booster and backing pump combination with interstage refrigerated condenser and cooled vapor trap at the high-vacuum inlet for extreme freedom from vapor contamination
0.0005 and lower (0.076 Pa and lower)	Single-stage pumps backing diffusion pumps, with refrigerated traps on the high-vacuum side of the diffusion pumps and possibly between the single-stage and diffusion pumps if evolution of condensable vapor is expected

VACUUM-SYSTEM PUMPING SPEED AND PIPE SIZE

A laboratory vacuum system has a volume of 500 ft^3 (14.2 m^3). Leakage into the system is expected at the rate of 0.00035 ft^3/min (0.00001 m^3/min). What backing pump speed, i.e., displacement, should an oil-sealed vacuum pump serving this system have if the pump blocking pressure is 0.150 mmHg and the desired operating pressure is 0.0002 mmHg? What should the speed of the diffusion pump be? What pipe size is needed for the connecting pipe of the backing pump if it has a displacement or pumping speed of 380 ft^3/min (10.8 m^3/min) at 0.150 mmHg and a length of 15 ft (4.6 m)?

Calculation Procedure:

1. *Compute the required backing pump speed*

Use the relation $d_b = G/P_b$, where d_b = backing pump speed or pump displacement, ft^3/min; G = gas leakage or flow rate, mm·min/ft^3. To convert the gas or leakage flow rate to mm·min/ft^3, multiply the ft^3/min by 760 mm, the standard atmospheric pressure, mmHg. Thus, $d_b = 760(0.00035)/0.150 = 1.775$ ft^3/min (0.05 m^3/min).

2. *Select the actual backing pump speed*

For practical purposes, since gas leakage and outgassing are impossible to calculate accurately, a backing pump speed or displacement of at least twice the computed value, or 2(1.775) = 3.550 ft^3/min (0.1 m^3/min), say 4 ft^3/min (0.11 m^3/min), would probably be used.

If this backing pump is to be used for pumping down the system, compute the pump-down

[1] C. M. Van Atta—*Vacuum Science and Engineering*, McGraw-Hill, New York, 1965.

time as shown in the previous calculation procedure. Should the pump-down time be excessive, increase the pump displacement until a suitable pump-down time is obtained.

3. *Compute the diffusion pump speed*

The diffusion pump reduces the system pressure from the blocking point, 0.150 mmHg, to the system operating pressure of 0.0002 mmHg. (*Note:* 1 torr = 1 mmHg.) Compute the diffusion pump speed from $d_d = G/P_d$, where d_d = diffusion pump speed, ft^3/min; P_d = diffusion-pump operating pressure, mmHg. Or, $d_d = 760(0.00035)/0.0002 = 1330$ ft^3/min (37.7 m^3/min). To allow for excessive leaks, outgassing, and manifold pressure loss, a 3000- or 4000-ft^3/min (84.9- or 113.2-m^3/min) diffusion pump would be chosen. To ensure reliability of service, two diffusion pumps would be chosen so that one could operate while the other was being overhauled.

4. *Compute the size of the connecting pipe*

In usual vacuum-pump practice, the pressure drop in pipes serving mechanical pumps is not allowed to exceed 20 percent of the inlet pressure prevailing under steady operating conditions. A correctly designed vacuum system, where this pressure loss is not exceeded, will have a pump-down time which closely approximates that obtained under ideal conditions.

Compute the pressure drop in the high-pressure region of vacuum pumps from $p_d = 1.9d_b L/d^4$, where p_d = pipe pressure drop, μm; d_b = backing pump displacement or speed, ft^3/min; L = pipe length, ft; d = inside diameter of pipe, in. Since the pressure drop should not exceed 20 percent of the inlet or system operating pressure, the drop for a backing pump is based on its blocking pressure, or 0.150 mmHg, or 150 μm. Hence $p_d = 0.20(150) = 30$ μm. Then $30 = 1.9(380)(15)/d^4$, and $d = 4.35$ in (110.5 mm). Use a 5-in (127.0-mm) diameter pipe.

In the low-pressure region, the diameter of the converting pipe should equal, or be larger than, the pump inlet connection. Whenever the size of a pump is increased, the diameter of the pipe should also be increased to conform with the above guide.

Related Calculations: Use the general procedures given here for laboratory- and production-type high-vacuum systems.

Materials Handling

REFERENCES: Apple—*Materials Handling Systems Design*, Wiley; Wasp—*Slurry Pipeline Transportation*, Trans Tech; Machinery Studies—*Materials Handling Equipment*, Business Trends; Bolz—*Materials Handling Handbook*, Wiley; Reisner and Eisenhart—*Bins and Bunkers for Handling Bulk Materials*, Trans Tech; *Chemical Engineering* Magazine—*Pneumatic Conveying of Bulk Materials*, McGraw-Hill; Hudson—*Conveyors*, Wiley; Buffalo Forge Company—*Fan Engineering*; Stanier—*Plant Engineering Handbook*, McGraw-Hill; Baumeister and Marks—*Standard Handbook for Mechanical Engineers*, McGraw-Hill.

BULK MATERIAL ELEVATOR AND CONVEYOR SELECTION

Choose a bucket elevator to handle 150 tons/h (136.1 t/h) of abrasive material weighing 50 lb/ft^3 (800.5 kg/m^3) through a vertical distance of 75 ft (22.9 m) at a speed of 100 ft/min (30.5 m/min). What hp input is required to drive the elevator? The bucket elevator discharges onto a horizontal conveyor which must transport the material 1400 ft (426.7 m). Choose the type of conveyor to use, and determine the required power input needed to drive it.

Calculation Procedure:

1. *Select the type of elevator to use*

Table 1 summarizes the various characteristics of bucket elevators used to transport bulk materials vertically. This table shows that a continuous bucket elevator would be a good choice, because it is a recommended type for abrasive materials. The second choice would be a pivoted bucket elevator. However, the continuous bucket type is popular and will be chosen for this application.

2. *Compute the elevator height*

To allow for satisfactory loading of the bulk material, the elevator length is usually increased by about 5 ft (1.5 m) more than the vertical lift. Hence, the elevator height = 75 + 5 = 80 ft (24.4 m).

TABLE 1 Bucket Elevators

	Centrifugal discharge	Perfect discharge	Continuous bucket	Gravity discharge	Pivoted bucket
Carrying paths	Vertical	Vertical to inclination 15° from vertical	Vertical to inclination 15° from vertical	Vertical and horizontal	Vertical and horizontal
Capacity range, tons/h (t/h), material weighing 50 lb/ft³ (800.5 kg/m³)	78 (70.8)	34 (30.8)	345 (312.9)	191 (173.3)	255 (231.3)
Speed range, ft/min (m/min)	306 (93.3)	120 (36.6)	100 (30.5)	100 (30.5)	80 (24.4)
Location of loading point	Boot	Boot	Boot	On lower horizontal run	On lower horizontal run
Location of discharge point	Over head wheel	Over head wheel	Over head wheel	On horizontal run	On horizontal run
Handling abrasive materials	Not preferred	Not preferred	Recommended	Not recommended	Recommended

Source: Link-Belt Div. of FMC Corp.

3. *Compute the required power input to the elevator*

Use the relation $hp = 2CH/1000$, where C = elevator capacity, tons/h; H = elevator height, ft. Thus, for this elevator, $hp = 2(150)(80)/1000 = 24.0$ hp (17.9 kW).

The power input relation given above is valid for continuous-bucket, centrifugal-discharge, perfect-discharge, and super-capacity elevators. A 25-hp (18.7-kW) motor would probably be chosen for this elevator.

4. *Select the type of conveyor to use*

Since the elevator discharges onto the conveyor, the capacity of the conveyor should be the same, per unit time, as the elevator. Table 2 lists the characteristics of various types of conveyors. Study of the tabulation shows that a belt conveyor would probably be best for this application, based on the speed, capacity, and type of material it can handle. Hence, it will be chosen for this installation.

5. *Compute the required power input to the conveyor*

The power input to a conveyor is composed of two portions: the power required to move the empty belt conveyor and the power required to move the load horizontally.

Determine from Fig. 1 the power required to move the empty belt conveyor, after choosing the required belt width. Determine the belt width from Table 3.

Thus, for this conveyor, Table 3 shows that a belt width of 42 in (106.7 cm) is required to transport up to 150 tons/h (136.1 t/h) at a belt speed of 100 ft/min (30.5 m/min). [Note that the next *larger* capacity, 162 tons/h (146.9 t/h), is used when the exact capacity required is not tabulated.] Find the horsepower required to drive the empty belt by entering Fig. 1 at the belt distance between centers, 1400 ft (426.7 m), and projecting vertically upward to the belt width, 42 in (106.7 cm). At the left, read the required power input as 7.2 hp (5.4 kW).

Compute the power required to move the load horizontally from $hp = (C/100)(0.4 + 0.00345L)$, where L = distance between conveyor centers, ft; other symbols as before. For this conveyor, $hp = (150/100)(0.4 + 0.00325 \times 1400) = 6.83$ hp (5.1 kW). Hence, the total horsepower to drive this horizontal conveyor is $7.2 + 6.83 = 14.03$ hp (10.5 kW).

The total horsepower input to this conveyor installation is the sum of the elevator and conveyor belt horsepowers, or $14.03 + 24.0 = 38.03$ hp (28.4 kW).

TABLE 2 Conveyor Characteristics

	Belt conveyor	Apron conveyor	Flight conveyor	Drag chain	En masse conveyor	Screw conveyor	Vibratory conveyor
Carrying paths	Horizontal to 18°	Horizontal to 25°	Horizontal to 45°	Horizontal or slight incline, 10°	Horizontal to 90°	Horizontal to 15°; may be used up to 90° but capacity falls off rapidly	Horizontal or slight incline, 5° above or below horizontal
Capacity range, tons/h (t/h) material weighing 50 lb/ft^3 (800.5 kg/m^3)	2160 (1959.5)	100 (90.7)	360 (326.6)	20 (18.1)	100 (90.7)	150 (136.1)	100 (90.7)
Speed range, ft/min (m/min)	600 (182.9)	100 (30.5)	150 (45.7)	20 (6.1)	80 (24.4)	100 (30.5)	40 (12.2)
Location of loading point	Any point	Any point	Any point	Any point	On horizontal runs	Any point	Any point
Location of discharge point	Over end wheel and intermediate points by tripper or plow	Over end wheel	At end of trough and intermediate points by gates	At end of trough	Any point on horizontal runs by gate	At end of trough and intermediate points by gates	At end of trough
Handling abrasive materials	Recommended	Recommended	Not recommended	Recommended with special steels	Not recommended	Not preferred	Recommended

Source: Link-Belt Div. of FMC Corp.

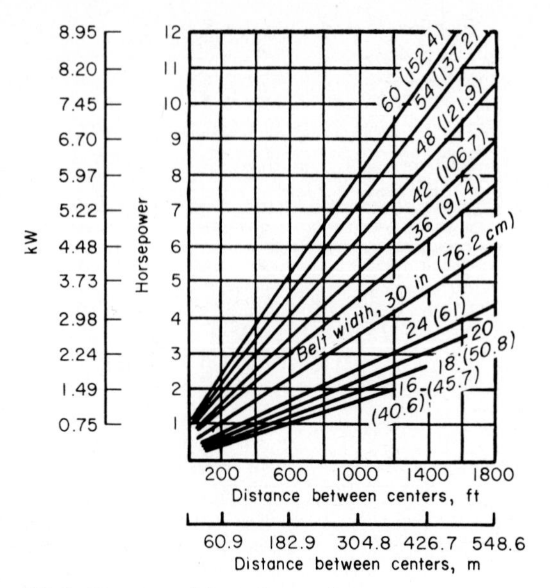

FIG. 1 Horsepower (kilowatts) required to move an empty conveyor belt at 100 ft/min (30.5 m/min).

Related Calculations: The procedure given here is valid for conveyors using rubber belts reinforced with cotton duck, open-mesh fabric, cords, or steel wires. It is also valid for stitched-canvas belts, balata belts, and flat-steel belts. The required horsepower input includes any power absorbed by idler pulleys.

Table 4 shows the minimum recommended belt widths for lumpy materials of various sizes. Maximum recommended belt speeds for various materials are shown in Table 5.

TABLE 3 Capacities of Troughed Rest [tons/h (t/h) with Belt Speed of 100 ft/min (30.5 m/min)]

Belt width, in (cm)	Weight of material, lb/ft³ (kg/m³)			
	30 (480.3)	50 (800.5)	100 (1601)	150 (2402)
30 (76.2)	47 (42.6)	79 (71.7)	158 (143.3)	237 (214.9)
36 (91.4)	69 (62.6)	114 (103.4)	228 (206.8)	342 (310.2)
42 (106.7)	97 (87.9)	162 (146.9)	324 (293.9)	486 (440.9)
48 (121.9)	130 (117.9)	215 (195.0)	430 (390.1)	645 (585.1)
60 (152.4)	207 (187.8)	345 (312.9)	690 (625.9)	1035 (938.9)

Source: United States Rubber Co.

TABLE 4 Minimum Belt Width for Lumps

Belt width, in (mm)	24 (609.6)	36 (914.4)	42 (1066.8)	48 (1219.2)
Sized materials, in (mm)	4½ (114.3)	8 (203.2)	10 (254)	12 (304.9)
Unsized material, in (mm)	8 (203.2)	14 (355.6)	20 (508)	35 (889)

TABLE 5 Maximum Belt Speeds for Various Materials

Width of belt		Light or free-flowing materials, grains dry sand, etc.		Moderately free-flowing sand, gravel, fine stone, etc.		Lump coal, coarse stone, crushed ore		Heavy sharp lumpy materials, heavy ores, lump coke	
in	mm	ft/min	m/min	ft/min	m/min	ft/min	m/min	ft/min	m/min
12–14	305–356	400	122	250	76	—	—	—	—
16–18	406–457	500	152	300	91	250	76	—	—
20–24	508–610	600	183	400	122	350	107	250	76
30–36	762–914	750	229	500	152	400	122	300	91

When a conveyor belt is equipped with a tripper, the belt must rise about 5 ft (1.5 m) above its horizontal plane of travel.

This rise must be included in the vertical-lift power input computation. When the tripper is driven by the belt, allow 1 hp (0.75 kW) for a 16-in (406.4-mm) belt, 3 hp (2.2 kW) for a 36-in (914.4-mm) belt, and 7 hp (5.2 kW) for a 60-in (1524-mm) belt. Where a rotary cleaning brush is driven by the conveyor shaft, allow about the same power input to the brush for belts of various widths.

SCREW CONVEYOR POWER INPUT AND CAPACITY

What is the required power input for a 100-ft (30.5-m) long screw conveyor handling dry coal ashes having a maximum density of 40 lb/ft^3 (640.4 kg/m^3) if the conveyor capacity is 30 tons/h (27.2 t/h)?

Calculation Procedure:

1. *Select the conveyor diameter and speed*

Refer to a manufacturer's engineering data or Table 6 for a listing of recommended screw conveyor diameters and speeds for various types of materials. Dry coal ashes are commonly rated as group 3 materials, Table 7, i.e., materials with small mixed lumps with fines.

To determine a suitable screw diameter, assume two typical values and obtain the recommended rpm from the sources listed above or Table 6. Thus, the maximum rpm recommended for a 6-in (152.4-mm) screw when handling group 3 material is 90, as shown in Table 6; for a 20-in (508.0-mm) screw, 60 r/min. Assume a 6-in (152.4-mm) screw as a trial diameter.

TABLE 6 Screw Conveyor Capacities and Speeds

Material group	Maximum material density		Maximum r/min for diameters of:	
	lb/ft^3	kg/m^3	6 in (152 mm)	20 in (508 mm)
1	50	801	170	110
2	50	801	120	75
3	75	1201	90	60
4	100	1601	70	50
5	125	2001	30	25

TABLE 7 Material Factors for Screw Conveyors

Material group	Material type	Material factor
1	Lightweight:	
	Barley, beans, flour, oats, pulverized coal, etc.	0.5
2	Fines and granular:	
	Coal—slack or fines	0.9
	Sawdust, soda ash	0.7
	Flyash	0.4
3	Small lumps and fines:	
	Ashes, dry alum	4.0
	Salt	1.4
4	Semiabrasives; small lumps:	
	Phosphate, cement	1.4
	Clay, limestone	2.0
	Sugar, white lead	1.0
5	Abrasive lumps:	
	Wet ashes	5.0
	Sewage sludge	6.0
	Flue dust	4.0

2. *Determine the material factor for the conveyor*

A material factor is used in the screw conveyor power input computation to allow for the character of the substance handled. Table 7 lists the material factor for dry ashes as $F = 4.0$. Standard references show that the average weight of dry coal ashes is 35 to 40 lb/ft^3 (640.4 kg/m^3).

3. *Determine the conveyor size factor*

A size factor that is a function of the conveyor diameter is also used in the power input computation. Table 8 shows that for a 6-in (152.4-mm) diameter conveyor the size factor $A = 54$.

4. *Compute the required power input to the conveyor*

Use the relation $hp = 10^{-6}(ALN + CWLF)$, where hp = hp input to the screw conveyor head shaft; A = size factor from step 3; L = conveyor length, ft; N = conveyor rpm; C = quantity of material handled, ft^3/h; W = density of material, lb/ft^3; F = material factor from step 2. For this conveyor, given the data listed above, $hp = 10^{-6}(54 \times 100 \times 60 + 1500 \times 40 \times 100 \times 4.0) = 24.3$ hp (18.1 kW). With a 90 percent motor efficiency, the required motor rating would be $24.3/0.90 = 27$ hp (20.1 kW). A 30-hp (22.4-kW) motor would be chosen to drive this conveyor. Since this is not an excessive power input, the 6-in (152.4-mm) conveyor is suitable for this application.

If the calculation indicates that an excessively large power input, say 50 hp (37.3 kW) or more, is required, then the larger-diameter conveyor should be analyzed. In general, a higher initial investment in conveyor size that reduces the power input will be more than recovered by the savings in power costs.

Related Calculations: Use the procedure given here for screw or spiral conveyors and feeders handling any material that will flow. The usual screw or spiral conveyor is suitable for conveying materials for distances up to about 200 ft (60.9 m), although special designs can be built

TABLE 8 Screw Conveyor Size Factors

Conveyor diameter, in (mm)	6 (152.4)	9 (228.6)	10 (254)	12 (304.8)	16 (406.4)	18 (457.2)	20 (508)	24 (609.6)
Size factor	54	96	114	171	336	414	510	690

for greater distances. Conveyors of this type can be sloped upward to angles of 35° with the horizontal. However, the capacity of the conveyor decreases as the angle of inclination is increased. Thus the reduction in capacity at a 10° inclination is 10 percent over the horizontal capacity; at 35° the reduction is 78 percent.

The capacities of screw and spiral conveyors are generally stated in ft^3/h (m^3/h) of various classes of materials at the maximum recommended shaft rpm. As the size of the lumps in the material conveyed increases, the recommended shaft rpm decreases. The capacity of a screw or spiral conveyor at a lower speed is found from (capacity at given speed, ft^3/h) [(lower speed, r/min)/(higher speed, r/min)]. Table 6 shows typical screw conveyor capacities at usual operating speeds.

Various types of screws are used for modern conveyors. These include short-pitch, variable-pitch, cut flights, ribbon, and paddle screws. The procedure given above also applies to these screws.

DESIGN AND LAYOUT OF PNEUMATIC CONVEYING SYSTEMS

A pneumatic conveying system for handling solids in an industrial exhaust installation contains two grinding-wheel booths and one lead each for a planer, sander, and circular saw. Determine the required duct sizes, resistance, and fan capacity for this pneumatic conveying system.

Calculation Procedure:

1. *Sketch the proposed exhaust system*

Make a freehand sketch, Fig. 2, of the proposed system. Show the main and branch ducts and the booths and hoods. Indicate all major structural interferences, such as building columns, deep girders, beams, overhead conveyors, piping, etc. Draw the layout approximately to scale.

Mark on the sketch the length of each duct run. Avoid, if possible, vertical drops or rises in the main exhaust duct between the hoods and the fan. Do this by locating the main duct centerline 10 ft (3 m) or so above the finished floor.

Number each hood or booth, and give each duct run an identifying letter. Although it is not absolutely necessary, it is more convenient during the design process to have the hoods in numerical order and the duct runs in alphabetical order.

2. *Determine the required air quantities and velocities*

Prepare a listing, columns 1 and 2, Table 9, of the booths, hoods, and duct runs. Enter the required air quantities and velocities for each booth or hood and duct in Table 9, columns 3 and 4. Select

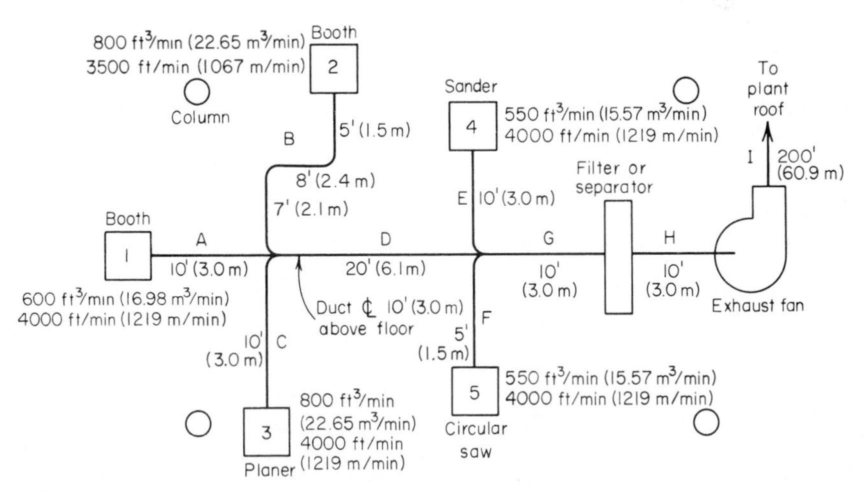

FIG. 2 Exhaust system layout.

the air quantities and velocities from the local code covering industrial exhaust systems, if such a code is available. If a code does not exist, use the ASHRAE *Guide* or Table 10.

Use extreme care in selecting the air quantities and velocities, because insufficient flow may cause dangerous atmospheric conditions. Harmful process wastes in the form of dust, gas, or moisture may injure plant personnel.

3. *Size the main and branch ducts*

Determine the required duct area by dividing the air quantity, ft^3/min (m^3/min), by the air velocity in the duct, or column 3/column 4, Table 9. Enter the result in column 5, Table 9.

Once the required duct area is known, find from Table 11 the nearest whole-number duct diameter corresponding to the required area. Avoid fractional diameters at this stage of the calculation, because ducts of these sizes are usually more expensive to fabricate. Later, if necessary, two or three duct sizes may be changed to fractional values. By selecting only whole-number diameters in the beginning, the cost of duct fabrication may be reduced somewhat. Enter the duct whole-number diameter in column 6, Table 9.

4. *Compute the actual air velocity in the duct*

Use Fig. 3 to determine the actual velocity in each duct. Enter the chart at the air quantity corresponding to that in the duct, and project vertically to the diameter curve representing the duct size. Read the actual velocity in the duct on the velocity scale, and enter the value in column 7 of Table 9.

The actual velocity in the duct should, in all cases, be equal to or greater than the design velocity shown in column 4, Table 9. If the actual velocity is less than the design velocity, decrease the duct diameter until the actual velocity is equal to or greater than the design velocity.

5. *Compute the duct velocity pressure*

With the actual velocity known, compute the corresponding velocity pressure in the duct from $h_v = (v/4005)^2$, where h_v = velocity pressure in the duct, inH_2O; v = air velocity in the duct, ft/min. Thus, for the duct run *A* in which the actual air velocity is 4300 ft/min (1310.6 m/min), $h_v = (4300/4005)^2 = 1.15$ in (29.2 mm) H_2O. Compute the actual velocity pressure in each duct run, and enter the result in column 8, Table 9.

6. *Compute the equivalent length of each duct*

Enter the total straight length of each duct, including any vertical drops, in column 9, Table 9. Use accurate lengths, because the system resistance is affected by the duct length.

Next list the equivalent length of each elbow in the duct runs in column 10, Table 9. For convenience, assume that the equivalent length of an elbow is 12 times the duct diameter in ft. Thus, an elbow in a 6-in (152.4-mm) diameter duct has an equivalent resistance of (6-in diameter/[(12 in/ft)(12)]) = 6 ft (1.83 m) of straight duct. When making this calculation, assume that all elbows have a radius equal to twice the diameter of the duct. Consider 45° bends as having the same resistance as 90° elbows. Note that branch ducts are usually arranged to enter the main duct at an angle of 45° or less. These assumptions are valid for all typical industrial exhaust systems and pneumatic conveying systems.

Find the total equivalent length of each duct by taking the sum of columns 9 and 10, Table 9, horizontally, for each duct run. Enter the result in column 11, Table 9.

7. *Determine the actual friction in each duct*

Using Fig. 3, determine the resistance, inH_2O (mmH_2O) per 100 ft (30.5 m) of each duct by entering with the air quantity and diameter of that duct. Enter the frictional resistance thus found in column 12, Table 9.

Compute actual friction in each duct by multiplying the friction per 100 ft (30.5 m) of duct, column 12, Table 9, by the total duct length, column 11 ÷ 100. Thus for duct run *A*, actual friction = 5.4(10/100) = 0.54 in (13.7 mm) H_2O. Compute the actual friction for the other duct runs in the same manner. Tabulate the results in column 13, Table 9.

8. *Compute the hood entrance losses*

Hoods are used in industrial exhaust systems to remove vapors, dust, fumes, and other undesirable airborne contaminants from the work area. The hood entrance loss, which depends upon the hood configuration, is usually expressed as a certain percentage of the velocity pressure in the branch

TABLE 9 Exhaust System Design Calculations

(1) Booth or hood	(2) Duct run	(3) ft^3/min (m^3/min) in duct	(4) Design velocity, ft/min (m/min)	(5) Duct area = column 3/column 4, ft^2 (m^2)	(6) Duct diameter, in (mm)	(7) Actual velocity, ft/min (m/min)	(8) Actual velocity pressure, inH_2O (mmH_2O)	(9) Length of straight duct, ft (m)	(10) Equivalent length of elbows, ft (m)	(11) Total duct length = column 9 + column 10, ft (m)	(12) Friction per 100 ft (30 m) of duct, inH_2O (mmH_2O)	(13) Actual friction, inH_2O (mmH_2O)
1	*A*	600	4000	0.150	5	4300	1.15	10	0	10	5.4	0.54
		(16.98)	(1219)	(0.014)	(127)	(1311)	(29.2)	(3.0)	(0)	(3.0)	(137.2)	(13.7)
2	*B*	800	3500	0.228	6	4200	1.0	20	18	38	4.0	1.57
		(22.65)	(1067)	(0.021)	(152)	(1280)	(25.4)	(6.1)	(5.5)	(11.6)	(101.6)	(39.9)
3	*C*	800	4000	0.200	6	4200	1.0	10	6	16	4.0	0.64
		(22.65)	(1219)	(0.019)	(152)	(1280)	(25.4)	(3.0)	(1.8)	(4.8)	(101.6)	(16.3)
	D	2200	4000	0.550	10	4000	1.0	20	0	20	2.1	0.42
		(62.28)	(1219)	(0.051)	(254)	(1219)	(25.4)	(6.1)	(0)	(6.1)	(53.3)	(10.7)
4	*E*	550	4000	0.137	5	4000	1.0	10	5	15	4.6	0.69
		(15.57)	(1219)	(0.013)	(127)	(1219)	(25.4)	(3.0)	(1.5)	(4.5)	(116.8)	(17.5)
5	*F*	550	4000	0.137	5	4000	1.0	5	5	10	4.6	0.46
		(15.57)	(1219)	(0.013)	(127)	(1219)	(25.4)	(1.5)	(1.5)	(3.0)	(116.8)	(11.7)
	G	3300	4000	0.825	12	4200	1.0	10	0	10	1.9	0.19
		(93.42)	(1219)	(0.077)	(305)	(1280)	(25.4)	(3.0)	(0)	(3.0)	(48.3)	(4.8)
	H	3300	3000	1.10	14	3000	0.55	10	14	24	0.84	0.20
		(93.42)	(1914)	(0.102)	(356)	(914)	(13.9)	(3.0)	(4.3)	(7.3)	(21.3)	(5.1)
	I	3300	2000	1.65	18	2000	0.25	200	0	200	0.25	0.50
		(93.42)	(610)	(0.153)	(457)	(610)	(6.4)	(60.9)	(0)	(60.9)	(6.4)	(12.7)

TABLE 9 Exhaust System Design Calculations *(Continued)*

System resistance						
		Hood number				
		1	2	3	4	5
Velocity pressure in hood branch, in (mm) H_2O		1.15 (29.2)	1.0 (25.4)	1.0 (25.4)	1.0 (25.4)	1.0 (25.4)
		50	11	50	60	60
Entrance loss (% of velocity pressure)		(50)	(11)	(50)	(60)	(60)
Entrance loss, in (mm) H_2O		0.58 (14.6)	0.11 (2.8)	0.50 (12.7)	0.60 (15.2)	0.60 (15.2)
Branch and main duct resistances	*A*	0.54 (13.7)				
	B	. . . . . .	1.57 (39.9)			
	C	. . . . . .	. . . . . .	0.64 (16.3)		
	D	0.42 (10.7)	0.42 (10.7)	0.42 (10.7)		
	E	. . . . . .	. . . . . .	. . . . . .	0.69 (17.5)	
	F	. . . . . .	. . . . . .	. . . . . .	. . . . . .	0.46 (11.7)
	G	0.19 (4.8)	0.19 (4.8)	0.19 (4.8)	0.19 (4.8)	0.19 (4.8)
	H	0.20 (5.1)	0.20 (5.1)	0.20 (5.1)	0.20 (5.1)	0.20 (5.1)
	I	0.50 (12.7)	0.50 (12.7)	0.50 (12.7)	0.50 (12.7)	0.50 (12.7)
Collector or filter resistance, in (mm) H_2O		2.00 (50.8)	2.00 (50.8)	2.00 (50.8)	2.00 (50.8)	2.00 (50.8)
Total resistance in each branch, in (mm) H_2O		4.43 (112.4)	4.99 (126.8)	4.45 (113.1)	4.18 (106.1)	3.95 (100.3)

TABLE 10 Recommended Exhaust Air Quantities

Operation	ft^3/min (m^3/min)	Branch duct velocity, ft/min (m/min)	Branch duct diameter, in (mm)
Sanding:			
Single drum, [10-in (25.4-cm) diameter]	400 (11.32)	4000 (1219)	4 (101.6)
Disk	550 (15.57)	4000 (1219)	5 (127)
Circular saws [16- to 24-in (40.6- to 60.9-cm) diameter]	450 (12.74)	4000 (1219)	4.5 (114.3)
Shoe machinery	550 (15.57)	4000 (1219)	5 (127)
Buffing and polishing wheels [16- to 24-in (40.6- to 60.9-cm) diameter]	600 (16.98)	4500 (1372)	5 (127)
Grinding wheels [16- to 20-in (40.6- to 50.8-cm) diameter]	600 (16.98)	4500 (1372)	5 (127)
Abrasive blast rooms	. . .	3500 (1067)	
Pharmaceuticals	. . .	3000 (1067)	

Conveying velocities

Material conveyed	Conveying velocity, ft/min (m/min)
Vapors, gases, fumes, fine dusts	1500 to 2000 (457 to 610)
Fine dry dusts	3000 (914)
Average industrial dusts	3500 (1067)
Coarse particles	3500 to 4500 (1067 to 1372)
Large particles, heavy loads, moist materials, pneumatic conveying	4500 and higher (1372 and higher)

TABLE 11 Duct Diameters and Areas

Diameter		Area	
in	mm	ft^2	m^2
4.0	102	0.0873	0.008
5.0	127	0.1364	0.013
6.0	152.4	0.1964	0.018
7.0	178	0.2673	0.025
8.0	203.2	0.3491	0.032
10.0	254	0.5454	0.051
12	305	0.7854	0.073
14	356	1.069	0.099
16	406.4	1.396	0.130
18	457.2	1.767	0.164
20	508	2.182	0.203
22	559	2.640	0.245
24	610	3.142	0.292

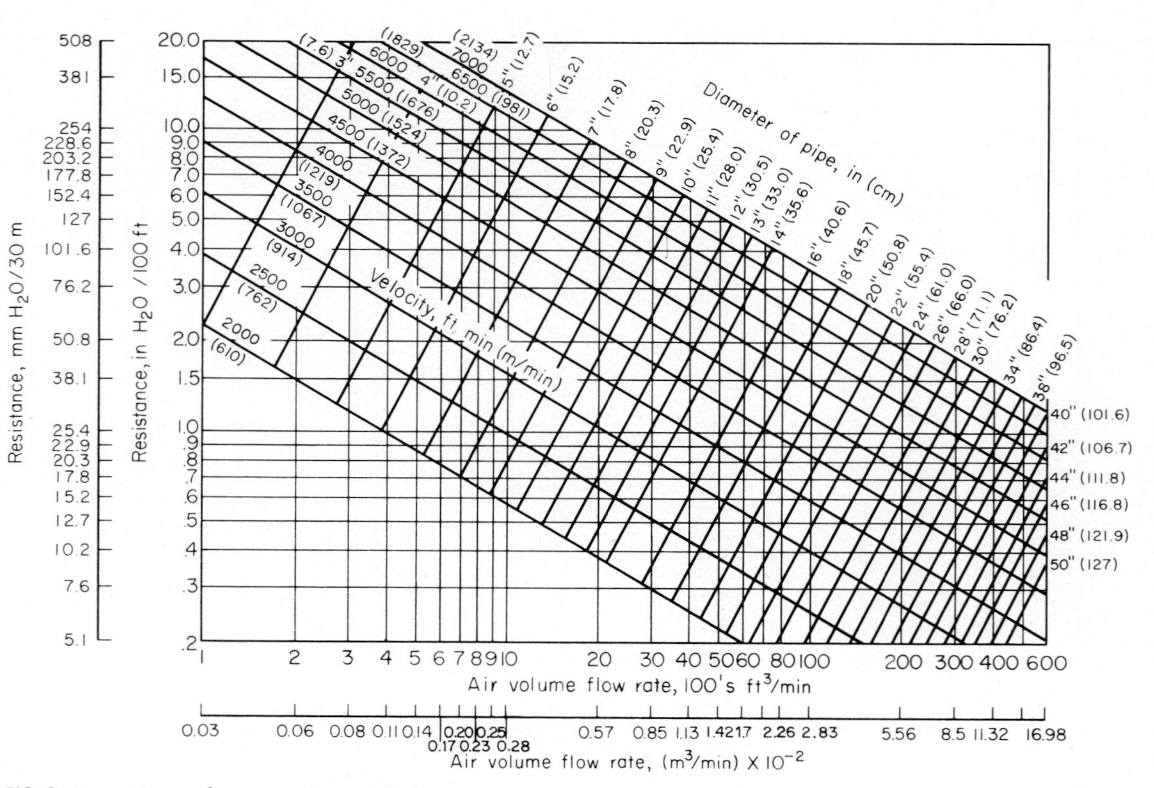

FIG. 3 Duct resistance chart. *(American Air Filter Co.)*

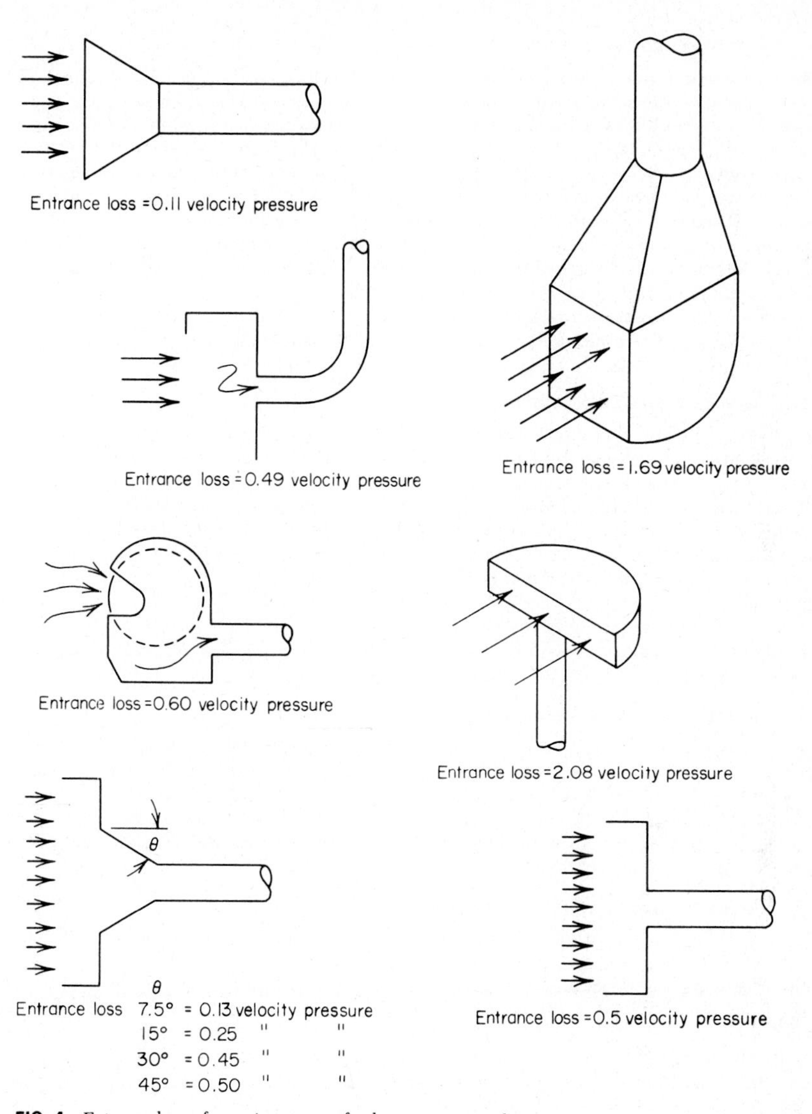

FIG. 4 Entrance losses for various types of exhaust-system intakes.

duct connected to the hood, Fig. 4. Since the hood entrance loss usually accounts for a large portion of the branch resistance, the entrance loss chosen should always be on the safe side.

List the hood designation number under the "System Resistance" heading, as shown in Table 9. Under each hood designation number, list the velocity pressure in the branch connected to that hood. Obtain this value from column 8, Table 9. List under the velocity pressure, the hood entrance loss from Fig. 4 for the particular type of hood used in that duct run. Take the product of these two values, and enter the result under the hood number on the "entrance loss, inH_2O" line. Thus, for hood 1, entrance loss = 1.15(0.50) = 0.58 in (14.7 mm) H_2O. Follow the same procedure for the other hoods listed.

9. *Find the resistance of each branch run*

List the main and branch runs, *A* through *F*, Table 9. Trace out each main and branch run in Fig. 2, and enter the actual friction listed in column 3 of Table 9. Thus for booth 1, the main and branch runs consist of *A*, *D*, *G*, *H*, and *I*. Insert the actual friction, in (mm) H_2O, as shown in Table 9, or $A = 9.54(242.3)$, $D = 0.42(10.7)$, $G = 0.19(4.8)$, $H = 0.20(5.1)$, $I = 0.50(12.7)$.

Determine the filter friction loss from the manufacturer's engineering data. It is common practice to design industrial exhaust systems on the basis of dirty filters or separators; i.e., the frictional resistance used in the design calculations is the resistance of a filter or separator containing the maximum amount of dust allowable under normal operating conditions. The frictional resistance of dirty filters can vary from 0.5 to 6 in (12.7 to 152.4 mm) H_2O or more. Assume that the frictional resistance of the filter used in this industrial exhaust system is 2.0 in (50.8 mm) H_2O.

Add the filter resistance to the main and branch duct resistance as shown in Table 9. Find the sum of each column in the table, as shown. This is the total resistance in each branch, inH_2O, Table 9.

10. *Balance the exhaust system*

Inspection of the lower part of Table 9 shows that the computed branch resistances are unequal. This condition is usually encountered during system design. To balance the system, certain duct sizes must be changed to produce equal resistance in all ducts. Or, if possible, certain ducts can be shortened. If duct shortening is not possible, as is often the case, an exhaust fan capable of operating against the largest resistance in a branch can be chosen. If this alternative is selected, special dampers must be fitted to the air inlets of the booths or ducts. For economical system operation, choose the balancing method that permits the exhaust fan to operate against the minimum resistance.

In the system being considered here, a fairly accurate balance can be obtained by decreasing the size of ducts *E* and *F* to 4.75 in (120.7 mm) and 4.375 in (111.1 mm), respectively. Duct *B* would be increased to 6.5 in (165.1 mm) in diameter.

11. *Choose the exhaust fan capacity and static pressure*

Find the required exhaust fan capacity in ft^3/min from the sum of the airflows in the ducts, *A* through *H*, column 3, Table 9, or 3300 ft^3/min (93.5 m^3/min). Choose a static pressure equal to or greater than the total resistance in the branch duct having the greatest resistance. Since this is slightly less than 4.5 in (114.3 mm) H_2O, a fan developing 4.5 in (114.3 mm) H_2O static pressure will be chosen. A 10 percent safety factor is usually applied to these values, giving a capacity of 3600 ft^3/min (101.9 m^3/min) and a static pressure of 5.0 in (127 mm) H_2O for this system.

12. *Select the duct material and thickness*

Galvanized sheet steel is popular for industrial exhaust systems, except where corrosive fumes and gases rule out galvanized material. Under these conditions, plastic, tile, stainless steel, or composition ducts may be substituted for galvanized ducts. Table 12 shows the recommended metal gage for galvanized ducts of various diameters. Do not use galvanized-steel ducts for gas temperatures higher than 400°F (204°C).

Hoods should be two gages heavier than the connected branch duct. Use supports not more than 12 ft (3.7 m) apart for horizontal ducts up to 8-in (203.2-mm) diameter. Supports can be spaced up to 20 ft (6.1 m) apart for larger ducts. Fit a duct cleanout opening every 10 ft (3 m). Where changes of diameter are made in the main duct, fit an eccentric taper with a length of at least 5 in (127 mm) for every 1-in (25.4-mm) change in diameter. The end of the main duct is usually extended 6 in (152.4 mm) beyond the last branch and closed with a removable cap. For additional data on industrial exhaust system design, see the newest issue of the ASHRAE *Guide*.

TABLE 12 Exhaust-System Duct Gages

Duct diameter, in (mm)	Metal gage
Up to 8 (203.2)	22
9 to 18 (228.6 to 457.2)	20
19 to 30 (482.6 to 762)	18
31 and larger (787.4 and larger)	16

Related Calculations: Use this procedure for any type of industrial exhaust system, such as those serving metalworking, woodworking, plating, welding, paint spraying, barrel filling, foundry, crushing, tumbling, and similar operations. Consult the local code or ASHRAE *Guide* for specific airflow requirements for these and other industrial operations.

This design procedure is also valid, in general, for industrial pneumatic conveying systems. For several comprehensive, worked-out designs of pneumatic conveying systems, see Hudson—*Conveyors*, Wiley.

Pumps and Pumping Systems

REFERENCES: Karassik—*Pump Handbook*, McGraw-Hill; Warring—*Pumps—Selection, Systems, and Applications*, Trade and Technical Press (England); Crawford—*Marine and Offshore Pumping and Piping Systems*, Butterworth; *Europump Terminology: Glossary of Pump Applications in English, German, Italian, and Spanish*, International Ideas; Isman—*Fire Service Pumps and Hydraulics*, Delmar; Pollak—*Pump User's Handbook*, Gulf Publishing; Anderson—*Centrifugal Pumps*, Trade and Technical Press (England); Walker—*Pump Selection*, Ann Arbor Science Press; Bartlett—*Pumping Stations for Water and Sewage*, Halsted Press; Koutitas—*Elememts of Computational Hydraulics*, Chapman and Hall; Blevins—*Applied Fluid Dynamics Handbook*, VNR; Herbich—*Offshore Pipeline Design Elements*, Dekker; Zienkiewicz—*Numerical Methods in Offshore Engineering*, Wiley; The Hydraulic Institute—*Standards of the Hydraulic Institute*; Allis-Chambers Manufacturing Company—*Pic-A-Pump*; Hicks and Edwards—*Pump Application Engineering*, McGraw-Hill; Stepanoff—*Centrifugal and Axial Flow Pumps*, Wiley; Karassik and Carter—*Centrifugal Pumps*, McGraw-Hill; Allen—*Using Centrifugal Pumps*, Oxford; Buffalo Pumps—*Centrifugal Pump Applications Manual*, Kristal and Annett—*Pumps*, McGraw-Hill; Economy Pumps, Inc.—*Pump Data*; Molloy—*Pumps and Pumping*, Chemical Publishing; Moore et al.—*The Vertical Pump*, Johnston Pump Company; Karassik—*Engineers' Guide to Centrifugal Pumps*, McGraw-Hill; Kovats and Desmur—*Pompes, Ventilateurs, Compresseurs*, Dunod, Paris; Fuchslocher and Schulz—*Die Pumpen*, Springer-Verlag, Berlin; Pfleiderer—*Die Kreiselpumpen*, Springer-Verlag, Berlin.

SIMILARITY OR AFFINITY LAWS FOR CENTRIFUGAL PUMPS

A centrifugal pump designed for a 1800-r/min operation and a head of 200 ft (60.9 m) has a capacity of 3000 gal/min (189.3 L/s) with a power input of 175 hp (130.6 kW). What effect will a speed reduction to 1200 r/min have on the head, capacity, and power input of the pump? What will be the change in these variables if the impeller diameter is reduced from 12 to 10 in (304.8 to 254 mm) while the speed is held constant at 1800 r/min?

Calculation Procedure:

1. *Compute the effect of a change in pump speed*

For any centrifugal pump in which the effects of fluid viscosity are negligible, or are neglected, the similarity or affinity laws can be used to determine the effect of a speed, power, or head change. For a *constant impeller diameter*, the laws are $Q_1/Q_2 = N_1/N_2$; $H_1/H_2 = (N_1/N_2)^2$; $P_1/P_2 = (N_1/N_2)^3$. For a *constant speed*, $Q_1/Q_2 = D_1/D_2$; $H_1/H_2 = (D_1/D_2)^2$; $P_1/P_2 = (D_1/D_2)^3$. In both sets of laws, Q = capacity, gal/min; N = impeller rpm; D = impeller diameter, in; H = total head, ft of liquid; P = bhp input. The subscripts 1 and 2 refer to the initial and changed conditions, respectively.

For this pump, with a constant impeller diameter, $Q_1/Q_2 = N_1/N_2$; $3000/Q_2 = 1800/1200$; $Q_2 = 2000$ gal/min (126.2 L/s). And, $H_1/H_2 = (N_1/N_2)^2 = 200/H_2 = (1800/1200)^2$; $H_2 = 88.9$ ft (27.1 m). Also, $P_1/P_2 = (N_1/N_2)^3 = 175/P_2 = (1800/1200)^3$; $P_2 = 51.8$ bhp (38.6 kW).

2. *Compute the effect of a change in impeller diameter*

With the speed constant, use the second set of laws. Or, for this pump, $Q_1/Q_2 = D_1/D_2$; $3000/Q_2 = 12/10$; $Q_2 = 2500$ gal/min (157.7 L/s). And $H_1/H_2 = (D_1/D_2)^2$; $200/H_2 = (12/10)^2$; $H_2 = 138.8$ ft (42.3 m). Also, $P_1/P_2 = (D_1/D_2)^3$; $175/P_2 = (12/10)^3$; $P_2 = 101.2$ bhp (75.5 kW).

Related Calculations: Use the similarity laws to extend or change the data obtained from centrifugal pump characteristic curves. These laws are also useful in field calculations when the pump head, capacity, speed, or impeller diameter is changed.

The similarity laws are most accurate when the efficiency of the pump remains nearly constant. Results obtained when the laws are applied to a pump having a constant impeller diameter are somewhat more accurate than for a pump at constant speed with a changed impeller diameter. The latter laws are more accurate when applied to pumps having a low specific speed.

If the similarity laws are applied to a pump whose impeller diameter is increased, be certain to consider the effect of the higher velocity in the pump suction line. Use the similarity laws for any liquid whose viscosity remains constant during passage through the pump. However, the accuracy of the similarity laws decreases as the liquid viscosity increases.

SIMILARITY OR AFFINITY LAWS IN CENTRIFUGAL PUMP SELECTION

A test-model pump delivers, at its best efficiency point, 500 gal/min (31.6 L/s) at a 350-ft (106.7-m) head with a required net positive suction head (NPSH) of 10 ft (3 m) a power input of 55 hp (41 kW) at 3500 r/min, when a 10.5-in (266.7-mm) diameter impeller is used. Determine the performance of the model at 1750 r/min. What is the performance of a full-scale prototype pump with a 20-in (50.4-cm) impeller operating at 1170 r/min? What are the specific speeds and the suction specific speeds of the test-model and prototype pumps?

Calculation Procedure:

1. ***Compute the pump performance at the new speed***

The similarity or affinity laws can be stated in general terms, with subscripts p and m for prototype and model, respectively, as $Q_p = K_d^3K_nQ_m$; $H_p = K_d^2K_n^2H_m$; $\text{NPSH}_p = K_2^2K_n^2\text{NPSH}_m$; $P_p = K_d^5K_n^3P_m$, where K_d = size factor = prototype dimension/model dimension. The usual dimension used for the size factor is the impeller diameter. Both dimensions should be in the same units of measure. Also, K_n = (prototype speed, r/min)/(model speed, r/min). Other symbols are the same as in the previous calculation procedure.

When the model speed is reduced from 3500 to 1750 r/min, the pump dimensions remain the same and $K_d = 1.0$; $K_n = 1750/3500 = 0.5$. Then $Q = (1.0)(0.5)(500) = 250$ r/min; $H = (1.0)^2(0.5)^2(350) = 87.5$ ft (26.7 m); $\text{NPSH} = (1.0)^2(0.5)^2(10) = 2.5$ ft (0.76 m); $P = (1.0)^5(0.5)^3(55) = 6.9$ hp (5.2 kW). In this computation, the subscripts were omitted from the equations because the same pump, the test model, was being considered.

2. ***Compute performance of the prototype pump***

First, K_d and K_n must be found: $K_d = 20/10.5 = 1.905$; $K_n = 1170/3500 = 0.335$. Then $Q_p = (1.905)^3(0.335)(500) = 1158$ gal/min (73.1 L/s); $H_p = (1.905)^2(0.335)^2(350) = 142.5$ ft (43.4 m); $\text{NPSH}_p = (1.905)^2(0.335)^2(10) = 4.06$ ft (1.24 m); $P_p = (1.905)^5(0.335)^3(55) = 51.8$ hp (38.6 kW).

3. ***Compute the specific speed and suction specific speed***

The specific speed or, as Horwitz[1] says, "more correctly, discharge specific speed," is $N_s = N(Q)^{0.5}/(H)^{0.75}$, while the suction specific speed $S = N(Q)^{0.5}/(\text{NPSH})^{0.75}$, where all values are taken at the best efficiency point of the pump.

For the model, $N_s = 3500(500)^{0.5}/(350)^{0.75} = 965$; $S = 3500(500)^{0.5}/(10)^{0.75} = 13{,}900$. For the prototype, $N_s = 1170(1158)^{0.5}/(142.5)^{0.75} = 965$; $S = 1170(1156)^{0.5}/(4.06)^{0.75} = 13{,}900$. The specific speed and suction specific speed of the model and prototype are equal because these units are geometrically similar or homologous pumps and both speeds are mathematically derived from the similarity laws.

Related Calculations: Use the procedure given here for any type of centrifugal pump where the similarity laws apply. When the term *model* is used, it can apply to a production test pump or to a standard unit ready for installation. The procedure presented here is the work of R. P. Horwitz, as reported in *Power* magazine.[1]

SPECIFIC-SPEED CONSIDERATIONS IN CENTRIFUGAL PUMP SELECTION

What is the upper limit of specific speed and capacity of a 1750-r/min single-stage double-suction centrifugal pump having a shaft that passes through the impeller eye if it handles clear water at 85°F (29.4°C) at sea level at a total head of 280 ft (85.3 m) with a 10-ft (3-m) suction lift? What is the efficiency of the pump and its approximate impeller shape?

[1]R. P. Horwitz, "Affinity Laws and Specific Speed Can Simplify Centrifugal Pump Selection," *Power*, November 1964.

Calculation Procedure:

1. *Determine the upper limit of specific speed*

Use the Hydraulic Institute upper specific-speed curve, Fig. 1, for centrifugal pumps or a similar curve, Fig. 2, for mixed- and axial-flow pumps. Enter Fig. 1 at the bottom at 280-ft (85.3-m) total head, and project vertically upward until the 10-ft (3-m) suction-lift curve is intersected. From here, project horizontally to the right to read the specific speed N_S = 2000. Figure 2 is used in a similar manner.

2. *Compute the maximum pump capacity*

For any centrifugal, mixed- or axial-flow pump, $N_S = (gpm)^{0.5}(rpm)/H_t^{0.75}$, where H_t = total head on the pump, ft of liquid. Solving for the maximum capacity, we get $gpm = (N_S H_t^{0.75}/rpm)^2 = (2000 \times 280^{0.75}/1750)^2 = 6040$ gal/min (381.1 L/s).

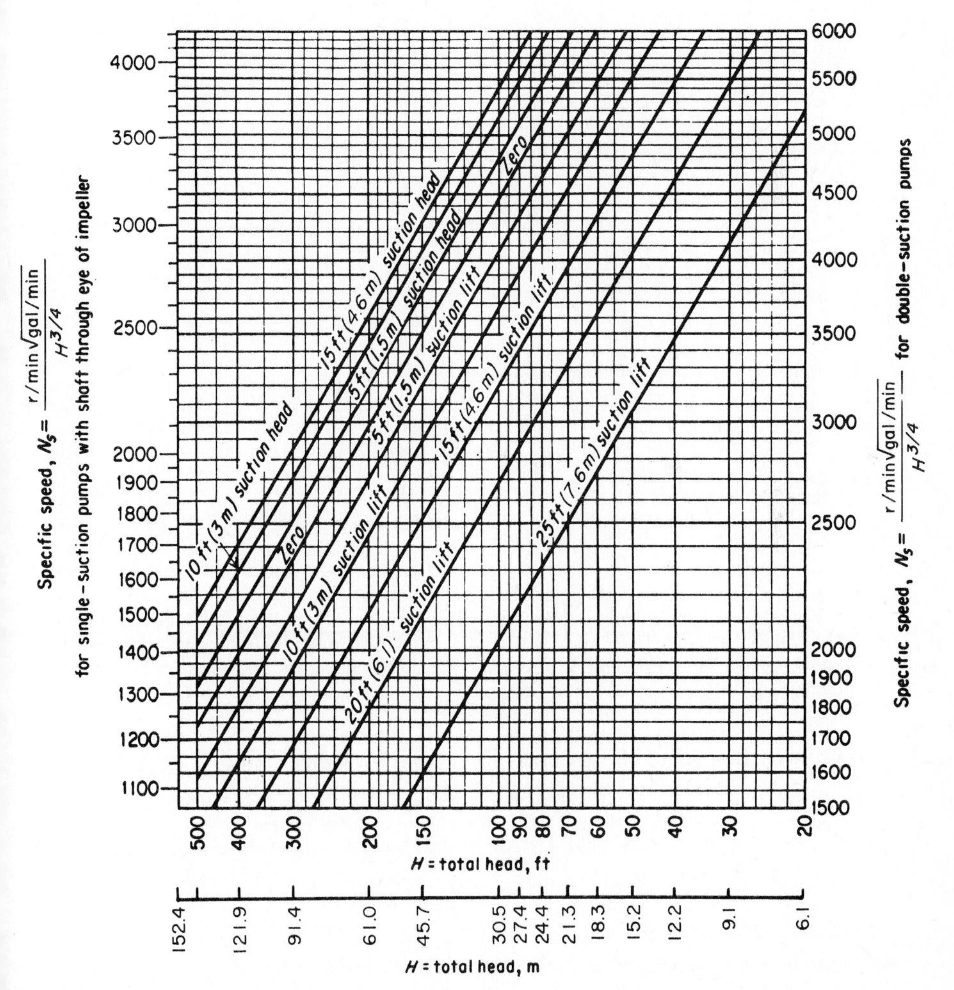

FIG. 1 Upper limits of specific speeds of single-stage, single- and double-suction centrifugal pumps handling clear water at 85°F (29.4°C) at sea level. *(Hydraulic Institute.)*

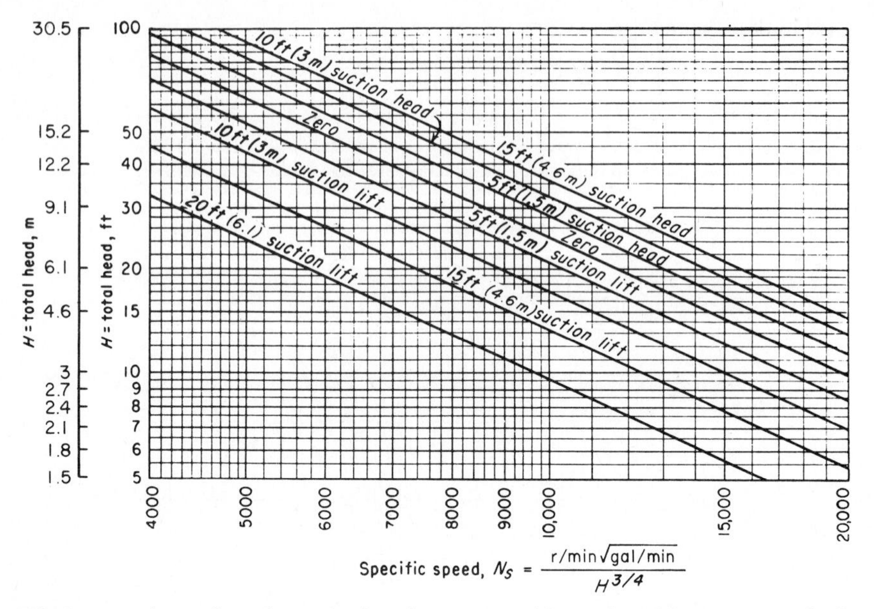

FIG. 2 Upper limits of specific speeds of single-suction mixed-flow and axial-flow pumps. *(Hydraulic Institute.)*

3. *Determine the pump efficiency and impeller shape*

Figure 3 shows the general relation between impeller shape, specific speed, pump capacity, efficiency, and characteristic curves. At $N_S = 2000$, efficiency = 87 percent. The impeller, as shown in Fig. 3, is moderately short and has a relatively large discharge area. A cross section of the impeller appears directly under the $N_S = 2000$ ordinate.

Related Calculations: Use the method given here for any type of pump whose variables are included in the Hydraulic Institute curves, Figs. 1 and 2, and in similar curves available from the same source. *Operating specific speed,* computed as above, is sometimes plotted on the performance curve of a centrifugal pump so that the characteristics of the unit can be better understood. *Type specific speed* is the operating specific speed giving maximum efficiency for a given pump and is a number used to identify a pump. Specific speed is important in cavitation and suction-lift studies. The Hydraulic Institute curves, Figs. 1 and 2, give upper limits of speed, head, capacity and suction lift for cavitation-free operation. When making actual pump analyses, be certain to use the curves (Figs. 1 and 2) in the latest edition of the *Standards of the Hydraulic Institute.*

SELECTING THE BEST OPERATING SPEED FOR A CENTRIFUGAL PUMP

A single-suction centrifugal pump is driven by a 60-Hz ac motor. The pump delivers 10,000 gal/min (630.9 L/s) of water at a 100-ft (30.5-m) head. The available net positive suction head = 32 ft (9.7 m) of water. What is the best operating speed for this pump if the pump operates at its best efficiency point?

Calculation Procedure:

1. *Determine the specific speed and suction specific speed*

Ac motors can operate at a variety of speeds, depending on the number of poles. Assume that the motor driving this pump might operate at 870, 1160, 1750, or 3500 r/min. Compute the specific

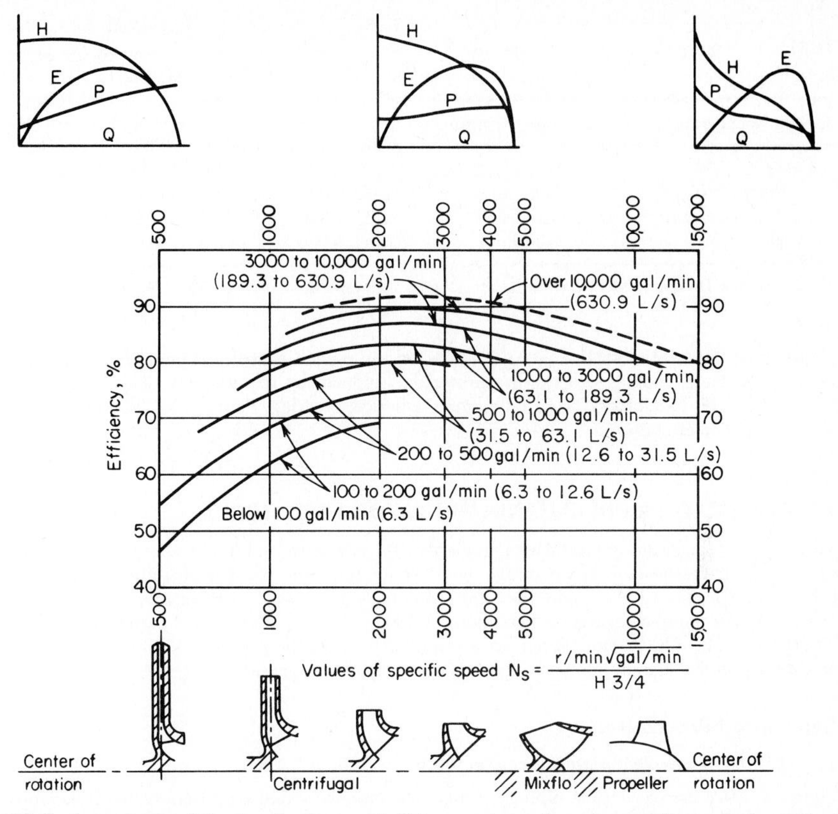

FIG. 3 Approximate relative impeller shapes and efficiency variations for various specific speeds of centrifugal pumps. *(Worthington Corporation.)*

speed $N_S = N(Q)^{0.5}/(H)^{0.75} = N(10{,}000)^{0.5}/(100)^{0.75} = 3.14N$ and the suction specific speed $S = N(Q)^{0.5}/(\text{NPSH})^{0.75} = N(10{,}000)^{0.5}/(32)^{0.75} = 7.43N$ for each of the assumed speeds. Tabulate the results as follows:

Operating speed, r/min	Required specific speed	Required suction specific speed
870	2,740	6,460
1,160	3,640	8,620
1,750	5,500	13,000
3,500	11,000	26,000

2. *Choose the best speed for the pump*

Analyze the specific speed and suction specific speed at each of the various operating speeds, using the data in Tables 1 and 2. These tables show that at 870 and 1160 r/min, the suction specific-speed rating is poor. At 1750 r/min, the suction specific-speed rating is excellent, and a turbine or mixed-flow type pump will be suitable. Operation at 3500 r/min is unfeasible because a suction specific speed of 26,000 is beyond the range of conventional pumps.

TABLE 1 Pump Types Listed by Specific Speed*

Specific speed range	Type of pump
Below 2,000	Volute, diffuser
2,000–5,000	Turbine
4,000–10,000	Mixed-flow
9,000–15,000	Axial-flow

*Peerless Pump Division, FMC Corporation.

TABLE 2 Suction Specific-Speed Ratings*

Single-suction pump	Double-suction pump	Rating
Above 11,000	Above 14,000	Excellent
9,000–11,000	11,000–14,000	Good
7,000–9,000	9,000–11,000	Average
5,000–7,000	7,000–9,000	Poor
Below 5,000	Below 7,000	Very poor

*Peerless Pump Division, FMC Corporation.

Related Calculations: Use this procedure for any type of centrifugal pump handling water for plant services, cooling, process, fire protection, and similar requirements. This procedure is the work of R. P. Horwitz, Hydrodynamics Division, Peerless Pump, FMC Corporation, as reported in *Power* magazine.

TOTAL HEAD ON A PUMP HANDLING VAPOR-FREE LIQUID

Sketch three typical pump piping arrangements with static suction lift and submerged, free, and varying discharge head. Prepare similar sketches for the same pump with static suction head. Label the various heads. Compute the total head on each pump if the elevations are as shown in Fig. 4 and the pump discharges a maximum of 2000 gal/min (126.2 L/s) of water through 8-in (203.2-mm) schedule 40 pipe. What hp is required to drive the pump? A swing check valve is used on the pump suction line and a gate valve on the discharge line.

Calculation Procedure:

1. *Sketch the possible piping arrangements*

Figure 4 shows the six possible piping arrangements for the stated conditions of the installation. Label the total static head, i.e., the *vertical* distance from the surface of the source of the liquid supply to the free surface of the liquid in the discharge receiver, or to the point of free discharge from the discharge pipe. When both the suction and discharge surfaces are open to the atmosphere, the total static head equals the vertical difference in elevation. Use the free-surface elevations that cause the maximum suction lift and discharge head, i.e., the *lowest* possible level in the supply tank and the *highest* possible level in the discharge tank or pipe. When the supply source is *below* the pump centerline, the vertical distance is called the *static suction lift;* with the supply *above* the pump centerline, the vertical distance is called *static suction head.* With variable static suction head, use the lowest liquid level in the supply tank when computing total static head. Label the diagrams as shown in Fig. 4.

2. *Compute the total static head on the pump*

The total static head H_{ts} ft = static suction lift, h_{sl} ft + static discharge head h_{sd} ft, where the pump has a suction lift, *s* in Fig. 4*a*, *b*, and *c*. In these installations, $H_{ts} = 10 + 100 = 110$ ft (33.5 m). Note that the static discharge head is computed between the pump centerline and the water level with an underwater discharge, Fig. 4*a;* to the pipe outlet with a free discharge, Fig. 4*b;* and to the maximum water level in the discharge tank, Fig. 4*c*. When a pump is discharging into a closed compression tank, the total discharge head equals the static discharge head plus the head equivalent, ft of liquid, of the internal pressure in the tank, or 2.31 × tank pressure, lb/in^2.

Where the pump has a static suction head, as in Fig. 4*d*, *e*, and *f*, the total static head H_{ts} ft $= h_{sd}$ − static suction head h_{sh} ft. In these installations, $H_t = 100 - 15 = 85$ ft (25.9 m).

The total static head, as computed above, refers to the head on the pump without liquid flow. To determine the total head on the pump, the friction losses in the piping system during liquid flow must be also determined.

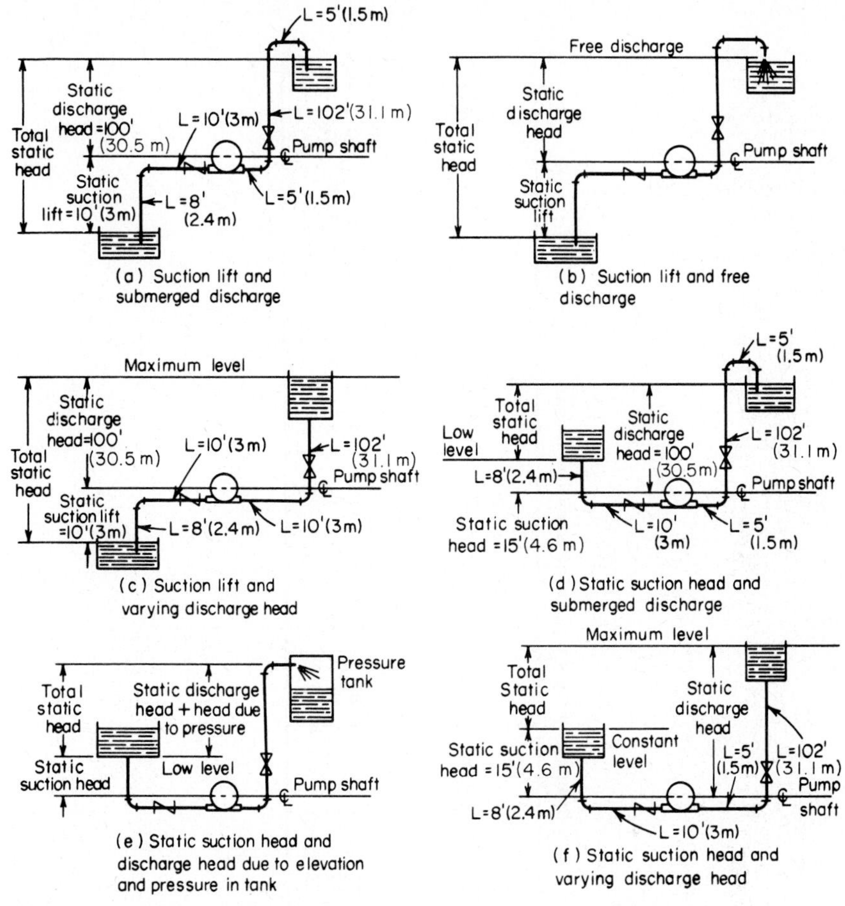

FIG. 4 Typical pump suction and discharge piping arrangements.

3. *Compute the piping friction losses*

Mark the length of each piece of straight pipe on the piping drawing. Thus, in Fig. 4*a*, the total length of straight pipe L_t ft = 8 + 10 + 5 + 102 + 5 = 130 ft (39.6 m), if we start at the suction tank and add each length until the discharge tank is reached. To the total length of straight pipe must be added the *equivalent* length of the pipe fittings. In Fig. 4*a* there are four long-radius elbows, one swing check valve, and one globe valve. In addition, there is a minor head loss at the pipe inlet and at the pipe outlet.

The equivalent length of one 8-in (203.2-mm) long-radius elbow is 14 ft (4.3 m) of pipe, from Table 3. Since the pipe contains four elbows, the total equivalent length = 4(14) = 56 ft (17.1 m) of straight pipe. The open gate valve has an equivalent resistance of 4.5 ft (1.4 m); and the open swing check valve has an equivalent resistance of 53 ft (16.2 m).

The entrance loss h_e ft, assuming a basket-type strainer is used at the suction-pipe inlet, is h_e ft = $Kv^2/2g$, where K = a constant from Fig. 5; v = liquid velocity, ft/s; g = 32.2 ft/s² (980.67 cm/s²). The exit loss occurs when the liquid passes through a sudden enlargement, as from a pipe to a tank. Where the area of the tank is large, causing a final velocity that is zero, $h_{ex} = v^2/2g$.

The velocity v ft/s in a pipe = $gpm/2.448d^2$. For this pipe, $v = 2000/[(2.448)(7.98)^2]$ = 12.82 ft/s (3.91 m/s). Then $h_e = 0.74(12.82)^2/[2(32.2)]$ = 1.89 ft (0.58 m), and $h_{ex} = (12.82)^2/$

TABLE 3 Resistance of Fittings and Valves (length of straight pipe giving equivalent resistance)

Pipe size		Standard ell		Medium-radius ell		Long-radius ell		45° Ell		Tee		Gate valve, open		Globe valve, open		Swing check, open	
in	mm	ft	m	ft	m	ft	m	ft	m	ft	m	ft	m	ft	m	ft	m
6	152.4	16	4.9	14	4.3	11	3.4	7.7	2.3	33	10.1	3.5	1.1	160	48.8	40	12.2
8	203.2	21	6.4	18	5.5	14	4.3	10	3.0	43	13.1	4.5	1.4	220	67.0	53	16.2
10	254.0	26	7.9	22	6.7	17	5.2	13	3.9	56	17.1	5.7	1.7	290	88.4	67	20.4
12	304.8	32	9.8	26	7.9	20	6.1	15	4.6	66	20.1	6.7	2.0	340	103.6	80	24.4

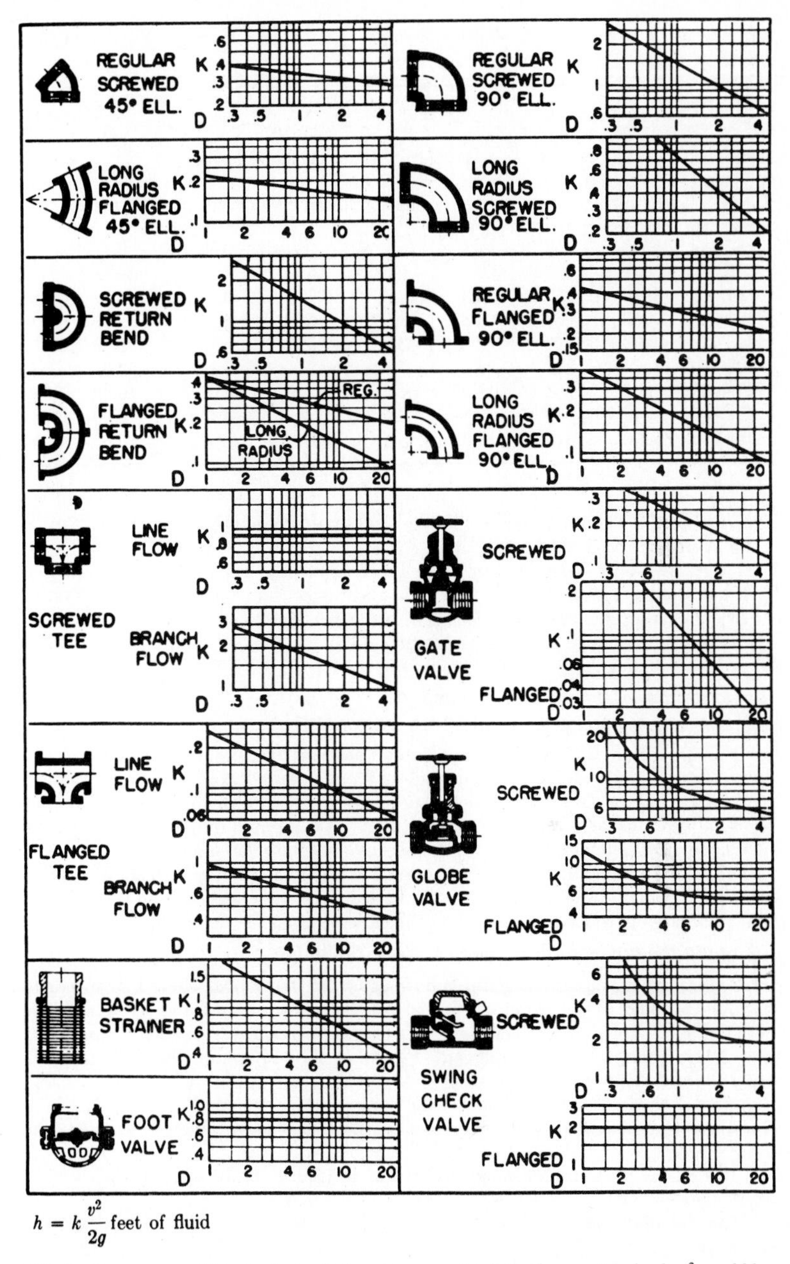

$$h = k \frac{v^2}{2g} \text{ feet of fluid}$$

FIG. 5 Resistance coefficients of pipe fittings. To convert to SI in the equation for h, v^2 would be measured in m/s and feet would be changed to meters. The following values would also be changed from inches to millimeters: 0.3 to 7.6, 0.5 to 12.7, 1 to 25.4, 2 to 50.8, 4 to 101.6, 6 to 152.4, 10 to 254, and 20 to 508. *(Hydraulic Institute.)*

TABLE 4 Pipe Friction Loss for Water (wrought-iron or steel schedule 40 pipe in good condition)

Diameter		Flow		Velocity		Velocity head		Friction loss per 100 ft (30.5 m) of pipe	
in	mm	gal/min	L/s	ft/s	m/s	ft water	m water	ft water	m water
6	152.4	1000	63.1	11.1	3.4	1.92	0.59	6.17	1.88
6	152.4	2000	126.2	22.2	6.8	7.67	2.3	23.8	7.25
6	152.4	4000	252.4	44.4	13.5	30.7	9.4	93.1	28.4
8	203.2	1000	63.1	6.41	1.9	0.639	0.195	1.56	0.475
8	203.2	2000	126.2	12.8	3.9	2.56	0.78	5.86	1.786
8	203.2	4000	252.4	25.7	7.8	10.2	3.1	22.6	6.888
10	254.0	1000	63.1	3.93	1.2	0.240	0.07	0.497	0.151
10	254.0	3000	189.3	11.8	3.6	2.16	0.658	4.00	1.219
10	254.0	5000	315.5	19.6	5.9	5.99	1.82	10.8	3.292

[(2)(32.2)] = 2.56 ft (0.78 m). Hence, the total length of the piping system in Fig. 4*a* is 130 + 56 + 4.5 + 53 + 1.89 + 2.56 = 247.95 ft (75.6 m), say 248 ft (75.6 m).

Use a suitable head-loss equation, or Table 4, to compute the head loss for the pipe and fittings. Enter Table 4 at an 8-in (203.2-mm) pipe size, and project horizontally across to 2000 gal/min (126.2 L/s) and read the head loss as 5.86 ft of water per 100 ft (1.8 m/30.5 m) of pipe.

The total length of pipe and fittings computed above is 248 ft (75.6 m). Then total friction-head loss with a 2000 gal/min (126.2-L/s) flow is H_f ft = (5.86)(248/100) = 14.53 ft (4.5 m).

4. *Compute the total head on the pump*

The total head on the pump $H_t = H_{ts} + H_f$. For the pump in Fig. 4*a*, $H_t = 110 + 14.53 = 124.53$ ft (37.95 m), say 125 ft (38.1 m). The total head on the pump in Fig. 4*b* and *c* would be the same. Some engineers term the total head on a pump the *total dynamic head* to distinguish between static head (no-flow vertical head) and operating head (rated flow through the pump).

The total head on the pumps in Fig. 4*d*, *c*, and *f* is computed in the same way as described above, except that the total static head is less because the pump has a static suction head. That is, the elevation of the liquid on the suction side reduces the total distance through which the pump must discharge liquid; thus the total static head is less. The static suction head is *subtracted* from the static discharge head to determine the total static head on the pump.

5. *Compute the horsepower required to drive the pump*

The brake horsepower input to a pump $bhp_i = (gpm)(H_t)(s)/3960e$, where s = specific gravity of the liquid handled; e = hydraulic efficiency of the pump, expressed as a decimal. The usual hydraulic efficiency of a centrifugal pump is 60 to 80 percent; reciprocating pumps, 55 to 90 percent; rotary pumps, 50 to 90 percent. For each class of pump, the hydraulic efficiency decreases as the liquid viscosity increases.

Assume that the hydraulic efficiency of the pump in this system is 70 percent and the specific gravity of the liquid handled is 1.0. Then $bhp_i = (2000)(127)(1.0)/(3960)(0.70) = 91.6$ hp (68.4 kW).

The theoretical or *hydraulic horsepower* $hp_h = (gpm)(H_t)(s)/3960$, or $hp_h = (2000) \times (127)(1.0)/3900 = 64.1$ hp (47.8 kW).

Related Calculations: Use this procedure for any liquid—water, oil, chemical, sludge, etc.—whose specific gravity is known. When liquids other than water are being pumped, the specific gravity and viscosity of the liquid, as discussed in later calculation procedures, must be taken into consideration. The procedure given here can be used for any class of pump—centrifugal, rotary, or reciprocating.

Note that Fig. 5 can be used to determine the equivalent length of a variety of pipe fittings. To use Fig. 5, simply substitute the appropriate K value in the relation $h = Kv^2/2g$, where h = equivalent length of straight pipe; other symbols as before.

PUMP SELECTION FOR ANY PUMPING SYSTEM

Give a step-by-step procedure for choosing the class, type, capacity, drive, and materials for a pump that will be used in an industrial pumping system.

Calculation Procedure:

1. *Sketch the proposed piping layout*

Use a single-line diagram, Fig. 6, of the piping system. Base the sketch on the actual job conditions. Show all the piping, fittings, valves, equipment, and other units in the system. Mark the *actual* and *equivalent* pipe length (see the previous calculation procedure) on the sketch. Be certain to include all vertical lifts, sharp bends, sudden enlargements, storage tanks, and similar equipment in the proposed system.

2. *Determine the required capacity of the pump*

The required capacity is the flow rate that must be handled in gal/min, million gal/day, ft^3/s, gal/h, bbl/day, lb/h, acre·ft/day, mil/h, or some similar measure. Obtain the required flow rate from the process conditions, for example, boiler feed rate, cooling-water flow rate, chemical feed

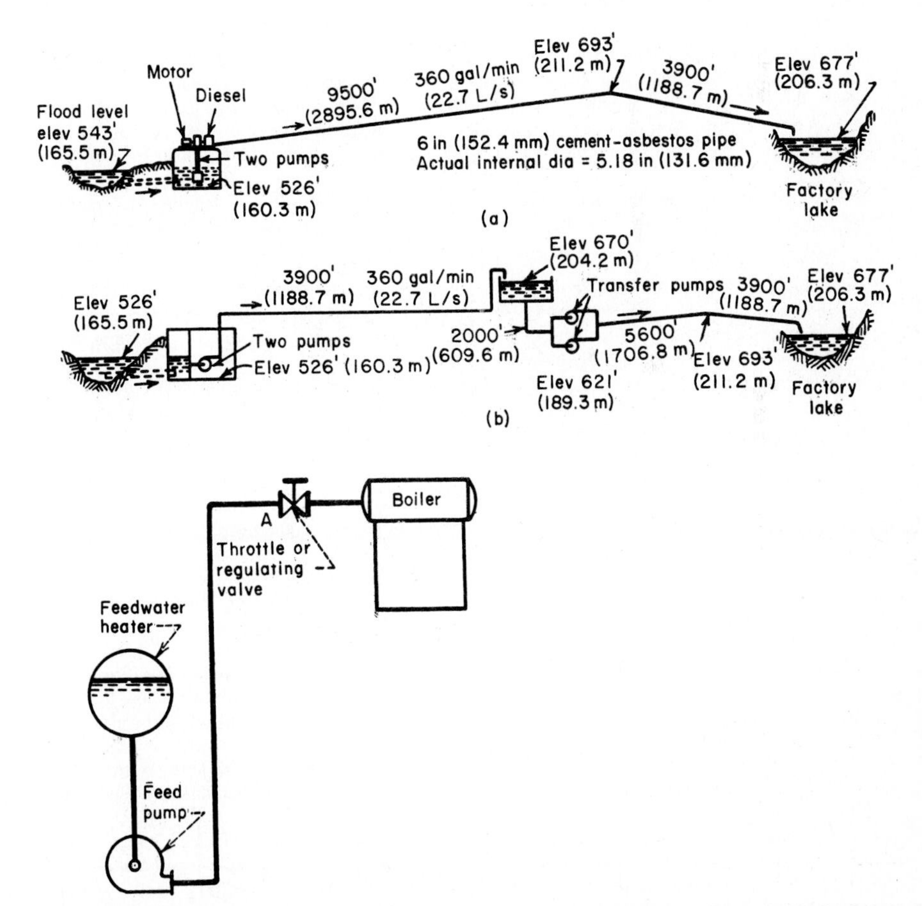

FIG. 6 (*a*) Single-line diagrams for an industrial pipeline; (*b*) single-line diagram of a boiler-feed system. *(Worthington Corporation.)*

rate, etc. The required flow rate for any process unit is usually given by the manufacturer or can be computed by using the calculation procedures given throughout this handbook.

Once the required flow rate is determined, apply a suitable factor of safety. The value of this factor of safety can vary from a low of 5 percent of the required flow to a high of 50 percent or more, depending on the application. Typical safety factors are in the 10 percent range. With flow rates up to 1000 gal/min (63.1 L/s), and in the selection of process pumps, it is common practice to round a computed required flow rate to the next highest round-number capacity. Thus, with a required flow rate of 450 gal/min (28.4 L/s) and a 10 percent safety factor, the flow of 450 + 0.10(450) = 495 gal/min (31.2 L/s) would be rounded to 500 gal/min (31.6 L/s) *before* the pump was selected. A pump of 500-gal/min (31.6-L/s), or larger, capacity would be selected.

3. *Compute the total head on the pump*

Use the steps given in the previous calculation procedure to compute the total head on the pump. Express the result in ft (m) of water—this is the most common way of expressing the head on a pump. Be certain to use the exact specific gravity of the liquid handled when expressing the head in ft (m) of water. A specific gravity less than 1.00 *reduces* the total head when expressed in ft (m) of water; whereas a specific gravity greater than 1.00 *increases* the total head when expressed in ft (m) of water. Note that variations in the suction and discharge conditions can affect the total head on the pump.

4. *Analyze the liquid conditions*

Obtain complete data on the liquid pumped. These data should include the name and chemical formula of the liquid, maximum and minimum pumping temperature, corresponding vapor pressure at these temperatures, specific gravity, viscosity at the pumping temperature, pH, flash point, ignition temperature, unusual characteristics (such as tendency to foam, curd, crystallize, become gelatinous or tacky), solids content, type of solids and their size, and variation in the chemical analysis of the liquid.

Enter the liquid conditions on a pump selection form like that in Fig. 7. Such forms are available from many pump manufacturers or can be prepared to meet special job conditions.

5. *Select the class and type of pump*

Three *classes* of pumps are used today—centrifugal, rotary, and reciprocating, Fig. 8. Note that these terms apply only to the mechanics of moving the liquid—not to the service for which the pump was designed. Each class of pump is further subdivided into a number of *types*, Fig. 8.

Use Table 5 as a general guide to the class and type of pump to be used. For example, when a large capacity at moderate pressure is required, Table 5 shows that a centrifugal pump would probably be best. Table 5 also shows the typical characteristics of various classes and types of pumps used in industrial process work.

Consider the liquid properties when choosing the class and type of pump, because exceptionally severe conditions may rule out one or another class of pump at the start. Thus, screw- and gear-type rotary pumps are suitable for handling viscous, nonabrasive liquid, Table 5. When an abrasive liquid must be handled, either another class of pump or another type of rotary pump must be used.

Also consider all the operating factors related to the particular pump. These factors include the type of service (continuous or intermittent), operating-speed preferences, future load expected and its effect on pump head and capacity, maintenance facilities available, possibility of parallel or series hookup, and other conditions peculiar to a given job.

Once the class and type of pump is selected, consult a rating table (Table 6) or rating chart, Fig. 9, to determine whether a suitable pump is available from the manufacturer whose unit will be used. When the hydraulic requirements fall between two standard pump models, it is usual practice to choose the next larger size of pump, unless there is some reason why an exact head and capacity are required for the unit. When one manufacturer does not have the desired unit, refer to the engineering data of other manufacturers. Also keep in mind that some pumps are custom-built for a given job when precise head and capacity requirements must be met.

Other pump data included in manufacturer's engineering information include characteristic curves for various diameter impellers in the same casing, Fig. 10, and variable-speed head-capacity curves for an impeller of given diameter, Fig. 11. Note that the required power input is given in Figs. 9 and 10 and may also be given in Fig. 11. Use of Table 6 is explained in the table.

Performance data for rotary pumps are given in several forms. Figure 12 shows a typical plot

Summary of Essential Data Required in Selection of Centrifugal Pumps

1. Number of Units Required

2. Nature of the Liquid to Be Pumped
   Is the liquid:
   a. Fresh or salt water, acid or alkali, oil, gasoline, slurry, or paper stock?
   b. Cold or hot and if hot, at what temperature? What is the vapor pressure of the liquid at the pumping temperature?
   c. What is its specific gravity?
   d. Is it viscous or nonviscous?
   e. Clear and free from suspended foreign matter or dirty and gritty? If the latter, what is the size and nature of the solids, and are they abrasive? If the liquid is of a pulpy nature, what is the consistency expressed either in percentage or in lb per cu ft of liquid? What is the suspended material?
   f. What is the chemical analysis, pH value, etc.? What are the expected variations of this analysis? If corrosive, what has been the past experience, both with successful materials and with unsatisfactory materials?

3. Capacity
   What is the required capacity as well as the minimum and maximum amount of liquid the pump will ever be called upon to deliver?

4. Suction Conditions
   Is there:
   a. A suction lift?
   b. Or a suction head?
   c. What are the length and diameter of the suction pipe?

5. Discharge Conditions
   a. What is the static head? Is it constant or variable?
   b. What is the friction head?
   c. What is the maximum discharge pressure against which the pump must deliver the liquid?

6. Total Head
   Variations in items 4 and 5 will cause variations in the total head.

7. Is the service continuous or intermittent?

8. Is the pump to be installed in a horizontal or vertical position? If the latter,
   a. In a wet pit?
   b. In a dry pit?

9. What type of power is available to drive the pump and what are the characteristics of this power?

10. What space, weight, or transportation limitations are involved?

11. Location of installation
    a. Geographical location
    b. Elevation above sea level
    c. Indoor or outdoor installation
    d. Range of ambient temperatures

12. Are there any special requirements or marked preferences with respect to the design, construction, or performance of the pump?

FIG. 7 Typical selection chart for centrifugal pumps. *(Worthington Corporation.)*

of the head and capacity ranges of different types of rotary pumps. Reciprocating-pump capacity data are often tabulated, as in Table 7.

6. *Evaluate the pump chosen for the installation*

Check the specific speed of a centrifugal pump, using the method given in an earlier calculation procedure. Once the specific speed is known, the impeller type and approximate operating efficiency can be found from Fig. 3.

Check the piping system, using the method of an earlier calculation procedure, to see whether the available net positive suction head equals, or is greater than, the required net positive suction head of the pump.

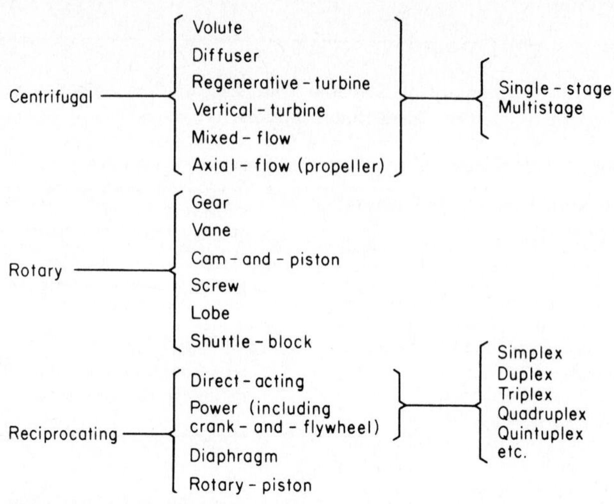

FIG. 8 Modern pump classes and types.

TABLE 5 Characteristics of Modern Pumps

	Centrifugal		Rotary	Reciprocating		
	Volute and diffuser	Axial flow	Screw and gear	Direct acting steam	Double acting power	Triplex
Discharge flow Usual maximum suction lift, ft (m)	Steady 15 (4.6)	Steady 15 (4.6)	Steady 22 (6.7)	Pulsating 22 (6.7)	Pulsating 22 (6.7)	Pulsating 22 (6.7)
Liquids handled	Clean, clear; dirty, abrasive; liquids with high solids content		Viscous; non-abrasive	Clean and clear		
Discharge pressure range	Low to high		Medium	Low to highest produced		
Usual capacity range	Small to largest available		Small to medium	Relatively small		
How increased head affects: Capacity Power input	Decrease Depends on specific speed		None Increase	Decrease Increase	None Increase	None Increase
How decreased head affects: Capacity Power input	Increase Depends on specific speed		None Decrease	Small increase Decrease	None Decrease	None Decrease

TABLE 6 Typical Centrifugal-Pump Rating Table

Size		Total head			
gal/min	L/s	20 ft, r/min—hp	6.1 m, r/min—kW	25 ft, r/min—hp	7.6 m, r/min—kW
3 CL:					
200	12.6	910—1.3	910-0.97	1010—1.6	1010—1.19
300	18.9	1000—1.9	1000—1.41	1100—2.4	1100—1.79
400	25.2	1200—3.1	1200—2.31	1230—3.7	1230—2.76
500	31.5	—	—	—	—
4 C:					
400	25.2	940—2.4	940—1.79	1040—3	1040—2.24
600	37.9	1080—4	1080—2.98	1170—4.6	1170—3.43
800	50.5	—	—	—	—

Example: 1080—4 indicates pump speed is 1080 r/min; actual input required to operate the pump is 4 hp (2.98 kW).
Source: Condensed from data of Goulds Pumps, Inc.; SI values added by handbook editor.

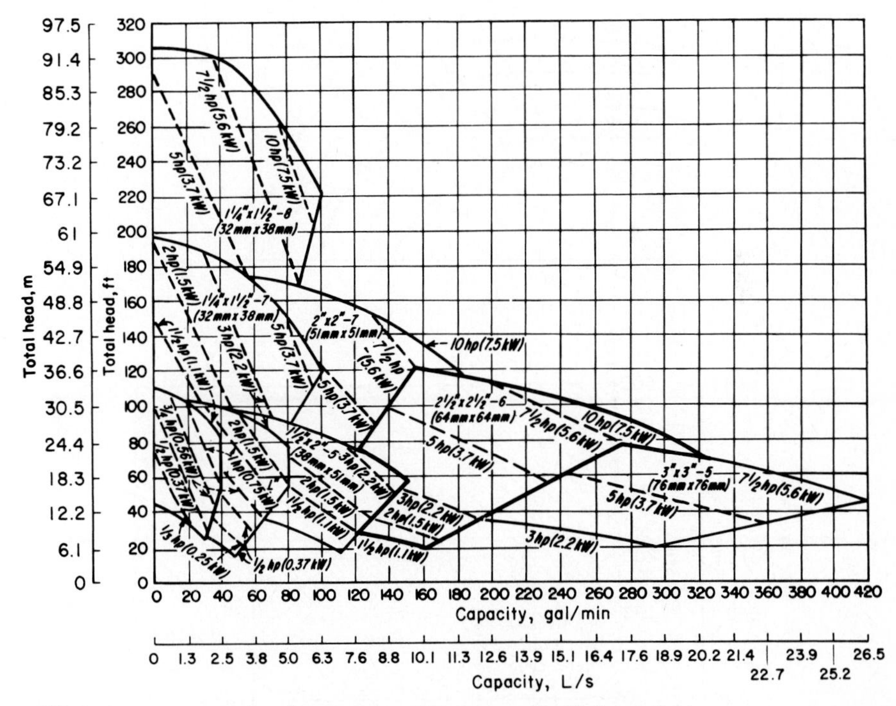

FIG. 9 Composite rating chart for a typical centrifugal pump. *(Goulds Pumps, Inc.)*

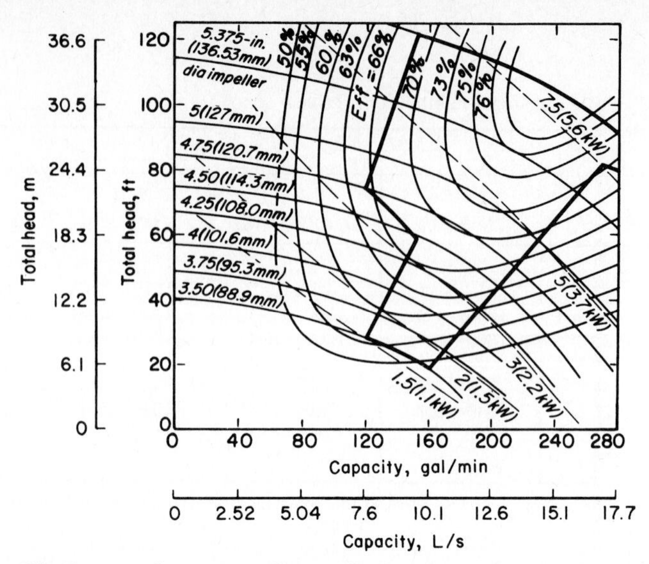

FIG. 10 Pump characteristics when impeller diameter is varied within the same casing.

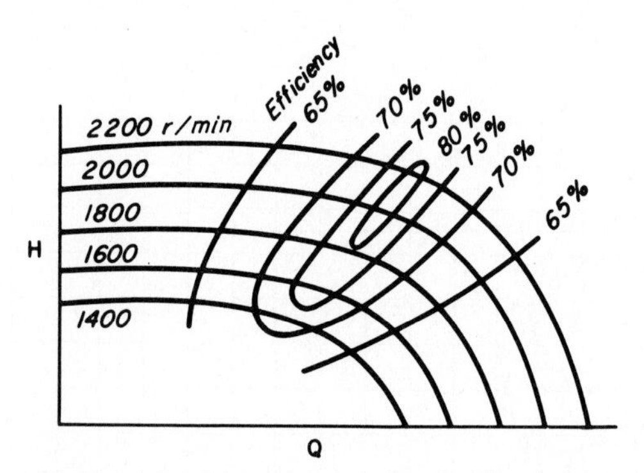

FIG. 11 Variable-speed head-capacity curves for a centrifugal pump.

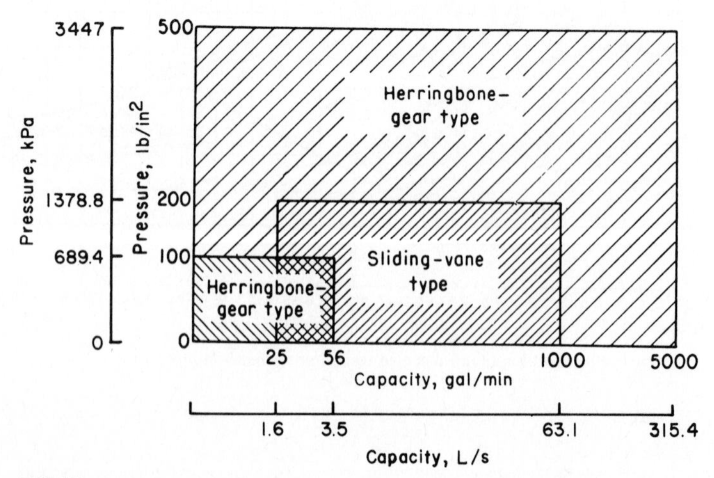

FIG. 12 Capacity ranges of some rotary pumps. *(Worthington Corporation.)*

TABLE 7 Capacities of Typical Horizontal Duplex Plunger Pumps

Size		Cold-water pressure service			
				Piston speed	
in	cm	gal/min	L/s	ft/min	m/min
6 × 3½ × 6	15.2 × 8.9 × 15.2	60	3.8	60	18.3
7½ × 4½ × 10	19.1 × 11.4 × 25.4	124	7.8	75	22.9
9 × 5 × 10	22.9 × 12.7 × 25.4	153	9.7	75	22.9
10 × 6 × 12	25.4 × 15.2 × 30.5	235	14.8	80	24.4
12 × 7 × 12	30.5 × 17.8 × 30.5	320	20.2	80	24.4

Size		Boiler-feed service					
				Boiler		Piston speed	
in	cm	gal/min	L/s	hp	kW	ft/min	m/min
6 × 3½ × 6	15.2 × 8.9 × 15.2	36	2.3	475	354.4	36	10.9
7½ × 4½ × 10	19.1 × 11.4 × 25.4	74	4.7	975	727.4	45	13.7
9 × 5 × 10	22.9 × 12.7 × 25.4	92	5.8	1210	902.7	45	13.7
10 × 6 × 12	25.4 × 15.2 × 30.5	141	8.9	1860	1387.6	48	14.6
12 × 7 × 12	30.5 × 17.8 × 30.5	192	12.1	2530	1887.4	48	14.6

Source: Courtesy of Worthington Corporation.

Determine whether a vertical or horizontal pump is more desirable. From the standpoint of floor space occupied, required NPSH, priming, and flexibility in changing the pump use, vertical pumps may be preferable to horizontal designs in some installations. But where headroom, corrosion, abrasion, and ease of maintenance are important factors, horizontal pumps may be preferable.

As a general guide, single-suction centrifugal pumps handle up to 50 gal/min (3.2 L/s) at total heads up to 50 ft (15.2 m); either single- or double-suction pumps are used for the flow rates to 1000 gal/min (63.1 L/s) and total heads to 300 ft (91.4 m); beyond these capacities and heads, double-suction or multistage pumps are generally used.

Mechanical seals are becoming more popular for all types of centrifugal pumps in a variety of services. Although they are more costly than packing, the mechanical seal reduces pump maintenance costs.

Related Calculations: Use the procedure given here to select any class of pump—centrifugal, rotary, or reciprocating—for any type of service—power plant, atomic energy, petroleum processing, chemical manufacture, paper mills, textile mills, rubber factories, food processing, water supply, sewage and sump service, air conditioning and heating, irrigation and flood control, mining and construction, marine services, industrial hydraulics, iron and steel manufacture.

ANALYSIS OF PUMP AND SYSTEM CHARACTERISTIC CURVES

Analyze a set of pump and system characteristic curves for the following conditions: friction losses without static head; friction losses with static head; pump without lift; system with little friction, much static head; system with gravity head; system with different pipe sizes; system with two discharge heads; system with diverted flow; and effect of pump wear on characteristic curve.

Calculation Procedure:

1. *Plot the system-friction curve*

Without static head, the system-friction curve passes through the origin (0,0), Fig. 13, because when no head is developed by the pump, flow through the piping is zero. For most piping systems, the friction-head loss varies as the square of the liquid flow rate in the system. Hence, a system-friction curve, also called a friction-head curve, is parabolic—the friction head increases as the flow rate or capacity of the system increases. Draw the curve as shown in Fig. 13.

2. *Plot the piping system and system-head curve*

Figure 14*a* shows a typical piping system with a pump operating against a static discharge head. Indicate the total static head, Fig. 14*b*, by a dashed line—in this installation H_{ts} = 110 ft. Since static head is a physical dimension, it does not vary with flow rate and is a constant for all flow rates. Draw the dashed line parallel to the abscissa, Fig. 14*b*.

From the point of no flow—zero capacity—plot the friction-head loss at various flow rates—100, 200, 300 gal/min (6.3, 12.6, 18.9 L/s), etc. Determine the friction-head loss by computing it as shown in an earlier calculation procedure. Draw a curve through the points obtained. This is called the *system-head curve*.

Plot the pump head-capacity (*H-Q*) curve of the pump on Fig. 14*b*. The *H-Q* curve can be obtained from the pump manufacturer or from a tabulation of *H* and *Q* values for the

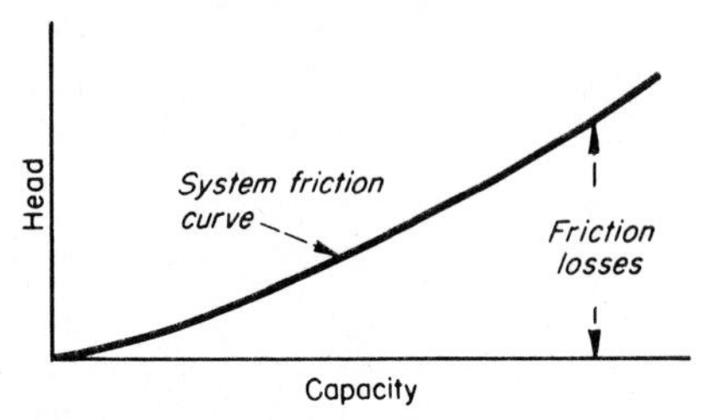

FIG. 13 Typical system-friction curve.

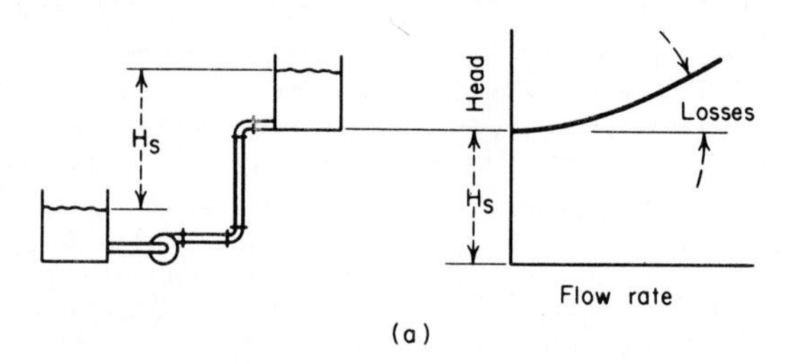

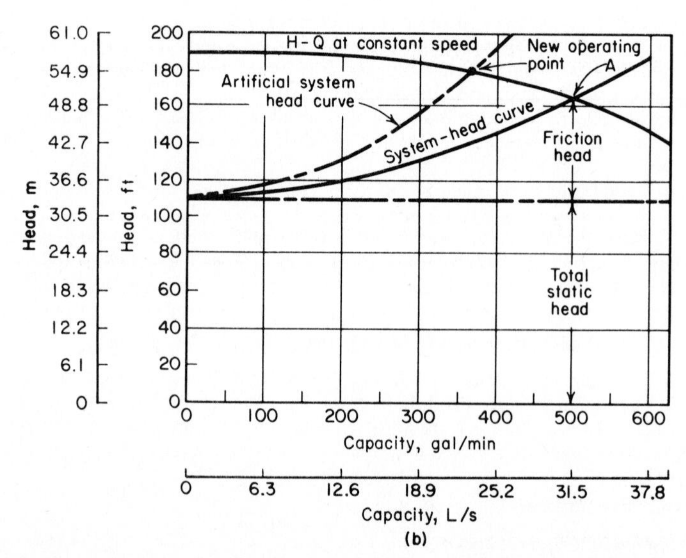

FIG. 14 (*a*) Significant friction loss and lift; (*b*) system-head curve superimposed on pump head-capacity curve. *(Peerless Pumps.)*

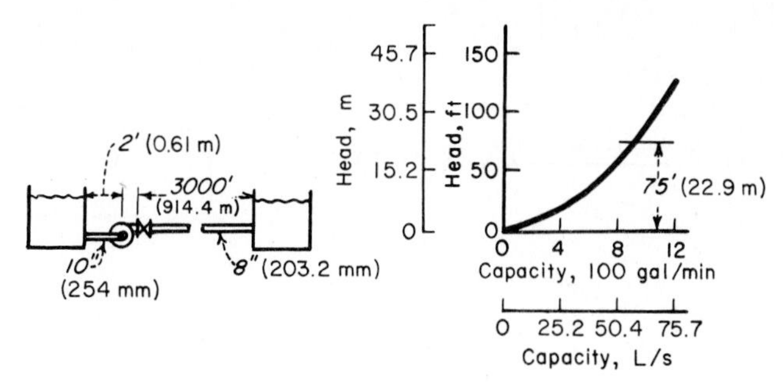

FIG. 15 No lift; all friction head. *(Peerless Pumps.)*

pump being considered. The point of intersection A between the H-Q and system-head curves is the operating point of the pump.

Changing the resistance of a given piping system by partially closing a valve or making some other change in the friction alters the position of the system-head curve and pump operating point. Compute the frictional resistance as before, and plot the artificial system-head curve as shown. Where this curve intersects the H-Q curve is the new operating point of the pump. System-head curves are valuable for analyzing the suitability of a given pump for a particular application.

3. *Plot the no-lift system-head curve and compute the losses*

With no static head or lift, the system-head curve passes through the origin (0,0), Fig. 15. For a flow of 900 gal/min (56.8 L/s) in this system, compute the friction loss as follows, using the Hydraulic Institute *Pipe Friction Manual* tables or the method of earlier calculation procedures:

	ft	m
Entrance loss from tank into 10-in (254-mm) suction pipe, $0.5v^2/2g$	0.10	0.03
Friction loss in 2 ft (0.61 m) of suction pipe	0.02	0.01
Loss in 10-in (254-mm) 90° elbow at pump	0.20	0.06
Friction loss in 3000 ft (914.4 m) of 8-in (203.2-mm) discharge pipe	74.50	22.71
Loss in fully open 8-in (203.2-mm) gate valve	0.12	0.04
Exit loss from 8-in (203.2-mm) pipe into tank, $v^2/2g$	0.52	0.16
Total friction loss	75.46	23.01

Compute the friction loss at other flow rates in a similar manner, and plot the system-head curve, Fig. 15. Note that if all losses in this system except the friction in the discharge pipe were ignored, the total head would not change appreciably. However, for the purposes of accuracy, all losses should always be computed.

4. *Plot the low-friction, high-head system-head curve*

The system-head curve for the vertical pump installation in Fig. 16 starts at the total static head, 15 ft (4.6 m), and zero flow. Compute the friction head for 15,000 gal/min as follows:

	ft	m
Friction in 20 ft (6.1 m) of 24-in (609.6-mm) pipe	0.40	0.12
Exit loss from 24-in (609.6-mm) pipe into tank, $v^2/2g$	1.60	0.49
Total friction loss	2.00	0.61

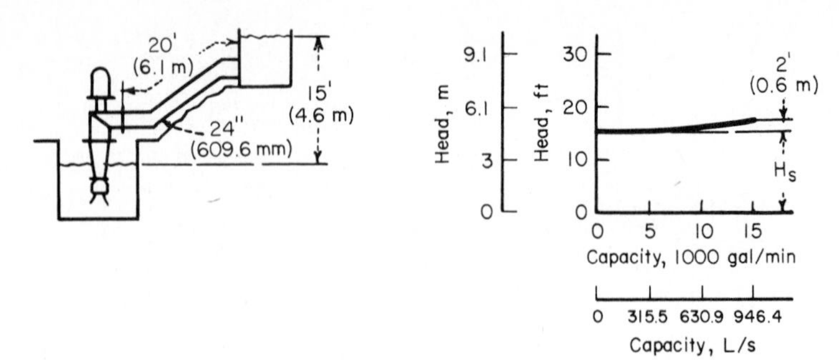

FIG. 16 Mostly lift; little friction head. *(Peerless Pumps.)*

Hence, almost 90 percent of the total head of 15 + 2 = 17 ft (5.2 m) at 15,000-gal/min (946.4-L/s) flow is static head. But neglect of the pipe friction and exit losses could cause appreciable error during selection of a pump for the job.

5. *Plot the gravity-head system-head curve*

In a system with gravity head (also called negative lift), fluid flow will continue until the system friction loss equals the available gravity head. In Fig. 17 the available gravity head is 50 ft (15.2 m). Flows up to 7200 gal/min (454.3 L/s) are obtained by gravity head alone. To obtain larger flow rates, a pump is needed to overcome the friction in the piping between the tanks. Compute the friction loss for several flow rates as follows:

	ft	m
At 5000 gal/min (315.5 L/s) friction loss in 1000 ft (305 m) of 16-in (406.4-mm) pipe	25	7.6
At 7200 gal/min (454.3 L/s), friction loss = available gravity head	50	15.2
At 13,000 gal/min (820.2 L/s), friction loss	150	45.7

Using these three flow rates, plot the system-head curve, Fig. 17.

6. *Plot the system-head curves for different pipe sizes*

When different diameter pipes are used, the friction loss vs. flow rate is plotted independently for the two pipe sizes. At a given flow rate, the total friction loss for the system is the sum of the loss for the two pipes. Thus, the combined system-head curve represents the sum of the static head and the friction losses for all portions of the pipe.

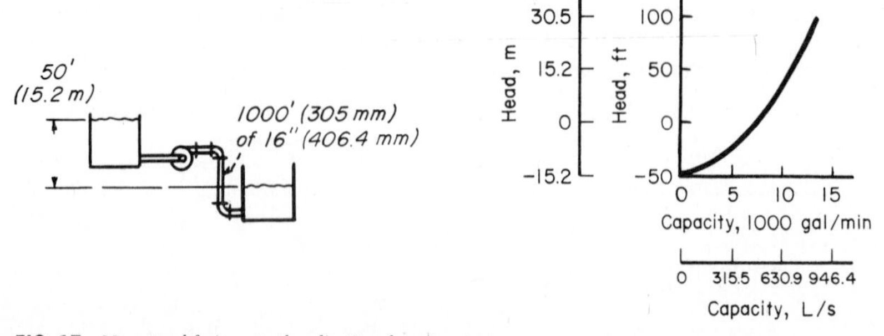

FIG. 17 Negative lift (gravity head). *(Peerless Pumps.)*

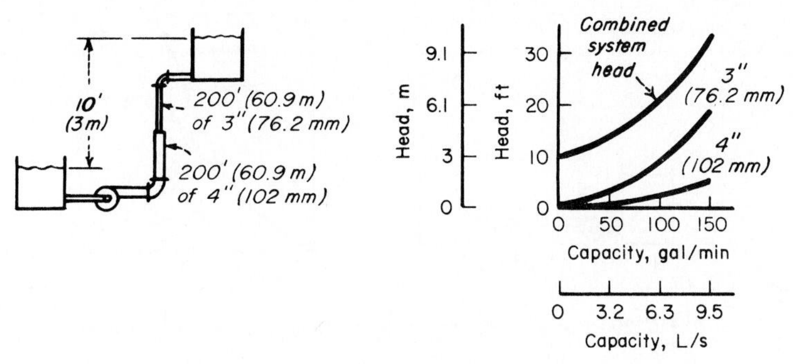

FIG. 18 System with two different pipe sizes. *(Peerless Pumps.)*

Figure 18 shows a system with two different pipe sizes. Compute the friction losses as follows:

	ft	m
At 150 gal/min (9.5 L/s), friction loss in 200 ft (60.9 m) of 4-in (102-mm) pipe	5	1.52
At 150 gal/min (9.5 L/s), friction loss in 200 ft (60.9 m) of 3-in (76.2-mm) pipe	19	5.79
Total static head for 3- (76.2-) and 4-in (102-mm) pipes	10	3.05
Total head at 150-gal/min (9.5-L/s) flow	34	10.36

Compute the total head at other flow rates, and then plot the system-head curve as shown in Fig. 18.

7. *Plot the system-head curve for two discharge heads*

Figure 19 shows a typical pumping system having two different discharge heads. Plot separate system-head curves when the discharge heads are different. Add the flow rates for the two pipes at the same head to find points on the combined system-head curve, Fig. 19. Thus,

	ft	m
At 550 gal/min (34.7 L/s), friction loss in 1000 ft (305 m) of 8-in (203.2-mm) pipe	10	3.05
At 1150 gal/min (72.6 L/s), friction	38	11.6
At 1150 gal/min (72.6 L/s), friction + lift in pipe 1	88	26.8
At 550 gal/min (34.7 L/s), friction + lift in pipe 2	88	26.8

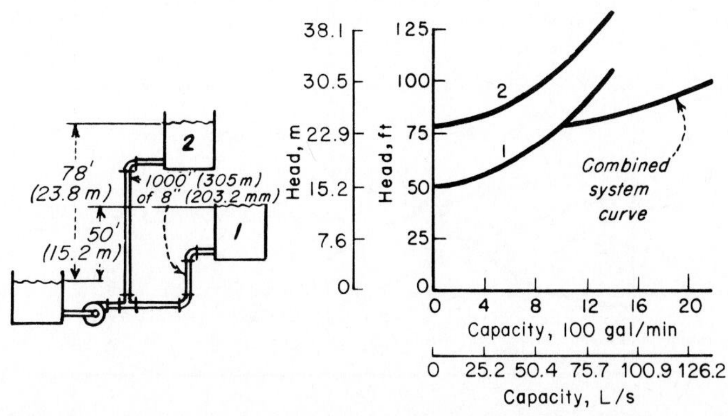

FIG. 19 System with two different discharge heads. *(Peerless Pumps.)*

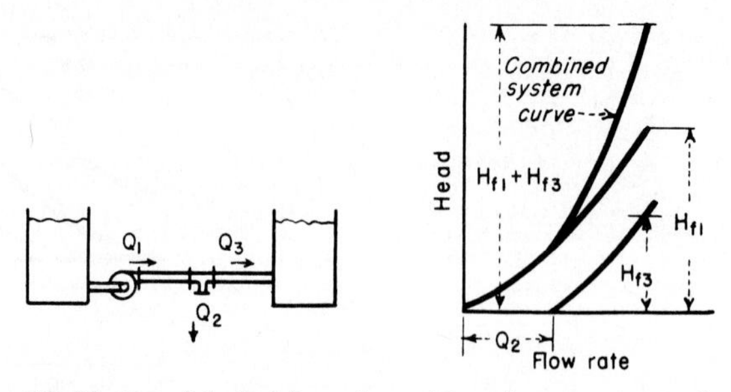

FIG. 20 Part of the fluid flow is diverted from the main pipe. *(Peerless Pumps.)*

The flow rate for the combined system at a head of 88 ft (26.8 m) is 1150 + 550 = 1700 gal/min (107.3 L/s). To produce a flow of 1700 gal/min (107.3 L/s) through this system, a pump capable of developing an 88-ft (26.8-m) head is required.

8. *Plot the system-head curve for diverted flow*

To analyze a system with diverted flow, assume that a constant quantity of liquid is tapped off at the intermediate point. Plot the friction loss vs. flow rate in the normal manner for pipe 1, Fig. 20. Move the curve for pipe 3 to the right at zero head by an amount equal to Q_2, since this represents the quantity passing through pipes 1 and 2 but not through pipe 3. Plot the combined system-head curve by adding, at a given flow rate, the head losses for pipes 1 and 3. With Q = 300 gal/min (18.9 L/s), pipe 1 = 500 ft (152.4 m) of 10-in (254-mm) pipe, and pipe 3 = 50 ft (15.2 m) of 6-in (152.4-mm) pipe.

	ft	m
At 1500 gal/min (94.6 L/s) through pipe 1, friction loss	11	3.35
Friction loss for pipe 3 (1500 − 300 = 1200 gal/min) (75.7 L/s)	8	2.44
Total friction loss at 1500-gal/min (94.6-L/s) delivery	19	5.79

9. *Plot the effect of pump wear*

When a pump wears, there is a loss in capacity and efficiency. The amount of loss depends, however, on the shape of the system-head curve. For a centrifugal pump, Fig. 21, the capacity loss is greater for a given amount of wear if the system-head curve is flat, as compared with a steep system-head curve.

Determine the capacity loss for a worn pump by plotting its H-Q curve. Find this curve by testing the pump at different capacities and plotting the corresponding head. On the same chart, plot the H-Q curve for a new pump of the same size, Fig. 21. Plot the system-head curve, and determine the capacity loss as shown in Fig. 21.

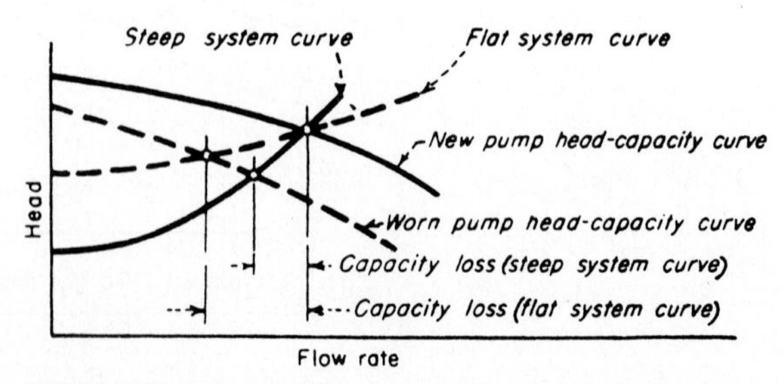

FIG. 21 Effect of pump wear on pump capacity. *(Peerless Pumps.)*

Related Calculations: Use the techniques given here for any type of pump—centrifugal, reciprocating, or rotary—handling any type of liquid—oil, water, chemicals, etc. The methods given here are the work of Melvin Mann, as reported in *Chemical Engineering,* and Peerless Pump Division of FMC Corp.

NET POSITIVE SUCTION HEAD FOR HOT-LIQUID PUMPS

What is the maximum capacity of a double-suction condensate pump operating at 1750 r/min if it handles 100°F (37.8°C) water from a hot well in a condenser having an absolute pressure of 2.0 in (50.8 mm) Hg if the pump centerline is 10 ft (30.5 m) below the hot-well liquid level and the friction-head loss in the suction piping and fitting is 5 ft (1.52 m) of water?

Calculation Procedure:

1. *Compute the net positive suction head on the pump*

The net positive suction head h_n on a pump when the liquid supply is *above* the pump inlet = pressure on liquid surface + static suction head − friction-head loss in suction piping and pump inlet − vapor pressure of the liquid, all expressed in ft absolute of liquid handled. When the liquid supply is *below* the pump centerline—i.e., there is a static suction lift—the vertical distance of the lift is *subtracted* from the pressure on the liquid surface instead of added as in the above relation.

The density of 100°F (37.8°C) water is 62.0 lb/ft^3 (992.6 kg/m^3), computed as shown in earlier calculation procedures in this handbook. The pressure on the liquid surface, in absolute ft of liquid = (2.0 inHg)(1.133)(62.4/62.0) = 2.24 ft (0.68 m). In this calculation, 1.133 = ft of 39.2°F (4°C) water = 1 inHg; 62.4 = lb/ft^3 (999.0 kg/m^3) of 39.2°F (4°C) water. The temperature of 39.2°F (4°C) is used because at this temperature water has its maximum density. Thus, to convert inHg to ft absolute of water, find the product of (inHg)(1.133)(water density at 39.2°F)/(water density at operating temperature). Express both density values in the same unit, usually lb/ft^3.

The static suction head is a physical dimension that is measured in ft (m) of liquid at the operating temperature. In this installation, h_{sh} = 10 ft (3 m) absolute.

The friction-head loss is 5 ft (1.52 m) of water. When it is computed by using the methods of earlier calculation procedures, this head loss is in ft (m) of water at maximum density. To convert to ft absolute, multiply by the ratio of water densities at 39.2°F (4°C) and the operating temperature, or (5)(62.4/62.0) = 5.03 ft (1.53 m).

The vapor pressure of water at 100°F (37.8°C) is 0.949 lb/in^2 (abs) (6.5 kPa) from the steam tables. Convert any vapor pressure to ft absolute by finding the result of [vapor pressure, lb/in^2 (abs)] (144 in^2/ft^2)/liquid density at operating temperature, or (0.949)(144)/62.0 = 2.204 ft (0.67 m) absolute.

With all the heads known, the net positive suction head is h_n = 2.24 + 10 − 5.03 − 2.204 = 5.01 ft (1.53 m) absolute.

2. *Determine the capacity of the condensate pump*

Use the Hydraulic Institute curve, Fig. 22, to determine the maximum capacity of the pump. Enter at the left of Fig. 22 at a net positive suction head of 5.01 ft (1.53 m), and project horizontally to the right until the 3500-r/min curve is intersected. At the top, read the capacity as 278 gal/min (17.5 L/s).

Related Calculations: Use this procedure for any condensate or boiler-feed pump handling water at an elevated temperature. Consult the *Standards of the Hydraulic Institute* for capacity curves of pumps having different types of construction. In general, pump manufacturers who are members of the Hydraulic Institute rate their pumps in accordance with the *Standards,* and a pump chosen from a catalog capacity table or curve will deliver the stated capacity. A similar procedure is used for computing the capacity of pumps handling volatile petroleum liquids. When you use this procedure, be certain to refer to the latest edition of the *Standards.*

CONDENSATE PUMP SELECTION FOR A STEAM POWER PLANT

Select the capacity for a condensate pump serving a steam power plant having a 140,000 lb/h (63,000 kg/h) exhaust flow to a condenser that operates at an absolute pressure of 1.0 in (25.4

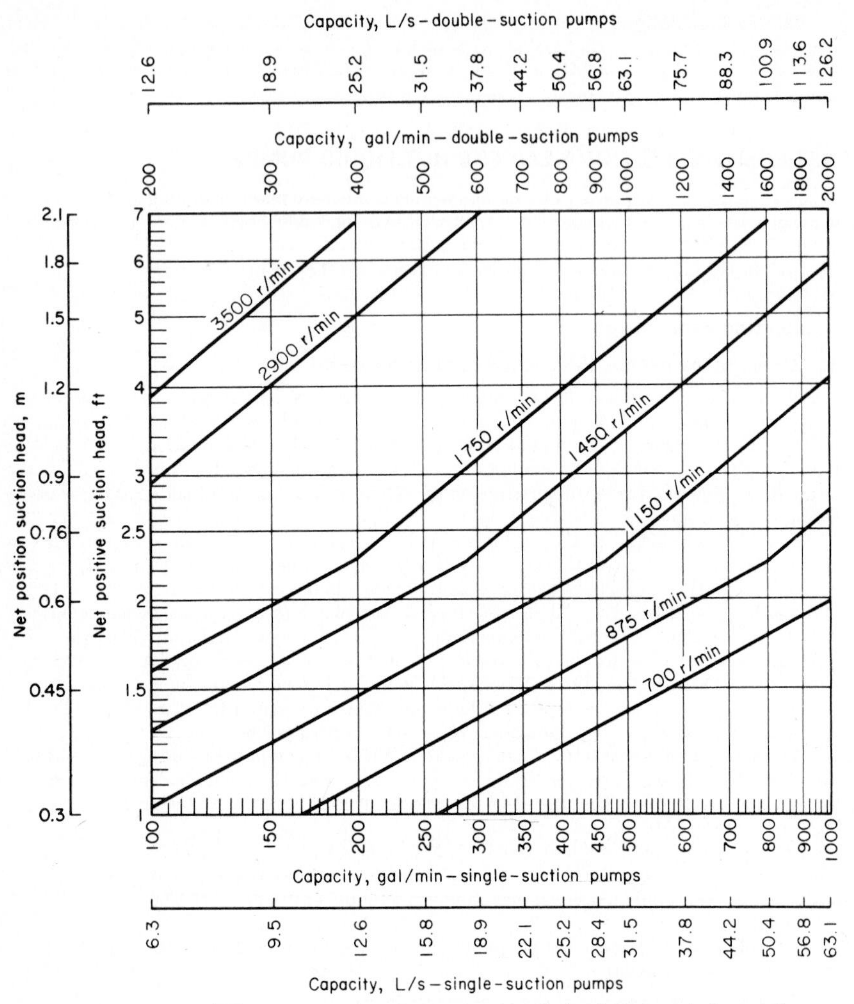

FIG. 22 Capacity and speed limitations of condensate pumps with the shaft through the impeller eye. *(Hydraulic Institute.)*

mm) Hg. The condensate pump discharges through 4-in (101.6-mm) schedule 40 pipe to an air-ejector condenser that has a frictional resistance of 8 ft (2.4 m) of water. From here, the condensate flows to and through a low-pressure heater that has a frictional resistance of 12 ft (3.7 m) of water and is vented to the atmosphere. The total equivalent length of the discharge piping, including all fittings and bends, is 400 ft (121.9 m), and the suction piping total equivalent length is 50 ft (15.2 m). The inlet of the low pressure heater is 75 ft (22.9 m) above the pump centerline, and the condenser hot-well water level is 10 ft (3 m) above the pump centerline. How much power is required to drive the pump if its efficiency is 70 percent?

Calculation Procedure:

1. *Compute the static head on the pump*

Sketch the piping system as shown in Fig. 23. Mark the static elevations and equivalent lengths as indicated.

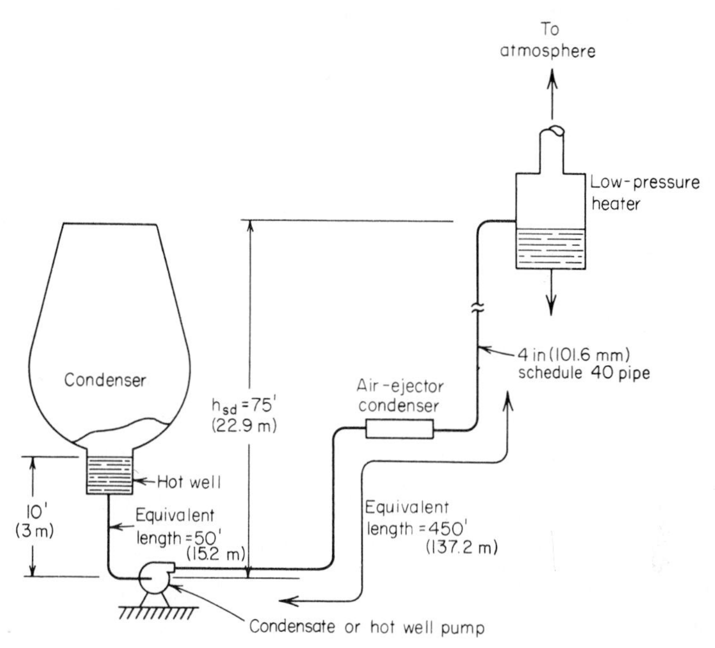

FIG. 23 Condensate pump serving a steam power plant.

The total head on the pump $H_t = H_{ts} + H_f$, where the symbols are the same as in earlier calculation procedures. The total static head $H_{ts} = h_{sd} - h_{sh}$. In this installation, $h_{sd} = 75$ ft (22.9 m). To make the calculation simpler, convert all the heads to absolute values. Since the heater is vented to the atmosphere, the pressure acting on the surface of the water in it = 14.7 lb/in^2 (abs) (101.3 kPa), or 34 ft (10.4 m) of water. The pressure acting on the condensate in the hot well is 1 in (25.4 mm) Hg = 1.133 ft (0.35 m) of water. [An absolute pressure of 1 in (25.4 mm) Hg = 1.133 ft (0.35 m) of water.] Thus, the absolute discharge static head = 75 + 34 = 109 ft (33.2 m), whereas the absolute suction head = 10 + 1.13 = 11.13 ft (3.39 m). Then $H_{ts} = h_{hd} - h_{sh}$ = 109.00 − 11.13 = 97.87 ft (29.8 m), say 98 ft (29.9 m) of water.

2. *Compute the friction head in the piping system*

The total friction head H_f = pipe friction + heater friction. The pipe friction loss is found first, as shown below. The heater friction loss, obtained from the manufacturer or engineering data, is then added to the pipe-friction loss. Both must be expressed in ft (m) of water.

To determine the pipe friction, use Fig. 24 of this section and Table 22 and Fig. 13 of the *Piping* section of this handbook in the following manner. Find the product of the liquid velocity, ft/s, and the pipe internal diameter, in, or vd. With an exhaust flow of 140,000 lb/h (63,636 kg/h) to the condenser, the condensate flow is the same, or 140,000 lb/h (63,636 kg/h) at a temperature of 79.03°F (21.6°C), corresponding to an absolute pressure in the condenser of 1 in (25.4 mm) Hg, obtained from the steam tables. The specific volume of the saturated liquid at this temperature and pressure is 0.01608 ft^3/lb (0.001 m^3/kg). Since 1 gal (0.26 L) of liquid occupies 0.13368 ft^3 (0.004 m^3), specific volume, gal/lb, is (0.01608/0.13368) = 0.1202 (1.01 L/kg). Therefore, a flow of 140,000 lb/h (63,636 kg/h) = a flow of (140,000)(0.1202) = 16,840 gal/h (63,739.4 L/h), or 16,840/60 = 281 gal/min (17.7 L/s). Then the liquid velocity $v = gpm/2.448d^2$ = $281/2.448(4.026)^2$ = 7.1 ft/s (2.1 m/s), and the product vd = (7.1)(4.026) = 28.55.

Enter Fig. 24 at a temperature of 79°F (26.1°C), and project vertically upward to the water curve. From the intersection, project horizontally to the right to vd = 28.55 and then vertically upward to read R = 250,000. Using Table 22 and Fig. 13 of the *Piping* section and R = 250,000, find the friction factor f = 0.0185. Then the head loss due to pipe friction $H_f = (L/D)(v^2/2g)$

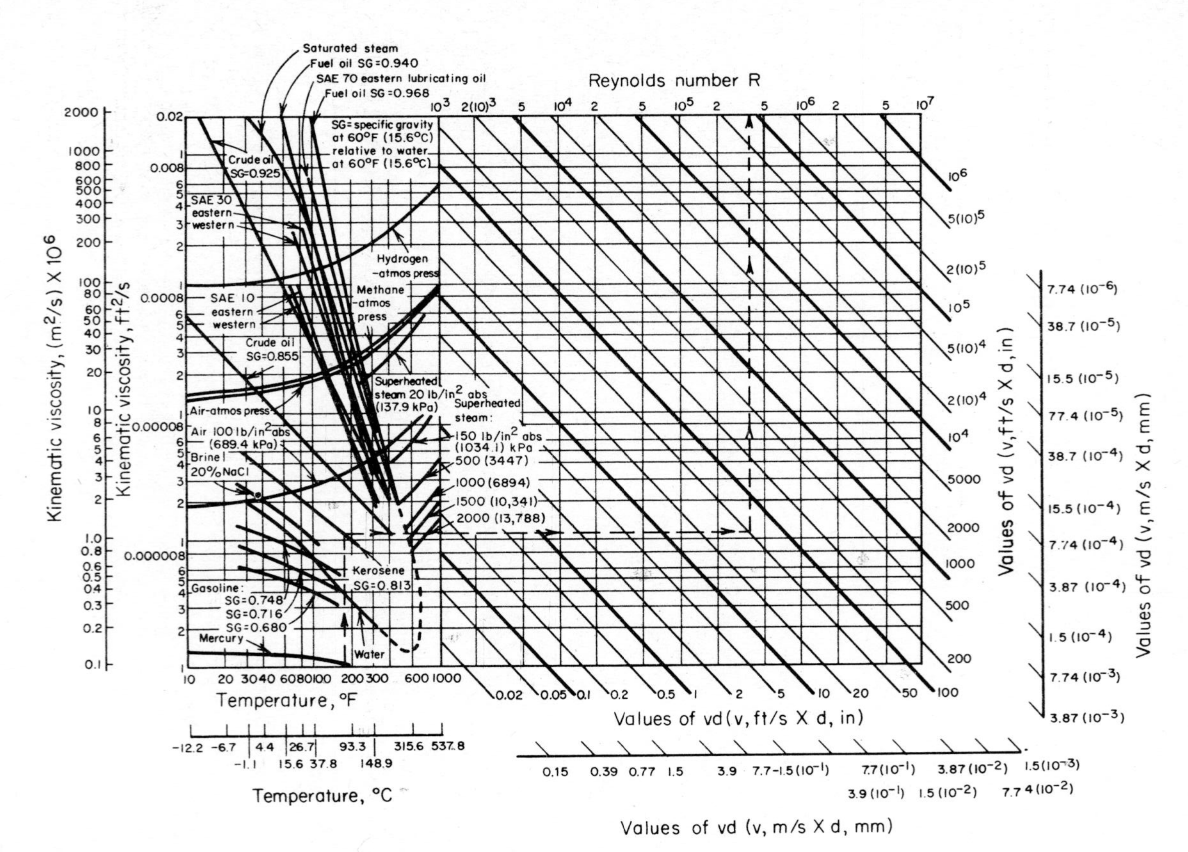

FIG. 24 Kinematic viscosity and Reynolds number chart. *(Hydraulic Institute.)*

$= 0.0185\ (450/4.026/12)/[(7.1)^2/2(32.2)] = 19.18$ ft (5.9 m). In this computation, L = total equivalent length of the pipe, pipe fittings, and system valves, or 450 ft (137.2 m).

3. *Compute the other head losses in the system*

There are two other head losses in this piping system: the entrance loss at the square-edged hot-well pipe leading to the pump and the sudden enlargement in the low-pressure heater. The velocity head $v^2/2g = (7.1)^2/2(32.2) = 0.784$ ft (0.24 m). Using k values from Fig. 5 in this section, $h_e = kv^2/2g = (0.5)(0.784) = 0.392$ ft (0.12 m); $h_{ex} = v^2/2g = 0.784$ ft (0.24 m).

4. *Find the total head on the pump*

The total head on the pump $H_t = H_{ts} + H_f = 97.87 + 19.18 + 8 + 12 + 0.392 + 0.784 =$ 138.226 ft (42.1 m), say 140 ft (42.7 m) of water. In this calculation, the 8- (2.4-m) and 12-ft (3.7-m) head losses are those occurring in the heaters. With a 25 percent safety factor, total head = $(1.25)(140) = 175$ ft (53.3 m).

5. *Compute the horsepower required to drive the pump*

The brake horsepower input $bhp_i = (gpm)(H_t)(s)/3960e$, where the symbols are the same as in earlier calculation procedures. At 1 in (25.4 mm) Hg, 1 lb (0.45 kg) of the condensate has a volume of 0.01608 ft^3 (0.000455 m^3). Since density = 1/specific volume, the density of the condensate = $1/0.01608 = 62.25$ ft^3/lb (3.89 m^3/kg). Water having a specific gravity of unity weighs 62.4 lb/ft^3 (999 kg/m^3). Hence, the specific gravity of the condensate is $62.25/62.4 = 0.997$. Then, assuming that the pump has an operating efficiency of 70 percent, we get $bhp_i = (281)(175) \times (0.997)/[3960(0.70)] = 17.7$ bhp (13.2 kW).

6. *Select the condensate pump*

Condensate or hot-well pumps are usually centrifugal units having two or more stages, with the stage inlets opposed to give better axial balance and to subject the sealing glands to positive internal pressure, thereby preventing air leakage into the pump. In the head range developed by this pump, 175 ft (53.3 m), two stages are satisfactory. Refer to a pump manufacturer's engineering data for specific stage head ranges. Either a turbine or motor drive can be used.

Related Calculations: Use this procedure to choose condensate pumps for steam plants of any type—utility, industrial, marine, portable, heating, or process—and for combined steam-diesel plants.

MINIMUM SAFE FLOW FOR A CENTRIFUGAL PUMP

A centrifugal pump handles 220°F (104.4°C) water and has a shutoff head (with closed discharge valve) of 3200 ft (975.4 m). At shutoff, the pump efficiency is 17 percent and the input brake horsepower is 210 (156.7 kW). What is the minimum safe flow through this pump to prevent overheating at shutoff? Determine the minimum safe flow if the NPSH is 18.8 ft (5.7 m) of water and the liquid specific gravity is 0.995. If the pump contains 500 lb (225 kg) of water, determine the rate of the temperature rise at shutoff.

Calculation Procedure:

1. *Compute the temperature rise in the pump*

With the discharge valve closed, the power input to the pump is converted to heat in the casing and causes the liquid temperature to rise. The temperature rise $t = (1 - e) \times H_s/778e$, where t = temperature rise during shutoff, °F; e = pump efficiency, expressed as a decimal; H_s = shutoff head, ft. For this pump, $t = (1 - 0.17)(3200)/[778(0.17)] = 20.4$°F (36.7°C).

2. *Compute the minimum safe liquid flow*

For general-service pumps, the minimum safe flow M gal/min = 6.0(bhp input at shutoff)/t. Or, $M = 6.0(210)/20.4 = 62.7$ gal/min (3.96 L/s). This equation includes a 20 percent safety factor.

Centrifugal boiler-feed pumps usually have a maximum allowable temperature rise of 15°F (27°C). The minimum allowable flow through the pump to prevent the water temperature from rising more than 15°F (27°C) is 30 gal/min (1.89 L/s) for each 100-bhp (74.6-kW) input at shutoff.

3. *Compute the temperature rise for the operating NPSH*

An NPSH of 18.8 ft (5.73 m) is equivalent to a pressure of 18.8(0.433)(0.995) = 7.78 lb/in^2 (abs) (53.6 kPa) at 220°F (104.4°C), where the factor 0.433 converts ft of water to lb/in^2. At 220°F (104.4°C), the vapor pressure of the water is 17.19 lb/in^2 (abs) (118.5 kPa), from the steam tables. Thus, the total vapor pressure the water can develop before flashing occurs = NPSH pressure + vapor pressure at operating temperature = 7.78 + 17.19 = 24.97 lb/in^2 (abs) (172.1 kPa). Enter the steam tables at this pressure, and read the corresponding temperature as 240°F (115.6°C). The allowable temperature rise of the water is then 240 − 220 = 20°F (36.0°C). Using the safe-flow relation of step 2, we find the minimum safe flow is 62.9 gal/min (3.97 L/s).

4. *Compute the rate of temperature rise*

In any centrifugal pump, the rate of temperature rise t_r °F/min = 42.4(bhp input at shutoff)/wc, where w = weight of liquid in the pump, lb; c = specific heat of the liquid in the pump, Btu/(lb·°F). For this pump containing 500 lb (225 kg) of water with a specific heat, c = 1.0, t_r = 42.4(210)/[500(1.0)] = 17.8°F/min (32°C/min). This is a very rapid temperature rise and could lead to overheating in a few minutes.

Related Calculations: Use this procedure for any centrifugal pump handling any liquid in any service—power, process, marine, industrial, or commercial. Pump manufacturers can supply a temperature-rise curve for a given model pump if it is requested. This curve is superimposed on the pump characteristic curve and shows the temperature rise accompanying a specific flow through the pump.

SELECTING A CENTRIFUGAL PUMP TO HANDLE A VISCOUS LIQUID

Select a centrifugal pump to deliver 750 gal/min (47.3 L/s) of 1000-SSU oil at a total head of 100 ft (30.5 m). The oil has a specific gravity of 0.90 at the pumping temperature. Show how to plot the characteristic curves when the pump is handling the viscous liquid.

Calculation Procedure:

1. *Determine the required correction factors*

A centrifugal pump handling a viscous liquid usually must develop a greater capacity and head, and it requires a larger power input than the same pump handling water. With the water performance of the pump known—from either the pump characteristic curves or a tabulation of pump performance parameters—Fig. 25, prepared by the Hydraulic Institute, can be used to find suitable correction factors. Use this chart only within its scale limits; do not extrapolate. Do not use the chart for mixed-flow or axial-flow pumps or for pumps of special design. Use the chart only for pumps handling uniform liquids; slurries, gels, paper stock, etc., may cause incorrect results. In using the chart, the available net positive suction head is assumed adequate for the pump.

To use Fig. 25, enter at the bottom at the required capacity, 750 gal/min (47.3 L/s), and project vertically to intersect the 100-ft (30.5-m) head curve, the required head. From here project horizontally to the 1000-SSU viscosity curve, and then vertically upward to the correction-factor curves. Read C_E = 0.635; C_Q = 0.95; C_H = 0.92 for $1.0Q_{NW}$. The subscripts E, Q, and H refer to correction factors for efficiency, capacity, and head, respectively; and NW refers to the water capacity at a particular efficiency. At maximum efficiency, the water capacity is given as $1.0Q_{NW}$; other efficiencies, expressed by numbers equal to or less than unity, give different capacities.

2. *Compute the water characteristics required*

The water capacity required for the pump $Q_w = Q_v/C_Q$ where Q_v = viscous capacity, gal/min. For this pump, Q_w = 750/0.95 = 790 gal/min (49.8 L/s). Likewise, water head $H_w = H_v/C_H$, where H_v = viscous head. Or, H_w = 100/0.92 = 108.8 (33.2 m), say 109 ft (33.2 m) of water.

Choose a pump to deliver 790 gal/min (49.8 L/s) of water at 109-ft (33.2-m) head of water, and the required viscous head and capacity will be obtained. Pick the pump so that it is operating at or near its maximum efficiency on water. If the water efficiency E_w = 81 percent at 790 gal/min (49.8 L/s) for this pump, the efficiency when handling the viscous liquid $E_v = E_wC_E$. Or, E_v = 0.81(0.635) = 0.515, or 51.5 percent.

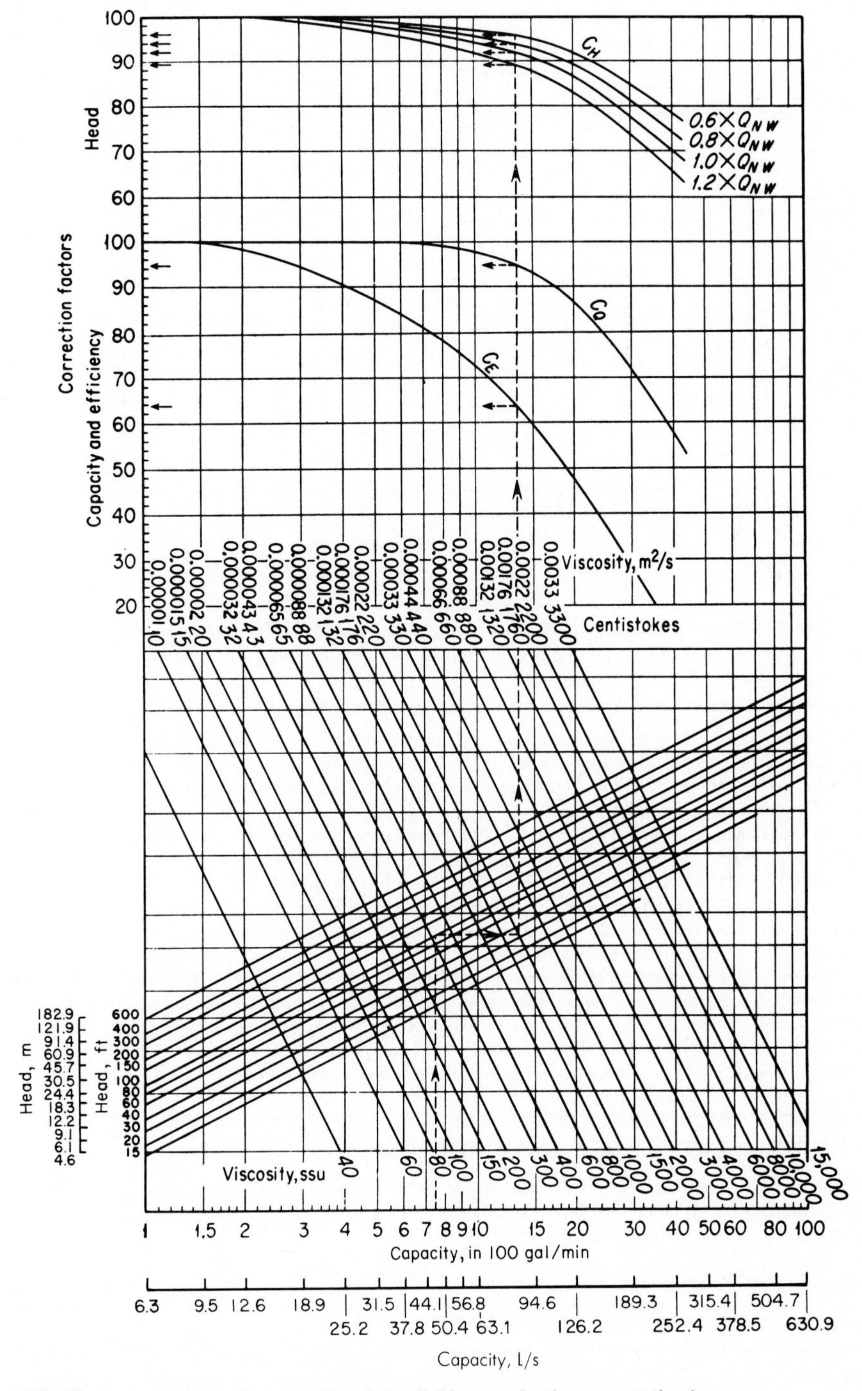

FIG. 25 Correction factors for viscous liquids handled by centrifugal pumps. *(Hydraulic Institute.)*

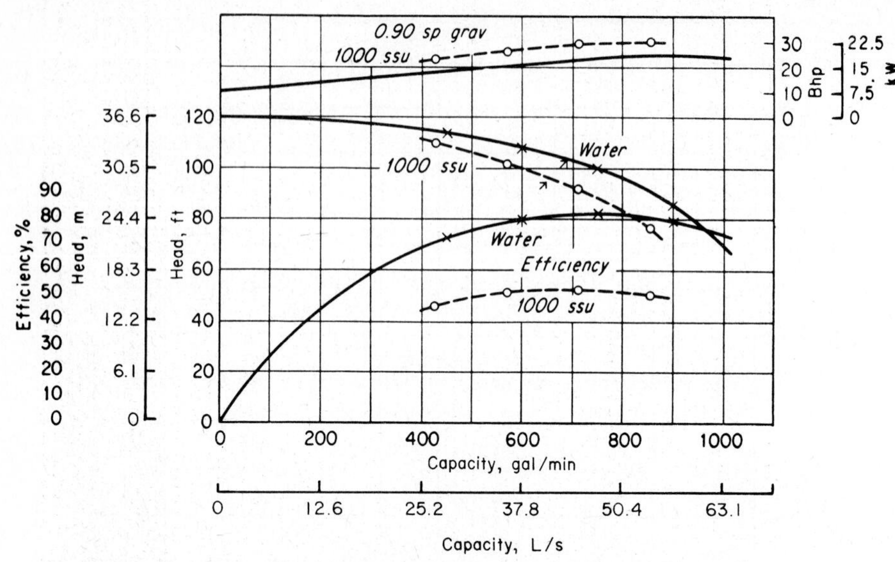

FIG. 26 Characteristics curves for water (solid line) and oil (dashed line). *(Hydraulic Institute.)*

The power input to the pump when handling viscous liquids is given by $P_v = Q_v H_v s / 3960 E_v$, where s = specific gravity of the viscous liquid. For this pump, P_v = (750) × (100)(0.90)/[3960(0.515)] = 33.1 hp (24.7 kW).

3. *Plot the characteristic curves for viscous-liquid pumping*

Follow these eight steps to plot the complete characteristic curves of a centrifugal pump handling a viscous liquid when the water characteristics are known: (*a*) Secure a complete set of characteristic curves (H, Q, P, E) for the pump to be used. (*b*) Locate the point of maximum efficiency for the pump when handling water. (*c*) Read the pump capacity, Q gal/min, at this point. (*d*) Compute the values of $0.6Q$, $0.8Q$, and $1.2Q$ at the maximum efficiency. (*e*) Using Fig. 25, determine the correction factors at the capacities in steps *c* and *d*. Where a multistage pump is being considered, use the head per stage (= total pump head, ft/number of stages), when entering Fig. 25. (*f*) Correct the head, capacity, and efficiency for each of the flow rates in *c* and *d*, using the correction factors from Fig. 25. (*g*) Plot the corrected head and efficiency against the corrected capacity, as in Fig. 26. (*h*) Compute the power input at each flow rate and plot. Draw smooth curves through the points obtained, Fig. 26.

Related Calculations: Use the method given here for any uniform viscous liquid—oil, gasoline, kerosene, mercury, etc—handled by a centrifugal pump. Be careful to use Fig. 25 only within its scale limits; *do not extrapolate*. The method presented here is that developed by the Hydraulic Institute. For new developments in the method, be certain to consult the latest edition of the Hydraulic Institute *Standards*.

PUMP SHAFT DEFLECTION AND CRITICAL SPEED

What are the shaft deflection and approximate first critical speed of a centrifugal pump if the total combined weight of the pump impellers is 23 lb (10.4 kg) and the pump manufacturer supplies the engineering data in Fig. 27?

Calculation Procedure:

1. *Determine the deflection of the pump shaft*

Use Fig. 27 to determine the shaft deflection. Note that this chart is valid for only one pump or series of pumps and must be obtained from the pump builder. Such a chart is difficult to prepare from test data without extensive test facilities.

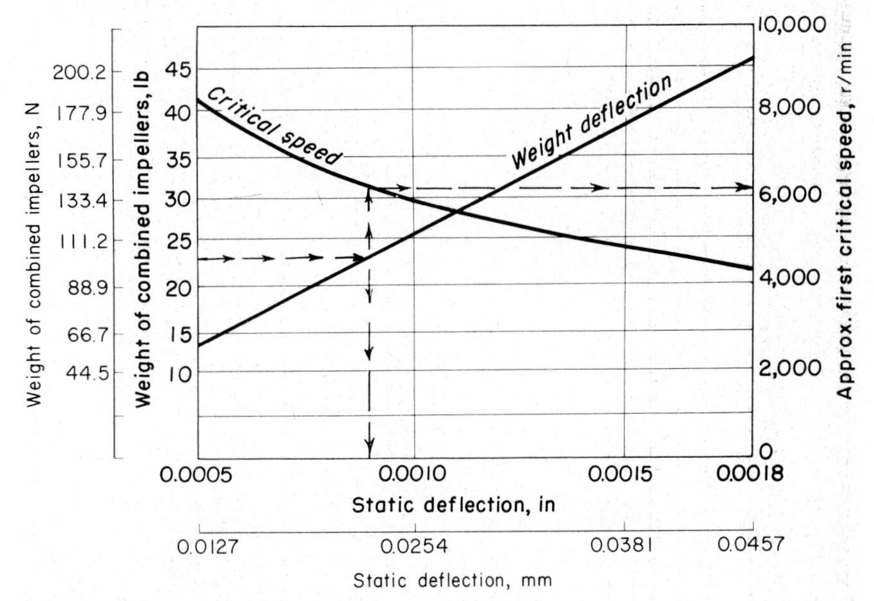

FIG. 27 Pump shaft deflection and critical speed. *(Goulds Pumps, Inc.)*

Enter Fig. 27 at the left at the total combined weight of the impellers, 23 lb (10.4 kg), and project horizontally to the right until the weight-deflection curve is intersected. From the intersection, project vertically downward to read the shaft deflection as 0.009 in (0.23 mm) at full speed.

2. *Determine the critical speed of the pump*

From the intersection of the weight-deflection curve in Fig. 27 project vertically upward to the critical-speed curve. Project horizontally right from this intersection and read the first critical speed as 6200 r/min.

Related Calculations: Use this procedure for any class of pump—centrifugal, rotary, or reciprocating—for which the shaft-deflection and critical-speed curves are available. These pumps can be used for any purpose—process, power, marine, industrial, or commercial.

EFFECT OF LIQUID VISCOSITY ON REGENERATIVE-PUMP PERFORMANCE

A regenerative (turbine) pump has the water head-capacity and power-input characteristics shown in Fig. 28. Determine the head-capacity and power-input characteristics for four different viscosity oils to be handled by the pump—400, 600, 900, and 1000 SSU. What effect does increased viscosity have on the performance of the pump?

Calculation Procedure:

1. *Plot the water characteristics of the pump*

Obtain a tabulation or plot of the water characteristics of the pump from the manufacturer or from their engineering data. With a tabulation of the characteristics, enter the various capacity and power points given, and draw a smooth curve through them, Fig. 28.

2. *Plot the viscous-liquid characteristics of the pump*

The viscous-liquid characteristics of regenerative-type pumps are obtained by test of the actual unit. Hence, the only source of this information is the pump manufacturer. Obtain these characteristics from the pump manufacturer or their test data, and plot them on Fig. 28, as shown, for each oil or other liquid handled.

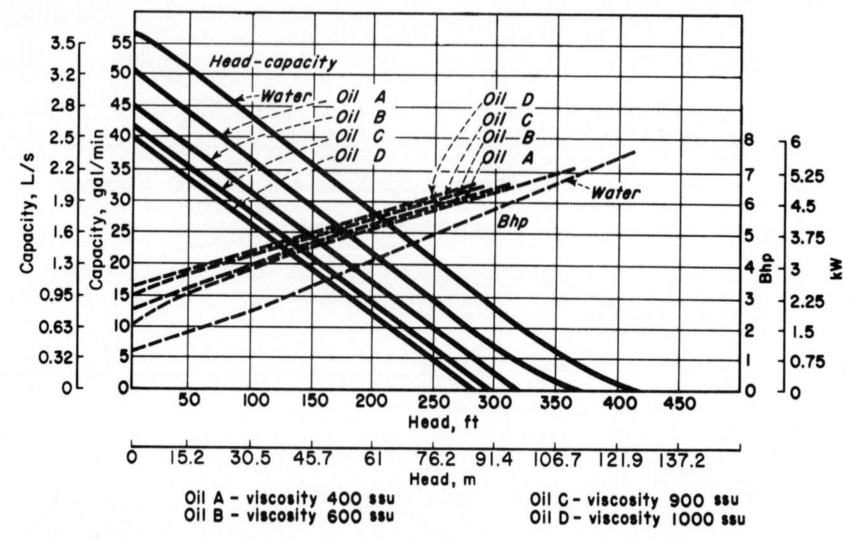

FIG. 28 Regenerative pump performance when handling water and oil. *(Aurora Pump Division, The New York Air Brake Company.)*

3. *Evaluate the effect of viscosity on pump performance*

Study Fig. 28 to determine the effect of increased liquid viscosity on the performance of the pump. Thus at a given head, say 100 ft (30.5 m), the capacity of the pump decreases as the liquid viscosity increases. At 100-ft (30.5-m) head, this pump has a water capacity of 43.5 gal/min (2.74 L/s), Fig. 28. The pump capacity for the various oils at 100-ft (30.5-m) head is 36 gal/min (2.27 L/s) for 400 SSU; 32 gal/min (2.02 L/s) for 600 SSU; 28 gal/min (1.77 L/s) for 900 SSU; and 26 gal/min (1.64 L/s) for 1000 SSU, respectively. There is a similar reduction in capacity of the pump at the other heads plotted in Fig. 28. Thus, as a general rule, the capacity of a regenerative pump decreases with an increase in liquid viscosity at constant head. Or conversely, at constant capacity, the head developed decreases as the liquid viscosity increases.

Plots of the power input to this pump show that the input power increases as the liquid viscosity increases.

Related Calculations: Use this procedure for a regenerative-type pump handling any liquid—water, oil, kerosene, gasoline, etc. A decrease in the viscosity of a liquid, as compared with the viscosity of water, will produce the opposite effect from that of increased viscosity.

EFFECT OF LIQUID VISCOSITY ON RECIPROCATING-PUMP PERFORMANCE

A direct-acting steam-driven reciprocating pump delivers 100 gal/min (6.31 L/s) of 70°F (21.1°C) water when operating at 50 strokes per minute. How much 2000-SSU crude oil will this pump deliver? How much 125°F (51.7°C) water will this pump deliver?

Calculation Procedure:

1. *Determine the recommended change in pump performance*

Reciprocating pumps of any type—direct-acting or power—having any number of liquid-handling cylinders—one to five or more—are usually rated for maximum delivery when handling 250-SSU liquids or 70°F (21.1°C) water. At higher liquid viscosities or water temperatures, the speed—strokes or rpm—is reduced. Table 8 shows typical recommended speed-correction factors for reciprocating pumps for various liquid viscosities and water temperatures. This table shows that with a liquid viscosity of 2000 SSU the pump speed should be reduced 20 percent. When 125°F (51.7°C) water is handled, the pump speed should be reduced 25 percent, as shown in Table 8.

TABLE 8 Speed-Correction Factors

Liquid viscosity, SSU	Speed reduction, %	Water temperature °F	°C	Speed reduction, %
250	0	70	21.1	0
500	4	80	26.7	9
1000	11	100	37.8	18
2000	20	125	51.7	25
3000	26	150	65.6	29
4000	30	200	93.3	34
5000	35	250	121.1	38

2. *Compute the delivery of the pump*

The delivery capacity of any reciprocating pump is directly proportional to the number of strokes per minute it makes or to its rpm.

When 2000-SSU oil is used, the pump strokes per minute must be reduced 20 percent, or (50)(0.20) = 10 strokes/min. Hence, the pump speed will be 50 − 10 = 40 strokes/min. Since the delivery is directly proportional to speed, the delivery of 2000-SSU oil = (40/50)(100) = 80 gal/min (5.1 L/s).

When handling 125°F (51.7°C) water, the pump strokes/min must be reduced 25 percent, or (50)(0.5) = 12.5 strokes/min. Hence, the pump speed will be 50.0 − 12.5 = 37.5 strokes/min. Since the delivery is directly proportional to speed, the delivery of 125°F (51.7°C) water = (37.5/50)(100) = 75 gal/min (4.7 L/s).

Related Calculations: Use this procedure for any type of reciprocating pump handling liquids falling within the range of Table 8. Such liquids include oil, kerosene, gasoline, brine, water, etc.

EFFECT OF VISCOSITY AND DISSOLVED GAS ON ROTARY PUMPS

A rotary pump handles 8000-SSU liquid containing 5 percent entrained gas and 10 percent dissolved gas at a 20-in (508-mm) Hg pump inlet vacuum. The pump is rated at 1000 gal/min (63.1 L/s) when handling gas-free liquids at viscosities less than 600 SSU. What is the output of this pump without slip? With 10 percent slip?

TABLE 9 Rotary Pump Speed Reduction for Various Liquid Viscosities

Liquid viscosity, SSU	Speed reduction, percent of rated pump speed
600	2
800	6
1,000	10
1,500	12
2,000	14
4,000	20
6,000	30
8,000	40
10,000	50
20,000	55
30,000	57
40,000	60

Calculation Procedure:

1. *Compute the required speed reduction of the pump*

When the liquid viscosity exceeds 600 SSU, many pump manufacturers recommend that the speed of a rotary pump be reduced to permit operation without excessive noise or vibration. The speed reduction usually recommended is shown in Table 9.

With this pump handling 8000-SSU liquid, a speed reduction of 40 percent is necessary, as shown in Table 9. Since the capacity of a rotary pump varies directly with its speed, the output of this pump when handling 8000-SSU liquid = (1000 gal/min) × (1.0 − 0.40) = 600 gal/min (37.9 L/s)

2. *Compute the effect of gas on the pump output*

Entrained or dissolved gas reduces the output

TABLE 10 Effect of Entrained or Dissolved Gas on the Liquid Displacement of Rotary Pumps (liquid displacement: percent of displacement)

Vacuum at pump inlet, inHg (mmHg)	Gas entrainment					Gas solubility					Gas entrainment and gas solubility combined				
	1%	2%	3%	4%	5%	2%	4%	6%	8	10%	1% 2%	2% 4%	3% 6%	4% 8%	5% 10%
5 (127)	99	97½	96½	95	93½	99½	99	98½	97	97½	98½	96½	96	92	91
10 (254)	98½	97¼	95½	94	92	99	97½	97	95	95	97½	95	90	90	88¼
15 (381)	98	96½	94½	92½	90½	97	96	94	92	90½	96	93	89½	86½	83¼
20 (508)	97½	94½	92	89	86½	96	92	89	86	83	94	88	83	78	74
25 (635)	94	89	84	79	75½	90	83	76½	71	66	85½	75½	68	61	55

For example, with 5 percent gas entrainment at 15 inHg (381 mmHg) vacuum, the liquid displacement will be 90½ percent of the pump displacement, neglecting slip, or with 10 percent dissolved gas liquid displacement will be 90½ percent of the pump displacement; and with 5 percent entrained gas combined with 10 percent dissolved gas, the liquid displacement will be 83¼ percent of pump replacement.

Source: Courtesy of Kinney Mfg. Div., The New York Air Brake Co.

of a rotary pump, as shown in Table 10. The gas in the liquid expands when the inlet pressure of the pump is below atmospheric and the gas occupies part of the pump chamber, reducing the liquid capacity.

With a 20-in (508-mm) Hg inlet vacuum, 5 percent entrained gas, and 10 percent dissolved gas, Table 10 shows that the liquid displacement is 74 percent of the rated displacement. Thus, the output of the pump when handling this viscous, gas-containing liquid will be (600 gal/min)(0.74) = 444 gal/min (28.0 L/s) without slip.

3. *Compute the effect of slip on the pump output*

Slip reduces rotary-pump output in direct proportion to the slip. Thus, with 10 percent slip, the output of this pump = (444 gal/min)(1.0 − 0.10) = 369.6 gal/min (23.3 L/s).

Related Calculations: Use this procedure for any type of rotary pump—gear, lobe, screw, swinging-vane, sliding-vane, or shuttle-block, handling any clear, viscous liquid. Where the liquid is gas-free, apply only the viscosity correction. Where the liquid viscosity is less than 600 SSU but the liquid contains gas or air, apply the entrained or dissolved gas correction, or both corrections.

SELECTION OF MATERIALS FOR PUMP PARTS

Select suitable materials for the principal parts of a pump handling cold ethylene chloride. Use the Hydraulic Institute recommendations for materials of construction.

Calculation Procedure:

1. *Determine which materials are suitable for this pump*

Refer to the data section of the Hydraulic Institute *Standards*. This section contains a tabulation of hundreds of liquids and the pump construction materials that have been successfully used to handle each liquid.

The table shows that for cold ethylene chloride having a specific gravity of 1.28, an all-bronze pump is satisfactory. In lieu of an all-bronze pump, the principal parts of the pump—casing, impeller, cylinder, and shaft—can be made of one of the following materials: austenitic steels (low-carbon 18-8; 18-8/Mo; highly alloyed stainless); nickel-base alloys containing chromium, molybdenum, and other elements, and usually less than 20 percent iron; or nickel-copper alloy (Monel metal). The order of listing in the *Standards* does not necessarily indicate relative superiority, since certain factors predominating in one instance may be sufficiently overshadowed in others to reverse the arrangement.

2. *Choose the most economical pump*

Use the methods of earlier calculation procedures to select the most economical pump for the installation. Where the corrosion resistance of two or more pumps is equal, the standard pump, in this instance an all-bronze unit, will be the most economical.

Related Calculations: Use this procedure to select the materials of construction for any class of pump—centrifugal, rotary, or reciprocating—in any type of service—power, process, marine, or commercial. Be certain to use the latest edition of the Hydraulic Institute *Standards*, because the recommended materials may change from one edition to the next.

SIZING A HYDROPNEUMATIC STORAGE TANK

A 200-gal/min (12.6-L/s) water pump serves a pumping system. Determine the capacity required for a hydropneumatic tank to serve this system if the allowable high pressure in the tank and system is 60 lb/in^2 (gage) (413.6 kPa) and the allowable low pressure is 30 lb/in^2 (gage) (206.8 kPa). How many starts per hour will the pump make if the system draws 3000 gal/min (189.3 L/s) from the tank?

Calculation Procedure:

1. *Compute the required tank capacity*

In the usual hydropneumatic system, a storage-tank capacity in gal of 10 times the pump capacity in gal/min is used, if this capacity produces a moderate running time for the pump. Thus, this system would have a tank capacity of (10)(200) = 2000 gal (7570.8 L).

2. *Compute the quantity of liquid withdrawn per cycle*

For any hydropneumatic tank the withdrawal, expressed as the number of gallons (liters) withdrawn per cycle, is given by $W = (v_L - v_H)/C$, where v_L = air volume in tank at the lower pressure, ft^3 (m^3); v_H = volume of air in tank at higher pressure, ft^3 (m^3); C = conversion factor to convert ft^3 (m^3) to gallons (liters), as given below.

Compute V_L and V_H using the gas law for v_H and either the gas law or the reserve percentage for v_L Thus, for v_H, the gas law gives $v_H = p_L v_L / p_H$, where p_L = lower air pressure in tank, lb/in² (abs) (kPa); p_H = higher air pressure in tank lb/in² (abs) (kPa); other symbols as before.

In most hydropneumatic tanks a liquid reserve of 10 to 20 percent of the total tank volume is kept in the tank to prevent the tank from running dry and damaging the pump. Assuming a 10 percent reserve for this tank, $v_L = 0.1\ V$, where V = tank volume in ft^3 (m^3). Since a 2000-gal (7570-L) tank is being used, the volume of the tank is 2000/7.481 ft³/gal = 267.3 ft³ (7.6 m³). With the 10 percent reserve at the 44.7 lb/in² (abs) (308.2-kPa) lower pressure, v_L = 0.9 (267.3) = 240.6 ft³ (6.3 m³), where 0.9 = V − 0.1 V.

At the higher pressure in the tank, 74.7 lb/in² (abs) (514.9 kPa), the volume of the air will be, from the gas law, $v_H = p_L v_L / p_H$ = 44.7 (240.6)/74.7 = 143.9 ft³ (4.1 m³). Hence, during withdrawal, the volume of liquid removed from the tank will be W_g = (240.6 − 143.9)/0.1337 = 723.3 gal (2738 L). In this relation the constant converts from cubic feet to gallons and is 0.1337. To convert from cubic meters to liters, use the constant 1000 in the denominator.

3. *Compute the pump running time*

The pump has a capacity of 200 gal/min (12.6 L/s). Therefore, it will take 723/200 = 3.6 min to replace the withdrawn liquid. To supply 3000 gal/h (11,355 L/h) to the system, the pump must start 3000/723 = 4.1, or 5 times per hour. This is acceptable because a system in which the pump starts six or fewer times per hour is generally thought satisfactory.

Where the pump capacity is insufficient to supply the system demand for short periods, use a smaller reserve. Compute the running time using the equations in steps 2 and 3. Where a larger reserve is used—say 20 percent—use the value 0.8 in the equations in step 2. For a 30 percent reserve, the value would be 0.70, and so on.

Related Calculations: Use this procedure for any liquid system having a hydropneumatic tank—well drinking water, marine, industrial, or process.

USING CENTRIFUGAL PUMPS AS HYDRAULIC TURBINES

Select a centrifugal pump to serve as a hydraulic turbine power source for a 1500-gal/min (5677.5-L/min) flow rate with 1290 ft (393.1 m) of head. The power application requires a 3600-r/min speed, the specific gravity of the liquid is 0.52, and the total available exhaust head is 20 ft (6.1 m). Analyze the cavitation potential and operating characteristics at an 80 percent flow rate.

Calculation Procedure:

1. *Choose the number of stages for the pump*

Search of typical centrifugal-pump data shows that a head of 1290 ft (393.1 m) is too large for a single-stage pump of conventional design. Hence, a two-stage pump will be the preliminary choice for this application. The two-stage pump chosen will have a design head of 645 ft (196.6 m) per stage.

2. *Compute the specific speed of the pump chosen*

Use the relation $N_s = \text{pump } rpm(Q)^{0.5}/H^{0.75}$, where N_s = specific speed of the pump; rpm = r/min of pump shaft; Q = pump capacity or flow rate, gal/min; H = pump head per stage, ft. Substituting, we get $N_s = 3600(1500)^{0.5}/(645)^{0.75} = 1090$. Note that the specific speed value is the same regardless of the system of units used—USCS or SI.

3. *Convert turbine design conditions to pump design conditions*

To convert from turbine design conditions to pump design conditions, use the pump manufacturer's conversion factors that relate turbine best efficiency point (bep) performance with pump bep performance. Typically, as specific speed N_s varies from 500 to 2800, these bep factors generally vary as follows: the conversion factor for capacity (gal/min or L/min) C_Q, from 2.2 to 1.1;

the conversion factor for head (ft or m) C_H, from 2.2 to 1.1; the conversion factor for efficiency C_E, from 0.92 to 0.99. Applying these conversion factors to the turbine design conditions yields the pump design conditions sought.

At the specific speed for this pump, the values of these conversion factors are determined from the manufacturer to be $C_Q = 1.24$; $C_H = 1.42$; $C_E = 0.967$.

Given these conversion factors, the turbine design conditions can be converted to the pump design conditions thus: $Q_p = Q_t/C_Q$, where Q_p = pump capacity or flow rate, gal/min or L/min; Q_t = turbine capacity or flow rate in the same units; other symbols are as given earlier. Substituting gives $Q_p = 1500/1.24 = 1210$ gal/min (4580 L/min).

Likewise, the pump discharge head, in feet of liquid handled, is $H_p = H_t/C_H$. So $H_p = 645/1.42 = 454$ ft (138.4 m).

4. *Select a suitable pump for the operating conditions*

Once the pump capacity, head, and rpm are known, a pump having its best bep at these conditions can be selected. Searching a set of pump characteristic curves and capacity tables shows that a two-stage 4-in (10-cm) unit with an efficiency of 77 percent would be suitable.

5. *Estimate the turbine horsepower developed*

To predict the developed horsepower, convert the pump efficiency to turbine efficiency. Use the conversion factor developed above. Or, the turbine efficiency $E_t = E_pC_E = (0.77)(0.967) = 0.745$, or 74.5 percent.

With the turbine efficiency known, the output brake horsepower can be found from bhp = $Q_tH_tE_ts/3960$, where s = fluid specific gravity; other symbols as before. Substituting, we get bhp = 1500(1290)(0.745)(0.52)/3960 = 198 hp (141 kW).

6. *Determine the cavitation potential of this pump*

Just as pumping requires a minimum net positive suction head, turbine duty requires a net positive exhaust head. The relation between the total required exhaust head (TREH) and turbine head per stage is the cavitation constant σ_r = TREH/H. Figure 29 shows σ_r vs. N_s for hydraulic turbines. Although a pump used as a turbine will not have exactly the same relationship, this curve provides a good estimate of σ_r for turbine duty.

To prevent cavitation, the total available exhaust head (TAEH) must be greater than the TREH. In this installation, $N_s = 1090$ and TAEH = 20 ft (6.1 m). From Fig. 29, $\sigma_r = 0.028$ and TREH = 0.028(645) = 18.1 ft (5.5 m). Because TAEH > TREH, there is enough exhaust head to prevent cavitation.

7. *Determine the turbine performance at 80 percent flow rate*

In many cases, pump manufacturers treat conversion factors as proprietary information. When this occurs, the performance of the turbine under different operating conditions can be predicted from the general curves in Figs. 30 and 31.

At the 80 percent flow rate for the turbine, or 1200 gal/min (4542 L/min), the operating point is 80 percent of bep capacity. For a specific speed of 1090, as before, the percentages of bep head and efficiency are shown in Figs. 30 and 31: 79.5 percent of bep head and 91 percent of bep efficiency. To find the actual performance, multiply by the bep values. Or, $H_t = 0.795(1290) = 1025$ ft (393.1 m); $E_t = 0.91(74.5) = 67.8$ percent.

The bhp at the new operating condition is then bhp = 1200(1025)(0.678)(0.52)/3960 = 110 hp (82.1 kW).

In a similar way, the constant-head curves in Figs. 32 and 33 predict turbine performance at different speeds. For example, speed is 80 percent of bep speed at 2880 r/min. For a specific speed of 1090, the percentages of bep capacity, efficiency, and power are 107 percent of the capacity, 94 percent of the efficiency, and 108 percent of the bhp. To get the actual performance, convert as before: $Q_t = 107(1500) = 1610$ gal/min (6094 L/min); $E_t = 0.94(74.5) = 70.0$ percent; bhp = 1.08(189) = 206 hp (153.7 kW).

Note that the bhp in this last instance is higher than the bhp at the best efficiency point. Thus more horsepower can be obtained from a given unit by reducing the speed and increasing the flow rate. When the speed is fixed, more bhp cannot be obtained from the unit, but it may be possible to select a smaller pump for the same application.

Related Calculations: Use this general procedure for choosing a centrifugal pump to drive—as a hydraulic turbine—another pump, a fan, a generator, or a compressor, where high-pressure liquid is available as a source of power. Because pumps are designed as fluid movers,

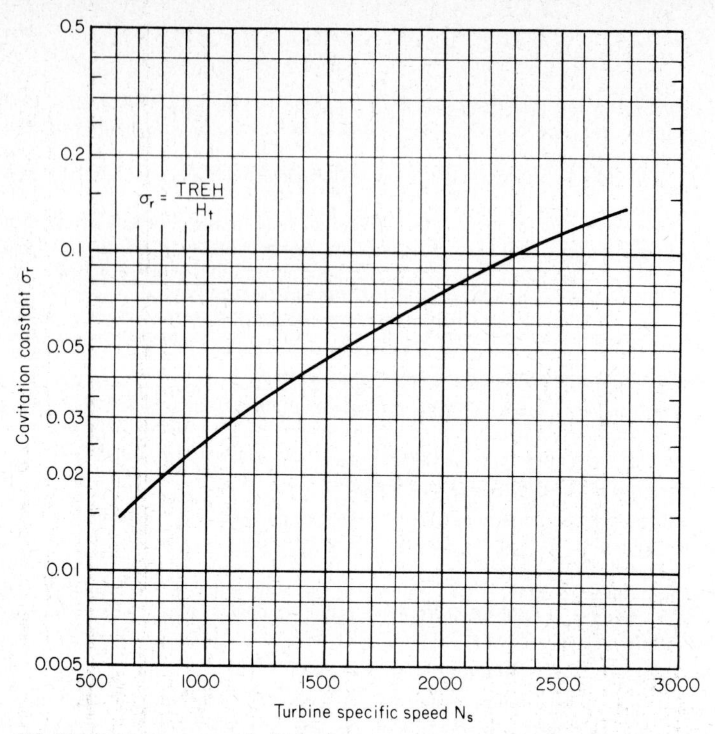

FIG. 29 Cavitation constant for hydraulic turbines. *(Chemical Engineering.)*

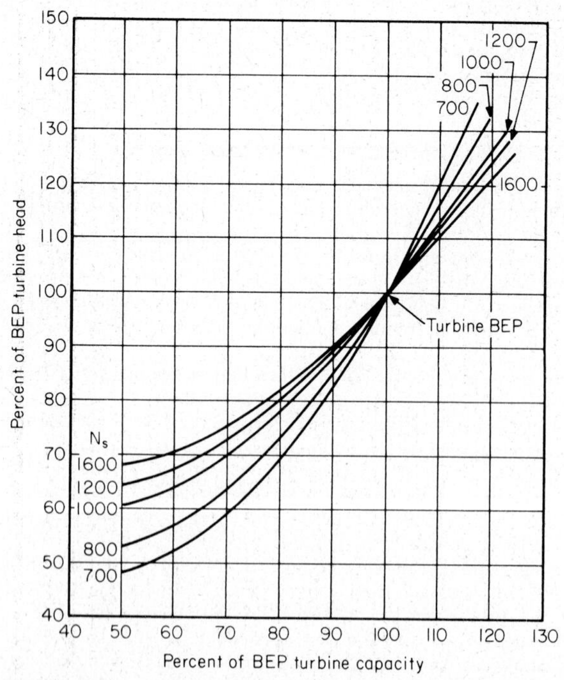

FIG. 30 Constant-speed curves for turbine duty. *(Chemical Engineering.)*

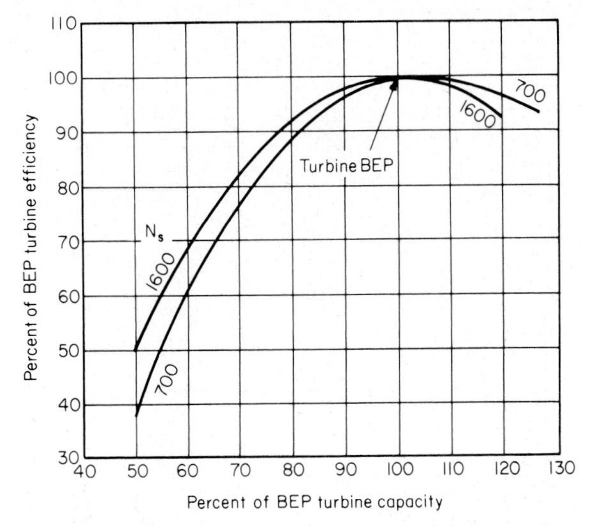

FIG. 31 Constant-speed curves for turbine duty. *(Chemical Engineering.)*

they may be less efficient as hydraulic turbines than equipment designed for that purpose. Steam turbines and electric motors are more economical where steam or electricity is available.

But using a pump as a turbine can pay off in remote locations where steam or electric power would require additional wiring or piping, in hazardous locations that require nonsparking equipment, where energy may be recovered from a stream that otherwise would be throttled, and when a radial-flow centrifugal pump is immediately available but a hydraulic turbine is not.

In the most common situation, there is a liquid stream with fixed head and flow rate and an application requiring a fixed rpm; these are the turbine design conditions. The objective is to pick a pump with a turbine bep at these conditions. With performance curves such as Fig. 34, turbine design conditions can be converted to pump design conditions. Then you select from a manufacturer's catalog a model that has its pump bep at those values.

The most common error in pump selection is using the turbine design conditions in choosing a pump from a catalog. Because catalog performance curves describe pump duty, not turbine duty, the result is an oversized unit that fails to work properly.

This procedure is the work of Fred Buse, Chief Engineer, Standard Pump Aldrich Division of Ingersoll-Rand Co., as reported in *Chemical Engineering* magazine.

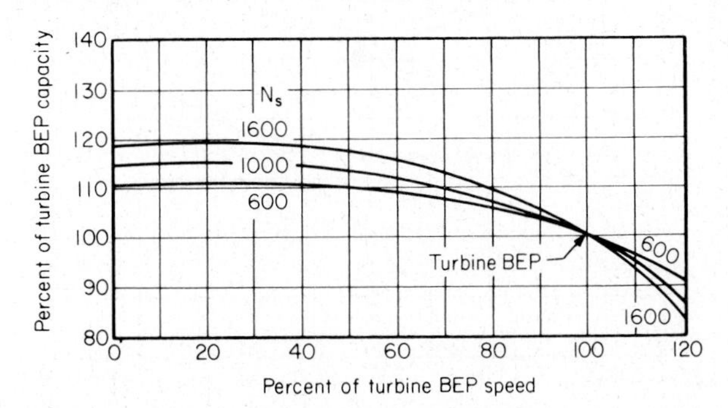

FIG. 32 Constant-head curves for turbine duty. *(Chemical Engineering.)*

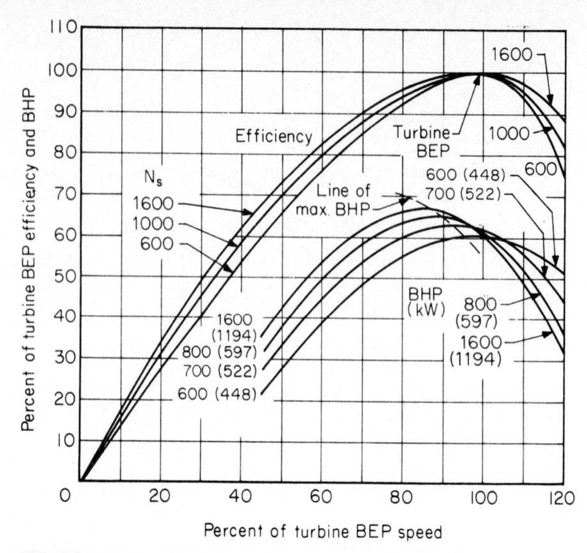

FIG. 33 Constant-head curves for turbine only. *(Chemical Engineering.)*

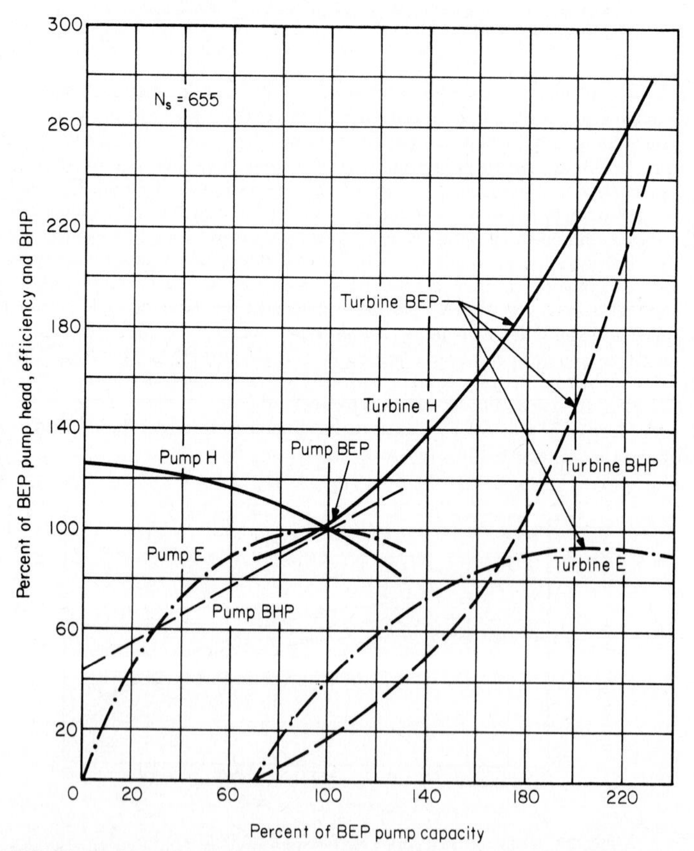

FIG. 34 Performance of a pump at constant speed in pump duty and turbine duty. *(Chemical Engineering.)*

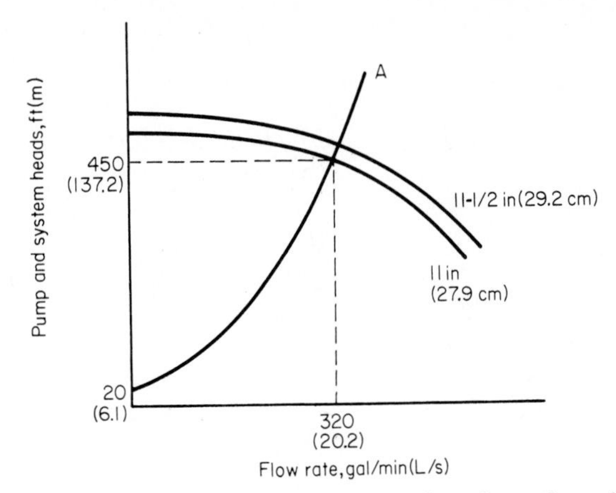

FIG. 35 System-head curves without recirculation flow. *(Chemical Engineering.)*

SIZING CENTRIFUGAL-PUMP IMPELLERS FOR SAFETY SERVICE

Determine the impeller size of a centrifugal pump that will provide a safe continuous-recirculation flow to prevent the pump from overheating at shutoff. The pump delivers 320 gal/min (20.2 L/s) at an operating head of 450 ft (137.2 m). The inlet water temperature is 220°F (104.4°C), and the system has an NPSH of 5 ft (1.5 m). Pump performance curves and the system-head characteristic curve for the discharge flow (without recirculation) are shown in Fig. 35, and the piping layout is shown in Fig. 36. The brake horsepower (bhp) for an 11-in (27.9-cm) and an 11.5-in (29.2-cm) impeller at shutoff is 53 and 60, respectively. Determine the permissible water temperature rise for this pump.

Calculation Procedure:

1. *Compute the actual temperature rise of the water in the pump*

Use the relation $P_0 = P_v + P_{\text{NPSH}}$, where P_0 = pressure corresponding to the actual liquid temperature in the pump during operation, lb/in² (abs) (kPa); P_v = vapor pressure in the pump at the inlet water temperature, lb/in² (abs) (kPa); P_{NPSH} = pressure created by the net positive suction head on the pumps, lb/in² (abs) (kPa). The head in feet (meters) must be converted to lb/in² (abs) (kPa) by the relation lb/in² (abs) = (NPSH, ft) (liquid density at the pumping temperature, lb/ft³)/(144 in²/ft²). Substituting yields P_0 = 17.2 lb/in² (abs) + 5(59.6)/144 = 19.3 lb/in² (abs) (133.1 kPa).

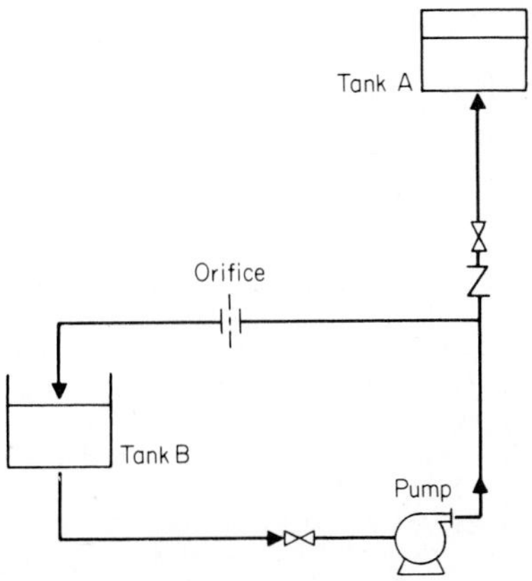

Using the steam tables, find the saturation temperature T_s corresponding to this absolute pressure as T_s = 226.1°F (107.8°C). Then the permissible temperature rise is $T_p = T_s - T_{op}$, where T_{op} = water temperature in the pump inlet. Or, $T_p = 226.1 - 220 = 6.1$°F (3.4°C).

FIG. 36 Pumping system with a continuous-recirculation line. *(Chemical Engineering.)*

2. *Compute the recirculation flow rate at the shutoff head*

From the pump characteristic curve with recirculation, Fig. 37, the continuous-recircu-

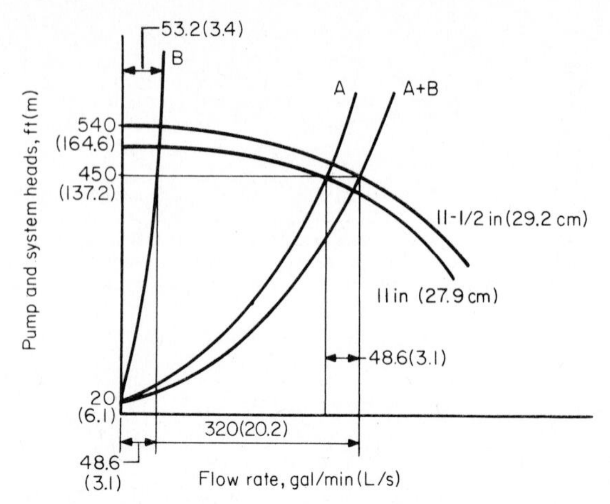

FIG. 37 System-head curves with recirculation flow. *(Chemical Engineering.)*

lation flow Q_B for an 11.5-in (29.2-cm) impeller at an operating head of 450 ft (137.2 m) is 48.6 gal/min (177.1 L/min). Find the continuous-recirculation flow at shutoff head H_s ft (m) of 540 ft (164.6 m) from $Q_s = Q_B(H_s/H_{op})^{0.5}$, where H_{op} = operating head, ft (m). Or, Q_s = 48.6(540/450) = 53.2 gal/min (201.4 L/min).

3. *Find the minimum safe flow for this pump*

The minimum safe flow, lb/h, is given by $w_{min} = 2545bhp/[C_pT_p + (1.285 \times 10^{-3})H_s]$, where C_p = specific head of the water; other symbols as before. Substituting, we find w_{min} = 2545(60)/[1.0(6.1) + (1.285 × 10^{-3})(540)] = 22,476 lb/h (2.83 kg/s). Converting to gal/min yields $Q_{min} = w_{min}/[(ft^3/h)(gal/min)(lb/ft^3)]$ for the water flowing through the pump. Or, Q_{min} = 22,476/[(8.021)(59.6)] = 47.1 gal/min (178.3 L/min).

4. *Compare the shutoff recirculation flow with the safe recirculation flow*

Since the shutoff recirculation flow Q_s = 53.2 gal/min (201.4 L/min) is greater than Q_{min} = 47.1 gal/min (178.3 L/min), the 11.5-in (29.2-cm) impeller is adequate to provide safe continuous recirculation. An 11.25-in (28.6-cm) impeller would not be adequate because Q_{min} = 45 gal/min (170.3 L/min) and Q_s = 25.6 gal/min (96.9 L/min).

Related Calculations: Safety-service pumps are those used for standby service in a variety of industrial plants serving the chemical, petroleum, plastics, aircraft, auto, marine, manufacturing, and similar businesses. Such pumps may be used for fire protection, boiler feed, condenser cooling, and related tasks. In such systems the pump is usually oversized and has a recirculation loop piped in to prevent overheating by maintaining a minimum safe flow. Figure 35 shows a schematic of such a system. Recirculation is controlled by a properly sized orifice rather than by valves because an orifice is less expensive and highly reliable.

The general procedure for sizing centrifugal pumps for safety service, using the symbols given earlier, is this: (1) Select a pump that will deliver the desired flow Q_A, using the head-capacity characteristic curves of the pump and system. (2) Choose the next larger diameter pump impeller to maintain a discharge flow of Q_A to tank *A*, Fig. 35, and a recirculation flow Q_B to tank *B*, Fig. 35. (3) Compute the recirculation flow Q_s at the pump shutoff point from $Q_s = Q_B(H_s/H_{op})^{0.5}$. (4) Calculate the minimum safe flow Q_{min} for the pump with the larger impeller diameter. (5) Compare the recirculation flow Q_s at the pump shutoff point with the minimum safe flow Q_{min}. If $Q_s \geq Q_{min}$, the selection process has been completed. If $Q_s < Q_{min}$, choose the next larger size impeller and repeat steps 3, 4, and 5 above until the impeller size that will provide the minimum safe recirculation flow is determined.

This procedure is the work of Mileta Mikasinovic and Patrick C. Tung, design engineers, Ontario Hydro, as reported in *Chemical Engineering* magazine.

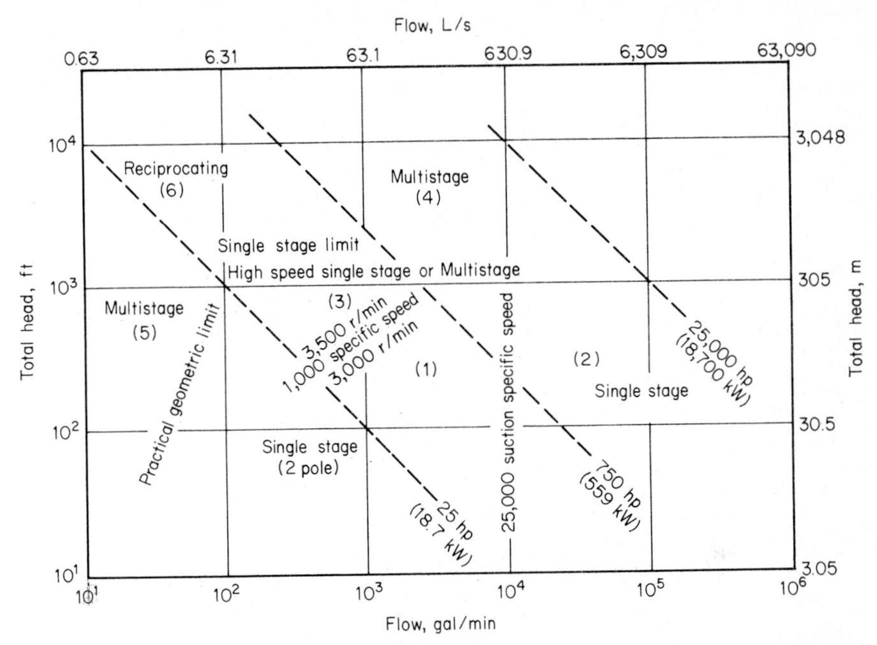

FIG. 38 Selection guide is based mainly on specific speed, which indicates impeller geometry. *(Chemical Engineering.)*

PUMP CHOICE TO REDUCE ENERGY CONSUMPTION AND LOSS

Choose an energy-efficient pump to handle 1000 gal/min (3800 L/min) of water at 60°F (15.6°C) at a total head of 150 ft (45.5 m). A readily commercially available pump is preferred for this application.

Calculation Procedure:

1. *Compute the pump horsepower required*

For any pump, $bhp_i = (gpm)(H_t)(s)/3960e$, where bhp_i = input brake (motor) horsepower to the pump; H_t = total head on the pump, ft; s = specific gravity of the liquid handled; e = hydraulic efficiency of the pump. For this application where $s = 1.0$ and a hydraulic efficiency of 70 percent can be safely assumed, $bhp_i = (1000)(150)(1)/(3960)(0.70) = 54.1$ bhp (40.3 kW).

2. *Choose the most energy-efficient pump*

Use Fig. 38, entering at the bottom at 1000 gal/min (3800 L/min) and projecting vertically upward to a total head of 150 ft (45.5 m). The resulting intersection is within area 1, showing from Table 11 that a single-stage 3500-r/min electric-motor-driven pump would be the most energy-efficient.

Related Calculations: The procedure given here can be used for pumps in a variety of applications—chemical, petroleum, commercial, industrial, marine, aeronautical, air-conditioning, cooling-water, etc., where the capacity varies from 10 to 1,000,000 gal/min (38 to 3,800,000 L/min) and the head varies from 10 to 10,000 ft (3 to 3300 m). Figure 38 is based primarily on the characteristic of pump specific speed $N_s = NQ^2/H^{3/4}$, where N = pump rotating speed, r/min; Q = capacity, gal/min (L/min); H = total head, ft (m).

When N_s is less than 1000, the operating efficiency of single-stage centrifugal pumps falls off dramatically; then either multistage or higher-speed pumps offer the best efficiency.

Area 1 of Fig. 38 is the densest, crowded both with pumps operating at 1750 and 3500 r/min,

TABLE 11 Type of Pump for Highest Energy Efficiency*

Area	Type
Area 1:	Single-stage, 3500 r/min
Area 2:	Single-stage, 1750 r/min or lower
Area 3:	Single-stage, above 3500 r/min, or multistage, 3500 r/min
Area 4:	Multistage
Area 5:	Multistage
Area 6:	Reciprocating

*Includes ANSI B73.1 standards; see area number in Fig. 38.

because years ago 3500-r/min pumps were not thought to be as durable as 1750-r/min ones. Since the adoption of the AVS standard in 1960 (superseded by ANSI B73.1), pumps with stiffer shafts have been proved reliable.

Also responsible for many 1750-r/min pumps in area 1 has been the impression that the higher (3500-r/min) speed causes pumps to wear out faster. However, because impeller tip speed is the same at both 3500 and 1750 r/min [as, for example, a 6-in (15-cm) impeller at 3500 r/min and a 12-in (30-cm) one at 1750 r/min], so is the fluid velocity, and so should be the erosion of metal surface. Another reason for not limiting operating speed is that improved impeller inlet design allows operation at 3500 r/min to capacities of 5000 gal/min (19,000 L/min) and higher.

Choice of operating speed also may be indirectly limited by specifications pertaining to suction performance, such as that fixing the top suction specific speed S directly or indirectly by choice of the sigma constant or by reliance on Hydraulic Institute charts.

Values of S below 8000 to 10,000 have long been accepted for avoiding cavitation. However, since the development of the inducer, S values in the range of 20,000 to 25,000 have become commonplace, and values as high as 50,000 have become practical.

The sigma constant, which relates NPSH to total head, is little used today, and Hydraulic Institute charts (which are being revised) are conservative.

In light of today's designs and materials, past restrictions resulting from suction performance limitations should be reevaluated or eliminated entirely.

Even if the most efficient pump has been selected, there are a number of circumstances in which it may not operate at peak efficiency. Today's cost of energy has made these considerations more important.

A centrifugal pump, being a hydrodynamic machine, is designed for a single peak operating-point capacity and total head. Operation at other than this best efficiency point (bep) reduces efficiency. Specifications now should account for such factors as these:

1. A need for a larger number of smaller pumps. When a process operates over a wide range of capacities, as many do, pumps will often work at less than full capacity, hence at lower efficiency. This can be avoided by installing two or three pumps in parallel, in place of a single large one, so that one of the smaller pumps can handle the flow when operations are at a low rate.

2. Allowance for present capacity. Pump systems are frequently designed for full flow at some time in the future. Before this time arrives, the pumps will operate far from their best efficiency points. Even if this interim period lasts only 2 or 3 years, it may be more economical to install a smaller pump initially and to replace it later with a full-capacity one.

3. Inefficient impeller size. Some specifications call for pump impeller diameter to be no larger than 90 or 95 percent of the size that a pump could take, so as to provide reserve head. If this reserve is used only 5 percent of the time, all such pumps will be operating at less than full efficiency most of the time.

4. Advantages of allowing operation to the right of the best efficiency point. Some specifications, the result of such thinking as that which provides reserve head, prohibit the selection of pumps that would operate to the right of the best efficiency point. This eliminates half of the pumps that might be selected and results in oversized pumps operating at lower efficiency.

This procedure is the work of John H. Doolin, Director of Product Development, Worthington Pumps, Inc., as reported in *Chemical Engineering* magazine.

Piping and Fluid Flow

REFERENCES: Severud and Marr—*Elevated Temperature Piping Design*, ASME; Jeppson—*Analysis of Flow in Pipe Networks*, Butterworths/Ann Arbor Science Press; Sherwood and Whistance—*Piping Guide*, Syentek Books; Williams—*Pipelines and Permafrost*, Longmans; Watters—*Modern Analysis and Control of Unsteady Flow in Pipelines*, Butterworths/Ann Arbor Science Press; Lambert—*Pipeline Instrumentation and Controls Handbook*, Gulf Publishing; Marks—*Oceanic Pipeline Computations*, Penwell; Kentish—*Industrial Pipework*, McGraw-Hill (UK); Brebbia and Ferrante—*Computational Hydraulics*, Butterworths; King and Crocker—*Piping Handbook*, McGraw-Hill; ANSA—*Code for Pressure Piping* (commonly called the *Piping Code*); ASME—*Fluid Meters—Their Theory and Application;* King and Brater—*Handbook of Hydraulics*, McGraw-Hill; Ingersoll-Rand Company—*Cameron Hydraulic Data*; The Hydraulic Institute—*Standards of the Hydraulic Institute*; Baumeister and Marks—*Standard Handbook for Mechanical Engineers*, McGraw-Hill; Littleton—*Industrial Piping*, McGraw-Hill; The Hydraulic Institute—*Pipe Friction Manual*; Black, Sivalls, and Bryson—*Valve Sizing Book* and *Cv Book*; Fluid Controls Institute—*Recommended Voluntary Standard Formulas for Sizing Control Valves*; Bell—*Petroleum Transportation Handbook*, McGraw-Hill; Perry—*Chemical Engineers' Handbook*, McGraw-Hill; Spielvogel—*Piping Stress Calculations Simplified*, Spielvogel Publishing; Grinnell Company, Inc.—*Piping Design and Engineering*; M. W. Kellogg Co.—*Design of Piping Systems*, Wiley; National Valve and Manufacturing Co.—*Piping Catalog*; McClain—*Fluid Flow in Pipes*, Industrial Press; Tube Turns Division of Chemetron Corp.—*Piping Engineering*.

PIPE-WALL THICKNESS AND SCHEDULE NUMBER

Determine the minimum wall thickness t_m in (mm) and schedule number SN for a branch steam pipe operating at 900°F (482.2°C) if the internal steam pressure is 1000 lb/in^2 (abs) (6894 kPa). Use ANSA B31.1 *Code for Pressure Piping* and the ASME *Boiler and Pressure Vessel Code* valves and equations where they apply. Steam flow rate is 72,000 lb/h (32,400 kg/h).

Calculation Procedure:

1. *Determine the required pipe diameter*

When the length of pipe is not given or is as yet unknown, make a first approximation of the pipe diameter, using a suitable velocity for the fluid. Once the length of the pipe is known, the pressure loss can be determined. If the pressure loss exceeds a desirable value, the pipe diameter can be increased until the loss is within an acceptable range.

Compute the pipe cross-sectional area a in^2 (cm^2) from $a = 2.4Wv/V$, where W = steam flow rate, lb/h (kg/h); v = specific volume of the steam, ft^3/lb (m^3/kg); V = steam velocity, ft/min (m/min). The only unknown in this equation, other than the pipe area, is the steam velocity V. Use Table 1 to find a suitable steam velocity for this branch line.

Table 1 shows that the recommended steam velocities for branch steam pipes range from 6000 to 15,000 ft/min (1828 to 4572 m/min). Assume that a velocity of 12,000 ft/min (3657.6 m/min) is used in this branch steam line. Then, by using the steam table to find the specific volume of steam at 900°F (482.2°C) and 1000 lb/in^2 (abs) (6894 kPa), $a = 2.4(72{,}000)(0.7604)/12{,}000 = 10.98$ in^2 (70.8 cm^2). The inside diameter of the pipe is then $d = 2(a/\pi)^{0.5} = 2(10.98/\pi)^{0.5} = 3.74$ in (95.0 mm). Since pipe is not ordinarily made in this fractional internal diameter, round it to the next larger size, or 4-in (101.6-mm) inside diameter.

2. *Determine the pipe schedule number*

The ANSA *Code for Pressure Piping*, commonly called the *Piping Code*, defines schedule number as SN $= 1000\ P_i/S$, where P_i = internal pipe pressure, lb/in^2 (gage); S = allowable stress in the pipe, lb/in^2, from *Piping Code*. Table 2 shows typical allowable stress values for pipe in power piping systems. For this pipe, assuming that seamless ferritic alloy steel (1% Cr, 0.55% Mo) pipe is used with the steam at 900°F (482°C), SN = (1000)(1014.7)/13,100 = 77.5. Since pipe is not ordinarily made in this schedule number, use the next *highest* readily available schedule number, or SN = 80. [Where large quantities of pipe are required, it is sometimes economically wise to order pipe of the exact SN required. This is not usually done for orders of less than 1000 ft (304.8 m) of pipe.]

3. *Determine the pipe-wall thickness*

Enter a tabulation of pipe properties, such as in Crocker and King—*Piping Handbook*, and find the wall thickness for 4-in (101.6-mm) SN 80 pipe as 0.337 in (8.56 mm).

TABLE 1 Recommended Fluid Velocities in Piping

Service	Velocity of fluid	
	ft/min	m/s
Boiler and turbine leads	6,000–12,000	30.5–60.9
Steam headers	6,000–8,000	30.5–40.6
Branch steam lines	6,000–15,000	30.5–76.2
Feedwater lines	250–850	1.3–4.3
Exhaust and low-pressure steam lines	6,000–15,000	30.5–76.2
Pump suction lines	100–300	0.51–1.52
Bleed steam lines	4,000–6,000	20.3–30.5
Service water mains	120–300	0.61–1.52
Vacuum steam lines	20,000–40,000	101.6–203.2
Steam superheater tubes	2,000–5,000	10.2–25.4
Compressed-air lines	1,500–2,000	7.6–10.2
Natural-gas lines (large cross-country)	100–150	0.51–0.76
Economizer tubes (water)	150–300	0.76–1.52
Crude-oil lines [6 to 30 in (152.4 to 762.0 mm)]	50–350	0.25–1.78

Related Calculations: Use the method given here for any type of pipe—steam, water, oil, gas, or air—in any service—power, refinery, process, commercial, etc. Refer to the proper section of B31.1 *Code for Pressure Piping* when computing the schedule number, because the allowable stress S varies for different types of service.

The *Piping Code* contains an equation for determining the minimum required pipe-wall thickness based on the pipe internal pressure, outside diameter, allowable stress, a temperature coefficient, and an allowance for threading, mechanical strength, and corrosion. This equation is seldom used in routine piping-system design. Instead, the schedule number as given here is preferred by most designers.

PIPE-WALL THICKNESS DETERMINATION BY PIPING CODE FORMULA

Use the ANSA B31.1 *Code for Pressure Piping* wall-thickness equation to determine the required wall thickness for an 8.625-in (219.1-mm) OD ferritic steel plain-end pipe if the pipe is used in 900°F (482°C) 900-lb/in^2 (gage) (6205-kPa) steam service.

Calculation Procedure:

1. *Determine the constants for the thickness equation*

Pipe-wall thickness to meet ANSA *Code* requirements for power service is computed from $t_m = \{DP/[2(S + YP)]\} + C$, where t_m = minimum wall thickness, in; D = outside diameter of pipe, in; P = internal pressure in pipe, lb/in^2 (gage); S = allowable stress in pipe material, lb/in^2; Y = temperature coefficient; C = end-condition factor, in.

Values of S, Y, and C are given in tables in the *Code for Pressure Piping* in the section on Power Piping. Using values from the latest edition of the *Code*, we get S = 12,500 lb/in^2 (86.2 MPa) for ferritic-steel pipe operating at 900°F (482°C); Y = 0.40 at the same temperature; C = 0.065 in (1.65 mm) for plain-end steel pipe.

2. *Compute the minimum wall thickness*

Substitute the given and *Code* values in the equation in step 1, or $t_m = [(8.625)(900)]/[2(12{,}500 + 0.4 \times 900)] + 0.065 = 0.367$ in (9.32 mm).

Since pipe mills do not fabricate to precise wall thicknesses, a tolerance above or below the computed wall thickness is required. An allowance must be made in specifying the wall thickness found with this equation by *increasing* the thickness by 12½ percent. Thus, for this pipe, wall thickness = 0.367 + 0.125(0.367) = 0.413 in (10.5 mm).

Refer to the *Code* to find the schedule number of the pipe. Schedule 60 8-in (203-mm) pipe

TABLE 2 Allowable Stresses (S Values) for Alloy-Steel Pipe in Power Piping Systems°

(Abstracted from ASME Power Boiler Code and Code for Pressure Piping, ASA B31.1)

Material	ASTM specification	Grade or symbol	Minimum tensile strength		S values for metal temperatures not to exceed†					
			lb/in^2	MPa	850°F	454°C	900°F	482°C	950°F	510°C
Seamless ferritic steels:										
Carbon-molybdenum	A335	P1	55,000	379.2	13,150	90.7	12,500	86.2	. . .	. . .
0.65 Cr, 0.55 Mo	A335	P2	55,000	379.2	13,150	90.7	12,500	86.2	10,000	68.9
1.00 Cr, 0.55 Mo	A335	P12	60,000	413.6	14,200	97.9	13,100	90.3	11,000	75.8

°Crocker and King—*Piping Handbook*.

†Where welded construction is used, consideration should be given to the possibility of graphite formation in carbon-molybdenum steel above 875°F (468°C) or in chromium-molybdenum steel containing less than 0.60 percent chromium above 975°F (523.9°C).

has a wall thickness of 0.406 in (10.31 mm), and schedule 80 pipe has a wall thickness of 0.500 in (12.7 mm). Since the required thickness of 0.413 in (10.5 mm) is greater than schedule 60 but less than schedule 80, the higher schedule number, 80, should be used.

3. *Check the selected schedule number*

From the previous calculation procedure, SN = $1000P_i/S$. For this pipe, SN = 1000(900)/12,500 = 72. Since piping is normally fabricated for schedule numbers 10, 20, 30, 40, 60, 80, 100, 120, 140, and 160, the next larger schedule number higher than 72, that is 80, will be used. This agrees with the schedule number found in step 2.

Related Calculations: Use this method in conjunction with the appropriate *Code* equation to determine the wall thickness of pipe conveying air, gas, steam, oil, water, alcohol, or any other similar fluids in any type of service. Be certain to use the correct equation, which in some cases is simpler than that used here. Thus, for lead pipe, $t_m = Pd/2S$, where P = safe working pressure of the pipe, lb/in^2 (gage); d = inside diameter of pipe, in; other symbols as before.

When a pipe will operate at a temperature between two tabulated *Code* values, find the allowable stress by interpolating between the tabulated temperature and stress values. Thus, for a pipe operating at 680°F (360°C), find the allowable stress at 650°F (343°C) [= 9500 lb/in^2 (65.5 MPa)] and 700°F (371°C) [= 9000 lb/in^2 (62.0 MPa)]. Interpolate thus: allowable stress at 680°F (360°C) = [(700°F − 680°F)/(700°F − 650°F)](9500 − 9000) + 9000 = 200 + 9000 = 9200 lb/in^2 (63.4 MPa). The same result can be obtained by interpolating downward from 9500 lb/in^2 (65.5 MPa), or allowable stress at 680°F (360°C) = 9500 − [(680 − 650)/(700 − 650)](9500 − 9000) = 9200 lb/in^2 (63.4 MPa).

DETERMINING THE PRESSURE LOSS IN STEAM PIPING

Use a suitable pressure-loss chart to determine the pressure loss in 510 ft (155.5 m) of 4-in (101.6-mm) flanged steel pipe containing two 90° elbows and four 45° bends. The schedule 40 piping conveys 13,000 lb/h (5850 kg/h) of 40-lb/in^2 (gage) (275.8-kPa) 350°F (177°C) superheated steam. List other methods of determining the pressure loss in steam piping.

Calculation Procedure:

1. *Determine the equivalent length of the piping*

The equivalent length of a pipe L_e ft = length of straight pipe, ft + equivalent length of fittings, ft. Using data from the Hydraulic Institute, Crocker and King—*Piping Handbook,* earlier sections of this handbook, or Fig. 1, find the equivalent length of a 90° 4-in (101.6-mm) elbow as 10 ft (3 m) of straight pipe. Likewise, the equivalent length of a 45° bend is 5 ft (1.5 m) of straight pipe. Substituting in the above relation and using the straight lengths and the number of fittings of each type, we get L_e = 510 + (2)(10) + 4(5) = 550 ft (167.6 m) of straight pipe.

2. *Compute the pressure loss, using a suitable chart*

Figure 2 presents a typical pressure-loss chart for steam piping. Enter the chart at the top left at the superheated steam temperature of 350°F (177°C), and project vertically downward until the 40-lb/in^2 (gage) (275.8-kPa) superheated steam pressure curve is intersected. From here, project horizontally to the right until the outer border of the chart is intersected. Next, project through the steam flow rate, 13,000 lb/h (5900 kg/h) on scale *B*, Fig. 2, to the pivot scale *C*. From this point, project through 4-in (101.6-mm) schedule 40 pipe on scale *D*, Fig. 2. Extend this line to intersect the pressure-drop scale, and read the pressure loss as 7.25 lb/in^2 (50 kPa) per 100 ft (30.4 m) of pipe.

Since the equivalent length of this pipe is 550 ft (167.6 m), the total pressure loss in the pipe is (550/100)(7.25) = 39.875 lb/in^2 (274.9 kPa), say 40 lb/in^2 (275.8 kPa).

3. *List the other methods of computing pressure loss*

Numerous pressure-loss equations have been developed to compute the pressure drop in steam piping. Among the better known are those of Unwin, Fritzche, Spitzglass, Babcock, Gutermuth, and others. These equations are discussed in some detail in Crocker and King—*Piping Handbook* and in the engineering data published by valve and piping manufacturers.

Most piping designers use a chart to determine the pressure loss in steam piping because a

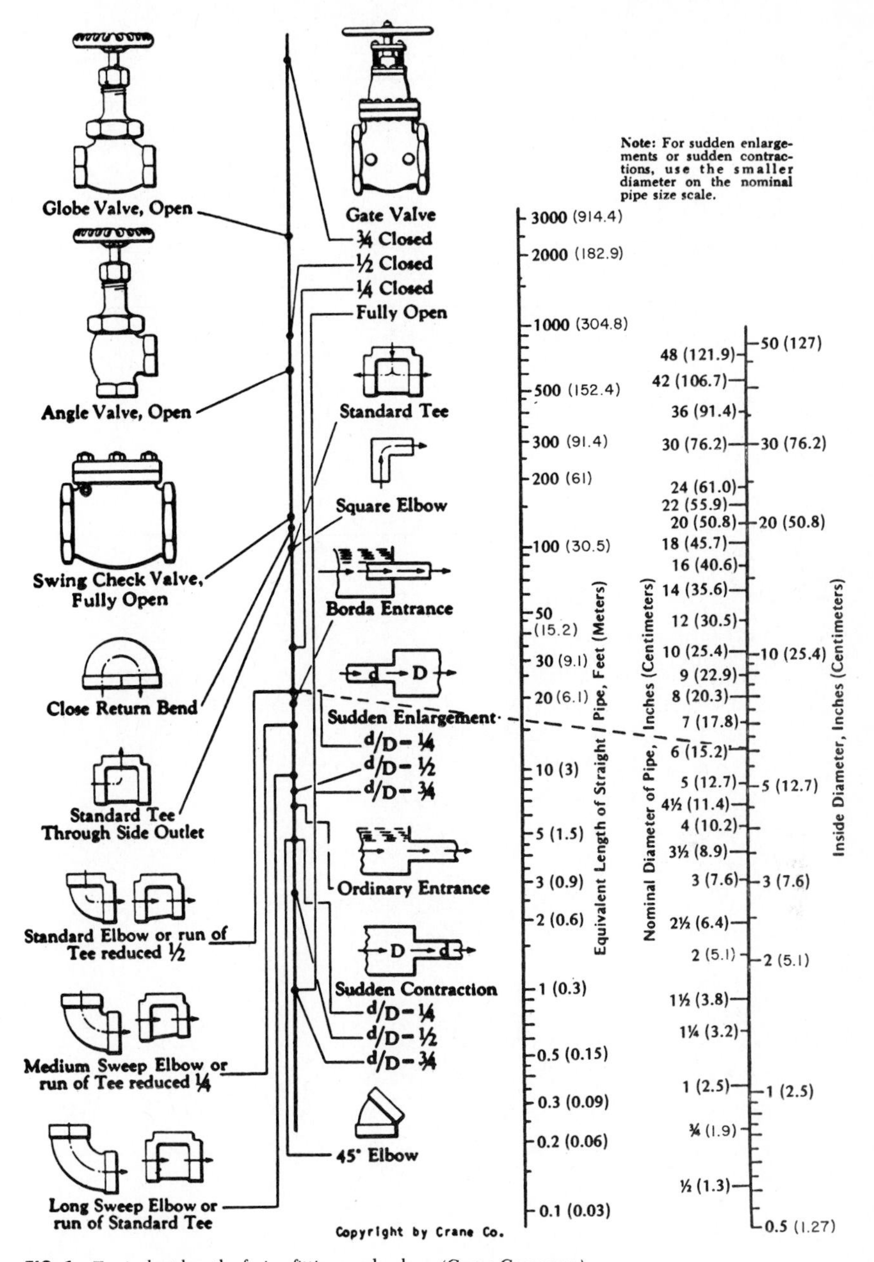

FIG. 1 Equivalent length of pipe fittings and valves. *(Crane Company.)*

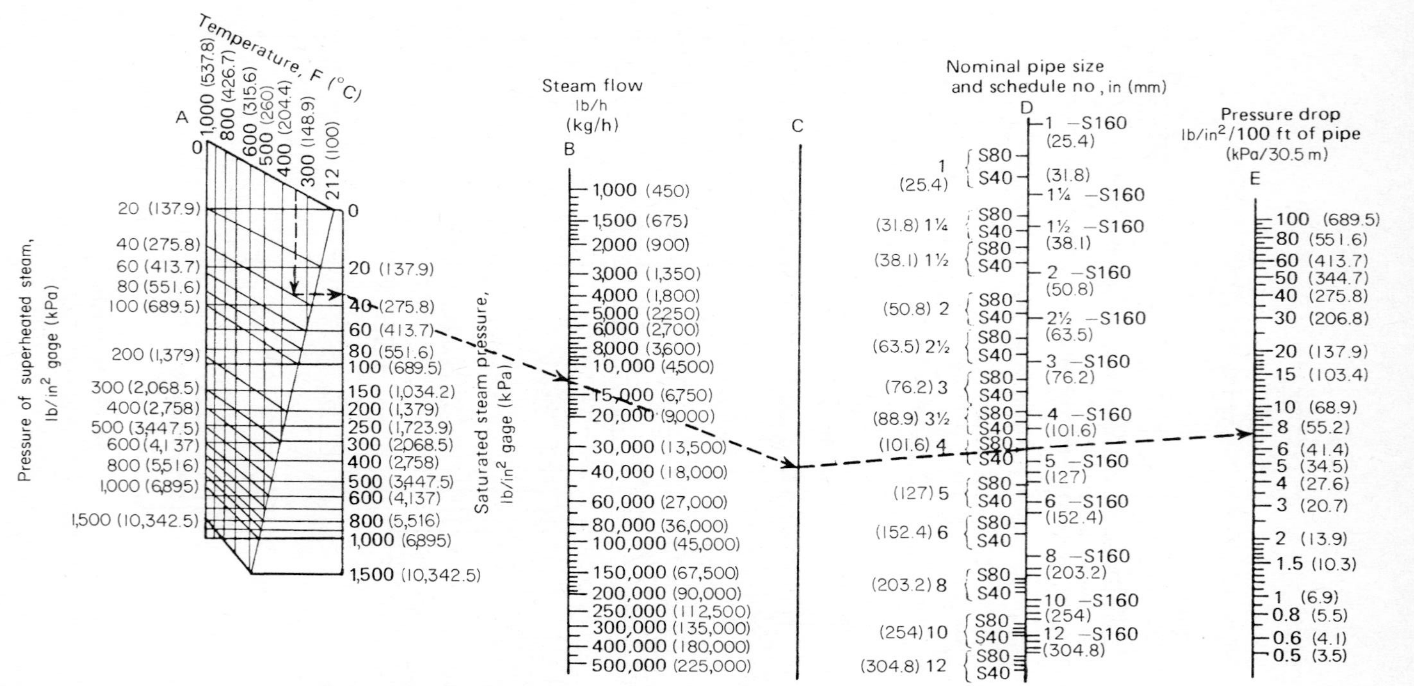

FIG. 2 Pressure loss in steam pipes based on the Fritzche formula. *(Power.)*

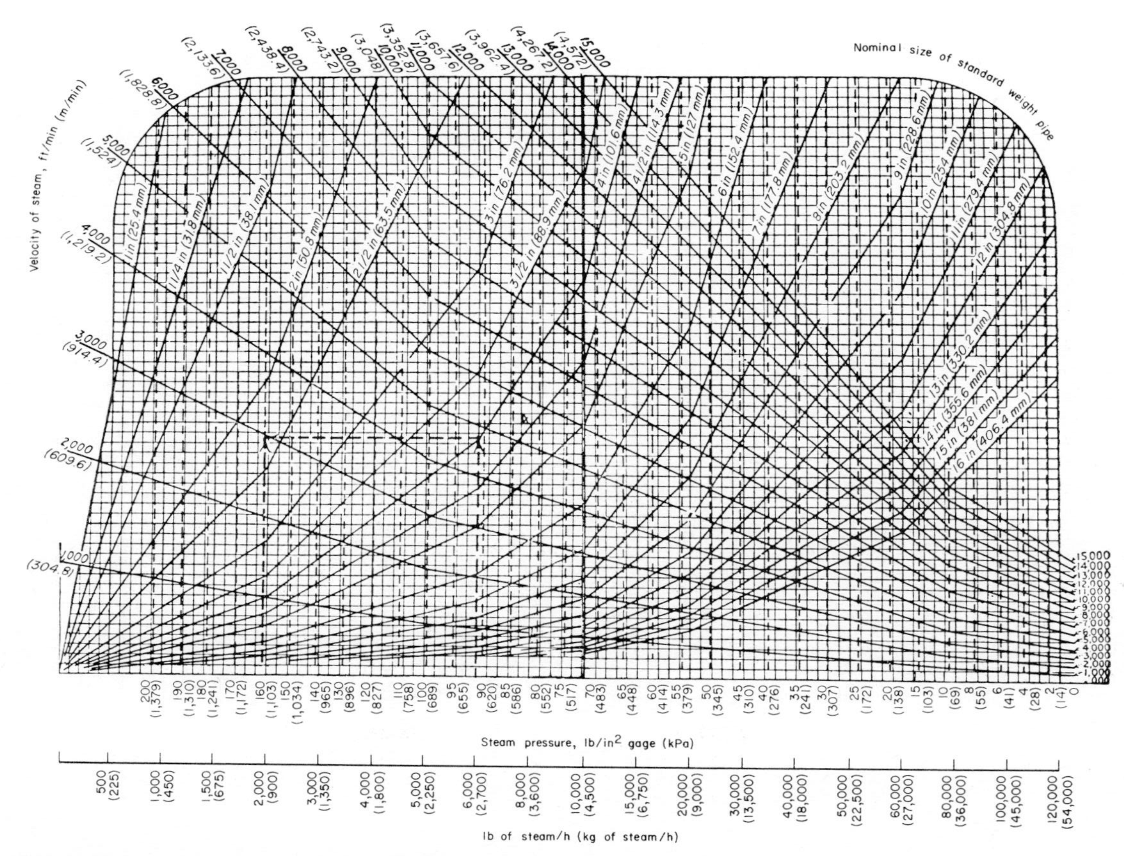

FIG. 3 Spitzglass chart for saturated steam flowing in schedule 40 pipe.

TABLE 3 Steam Velocities Used in Pipe Design

Steam condition	Steam pressure		Steam use	Steam velocity	
	lb/in^2	kPa		ft/min	m/min
Saturated	0–15	0–103.4	Heating	4,000–6,000	1,219.2–1,828.8
Saturated	50–150	344.7–1,034.1	Process	6,000–10,000	1,828.8–3,048.0
Superheated	200 and higher	1,378.8 and higher	Boiler leads	10,000–15,000	3,048.0–4,572.0

chart saves time and reduces the effort involved. Further, the accuracy obtained is sufficient for all usual design practice.

Figure 3 is a popular flowchart for determining steam flow rate, pipe size, steam pressure, or steam velocity in a given pipe. Using this chart, the designer can determine any one of the four variables listed above when the other three are known. In solving a problem on the chart in Fig. 3, use the steam-quantity lines to intersect pipe sizes and the steam-pressure lines to intersect steam velocities. Here are two typical applications of this chart.

Example: What size schedule 40 pipe is needed to deliver 8000 lb/h (3600 kg/h) of 120-lb/in^2 (gage) (827.3-kPa) steam at a velocity of 5000 ft/min (1524 m/min)?

Solution: Enter Fig. 3 at the upper left at a velocity of 5000 ft/min (1524 m/min), and project along this velocity line until the 120-lb/in^2 (gage) (827.3-kPa) pressure line is intersected. From this intersection, project horizontally until the 8000-lb/h (3600-kg/h) vertical line is intersected. Read the *nearest* pipe size as 4 in (101.6 mm) on the *nearest* pipe-diameter curve.

Example: What is the steam velocity in a 6-in (152.4-mm) pipe delivering 20,000 lb/h (9000 kg/h) of steam at 85 lb/in^2 (gage) (586 kPa)?

Solution: Enter the bottom of the chart, Fig. 3, at the flow rate of 20,000 lb/h (9000 kg/h), and project vertically upward until the 6-in (152.4-mm) pipe curve is intersected. From this point, project horizontally to the 85-lb/in^2 (gage) (586-kPa) curve. At the intersection, read the velocity as 7350 ft/min (2240.3 m/min).

Table 3 shows typical steam velocities for various industrial and commercial applications. Use the given values as guides when sizing steam piping.

PIPING WARM-UP CONDENSATE LOAD

How much condensate is formed in 5 min during warm-up of 500 ft (152.4 m) of 6-in (152.4-mm) schedule 40 steel pipe conveying 215-lb/in^2 (abs) (1482.2-kPa) saturated steam if the pipe is insulated with 2 in (50.8 mm) of 85 percent magnesia and the minimum external temperature is 35°F (1.7°C)?

Calculation Procedure:

1. *Compute the amount of condensate formed during pipe warm-up*

For any pipe, the condensate formed during warm-up C_h lb/h $= 60(W_p)(\Delta t)(s)/h_{fg}N$, where W_p = total weight of pipe, lb; Δt = difference between final and initial temperature of the pipe, °F; s = specific heat of pipe material, Btu/(lb·°F); h_{fg} = enthalpy of vaporization of the steam, Btu/lb; N = warm-up time, min.

A table of pipe properties shows that this pipe weighs 18.974 lb/ft (28.1 kg/m). The steam table shows that the temperature of 215-lb/in^2 (abs) (1482.2-kPa) saturated steam is 387.89°F (197.7°C), say 388°F (197.8°C); the enthalpy h_{fg} = 837.4 Btu/lb (1947.8 kJ/kg). The specific heat of steel pipe s = 0.144 Btu/(lb·°F) [0.6 kJ/(kg·°C)]. Then $C_h = 60(500 \times 18.974)(388 - 35)(0.114)/[(837.4)(5)]$ = 5470 lb/h (2461.5 kg/h).

2. *Compute the radiation-loss condensate load*

Condensate is also formed by radiation of heat from the pipe during warm-up and while the pipe is operating. The warm-up condensate load decreases as the radiation load increases, the peak occurring midway (2½ min in this case) through the warm-up period. For this reason, one-half the normal radiation load is added to the warm-up load. Where the radiation load is small, it is often disregarded. However, the load must be computed before its magnitude can be determined.

For any pipe, $C_r = (L)(A)(\Delta t)(H)/h_{fg}$, where L = length of pipe, ft; A = external area of pipe, ft^2 per ft of length; H = heat loss through bare pipe or pipe insulation, Btu/($ft^2 \cdot h \cdot °F$), from the piping or insulation tables. This 6-in (152.4-mm) schedule 40 pipe has an external area $A = 1.73\ ft^2/ft$ ($0.53\ m^2/m$) of length. The heat loss through 2 in (50.8 mm) of 85 percent magnesia, from insulation tables, is $H = 0.286$ Btu/($ft^2 \cdot h \cdot °F$) [1.62 W/($m^2 \cdot °C$)]. Then $C_r = (500) \times (1.73)(388 - 35)(0.286)/837.4 = 104.2$ lb/h (46.9 kg/h). Adding half the radiation load to the warm-up load gives 5470 + 52.1 = 5522.1 lb/h (2484.9 kg/h).

3. *Apply a suitable safety factory to the condensate load*

Trap manufacturers recommend a safety factor of 2 for traps installed between a boiler and the end of a steam main; traps at the end of a long steam main or ahead of pressure-regulating or shutoff valves usually have a safety factor of 3. With a safety factor of 3 for this pipe, the steam trap should have a capacity of at least 3(5522.1) = 16,566.3 lb/h (7454.8 kg/h), say 17,000 lb/h (7650.0 kg/h).

Related Calculations: Use this method to find the warm-up condensate load for any type of steam pipe—main or auxiliary— in power, process, heating, or vacuum service. The same method is applicable to other vapors that form condensate—Dowtherm, refinery vapors, process vapors, and others.

STEAM TRAP SELECTION FOR INDUSTRIAL APPLICATIONS

Select steam traps for the following four types of equipment: (1) the steam directly heats solid materials as in autoclaves, retorts, and sterilizers; (2) the steam indirectly heats a liquid through a metallic surface, as in heat exchangers and kettles, where the quantity of liquid heated is known and unknown; (3) the steam indirectly heats a solid through a metallic surface, as in dryers using cylinders or chambers and platen presses; and (4) the steam indirectly heats air through metallic surfaces, as in unit heaters, pipe coils, and radiators.

Calculation Procedure:

1. *Determine the condensate load*

The first step in selecting a steam trap for any type of equipment is determination of the condensate load. Use the following general procedure.

a. Solid materials in autoclaves, retorts, and sterilizers. How much condensate is formed when 2000 lb (900.0 kg) of solid material with a specific heat of 1.0 is processed in 15 min at 240°F (115.6°C) by 25-lb/in^2 (gage) (172.4-kPa) steam from an initial temperature of 60°F in an insulated steel retort?

For this type of equipment, use $C = WSP$, where C = condensate formed, lb/h; W = weight of material heated, lb; s = specific heat, Btu/(lb·°F); P = factor from Table 4. Thus, for this application, $C = (2000)(1.0)(0.193) = 386$ lb (173.7 kg) of condensate. Note that P is based on a temperature rise of 240 − 60 = 180°F (100°C) and a steam pressure of 25 lb/in^2 (gage) (172.4 kPa). For the retort, using the specific heat of steel from Table 5, $C = (4000)(0.12)(0.193) = 92.6$ lb of condensate, say 93 lb (41.9 kg). The total weight of condensate formed in 15 min is 386 + 93 = 479 lb (215.6 kg). In 1 h, 479(60/15) = 1916 lb (862.2 kg) of condensate is formed.

TABLE 4 Factors $P = (T - t)/L$ to Find Condensate Load

Pressure		Temperature		
lb/in^2 (abs)	kPa	160°F (71.1°C)	180°F (82.2°C)	200°F (93.3°C)
20	137.8	0.170	0.192	0.213
25	172.4	0.172	0.193	0.214
30	206.8	0.172	0.194	0.215

TABLE 5 Use These Specific Heats to Calculate Condensate Load

Solids	Btu/(lb·°F)	kJ/(kg·°C)	Liquids	Btu/(lb·°F)	kJ/(kg·°C)
Aluminum	0.23	0.96	Alcohol	0.65	2.7
Brass	0.10	0.42	Carbon tetrachloride	0.20	0.84
Copper	0.10	0.42	Gasoline	0.53	2.22
Glass	0.20	0.84	Glycerin	0.58	2.43
Iron	0.13	0.54	Kerosene	0.47	1.97
Steel	0.12	0.50	Oils	0.40–0.50	1.67–2.09

A safety factor must be applied to compensate for radiation and other losses. Typical safety factors used in selecting steam traps are as follows:

Steam mains and headers	2–3
Steam heating pipes	2–6
Purifiers and separators	2–3
Retorts for process	2–4
Unit heaters	3
Submerged pipe coils	2–4
Cylinder dryers	4–10

With a safety factor of 4 for this process retort, the trap capacity = (4)(1916) = 7664 lb/h (3449 kg/h), say 7700 lb/h (3465 kg/h).

b(1). Submerged heating surface and a known quantity of liquid. How much condensate forms in the jacket of a kettle when 500 gal (1892.5 L) of water is heated in 30 min from 72 to 212°F (22.2 to 100°C) with 50-lb/in^2 (gage) (344.7-kPa) steam?

For this type of equipment, $C = GwsP$, where G = gal of liquid heated; w = weight of liquid, lb/gal. Substitute the appropriate values as follows: $C = (500)(8.33)(1.0) \times (0.154) = 641$ lb (288.5 kg), or (641)(60/30) = 1282 lb/h (621.9 kg/h). With a safety factor of 3, the trap capacity = (3)(1282) = 3846 lb/h (1731 kg/h), say 3900 lb/h (1755 kg/h).

b(2). Submerged heating surface and an unknown quantity of liquid. How much condensate is formed in a coil submerged in oil when the oil is heated as quickly as possible from 50 to 250°F (10 to 121°C) by 25-lb/in^2 (gage) (172.4-kPa) steam if the coil has an area of 50 ft^2 (4.66 m^2) and the oil is free to circulate around the coil?

For this condition, $C = UAP$, where U = overall coefficient of heat transfer, Btu/(h·ft^2·°F), from Table 6; A = area of heating surface, ft^2. With free convection and a condensing-vapor-to-liquid type of heat exchanger, U = 10 to 30. With an average value of $U = 20$, C = (20)(50)(0.214) = 214 lb/h (96.3 kg/h) of condensate. Choosing a safety factor 3 gives trap capacity = (3)(214) = 642 lb/h (289 kg/h), say 650 lb/h (292.5 kg/h).

b(3). Submerged surfaces having more area than needed to heat a specified quantity of liquid in a given time with condensate withdrawn as rapidly as formed. Use Table 7 instead of step *b*(1) or *b*(2). Find the condensation rate by multiplying the submerged area by the appropriate factor from Table 7. Use this method for heating water, chemical solutions, oils, and other liquids. Thus, with steam at 100 lb/in^2 (gage) (689.4 kPa) and a temperature of 338°F (170°C) and heating oil from 50 to 226°F (10 to 108°C) with a submerged surface having an area of 500 ft^2 (46.5 m^2), the mean temperature difference = steam temperature minus the average liquid temperature = Mtd = 338 − (50 + 226/2) = 200°F (93.3°C). The factor from Table 7 for 100 lb/in^2 (gage) (689.4 kPa) steam and a 200°F (93.3°C) Mtd is 56.75. Thus, the condensation rate = (56.75)(500) = 28,375 lb/h (12,769 kg/h). With a safety factor of 2, the trap capacity = (2)(28,375) = 56,750 lb/h (25,538 kg/h).

c. Solids indirectly heated through a metallic surface. How much condensate is formed in a chamber dryer when 1000 lb (454 kg) of cereal is dried to 750 lb (338 kg) by 10-lb/in^2 (gage) (68.9-kPa) steam? The initial temperature of the cereal is 60°F (15.6°C), and the final temperature equals that of the steam.

TABLE 6 Ordinary Ranges of Overall Coefficients of Heat Transfer

Type of heat exchanger	State of controlling resistance: Free convection, U		State of controlling resistance: Forced convection, U		Typical fluid	Typical apparatus
Liquid to liquid	25–60	[141.9–340.7]	150–300	[851.7–1703.4]	Water	Liquid-to-liquid heat exchangers
Liquid to liquid	5–10	[28.4–56.8]	20–50	[113.6–283.9]	Oil	
Liquid to gas*	1–3	[5.7–17.0]	2–10	[11.4–56.8]	—	Hot-water radiators
Liquid to boiling liquid	20–60	[113.6–340.7]	50–150	[283.9–851.7]	Water	Brine coolers
Liquid to boiling liquid	5–20	[28.4–113.6]	25–60	[141.9–340.7]	Oil	
Gas* to liquid	1–3	[5.7–17.0]	2–10	[11.4–56.8]	—	Air coolers, economizers
Gas* to gas	0.6–2	[3.4–11.4]	2–6	[11.4–34.1]	—	Steam superheaters
Gas* to boiling liquid	1–3	[5.7–17.0]	2–10	[11.4–56.8]	—	Steam boilers
Condensing vapor to liquid	50–200	[283.9–1136]	150–800	[851.7–4542.4]	Steam to water	Liquid heaters and condensers
Condensing vapor to liquid	10–30	[56.8–170.3]	20–60	[113.6–340.7]	Steam to oil	
Condensing vapor to liquid	40–80	[227.1–454.2]	60–150	[340.7–851.7]	Organic vapor to water	
Condensing vapor to liquid	—	—	15–300	[85.2–1703.4]	Steam-gas mixture	
Condensing vapor to gas*	1–2	[5.7–11.4]	2–10	[11.4–56.8]	—	Steam pipes in air, air heaters
Condensing vapor to boiling liquid	40–100	[227.1–567.8]	—	—	—	Scale-forming evaporators
Condensing vapor to boiling liquid	300–800	[1703.4–4542.4	—	—	Steam to water	
Condensing vapor to boiling liquid	50–150	[283.9–851.7]	—	—	Steam to oil	

*At atmospheric pressure.

Note: U = Btu/(h·ft^2·°F) [W/(m^2·°C)]. Under many conditions, either higher or lower values may be realized.

TABLE 7 Condensate Formed in Submerged Steel* Heating Elements, $lb/(ft^2 \cdot h)$ [$kg/(m^2 \cdot min)$]

MTD†		Steam pressure				
°F	°C	75 lb/in² (abs) (517.1 kPa)	100 lb/in² (abs) (689.4 kPa)	150 lb/in² (abs) (1034.1 kPa)	Btu/(ft²·h)	kW/m²
175	97.2	44.3 (3.6)	45.4 (3.7)	46.7 (3.8)	40,000	126.2
200	111.1	54.8 (4.5)	56.8 (4.6)	58.3 (4.7)	50,000	157.7
250	138.9	90.0 (7.3)	93.1 (7.6)	95.7 (7.8)	82,000	258.6

*For copper, multiply table data by 2.0; for brass, by 1.6.
†Mean temperature difference, °F or °C, equals temperature of steam minus average liquid temperature. Heat-transfer data for calculating this table obtained from and used by permission of the American Radiator & Standard Sanitary Corp.

For this condition, $C = 970(W - D)/h_{fg} + WP$, where D = dry weight of the material, lb; h_{fg} = enthalpy of vaporization of the steam at the trap pressure, Btu/lb. From the steam tables and Table 4, $C = 970(1000 - 750)/952 + (1000)(0.189) = 443.5$ lb/h (199.6 kg/h) of condensate. With a safety factor of 4, the trap capacity = (4)(443.5) = 1774 lb/h (798.3 kg/h).

d. Indirect heating of air through a metallic surface. How much condensate is formed in a unit heater using 10-lb/in² (gage) (68.9-kPa) steam if the entering-air temperature is 30°F (−1.1°C) and the leaving-air temperature is 130°F (54.4°C)? Airflow is 10,000 ft³/min (281.1 m³/min).

Use Table 8, entering at a temperature difference of 100°F (37.8°C) and projecting to a steam pressure of 10 lb/in² (gage) (68.9 kPa). Read the condensate formed as 122 lb/h (54.9 kg/h) per 1000 ft³/min (28.3 m³/min). Since 10,000 ft³/min (283.1 m³/min) of air is being heated, the condensation rate = (10,000/1000)(122) = 1220 lb/h (549 kg/h). With a safety factor of 3, the trap capacity = (3)(1220) = 3660 lb/h (1647 kg/h), say 3700 lb/h (1665 kg/h).

Table 9 shows the condensate formed by radiation from bare iron and steel pipes in still air and with forced-air circulation. Thus, with a steam pressure of 100 lb/in² (gage) (689.4 kPa) and an initial air temperature of 75°F (23.9°C), 1.05 lb/h (0.47 kg/h) of condensate will be formed per ft² (0.09 m²) of heating surface in still air. With forced-air circulation, the condensate rate is (5)(1.05) = 5.25 (lb/(h·ft²) [25.4 kg/(h·m²)] of heating surface.

Unit heaters have a *standard rating* based on 2-lb/in² (gage) (13.8-kPa) steam with entering air at 60°F (15.6°C). If the steam pressure or air temperature is different from these standard conditions, multiply the heater Btu/h capacity rating by the appropriate correction factor form, Table 10. Thus, a heater rated at 10,000 Btu/h (2931 W) with 2-lb/in² (gage) (13.8-kPa) steam and 60°F (15.6°C) air would have an output of (1.290)(10,000) = 12,900 Btu/h (3781 W) with 40°F (4.4°C) inlet air and 10-lb/in² (gage) (68.9-kPa) steam. Trap manufacturers usually list heater Btu ratings and recommend trap model numbers and sizes in their trap engineering data. This allows easier selection of the correct trap.

TABLE 8 Steam Condensed by Air, lb/h at 1000 ft³/min (kg/h at 28.3 m³/min)*

Temperature difference		Pressure		
°F	°C	5 lb/in² (gage) (34.5 kPa)	10 lb/in² (gage) (68.9 kPa)	50 lb/in² (gage) (344.7 kPa)
50	27.8	61 (27.5)	61 (27.5)	63 (28.4)
100	55.6	120 (54.0)	122 (54.9)	126 (56.7)
150	83.3	180 (81.0)	183 (82.4)	189 (85.1)

*Based on 0.0192 Btu (0.02 kJ) absorbed per ft³ (0.028 m³) of saturated air per °F (0.556°C) at 32°F (0°C). For 0°F (−17.8°C), multiply by 1.1.

TABLE 9 Condensate Formed by Radiation from Bare Iron and Steel, lb/(ft²·h) [kg/(m²·h)]

Air temperature		Steam pressure			
°F	°C	50 lb/in² (gage) (344.7 kPa)	75 lb/in² (gage) (517.1 kPa)	100 lb/in² (gage) (689.5 kPa)	150 lb/in² (gage) (1034 kPa)
65	18.3	0.82 (3.97)	1.00 (5.84)	1.08 (5.23)	1.32 (6.39)
70	21.2	0.80 (3.87)	0.98 (4.74)	1.06 (5.13)	1.21 (5.86)
75	23.9	0.77 (3.73)	0.88 (4.26)	1.05 (5.08)	1.19 (5.76)

°Based on still air; for forced-air circulation, multiply by 5.

2. *Select the trap size based on the load and steam pressure*

Obtain a chart or tabulation of trap capacities published by the manufacturer whose trap will be used. Figure 4 is a capacity chart for one type of bucket trap manufactured by Armstrong Machine Works. Table 11 shows typical capacities of impulse traps manufactured by the Yarway Company.

To select a trap from Fig. 4, when the condensation rate is uniform and the pressure across the trap is constant, enter at the left at the condensation rate, say 8000 lb/h (3600 kg/h) (as obtained from step 1). Project horizontally to the right to the vertical ordinate representing the pressure across the trap [= Δp = steam-line pressure, lb/in² (gage) − return-line pressure with trap valve closed, lb/in² (gage)]. Assume Δp = 20 lb/in² (gage) (138 kPa) for this trap. The intersection of the horizontal 8000-lb/h (3600-kg/h) projection and the vertical 20-lb/in² (gage) (137.9-kPa) projection is on the sawtooth capacity curve for a trap having a 9/16-in (14.3-mm) diameter orifice. If these projections intersected beneath this curve, a 9/16-in (14.3-mm) orifice would still be used if the point were between the verticals for this size orifice.

The dashed lines extending downward from the sawtooth curves show the capacity of a trap at reduced Δp. Thus, the capacity of a trap with a 3/8-in (9.53-mm) orifice at Δp = 30 lb/in² (gage) (207 kPa) is 6200 lb/h (2790 kg/h), read at the intersection of the 30-lb/in² (gage) (207-kPa) ordinate and the dashed curve extended from the 3/8-in (9.53-mm) solid curve.

To select an impulse trap from Table 11, enter the table at the trap inlet pressure, say 125 lb/in² (gage) (862 kPa), and project to the desired capacity, say 8000 lb/h (3600 kg/h), determined from step 1. Table 11 shows that a 2-in (50.8-mm) trap having an 8530-lb/h (3839-kg/h) capacity must be used because the next smallest size has a capacity of 5165 lb/h (2324 kg/h). This capacity is less than that required.

Some trap manufacturers publish capacity tables relating various trap models to specific types of equipment. Such tables simplify trap selection, but the condensation rate must still be computed as given here.

Related Calculations: Use the procedure given here to determine the trap capacity required for any industrial, commercial, or domestic application including acid vats, air dryers, asphalt tanks, autoclaves, baths (dyeing), belt presses, bleach tanks, blenders, bottle washers, brewing kettles, cabinet dryers, calenders, can washers, candy kettles, chamber dryers, chambers (reaction), cheese kettles, coils (cooking, kettle, pipe, tank, tank-car), confectioners' kettles, continuous dryers,

TABLE 10 Unit-Heater Correction Factors

Steam pressure		Temperature of entering air		
lb/in² (gage)	kPa	20°F (−6.7°C)	40°F (4.4°C)	60°F (15.6°C)
5	34.5	1.370	1.206	1.050
10	68.9	1.460	1.290	1.131
15	103.4	1.525	1.335	1.194

Source: Yarway Corporation; SI values added by handbook editor.

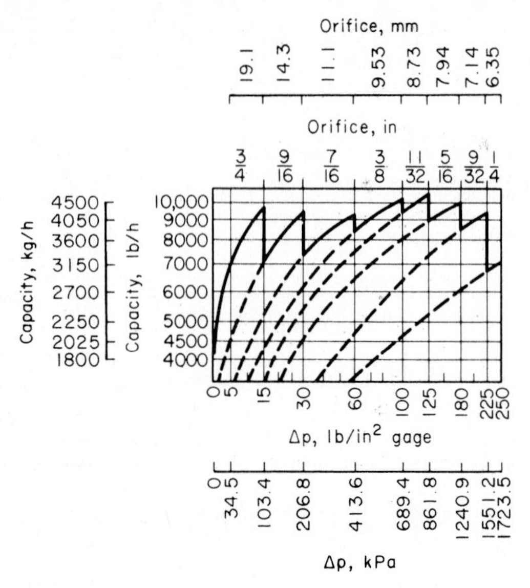

FIG. 4 Capacities of one type of bucket steam trap. *(Armstrong Machine Works.)*

conveyor dryers, cookers (nonpressure and pressure), cooking coils, cooking kettles, cooking tanks, cooking vats, cylinder dryers, cylinders (jacketed), double-drum dryers, drum dryers, drums (dyeing), dry cans, dry kilns, dryers (cabinet, chamber, continuous, conveyor, cylinder, drum, festoon, jacketed, linoleum, milk, paper, pulp, rotary, shelf, stretch, sugar, tray, tunnel), drying rolls, drying rooms, drying tables, dye vats, dyeing baths and drums, dryers (package), embossing-press platens, evaporators, feedwater heaters, festoon dryers, fin-type heaters, fourdriniers, fuel-oil pre-heaters, greenhouse coils, heaters (steam), heat exchangers, heating coils and kettles, hot-break tanks, hot plates, kettle coils, kettles (brewing, candy, cheese, confectioners', cooking, heating, process), kiers, kilns (dry), liquid heaters, mains (steam), milk-bottle washers, milk-can washers, milk dryers, mixers, molding-press platens, package dryers, paper dryers, percolators, phonograph-record press platens, pipe coils (still- and circulating-air), platens, plating tanks, plywood press platens, preheaters (fuel-oil), preheating tanks, press platens, pressure cookers, process kettles, pulp dryers, purifiers, reaction chambers, retorts, rotary dryers, steam mains (risers, separators), stocking boarders, storage-tank coils, storage water heaters, stretch dryers, sugar dryers, tank-car coils, tire-mold presses, tray dryers, tunnel dryers, unit heaters, vats, veneer press platens, vulcanizers, and water stills. Hospital equipment—such as autoclaves and sterilizers—can be analyzed in the same way, as can kitchen equipment—bain marie, compartment cooker, egg boiler,

TABLE 11 Capacities of Impulse Traps, lb/h (kg/h)
[*Maximum continuous discharge of condensate, based on condensate at 30°F (16.7°C) below steam temperature.*]

Pressure at trap inlet		Trap nominal size	
lb/in² (gage)	kPa	1.25 in (38.1 mm)	2.0 in (50.8 mm)
125	861.8	6165 (2774)	8530 (3839)
150	1034.1	6630 (2984)	9075 (4084)
200	1378.8	7410 (3335)	9950 (4478)

Source: Yarway Corporation.

kettles, steam table, and urns; and laundry equipment—blanket dryers, curtain dryers, flatwork ironers, presses (dry-cleaning, laundry), sock forms, starch cookers, tumblers, etc.

When using a trap capacity diagram or table, be sure to determine the basis on which it was prepared. Apply any necessary correction factors. Thus, *cold-water capacity ratings* must be corrected for traps operating at higher condensate temperatures. Correction factors are published in trap engineering data. The capacity of a trap is greater at condensate temperatures less than 212°F (100°C) because at or above this temperature condensate forms flash steam when it flows into a pipe or vessel at atmospheric [14.7 lb/in^2 (abs) (101.3 kPa)] pressure. At altitudes above sea level, condensate flashes into steam at a lower temperature, depending on the altitude.

The method presented here is the work of L. C. Campbell, Yarway Corporation, as reported in *Chemical Engineering*.

SELECTING HEAT INSULATION FOR HIGH-TEMPERATURE PIPING

Select the heat insulation for a 300-ft (91.4-m) long 10-in (254-mm) turbine lead operating at 570°F (299°C) for 8000 h/year in a 70°F (21.1°C) turbine room. How much heat is saved per year by this insulation? The boiler supplying the turbine has an efficiency of 80 percent when burning fuel having a heating value of 14,000 Btu/lb (32.6 MJ/kg). Fuel costs $6 per ton ($5.44 per metric ton). How much money is saved by the insulation each year? What is the efficiency of the insulation?

Calculation Procedure:

1. *Choose the type of insulation to use*

Refer to an insulation manufacturer's engineering data or Crocker and King—*Piping Handbook* for recommendations about a suitable insulation for a pipe operating in the 500 to 600°F (260 to 316°C) range. These references will show that calcium silicate is a popular insulation for this temperature range. Table 12 shows that a thickness of 3 in (76.2 mm) is usually recommended for 10-in (254-mm) pipe operating at 500 to 599°F (260 to 315°C).

2. *Determine heat loss through the insulation*

Refer to an insulation manufacturer's engineering data to find the heat loss through 3-in (76.2-mm) thick calcium silicate as 0.200 Btu/(h·ft^2·°F) [1.14 W/(m^2·°C)]. Since 10-in (254-mm) pipe has an area of 2.817 ft^2/ft (0.86 m^2/m) of length and since the temperature difference across the pipe is 570 − 70 = 500°F (260°C), the heat loss per hour = (0.200)(2.817)(500) = 281.7 Btu/(h·ft) (887.9 W/m^2). The heat loss from bare 10-in (254-mm) pipe with a 500°F (260°C) temperature difference is, from an insulation manufacturer's engineering data, 4.640 Btu/(h·ft^2·°F) [26.4 W/(m^2·°C)], or (4.64)(2.817)(500) = 6510 Btu/(h·ft) (6.3 kW/m).

3. *Determine annual heat saving*

The heat saved = bare-pipe loss, Btu/h − insulated-pipe loss, Btu/h = 6510 − 281.7 = 6228.3 Btu/(h·ft) (5989 W/m) of pipe. Since the pipe is 300 ft (91.4 m) long and operates 8000 h per year, the annual heat saving = (300)(8000)(6228.3) = 14,940,000,000 Btu/year (547.4 kW).

4. *Compute the money saved by the heat insulation*

The heat saved in fuel as fired = (annual heat saving, Btu/year)/(boiler efficiency) = 14,940,000,000/0.80 = 18,680,000,000 Btu/year (5473 MW). Weight of fuel saved = (annual heat saving, Btu/year)/(heating value of fuel, Btu/lb)(2000 lb/ton) = 18,680,000,000/[(14,000)(2000)] = 667 tons (605 t). At $6 per ton ($5.44 per metric ton), the monetary saving is ($6)(667) = $4002 per year.

5. *Determine the insulation efficiency*

Insulation efficiency = (bare-pipe loss − insulated-pipe loss)/bare pipe loss, all expressed in Btu/h, or bare-pipe loss = (6510.0 − 281.7)/6510.0 = 0.957, or 95.7 percent.

Related Calculations: Use this method for any type of insulation—magnesia, fiber-glass, asbestos, felt, diatomaceous, mineral wool, etc.—used for piping at elevated temperatures conveying steam, water, oil, gas, or other fluids or vapors. To coordinate and simplify calculations, become familiar with the insulation tables in a reliable engineering handbook or comprehensive insulation catalog. Such familiarity will simplify routine calculations.

TABLE 12 Recommended Insulation Thickness

Nominal pipe size		Pipe temperature			
		400–499°F	204–259°C	500–599°F	260–315°C
in	mm	in	mm	in	mm
6	152.4	2½[*]	63.5	2½	63.5
8	203.2	2½	63.5	3	76.2
10	254.0	2½	63.5	3	76.2
12	304.8	3	76.2	3	76.2
14 and over	355.6 and over	3	76.2	3½	88.9

[*] Available in single- or double-layer insulation.

ORIFICE METER SELECTION FOR A STEAM PIPE

Steam is metered with an orifice meter in a 10-in (254-mm) boiler lead having an internal diameter of d_p = 9.760 in (247.9 mm). Determine the maximum rate of steam flow that can be measured with a steel orifice plate having a diameter of d_0 = 5.855 in (148.7 mm) at 70°F (21.1°C). The upstream pressure tap is 1D ahead of the orifice, and the downstream tap is 0.5D past the orifice. Steam pressure at the orifice inlet p_p = 250 lb/in^2 (gage) (1724 kPa), temperature is 640°F (338°C). A differential gage fitted across the orifice has a maximum range of 120 in (304.8 cm) of water. What is the steam flow rate when the observed differential pressure is 40 in (101.6 cm) of water? Use the ASME Research Committee on Fluid Meters method in analyzing the meter. Atmospheric pressure is 14.696 lb/in^2 (abs) (101.3 kPa).

Calculation Procedure:

1. *Determine the diameter ratio and steam density*

For any orifice meter, diameter ratio = β = meter orifice diameter, in/pipe internal diameter, in = 5.855/9.760 = 0.5999.

Determine the density of the steam by entering the superheated steam table at 250 + 14.696 = 264.696 lb/in^2 (abs) (1824.8 kPa) and 640°F (338°C) and reading the specific volume as 2.387 ft^3/lb (0.15 m^3/kg). For steam, the density = 1/specific volume = d_s = 1/2.387 = 0.4193 lb/ft^3 (6.7 kg/m^3).

2. *Determine the steam viscosity and meter flow coefficient*

From the ASME publication, *Fluid Meters—Their Theory and Application*, the steam viscosity gu_1 for a steam system operating at 640°F (338°C) is gu_1 = 0.0000141 in·lb/(°F·s·ft^2) [0.000031 N·m/(°C·s·m^2)].

Find the flow coefficient K from the same ASME source by entering the 10-in (254-mm) nominal pipe diameter table at β = 0.5999 and projecting to the appropriate Reynolds number column. Assume that the Reynolds number = 10^7, approximately, for the flow conditions in this pipe. Then K = 0.6486. Since the Reynolds number for steam pressures above 100 lb/in^2 (689.4 kPa) ranges from 10^6 to 10^7, this assumption is safe because the value of K does not vary appreciably in this Reynolds number range. Also, the Reynolds number cannot be computed yet because the flow rate is unknown. Therefore, assumption of the Reynolds number is necessary. The assumption will be checked later.

3. *Determine the expansion factor and the meter area factor*

Since steam is a compressible fluid, the expansion factor Y_1 must be determined. For superheated steam, the ratio of the specific heat at constant pressure c_p to the specific heat at constant volume c_v is $k = c_p/c_v$ = 1.3. Also, the ratio of the differential maximum pressure reading h_w, in of water, to the maximum pressure in the pipe, lb/in^2 (abs) = 120/246.7 = 0.454. From the expansion-factor curve in the ASME *Fluid Meters*, Y_1 = 0.994 for β = 0.5999 and the pressure ratio =

0.454. And, from the same reference, the meter area factor F_a = 1.0084 for a steel meter operating at 640°F (338°C).

4. *Compute the rate of steam flow*

For square-edged orifices, the flow rate, lb/s = w = $0.0997F_aKd^2Y_1(h_wd_s)^{0.5}$ = $(0.0997)(1.0084)(0.6486)(5.855)^2(0.994)(120 \times 0.4188)^{0.5}$ = 15.75 lb/s (7.1 kg/s).

5. *Compute the Reynolds number for the actual flow rate*

For any steam pipe, the Reynolds number R = $48w/(d_p\,gu_1)$ = 48(15.75)/[(3.1416)(0.760)(0.0000141)] = 1,750,000.

6. *Adjust the flow coefficient for the actual Reynolds number*

In step 2, $R = 10^7$ was assumed and K = 0.6486. For R = 1,750,000, K = 0.6489, from ASME *Fluid Meters*, by interpolation. Then the actual flow rate w_h = (computed flow rate)(ratio of flow coefficients based on assumed and actual Reynolds numbers) = (15.75)(0.6489/0.6486)(3.600) = 56,700 lb/h (25,515 kg/h), closely, where the value 3600 is a conversion factor for changing lb/s to lb/h.

7. *Compute the flow rate for a specific differential gage deflection*

For a 40-in (101.6-cm) H_2O deflection, F_a is unchanged and equals 1.0084. The expansion factor changes because h_w/p_p = 40/264.7 = 0.151. From the ASME *Fluid Meters*, Y_1 = 0.998. By assuming again that $R = 10^7$, K = 0.6486, as before, w = $(0.0997)(1.0084)(0.6486)(5.855)^2(0.998)(40 \times 0.4188)^{0.5}$ = 9.132 lb/s (4.1 kg/s). Computing the Reynolds number as before, gives R = (40)(0.132)/[(3.1416)(0.76)(0.0000141)] = 1,014,000. The value of K corresponding to this value, as before, is from ASME—*Fluid Meters:* K = 0.6497. Therefore, the flow rate for a 40 in (101.6 cm) H_2O reading, in lb/h = w_h = (0.132)(0.6497/0.6486)(3600) = 32,940 lb/h (14,823 kg/h).

Related Calculations: Use these steps and the ASME *Fluid Meters* or comprehensive meter engineering tables giving similar data to select or check an orifice meter used in any type of steam pipe—main, auxiliary, process, industrial, marine, heating, or commercial, conveying wet, saturated, or superheated steam.

SELECTION OF A PRESSURE-REGULATING VALVE FOR STEAM SERVICE

Select a single-seat spring-loaded diaphragm-actuated pressure-reducing valve to deliver 350 lb/h (158 kg/h) of steam at 50 lb/in² (gage) (344.7 kPa) when the initial pressure is 225 lb/in² (gage) (1551 kPa). Also select an integral pilot-controlled piston-operated single-seat pressure-regulating valve to deliver 30,000 lb/h (13,500 kg/h) of steam at 40 lb/in² (gage) (275.8 kPa) with an initial pressure of 225 lb/in² (gage) (1551 kPa) saturated. What size pipe must be used on the downstream side of the valve to produce a velocity of 10,000 ft/min (3048 m/min)? How large should the pressure-regulating valve be if the steam entering the valve is at 225 lb/in² (gage) (1551 kPa) and 600°F (316°C)?

Calculation Procedure:

1. *Compute the maximum flow for the diaphragm-actuated valve*

For best results in service, pressure-reducing valves are selected so that they operate 60 to 70 percent open at normal load. To obtain a valve sized for this opening, divide the desired delivery, lb/h, by 0.7 to obtain the maximum flow expected. For this valve then, the maximum flow = 350/0.7 = 500 lb/h (225 kg/h).

2. *Select the diaphragm-actuated valve size*

Using a manufacturer's engineering data for an acceptable valve, enter the appropriate valve capacity table at the valve inlet steam pressure, 225 lb/in² (gage) (1551 kPa), and project to a capacity of 500 lb/h (225 kg/h), as in Table 13. Read the valve size as ¾ in (19.1 mm) at the top of the capacity column.

3. *Select the size of the pilot-controlled pressure-regulating valve*

Enter the capacity table in the engineering data of an acceptable pilot-controlled pressure-regulating valve, similar to Table 14, at the required capacity, 30,000 lb/h (13,500 kg/h). Project

TABLE 13 Pressure-Reducing-Valve Capacity, lb/h (kg/h)

Inlet pressure		Valve size		
lb/in^2 (gage)	kPa	½ in (12.7 mm)	¾ in (19.1 mm)	1 in (25.4 mm)
200	1379	420 (189)	460 (207)	560 (252)
225	1551	450 (203)	500 (225)	600 (270)
250	1724	485 (218)	560 (252)	650 (293)

Source: Clark-Reliance Corporation.

across until the correct inlet steam pressure column, 225 lb/in^2 (gage) (1551 kPa), is intercepted, and read the required valve size as 4 in (101.6 mm).

Note that it is not necessary to compute the maximum capacity before entering the table, as in step 1, for the pressure-reducing valve. Also note that a capacity table such as Table 14 can be used only for valves conveying saturated steam, unless the table notes state that the values listed are valid for other steam conditions.

4. *Determine the size of the downstream pipe*

Enter Table 14 at the required capacity, 30,000 lb/h (13,500 kg/h); project across to the valve *outlet pressure,* 40 lb/in^2 (gage) (275.8 kPa); and read the required pipe size as 8 in (203.2 mm) for a velocity of 10,000 ft/min (3048 m/min). Thus, the pipe immediately downstream from the valve must be enlarged from the valve size, 4 in (101.6 mm), to the required pipe size, 8 in (203.2 mm), to obtain the desired steam velocity.

5. *Determine the size of the valve handling superheated steam*

To determine the correct size of a pilot-controlled pressure-regulating valve handling superheated steam, a correction must be applied. Either a factor or a tabulation of corrected pressures, Table 15, may be used. To use Table 15, enter at the valve inlet pressure, 225 lb/in^2 (gage) (1551.2 kPa), and project across to the total temperature, 600°F (316°C), to read the corrected pressure, 165 lb/in^2 (gage) (1137.5 kPa). Enter Table 14 at the *next highest* saturated steam pressure, 175 lb/in^2 (gage) (1206.6 kPa); project down to the required capacity, 30,000 lb/h (13,500 kg/h); and read the required valve size as 5 in (127 mm).

Related Calculations: To simplify pressure-reducing and pressure-regulating valve selection, become familiar with two or three acceptable valve manufacturers' engineering data. Use the procedures given in the engineering data or those given here to select valves for industrial, marine, utility, heating, process, laundry, kitchen, or hospital service with a saturated or superheated steam supply.

Do not oversize reducing or regulating valves. Oversizing causes chatter and excessive wear.

When an anticipated load on the downstream side will not develop for several months after installation of a valve, fit to the valve a reduced-area disk sized to handle the present load. When the load increases, install a full-size disk. Size the valve for the ultimate load, not the reduced load.

Where there is a wide variation in demand for steam at the reduced pressure, consider installing two regulators piped in parallel. Size the smaller regulator to handle light loads and the larger

TABLE 14 Pressure-Regulating-Valve Capacity

Steam capacity		Initial steam pressure, saturated			
lb/h	kg/h	40 lb/in^2 (gage) (276 kPa)	175 lb/in^2 (gage) (1206 kPa)	225 lb/in^2 (gage) (1551 kPa)	300 lb/in^2 (gage) (2068 kPa)
20,000	9,000	6* (152.4)	4 (101.6)	4 (101.6)	3 (76.2)
30,000	13,500	8 (203.2)	5 (127.0)	4 (101.6)	4 (101.6)
40,000	18,000	—	5 (127.0)	5 (127.0)	4 (101.6)

*Valve diameter measured in inches (millimeters).
Source: Clark-Reliance Corporation.

TABLE 15 Equivalent Saturated Steam Values for Superheated Steam at Various Pressures and Temperatures

Steam pressure		Steam temperature		Total temperature					
				500°F	600°F	700°F	260.0°C	315.6°C	371.1°C
lb/in² (gage)	kPa	°F	°C	Steam values, lb/in² (gage)			Steam values, kPa		
205	1413.3	389	198	171	149	133	1178.9	1027.2	916.9
225	1551.2	397	203	190	165	147	1309.9	1137.5	1013.4
265	1826.9	411	211	227	200	177	1564.9	1378.8	1220.2

Source: Clark-Reliance Corporation.

regulator to handle the difference between 60 percent of the light load and the maximum heavy load. Set the larger regulator to open when the minimum allowable reduced pressure is reached. Then both regulators will be open to handle the heavy load. Be certain to use the actual regulator inlet pressure and not the boiler pressure when sizing the valve if this is different from the inlet pressure. Data in this calculation procedure are based on valves built by the Clark-Reliance Corporation, Cleveland, Ohio.

Some valve manufacturers use the valve flow coefficient C_v for valve sizing. This coefficient is defined as the flow rate, lb/h, through a valve of given size when the pressure loss across the valve is 1 lb/in² (6.89 kPa). Tabulations like Tables 13 and 14 incorporate this flow coefficient and are somewhat easier to use. These tables make the necessary allowances for downstream pressures less than the critical pressure (= 0.55 × absolute upstream pressure, lb/in², for superheated steam and hydrocarbon vapors; and 0.58 × absolute upstream pressure, lb/in², for saturated steam). The accuracy of these tabulations equals that of valve sizes determined by using the flow coefficient.

HYDRAULIC RADIUS AND LIQUID VELOCITY IN WATER PIPES

What is the velocity of 1000 gal/min (63.1 L/s) of water flowing through a 10-in (254-mm) inside-diameter cast-iron water main? What is the hydraulic radius of this pipe when it is full of water? When the water depth is 8 in (203.2 mm)?

Calculation Procedure:

1. *Compute the water velocity in the pipe*

For any pipe conveying water, the liquid velocity is v ft/s = gal/min/$(2.448d^2)$, where d = internal pipe diameter, in. For this pipe, $v = 1000/[2.448(100)] = 4.08$ ft/s (1.24 m/s), or $(60)(4.08) = 244.8$ ft/min (74.6 m/min).

2. *Compute the hydraulic radius for a full pipe*

For any pipe, the hydraulic radius is the ratio of the cross-sectional area of the pipe to the wetted perimeter, or $d/4$. For this pipe, when full of water, the hydraulic radius = $10/4 = 2.5$.

3. *Compute the hydraulic radius for a partially full pipe*

Use the hydraulic radius tables in King and Brater—*Handbook of Hydraulics*, or compute the wetted perimeter by using the geometric properties of the pipe, as in step 2. From the King and Brater table, the hydraulic radius = Fd, where F = table factor for the ratio of the depth of water, in/diameter of channel, in = $8/10 = 0.8$. For this ratio, $F = 0.304$. Then, hydraulic radius = $(0.304)(10) = 3.04$ in (77.2 mm).

Related Calculations: Use this method to determine the water velocity and hydraulic radius in any pipe conveying cold water—water supply, plumbing, process, drain, or sewer.

TABLE 16 Values of C in Hazen-Williams Formula

Type of pipe	C*	Type of pipe	C*
Cement-asbestos	140	Cast iron or wrought iron	100
Asphalt-lined iron or steel	140	Welded or seamless steel	100
Copper or brass	130	Concrete	100
Lead, tin, or glass	130	Corrugated steel	60
Wood stave	110		

*Values of C commonly used for design. The value of C for pipes made of corrosive materials decreases as the age of the pipe increases; the values given are those that apply at an age of 15 to 20 years. For example, the value of C for cast-iron pipes 30 in (762 mm) in diameter or greater at various ages is approximately as follows: new, 130; 5 years old, 120; 10 years old, 115; 20 years old, 100; 30 years old, 90; 40 years old, 80; and 50 years old, 75. The value of C for smaller-size pipes decreases at a more rapid rate.

FRICTION-HEAD LOSS IN WATER PIPING OF VARIOUS MATERIALS

Determine the friction-head loss in 2500 ft (762 m) of clean 10-in (254-mm) new tar-dipped cast-iron pipe when 2000 gal/min (126.2 L/s) of cold water is flowing. What is the friction-head loss 20 years later? Use the Hazen-Williams and Manning formulas, and compare the results.

Calculation Procedure:

1. *Compute the friction-head loss by the Hazen-Williams formula*

The Hazen-Williams formula is $h_f = [v/(1.318CR_h^{0.63})]^{1.85}$, where h_f = friction-head loss per ft of pipe, ft of water; v = water velocity, ft/s; C = a constant depending on the condition and kind of pipe; R_h = hydraulic radius of pipe, ft.

For a water pipe, v = gal/min/$(2.44d^2)$; for this pipe, $v = 2000/[2.448(10)^2] = 8.18$ ft/s (2.49 m/s). From Table 16 or Crocker and King—*Piping Handbook*, C for new pipe = 120; for 20-year-old pipe, $C = 90$; $R_h = d/4$ for a full-flow pipe = 10/4 = 2.5 in, or 2.5/12 = 0.208 ft (63.4 mm). Then $h_f = [8.18/(1.318 \times 120 \times 0.208^{0.63})]^{1.85} = 0.0263$ ft (8.0 mm) of water per ft (m) of pipe. For 2500 ft (762 m) of pipe, the total friction-head loss = 2500(0.0263) = 65.9 ft (20.1 m) of water for the new pipe.

For 20-year-old pipe and the same formula, except with $C = 90$, $h_f = 0.0451$ ft (13.8 mm) of water per ft (m) of pipe. For 2500 ft (762 m) of pipe, the total friction-head loss = 2500(0.0451) = 112.9 ft (34.4 m) of water. Thus, the friction-head loss nearly doubles [from 65.9 to 112.9 ft (20.1 to 34.4 m)] in 20 years. This shows that it is wise to design for future friction losses; otherwise, pumping equipment may become overloaded.

2. *Compute the friction-head loss from the Manning formula*

The Manning formula is $h_f = n^2v^2/2.208R_h^{4/3}$, where n = a constant depending on the condition and kind of pipe; other symbols as before.

Using $n = 0.011$ for new coated cast-iron pipe from Table 17 or Crocker and King—*Piping Handbook*, we find $h_f = (0.011)^2(8.18)^2/[2.208(0.208)^{4/3}] = 0.0295$ ft (8.9 mm) of water per ft (m) of pipe. For 2500 ft (762 m) of pipe, the total friction-head loss = 2500(0.0295) = 73.8 ft (22.5 m) of water, as compared with 65.9 ft (20.1 m) of water computed with the Hazen-Williams formula.

For coated cast-iron pipe in fair condition, $n = 0.013$, and $h_f = 0.0411$ ft (12.5 mm) of water. For 2500 ft (762 m) of pipe, the total friction-head loss = 2500(0.0411) = 102.8 ft (31.3 m) of water, as compared with 112.9 ft (34.4 m) of water computed with the Hazen-Williams formula. Thus, the Manning formula gives results higher than the Hazen-Williams in one case and lower in another. However, the differences in each case are not excessive; $(73.8 - 65.9)/65.9 = 0.12$, or 12 percent higher, and $(112.9 - 102.8)/102.8 = 0.0983$, or 9.83 percent lower. Both these differences are within the normal range of accuracy expected in pipe friction-head calculations.

TABLE 17 Roughness Coefficients (Manning's n) for Closed Conduits

Type of conduit			Manning's n Good construction*	Manning's n Fair construction*
Concrete pipe			0.013	0.015
Corrugated metal pipe or pipe arch, 2⅔ × ½ in (67.8 × 12.7 mm) corrugation, riveted:				
Plain			0.024	
Paved invert:				
Percent of circumference paved	25	50		
Depth of flow:				
Full	0.021	0.018		
0.8D	0.021	0.016		
0.6D	0.019	0.013		
Vitrified clay pipe			0.012	0.014
Cast-iron pipe, uncoated			0.013	
Steel pipe			0.011	
Brick			0.014	0.017
Monolithic concrete:				
Wood forms, rough			0.015	0.017
Wood forms, smooth			0.012	0.014
Steel forms			0.012	0.013
Cemented-rubble masonry walls:				
Concrete floor and top			0.017	0.022
Natural floor			0.019	0.025
Laminated treated wood			0.015	0.017
Vitrified-clay liner plates			0.015	

*For poor-quality construction, use larger values of n.

Related Calculations: The Hazen-Williams and Manning formulas are popular with many piping designers for computing pressure losses in cold-water piping. To simplify calculations, most designers use the precomputed tabulated solutions available in Crocker and King—*Piping Handbook*, King and Brater—*Handbook of Hydraulics*, and similar publications. In the rush of daily work these precomputed solutions are also preferred over the more complex Darcy-Weisbach equation used in conjunction with the friction factor f, the Reynolds number R, and the roughness-diameter ratio.

Use the method given here for sewer lines, water-supply pipes for commercial, industrial, or process plants, and all similar applications where cold water at temperatures of 33 to 90°F (0.6 to 32.2°C) flows through a pipe made of cast iron, riveted steel, welded steel, galvanized iron, brass, glass, wood-stove, concrete, vitrified, common clay, corrugated metal, unlined rock, or enameled steel. Thus, either of these formulas, used in conjunction with a suitable constant, gives the friction-head loss for a variety of piping materials. Suitable constants are given in Tables 16 and 17 and in the above references. For the Hazen-Williams formula, the constant C varies from about 70 to 140, while n in the Manning formula varies from about 0.017 for $C = 70$ to 0.010 for $C = 140$. Values obtained with these formulas have been used for years with satisfactory results. At present, the Manning formula appears the more popular.

CHART AND TABULAR DETERMINATION OF FRICTION HEAD

Figure 5 shows a process piping system supplying 1000 gal/min (63.1 L/s) of 70°F (21.1°C) water. Determine the total friction head, using published charts and pipe-friction tables. All the valves and fittings are flanged, and the piping is 10-in (254-mm) steel, schedule 40.

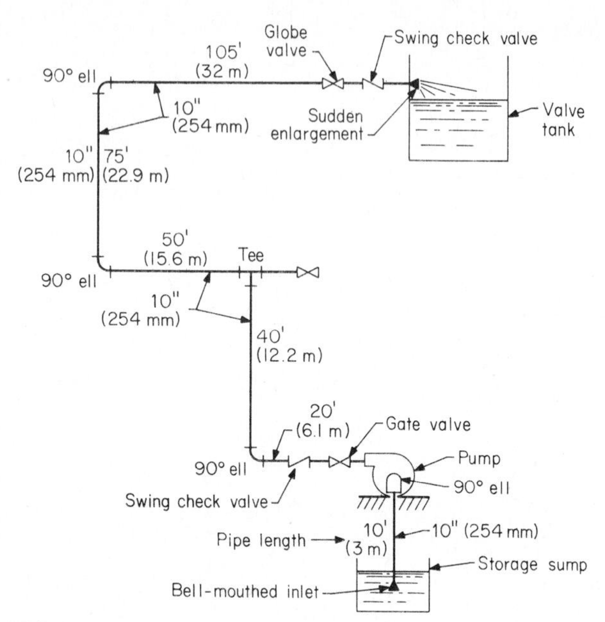

FIG. 5 Typical industrial piping system.

Calculation Procedure:

1. *Determine the total length of the piping*

Mark the length of each piping run on the drawing after scaling it or measuring it in the field. Determine the total length by adding the individual lengths, starting at the supply source of the liquid. In Fig. 5, beginning at the storage sump, the total length of piping = 10 + 20 + 40 + 50 + 75 + 105 = 300 ft (91.4 m). Note that the physical length of the fittings is included in the length of each run.

2. *Compute the equivalent length of each fitting*

The frictional resistance of pipe fittings (elbows, tees, etc.) and valves is greater than the actual length of each fitting. Therefore, the equivalent length of straight piping having a resistance equal to that of the fittings must be determined. This is done by finding the equivalent length of each fitting and taking the sum for all the fittings.

Use the equivalent length table in the pump section of this handbook or in Crocker and King—*Piping Handbook*, Baumeister and Marks—*Standard Handbook for Mechanical Engineers*, or *Standards of the Hydraulic Institute*. Equivalent length values will vary slightly from one reference to another.

Starting at the supply source, as in step 1, for 10-in (254-mm) flanged fittings throughout, we see the equivalent fitting lengths are: bell-mouth inlet, 2.9 ft (0.88 m); 90° ell at pump, 14 ft (4.3 m); gate valve, 3.2 ft (0.98 m); swing check valve, 120 ft (36.6 m); 90° ell, 14 ft (4.3 m); tee, 30 ft (9.1 m); 90° ell, 14 ft (4.3 m); 90° ell, 14 ft (4.3 m); globe valve, 310 ft (94.5 m); swing check valve, 120 ft (36.6 m); sudden enlargement = (liquid velocity, ft/s)$^2/2g$ = $(4.07)^2/2(32.2)$ = 0.257 ft (0.08 m), where the terminal velocity is zero, as in the tank. Find the liquid velocity as shown in a previous calculation procedure in this section. The sum of the fitting equivalent lengths is 2.9 + 14 + 3.2 + 120 + 14 + 30 + 14 + 14 + 310 + 120 + 0.257 = 642.4 ft (159.8 m). Adding this to the straight length gives a total length of 642.4 + 300 = 942.4 ft (287.2 m).

3. *Compute the friction-head loss by using a chart*

Figure 6 is a popular friction-loss chart for fairly rough pipe, which is any ordinary pipe after a few years' use. Enter at the left at a flow of 1000 gal/min (63.1 L/s), and project to the right until

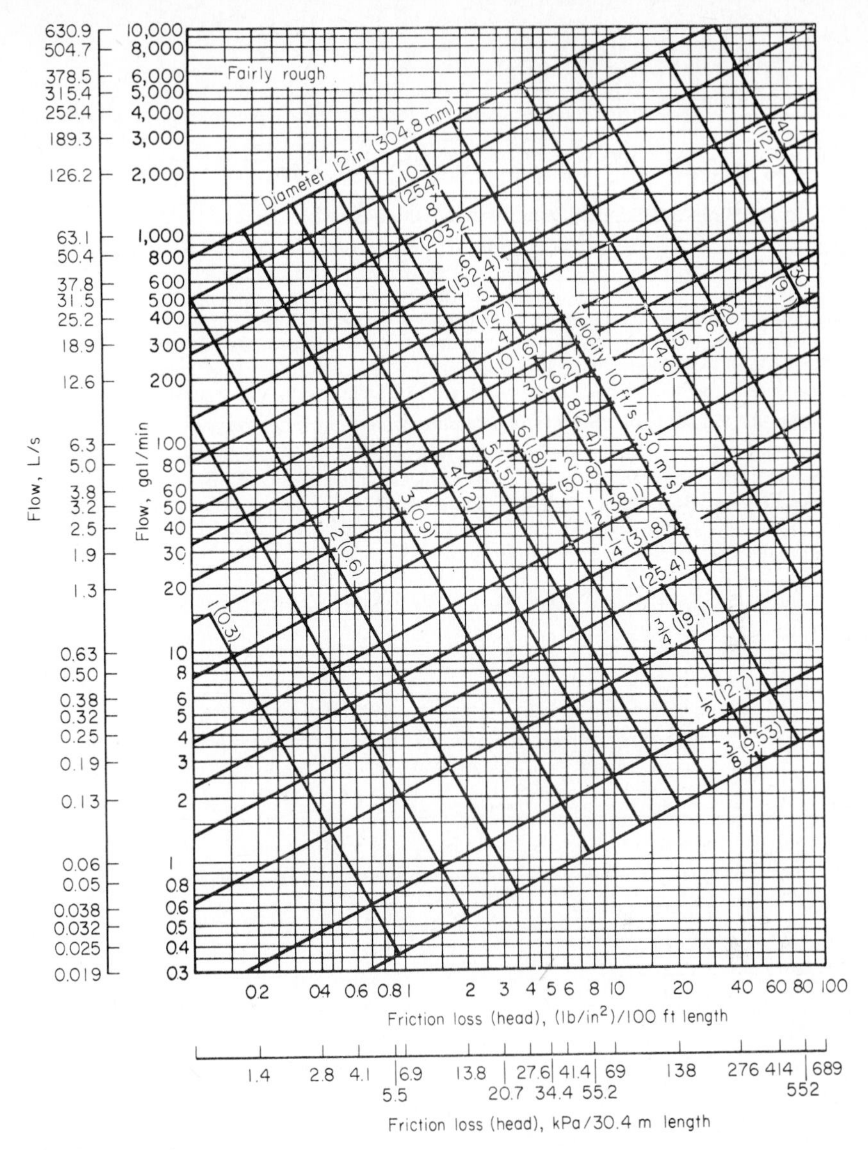

FIG. 6 Friction loss in water piping.

TABLE 18 Absolute Roughness Classification of Pipe Surfaces for Selection of Friction Factor *f* in Fig. 7

Commercial pipe surface (new)	Absolute roughness ϵ, ft	Absolute roughness ϵ, mm	Commercial pipe surface (new)	Absolute roughness ϵ, ft	Absolute roughness ϵ, mm
Glass, drawn brass, copper, lead	Smooth	Smooth	Cast iron	0.00085	0.26
Wrought iron, steel	0.00015	0.05	Wood stave	0.0006–0.003	0.18–0.91
Asphalted cast iron	0.0004	0.12	Concrete	0.001–0.01	0.30–3.05
Galvanized iron	0.0005	0.15	Riveted steel	0.003–0.03	0.91–9.14

the 10-in (254-mm) diameter curve is intersected. Read the friction-head loss at the top or bottom of the chart as 0.4 lb/in^2 (2.8 kPa), closely, per 100 ft (30.5 m) of pipe. Therefore, total friction-head loss = (0.4)(942.4/100) = 3.77 lb/in^2 (26 kPa). Converting gives (3.77)(2.31) = 8.71 ft (2.7 m) of water.

4. *Compute the friction-head loss from tabulated data*

Using the *Standards of the Hydraulic Institute* pipe-friction table, we find that the friction head h_f of water per 100 ft (30.5 m) of pipe = 0.500 ft (0.15 m). Hence, the total friction head = (0.500)(942.4/100) = 4.71 ft (1.4 m) of water. The Institute recommends that 15 percent be added to the tabulated friction head, or (1.15)(4.71) = 5.42 ft (1.66 m) of water.

Using the friction-head tables in Crocker and King—*Piping Handbook,* the friction head = 6.27 ft (1.9 m) per 1000 ft (304.8 m) of pipe with C = 130 for new, very smooth pipe. For this piping system, the friction-head loss = (942.4/1000)(6.27) = 5.91 ft (1.8 m) of water.

5. *Use the Reynolds number method to determine the friction head*

In this method, the friction factor is determined by using the Reynolds number R and the relative roughness of the pipe ϵ/D, where ϵ = pipe roughness, ft, and D = pipe diameter, ft.

For any pipe, $R = Dv/\nu$, where v = liquid velocity, ft/s, and ν = kinematic viscosity, ft^2/s. Using King and Brater—*Handbook of Hydraulics,* v = 4.07 ft/s (1.24 m/s), and ν = 0.00001059 ft^2/s (0.00000098 m^2/s) for water at 70°F (21.1°C). Then R = (10/12)(4.07)/0.00001059 = 320,500.

From Table 18 or the above reference, ϵ = 0.00015, and ϵ/D = 0.00015/(10/12) = 0.00018.

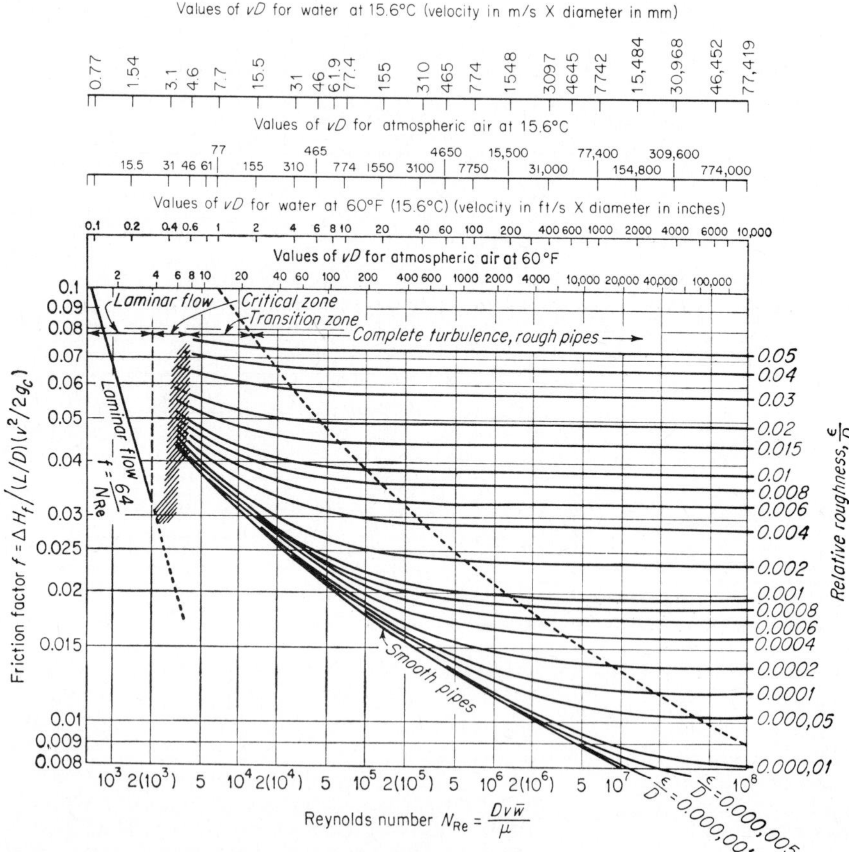

FIG. 7 Friction factors for laminar and turbulent flow.

From the Reynolds-number, relative-roughness, friction-factor curve in Fig. 7 or in Baumeister—*Standard Handbook for Mechanical Engineers*, the friction factor $f = 0.016$.

Apply the Darcy-Weisbach equation $h_f = f(l/D)(v^2/2g)$, where l = total pipe length, including the fittings' equivalent length, ft. Then $h_f = (0.016)(942.4/10/12)(4.07)^2/(2 \times 32.2) = 4.651$ ft (1.43 m) of water.

6. *Compare the results obtained*

Three different friction-head values were obtained: 8.71, 5.91, and 4.651 ft (2.7, 1.8, and 1.4 m) of water. The results show the variations that can be expected with the different methods. Actually, the Reynolds number method is probably the most accurate. As can be seen, the other two methods give safe results—i.e., the computed friction head is higher. The *Pipe Friction Manual*, published by the Hydraulic Institute, presents excellent simplified charts for use with the Reynolds number method.

Related Calculations: Use any of these methods to compute the friction-head loss for any type of pipe. The Reynolds number method is useful for a variety of liquids other than water—mercury, gasoline, brine, kerosene, crude oil, fuel oil, and lube oil. It can also be used for saturated and superheated steam, air, methane, and hydrogen.

RELATIVE CARRYING CAPACITY OF PIPES

What is the equivalent steam-carrying capacity of a 24-in (609.6-mm) inside-diameter pipe in terms of a 10-in (254-mm) inside-diameter pipe? What is the equivalent water-carrying capacity of a 23-in (584.2-mm) inside-diameter pipe in terms of a 13.25-in (336.6-mm) inside-diameter pipe?

Calculation Procedure:

1. *Compute the relative carrying capacity of the steam pipes*

For steam, air, or gas pipes, the number N of small pipes of inside diameter d_2 in equal to one pipe of larger inside diameter d_1 in is $N = (d_1^3\sqrt{d_2 + 3.6})/(d_2^3 + \sqrt{d_1 + 3.6})$. For this piping system, $N = (24^3 + \sqrt{10} + 3.6)/(10^3 + \sqrt{24} + 3.6) = 9.69$, say 9.7. Thus, a 24-in (609.6-mm) inside-diameter steam pipe has a carrying capacity equivalent to 9.7 pipes having a 10-in (254-mm) inside diameter.

2. *Compute the relative carrying capacity of the water pipes*

For water, $N = (d_2/d_1)^{2.5} = (23/13.25)^{2.5} = 3.97$. Thus, one 23-in (584-cm) inside-diameter pipe can carry as much water as 3.97 pipes of 13.25-in (336.6-mm) inside diameter.

Related Calculations: Crocker and King—*Piping Handbook* and certain piping catalogs (Crane, Walworth, National Valve and Manufacturing Company) contain tabulations of relative carrying capacities of pipes of various sizes. Most piping designers use these tables. However, the equations given here are useful for ranges not covered by the tables and when the tables are unavailable.

PRESSURE-REDUCING VALVE SELECTION FOR WATER PIPING

What size pressure-reducing valve should be used to deliver 1200 gal/h (1.26 L/s) of water at 40 lb/in^2 (275.8 kPa) if the inlet pressure is 140 lb/in^2 (965.2 kPa)?

Calculation Procedure:

1. *Determine the valve capacity required*

Pressure-reducing valves in water systems operate best when the nominal load is 60 to 70 percent of the maximum load. Using 60 percent, we see that the maximum load for this valve = 1200/0.6 = 2000 gal/h (2.1 L/s).

2. *Determine the valve size required*

Enter a valve capacity table in suitable valve engineering data at the valve inlet pressure, and project to the exact, or next higher, valve capacity. Thus, enter Table 19 at 140 lb/in^2 (965.2 kPa)

TABLE 19 Maximum Capacities of Water Pressure-Reducing Valves, gal/h (L/s)

Inlet pressure		Valve size		
lb/in² (gage)	kPa	¾ in (19.1 mm)	1 in (25.4 mm)	1¼ in (31.8 mm)
120	827.3	1550 (1.6)	2000 (2.1)	4500 (4.7)
140	965.2	1700 (1.8)	2200 (2.3)	5000 (5.3)
160	1103.0	1850 (1.9)	2400 (2.5)	5500 (5.8)

Source: Clark-Reliance Corporation.

and project to the next higher capacity, 2200 gal/h (2.3 L/s), since a capacity of 2000 gal/h (2.1 L/s) is not tabulated. Read at the top of the column the required valve size as 1 in (25.4 mm).

Some valve manufacturers present the capacity of their valves in graphical instead of tabular form. One popular chart, Fig. 8, is entered at the difference between the inlet and outlet pressures on the abscissa, or 140 − 40 = 100 lb/in² (689.4 kPa). Project vertically to the flow rate of 2000/60 = 33.3 gal/min (2.1 L/s). Read the valve size on the intersecting valve capacity curve, or on the next curve if there is no intersection with the curve. Figure 8 shows that a 1-in (25.4-mm) valve should be used. This agrees with the tabulated capacity.

Related Calculations: Use this method for pressure-reducing valves in any type of water piping—process, domestic, commercial—where the water temperature is 100°F (37.8°C) or less. Table 19 is from data prepared by the Clark-Reliance Corporation; Fig. 8 is from Foster Engineering Company data.

Some valve manufacturers use the valve flow coefficient C_v for valve sizing. This coefficient is defined as the flow rate, gal/min, through a valve of given size when the pressure loss across the valve is 1 lb/in² (6.9 kPa). Tabulations like Table 19 and flowcharts like Fig. 8 incorporate this flow coefficient and are somewhat easier to use. Their accuracy equals that of the flow coefficient method.

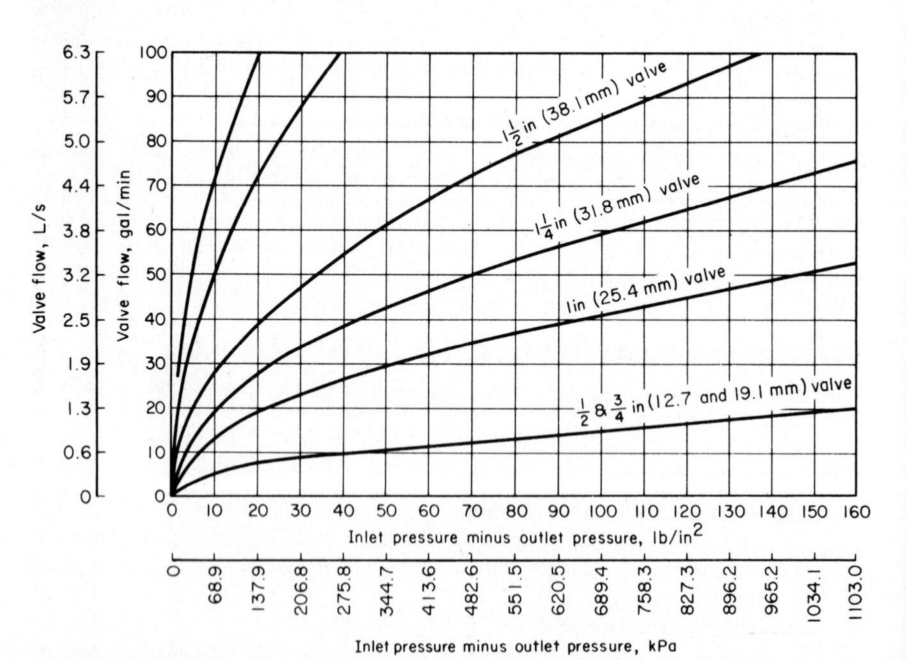

FIG. 8 Pressure-reducing valve flow capacity. *(Foster Engineering Company.)*

SIZING A WATER METER

A 6 × 4 in (152.4 × 101.6 mm) Venturi tube is used to measure water flow rate in a piping system. The dimensions of the meter are: inside pipe diameter d_p = 6.094 in (154.8 mm); throat diameter d = 4.023 in (102.2 mm). The differential pressure is measured with a mercury manometer having water on top of the mercury. The average manometer reading for 1 h is 10.1 in (256.5 mm) of mercury. The temperature of the water in the pipe is 41°F (5.0°C), and that of the room is 77°F (25°C). Determine the water flow rate in lb/h, gal/h, and gal/min. Use the ASME Research Committee on Fluid Meters method in analyzing the meter.

Calculation Procedure:

1. *Convert the pressure reading to standard conditions*

The ASME meter equation constant is based on a manometer liquid temperature of 68°F (20.0°C). Therefore, the water and mercury density at room temperature, 77°F (25°C), and the water density at 68°F (20.0°C), must be used to convert the manometer reading to standard conditions by the equation $h_w = h_m(m_d - w_d)/w_s$, where h_w = equivalent manometer reading, in (mm) H_2O at 68°F (20.0°C); h_m = manometer reading at room temperature, in mercury; m_d = mercury density at room temperature, lb/ft^3; w_d = water density at room temperature, lb/ft^3; w_s = water density at standard conditions, 68°F (20.0°C), lb/ft^3. From density values from the ASME publication *Fluid Meters: Their Theory and Application*, h_w = 10.1(844.88 − 62.244)/62.316 = 126.8 in (322.1 cm) of water at 68°F (20.0°C).

2. *Determine the throat-to-pipe diameter ratio*

The throat-to-pipe diameter ratio β = 4.023/6.094 = 0.6602. Then $1/(1 - \beta^4)^{0.5} = 1/(1 - 0.6602^4)^{0.5} = 1.1111$.

3. *Assume a Reynolds number value, and compute the flow rate*

The flow equation for a Venturi tube is w lb/h $= 359.0(Cd^2/\sqrt{1-\beta^4})(w_{dp}h_w)^{0.5}$, where C = meter discharge coefficient, expressed as a function of the Reynolds number; w_{dp} = density of the water at the pipe temperature, lb/ft^3. With a Reynolds number greater than 250,000, C is a constant. As a first trial, assume $R > 250{,}000$ and C = 0.984 from *Fluid Meters*. Then $w = 359.0(0.984)(4.023)^2(1.1111)(62.426 \times 126.8)^{0.5}$ = 565,020 lb/h (254,259 kg/h), or 565,020/8.33 lb/gal = 67,800 gal/h (71.3 L/s), or 67,800/60 min/h = 1129 gal/min (71.23 L/s).

4. *Check the discharge coefficient by computing the Reynolds numbers*

For a water pipe, $R = 48w_s/(\pi d_p g u)$, where w_s = flow rate, lb/s = w/3600; u = coefficient of absolute viscosity. Using *Fluid Meters* data for water at 41°F (5°C), we find $R = 48(156.95)/[(\pi \times 6.094)(0.001004)]$ = 391,900. Since C is constant for $R > 250{,}000$, use of C = 0.984 is correct, and no adjustment in the computations is necessary. Had the value of C been incorrect, another value would be chosen and the Reynolds number recomputed. Continue this procedure until a satisfactory value for C is obtained.

5. *Use an alternative solution to check the results*

Fluid Meters gives another equation for Venturi meter flow rate, that is, w lb/s $= 0.525(Cd^2/\sqrt{1-\beta^4})[w_{dp}(p_1 - p_2)]^{0.5}$, where $p_1 - p_2$ is the manometer differential pressure in lb/in^2. Using the conversion factor in *Fluid Meters* for converting in of mercury under water at 77°F (25°C) to lb/in^2 (kPa), we get $p_1 - p_2$ = (10.1)(0.4528) = 4.573 lb/in^2 (31.5 kPa). Then $w = (0.525)(0.984)(4.023)^2(1.1111)(62.426 \times 4.573)^{0.5}$ = 156.9 lb/s (70.6 kg/s), or (156.9)(3600 s/h) = 564,900 lb/h (254,205 kg/h), or 564,900/8.33 lb/gal = 67,800 gal/h (71.3 L/s), or 67,800/60 min/h = 1129 gal/min (71.2 L/s). This result agrees with that computed in step 3 within 1 part in 5600. This is much less than the probable uncertainties in the values of the discharge coefficient and the differential pressure.

Related Calculations: Use this method for any Venturi tube serving cold-water piping in process, industrial, water-supply, domestic, or commercial service.

EQUIVALENT LENGTH OF A COMPLEX SERIES PIPELINE

Figure 9 shows a complex series pipeline made up of four lengths of different size pipe. Determine the equivalent length of this pipe if each size of pipe has the same friction factor.

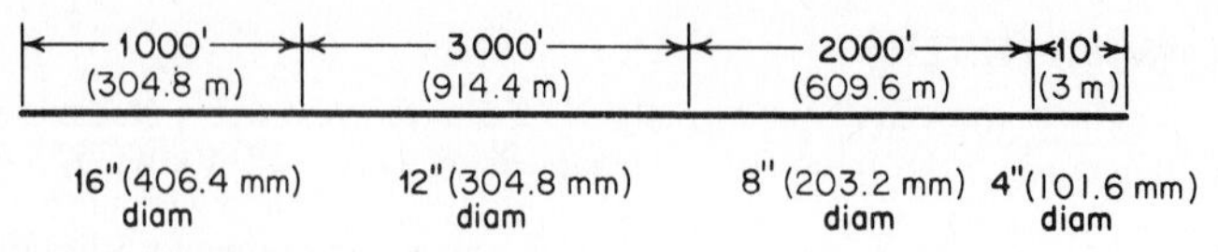

FIG. 9 Complex series pipeline.

Calculation Procedure:

1. *Select the pipe size for expressing the equivalent length*

The usual procedure in analyzing complex pipelines is to express the equivalent length in terms of the smallest, or next to smallest, diameter pipe. Choose the 8-in (203.2-mm) size as being suitable for expressing the equivalent length.

2. *Find the equivalent length of each pipe*

For any complex series pipeline having equal friction factors in all the pipes, L_e = equivalent length, ft, of a section of constant diameter = (actual length of section, ft) (inside diameter, in, of pipe used to express the equivalent length/inside diameter, in, of section under consideration)^5.

For the 16-in (406.4-mm) pipe, $L_e = (1000)(7.981/15.000)^5 = 42.6$ ft (12.9 m). The 12-in (304.8-mm) pipe is next; for it $L_e = (3000)(7.981/12.00)^5 = 390$ ft (118.9 m). For the 8-in (203.2-mm) pipe, the equivalent length = actual length = 2000 ft (609.6 m). For the 4-in (101.6-mm) pipe, $L_e = (10)(7.981/4.026)^5 = 306$ ft (93.3 m). Then the total equivalent length of 8-in (203.2-mm) pipe = sum of the equivalent lengths = 42.6 + 390 + 2000 + 306 = 2738.6 ft (834.7 m); or, by rounding off, 2740 ft (835.2 m) of 8-in (203.2-mm) pipe will have a frictional resistance equal to the complex series pipeline shown in Fig. 9. To compute the actual frictional resistance, use the methods given in previous calculation procedures.

Related Calculations: Use this general procedure for any complex series pipeline conveying water, oil, gas, steam, etc. See Crocker and King—*Piping Handbook* for derivation of the flow equations. Use the tables in Crocker and King to simplify finding the fifth power of the inside diameter of a pipe. The method of the next calculation procedure can also be used if a given flow rate is assumed.

Choosing a flow rate of 1000 gal/min (63.1 L/s) and using the tables in the Hydraulic Institute *Pipe Friction Manual* give an equivalent length of 2770 ft (844.3 m) for the 8-in (203.2-mm) pipe. This compares favorably with the 2740 ft (835.2 m) computed above. The difference of 30 ft (9.1 m) is negligible and can be accounted for by calculator variations.

The equivalent length is found by summing the friction-head loss for 1000-gal/min (63.1-L/s) flow for each length of the four pipes—16, 12, 8, and 4 in (406, 305, 203, and 102 mm)—and dividing this by the friction-head loss for 1000 gal/min (63.1 L/s) flowing through an 8-in (203.2-mm) pipe. Be careful to observe the units in which the friction-head loss is stated, because errors are easy to make if the units are ignored.

EQUIVALENT LENGTH OF A PARALLEL PIPING SYSTEM

Figure 10 shows a parallel piping system used to supply water for industrial needs. Determine the equivalent length of a single pipe for this system. All pipes in the system are approximately horizontal.

Calculation Procedure:

1. *Assume a total head loss for the system*

To determine the equivalent length of a parallel piping system, assume a total head loss for the system. Since this head loss is assumed for computation purposes only, its value need not be exact or even approximate. Assume a total head loss of 50 ft of water for each pipe in this system.

2. *Compute the flow rate in each pipe in the system*

Assume that the roughness coefficient C in the Hazen-Williams formula is equal for each of the pipes in the system. This is a valid assumption. Using the assumed value of C, compute the flow rate in each pipe. To allow for possible tuberculation of the pipe, assume that $C = 100$.

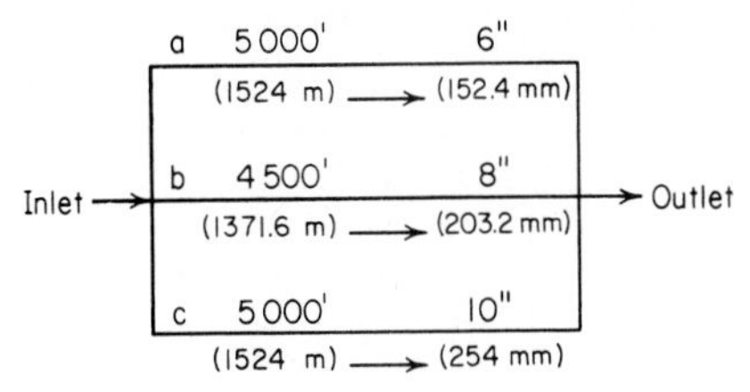

FIG. 10 Parallel piping system.

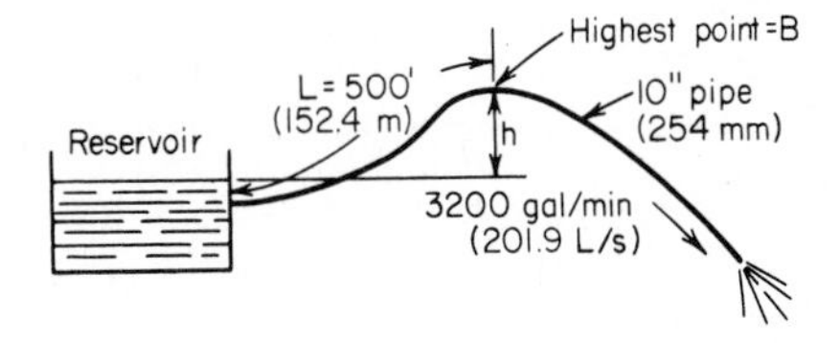

FIG. 11 Liquid siphon piping system.

The Hazen-Williams formula is given in a previous calculation procedure and can be used to solve for the flow rate in each pipe. A more rapid way to make the computation is to use the friction-loss tabulations for the Hazen-Williams formula in Crocker and King—*Piping Handbook*, the Hydraulic Institute—*Pipe Friction Manual*, or a similar set of tables.

Using such a set of tables, enter at the friction-head loss equal to 50 ft (15.2 m) per 5000 ft (1524 m) of pipe for the 6-in (152.4-mm) line. Find the corresponding flow rate Q gal/min. Using the Hydraulic Institute tables, Q_a = 380 gal/min (23.9 L/s); Q_b = 850 gal/min (53.6 L/s); Q_c = 1450 gal/min (91.5 L/s). Hence, the total flow = ΣQ = 380 + 835 + 1450 = 2665 gal/min (168.1 L/s).

3. *Find the equivalent size and length of the pipe*

Using the Hydraulic Institute tables again, look for a pipe having a 50-ft (15.2-m) head loss with a flow of 2665 gal/min (168.1 L/s). Any pipe having a discharge equal to the sum of the discharge rates for all the pipes, at the assumed friction head, is an equivalent pipe.

Interpolating friction-head values in the 14-in (355.6-mm) outside-diameter [13.126-in (333.4-mm) inside-diameter] table shows that 5970 ft (1820 m) of this pipe is equivalent to the system in Fig. 10. This equivalent size can be used in any calculations related to this system—selection of a pump, determination of head loss with longer or shorter mains, etc. If desired, another equivalent-size pipe could be found by entering a different pipe-size table. Thus, 5570 ft (1697.7 m) of 12-in (304.8-mm) pipe [11.938-in (303.2-mm) inside diameter] is also equivalent to this system.

Related Calculations: Use this procedure for any liquid—water, oil, gasoline, brine—flowing through a parallel piping system. The pipes are assumed to be full at all times.

MAXIMUM ALLOWABLE HEIGHT FOR A LIQUID SIPHON

What is the maximum height h ft (m), Fig. 11, that can be used for a siphon in a water system if the length of the pipe from the water source to its highest point is 500 ft (152.4 m), the water velocity is 13.0 ft/s (3.96 m/s), the pipe diameter is 10 in (254 mm), and the water temperature is 70°F (21.1°C) if 3200 gal/min (201.9 L/s) is flowing?

Calculation Procedure:

1. *Compute the velocity of the water in the pipe*

From an earlier calculation procedure, $v = gpm/(2.448d^2)$. With an internal diameter of 10.020 in (254.5 mm), $v = 3200/[(2.448)(10.02)^2]$ = 13.0 ft/s (3.96 m/s).

2. *Determine the vapor pressure of the water*

Using a steam table, we see that the vapor pressure of water at 70°F (21.1°C) is p_v = 0.3631 lb/in^2 (abs) (2.5 kPa), or (0.3631) (144 in^2/ft^2) = 52.3 lb/ft^2 (2.5 kPa). The specific volume of water at 70°F (21.1°C) is, from a steam table, 0.01606 ft^3/lb (0.001 m^3/kg). Converting this to density at 70°F (21.1°C), density = 1/0.01606 = 62.2 lb/ft^3 (995.8 kg/m^3). The vapor pressure in ft of 70°F (21.1°C) water is then f_v = (52.3 lb/ft^2)/(62.2 lb/ft^3) = 0.84 ft (0.26 m) of water.

3. *Compute or determine the friction-head loss and velocity head*

From the reservoir to the highest point of the siphon, B, Fig. 11, the friction head in the pipe must be overcome. Use the Hazen-Williams or a similar formula to determine the friction head, as given in earlier calculation procedures or a pipe-friction table. From the Hydraulic Institute

Pipe Friction Manual, h_f = 4.59 ft per 100 ft (1.4 m per 30.5 m), or (500/100)(4.59) = 22.95 ft (7.0 m). From the same table, velocity head = 2.63 ft/s (0.8 m/s).

4. *Determine the maximum height for the siphon*

For a siphon handling water, the maximum allowable height h at sea level with an atmospheric pressure of 14.7 lb/in² (abs) (101.3 kPa) = [14.7 × (144 in²/ft²)/(density of water at operating temperature, lb/ft³) − (vapor pressure of water at operating temperature, ft + 1.5 × velocity head, ft + friction head, ft)]. For this pipe, $h = 14.7 \times 144/62.2 - (0.84 + 1.5 \times 2.63 + 22.95) = 11.32$ ft (3.45 m). In actual practice, the value of h is taken as 0.75 to 0.8 the computed value. Using 0.75 gives $h = (0.75)(11.32) = 8.5$ ft (2.6 m).

Related Calculations: Use this procedure for any type of siphon conveying a liquid—water, oil, gasoline, brine, etc. Where the liquid has a specific gravity different from that of water, i.e., less than or greater than 1.0, proceed as above, expressing all heads in ft of liquid handled. Divide the resulting siphon height by the specific gravity of the liquid. At elevations above atmospheric, use the actual atmospheric pressure instead of 14.7 lb/in² (abs) (101.3 kPa).

WATER-HAMMER EFFECTS IN LIQUID PIPELINES

What is the maximum pressure developed in a 200-lb/in² (1378.8-kPa) water pipeline if a valve is closed nearly instantly or pumps discharging into the line are all stopped at the same instant? The pipe is 8-in (203.2-mm) schedule 40 steel, and the water flow rate is 2800 gal/min (176.7 L/s). What maximum pressure is developed if the valve closes in 5 s and the line is 5000 ft (1524 m) long?

Calculation Procedure:

1. *Determine the velocity of the pressure wave*

For any pipe, the velocity of the pressure wave during water hammer is found from $v_w = 4720/(1 + Kd/Et)^{0.5}$, where v_w = velocity of the pressure wave in the pipeline, ft/s; K = bulk modulus of the liquid in the pipeline = 300,000 for water; d = internal diameter of pipe, in; E = modulus of elasticity of pipe material, lb/in² = 30 × 10⁶ lb/in² (206.8 GPa) for steel; t = pipe-wall thickness, in. For 8-in (203.2-mm) schedule 40 steel pipe and data from a table of pipe properties, $v_w = 4720/[1 + 300{,}000 \times 7.981/(30 \times 10^6 \times 0.322)]^{0.5} = 4225.6$ ft/s (1287.9 m/s).

2. *Compute the pressure increase caused by water hammer*

The pressure increase p_1 lb/in² due to water hammer = $v_w v/[32.2(2.31)]$, where v = liquid velocity in the pipeline, ft/s; 32.2 = acceleration due to gravity, ft/s²; 2.31 ft of water = 1-lb/in² (6.9-kPa) pressure.

For this pipe, $v = 0.4085gpm/d^2 = 0.4085(2800)/(7.981)^2 = 18.0$ ft/s (5.5 m/s). Then $p_i = (4225.6)(18)/[32.2(2.31)] = 1022.56$ lb/in² (7049.5 kPa). The maximum pressure developed in the pipe is then p_i + pipe operating pressure = 1022.56 + 200 = 1222.56 lb/in² (8428.3 kPa).

3. *Compute the hammer pressure rise caused by valve closure*

The hammer pressure rise caused by valve closure p_v lb/in² = $2p_iL/v_wT$, where L = pipeline length, ft; T = valve closing time, s. For this pipeline, $p_v = 2(1022.56)(5000)/[(4225.6)(5)] = 484$ lb/in² (3336.7 kPa). Thus, the maximum pressure in the pipe will be 484 + 200 = 648 lb/in² (4467.3 kPa).

Related Calculations: Use this procedure for any type of liquid—water, oil, etc.—in a pipeline subject to sudden closure of a valve or stoppage of a pump or pumps. The effects of water hammer can be reduced by relief valves, slow-closing check valves on pump discharge pipes, air chambers, air spill valves, and air injection into the pipeline.

SPECIFIC GRAVITY AND VISCOSITY OF LIQUIDS

An oil has a specific gravity of 0.8000 and a viscosity of 200 SSU (Saybolt Seconds Universal) at 60°F (15.6°C). Determine the API gravity and Bé gravity of this oil at 70°F (21.1°C) and its weight in lb/gal (kg/L). What is the kinematic viscosity in cSt? What is the absolute viscosity in cP?

Calculation Procedure:

1. *Determine the API gravity of the liquid*

For any oil at 60°F (15.6°C), its specific gravity S, in relation to water at 60°F (15.6°C), is $S = 141.5/(131.5 + °API)$; or $°API = (141.5 - 131.5S)/S$. For this oil, °API = [141.5 − 131.5(0.80)]/0.80 = 45.4 °API.

2. *Determine the Bé gravity of the liquid*

For any liquid lighter than water, $S = 140/(130 + \text{Bé})$; or $\text{Bé} = (140 - 130S)/S$. For this oil, Bé = [140 − 130(0.80)]/0.80 = 45 Bé.

3. *Compute the weight per gal of liquid*

With a specific gravity of S, the weight of 1 ft^3 of oil = (S)[weight of 1 ft^3 (1 m^3) of fresh water at 60°F (15.6°C)] = (0.80)(62.4) = 49.92 lb/ft^3 (799.2 kg/m^3). Since 1 gal (3.8 L) of liquid occupies 0.13368 ft^3 the weight of this oil is (49.92)(0.13368) = 6.66 lb/gal (0.79 kg/L).

4. *Compute the kinematic viscosity of the liquid*

For any liquid having a viscosity between 32 and 99 SSU, the kinematic viscosity $k = 0.226 \text{ SSU} - 195/\text{SSU}$ cSt. For this oil, $k = 0.226(200) - 195/200 = 44.225$ cSt.

5. *Convert the kinematic viscosity to absolute viscosity*

For any liquid, the absolute viscosity, cP = (kinematic viscosity, cSt)(specific gravity). Thus, for this oil, the absolute viscosity = (44.225)(0.80) = 35.38 cP.

Related Calculations: For liquids *heavier* than water, $S = 145/(145 - \text{Bé})$. When the SSU viscosity is greater than 100 s, $k = 0.220$ (SSU)—135/SSU. Use these relations for any liquid—brine, gasoline, crude oil, kerosene, Bunker C, diesel oil, etc. Consult the *Pipe Friction Manual* and Crocker and King—*Piping Handbook* for tabulations of typical viscosities and specific gravities of various liquids.

PRESSURE LOSS IN PIPING HAVING LAMINAR FLOW

Fuel oil at 300°F (148.9°C) and having a specific gravity of 0.850 is pumped through a 30,000-ft (9144-m) long 24-in (609.6-mm) pipe at the rate of 500 gal/min (31.6 L/s). What is the pressure loss if the viscosity of the oil is 75 cP (0.075 Pa·s)?

Calculation Procedure:

1. *Determine the type of flow that exists*

Flow is laminar (also termed *viscous*) if the Reynolds number R for the liquid in the pipe is less than 1200. Tubulent flow exists if the Reynolds number is greater than 2500. Between these values is a zone in which either condition may exist, depending on the roughness of the pipe wall, entrance conditions, and other factors. Avoid sizing a pipe for flow in this critical zone because excessive pressure drops result without a corresponding increase in the pipe discharge.

Compute the Reynolds number from $R = 3.162G/kd$, where G = flow rate gal/min (L/s); k = kinematic viscosity of liquid, cSt = viscosity z, cP/specific gravity of the liquid S; d = inside diameter of pipe, in (cm). From a table of pipe properties, d = 22.626 in (574.7 mm). Also, $k = z/S = 75/0.85 = 88.2$ cSt. Then $R = 3162(500)/[88.2(22.626)] = 792$. Since $R < 1200$, laminar flow exists in this pipe.

2. *Compute the pressure loss by using the Poiseuille formula*

The Poiseuille formula gives the pressure drop p_d lb/in^2 (kPa) $= 2.73(10^{-4})luG/d^4$, where l = total length of pipe, including equivalent length of fittings, ft; u = absolute viscosity of liquid, cP (Pa·s); G = flow rate, gal/min (L/s); d = inside diameter of pipe, in (cm). For this pipe, $p_d = 2.73(10^{-4})(10{,}000)(75)(500)/262{,}078 = 1.17$ lb/in^2 (8.1 kPa).

Related Calculations: Use this procedure for any pipe in which there is laminar flow of the liquid. Other liquids for which this method can be used include water, molasses, gasoline, brine, kerosene, and mercury. Table 20 gives a quick summary of various ways in which the Reynolds number can be expressed. The symbols in Table 20, in the order of their appearance, are D = inside diameter of pipe, ft (m); v = liquid velocity, ft/s (m/s); ρ = liquid density, lb/ft^3 (kg/m^3); μ = absolute viscosity of liquid, lb mass/(ft·s) [kg/(m·s)]; d = inside diameter of pipe, in

TABLE 20 Reynolds Number

Reynolds number R	Numerator				Denominator	
	Coefficient	First symbol	Second symbol	Third symbol	Fourth symbol	Fifth symbol
$Dv\rho/\mu$	. . .	ft	ft/s	lb/ft^3	lb mass/(ft·s)	
$124dv\rho/z$	124	in	ft/s	lb/ft^3	cP	
$50.7G\rho/dz$	50.7	gal/min	lb/ft^3	. . .	in	cP
$6.32W/dz$	6.32	lb/h	. . .	. . .	in	cP
$35.5B\rho/dz$	35.5	bbl/h	lb/ft^3	. . .	in	cP
$7742dv/k$	7,742	in	ft/s	. . .	. . .	cP
$3162G/dk$	3,162	gal/min	. . .	. . .	in	cP
$2214B/dk$	2,214	bbl/h	. . .	. . .	in	cP
$22{,}735q\rho/dz$	22,735	ft^3/s	lb/ft^3	. . .	in	cP
$378.9Q\rho/dz$	378.9	ft^3/min	lb/ft^3	. . .	in	cP

(cm). From a table of pipe properties, d = 22.626 in (574.7 mm). Also, $k = z/S$ liquid flow rate, lb/h (kg/h); B = liquid flow rate, bbl/h (L/s); k = kinematic viscosity of the liquid, cSt; q = liquid flow rate, ft^3/s (m^3/s); Q = liquid flow rate, ft^3/min (m^3/min). Use Table 20 to find the Reynolds number for any liquid flowing through a pipe.

DETERMINING THE PRESSURE LOSS IN OIL PIPES

What is the pressure drop in a 5000-ft (1524-m) long 6-in (152.4-mm) oil pipe conveying 500 bbl/h (22.1 L/s) of kerosene having a specific gravity of 0.813 at 65°F (18.3°C), which is the temperature of the liquid in the pipe? The pipe is schedule 40 steel.

Calculation Procedure:

1. *Determine the kinematic viscosity of the oil*

Use Fig. 12 and Table 21 or the Hydraulic Institute—*Pipe Friction Manual* kinematic viscosity and Reynolds number chart to determine the kinematic viscosity of the liquid. Enter Table 12 at kerosene, and find the coordinates as X = 10.2, Y = 16.9. Using these coordinates, enter Fig. 12 and find the absolute viscosity of kerosene at 65°F (18.3°C) as 2.4 cP. By the method of a previous calculation procedure, the kinematic viscosity = absolute viscosity, cP/specific gravity of the liquid = 2.4/0.813 = 2.95 cSt. This value agrees closely with that given in the *Pipe Friction Manual.*

2. *Determine the Reynolds number of the liquid*

The Reynolds number can be found from the *Pipe Friction Manual* chart mentioned in step 1 or computed from $R = 2214B/(dk) = 2214(500)/[(6.065)(2.95)] = 61{,}900$.

To use the *Pipe Friction Manual* chart, compute the velocity of the liquid in the pipe by converting the flow rate to ft^3/s. Since there is 42 gal/bbl (0.16 L) and 1 gal (0.00379 L) = 0.13368 ft^3 (0.00378 m^3), 1 bbl = (42)(0.13368) = 5.6 ft^3 (0.16 m^3). With a flow rate of 500 bbl/h (79.5 m^3/h) the equivalent flow = (500)(5.6) = 2800 ft^3/h (79.3 m^3/h), or 2800/3600 s/h = 0.778 ft^3/s (0.02 m^3/s). Since 6-in (152.4-mm), schedule 40 pipe has a cross-sectional area of 0.2006 ft^2 (0.02 m^2) internally, the liquid velocity = 0.778/0.2006 = 3.88 ft/s (1.2 m/s). Then, the product (velocity, ft/s)(internal diameter, in) = (3.88)(6.065) = 23.75 ft/s. In the *Pipe Friction Manual,* project horizontally from the kerosene specific-gravity curve to the vd product of 23.75, and read the Reynolds number as 61,900, as before. In general, the Reynolds number can be found more quickly by computing it using the appropriate relation given in an earlier calculation procedure, unless the flow velocity is already known.

3. *Determine the friction factor of this pipe*

Enter Fig. 13 at the Reynolds number value of 61,900, and project to the curve 4 as indicated by Table 22. Read the friction factor as 0.0212 at the left. Alternatively, the *Pipe Friction Manual* friction-factor chart could be used, if desired.

VISCOSITIES

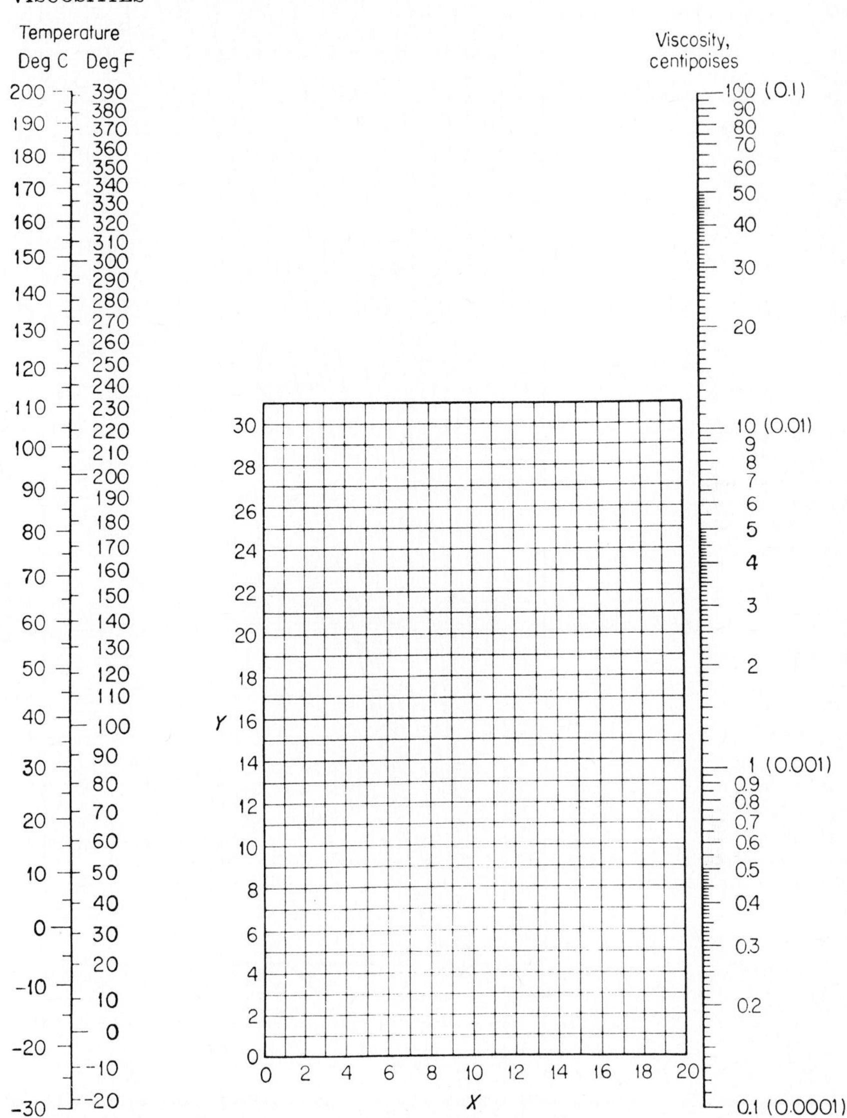

FIG. 12 Viscosities of liquids at 1 atm. For coordinates, see Table 21.

4. *Compute the pressure loss in the pipe*

Use the Fanning formula $p_d = 1.06(10^{-4})f\rho lB^2/d^5$. In this formula, ρ = density of the liquid, lb/ft^3. For kerosene, ρ = (density of water, lb/ft^3)(specific gravity of the kerosene) = (62.4)(0.813) = 50.6 lb/ft^3 (810.1 kg/m^3). Then $p_d = 1.06(10^{-4})(0.0212)(50.6)(5000)(500)^2/8206$ = 17.3 lb/in^2 (119.3 kPa).

Related Calculations: The Fanning formula is popular with oil-pipe designers and can be stated in various ways: (1) with velocity v ft/s, $p_d = 1.29(10^{-3})f\rho v^2 l/d$; (2) with velocity V ft/min, $p_d = 3.6(10^{-7})f\rho V^2 l/d$; (3) with flow rate in G gal/min, $p_d = 2.15(10^{-4})f\rho lG^2/d^2$; (4) with the flow rate in W lb/h, $p_d = 3.36(10^{-6})flW^2/d^5\rho$.

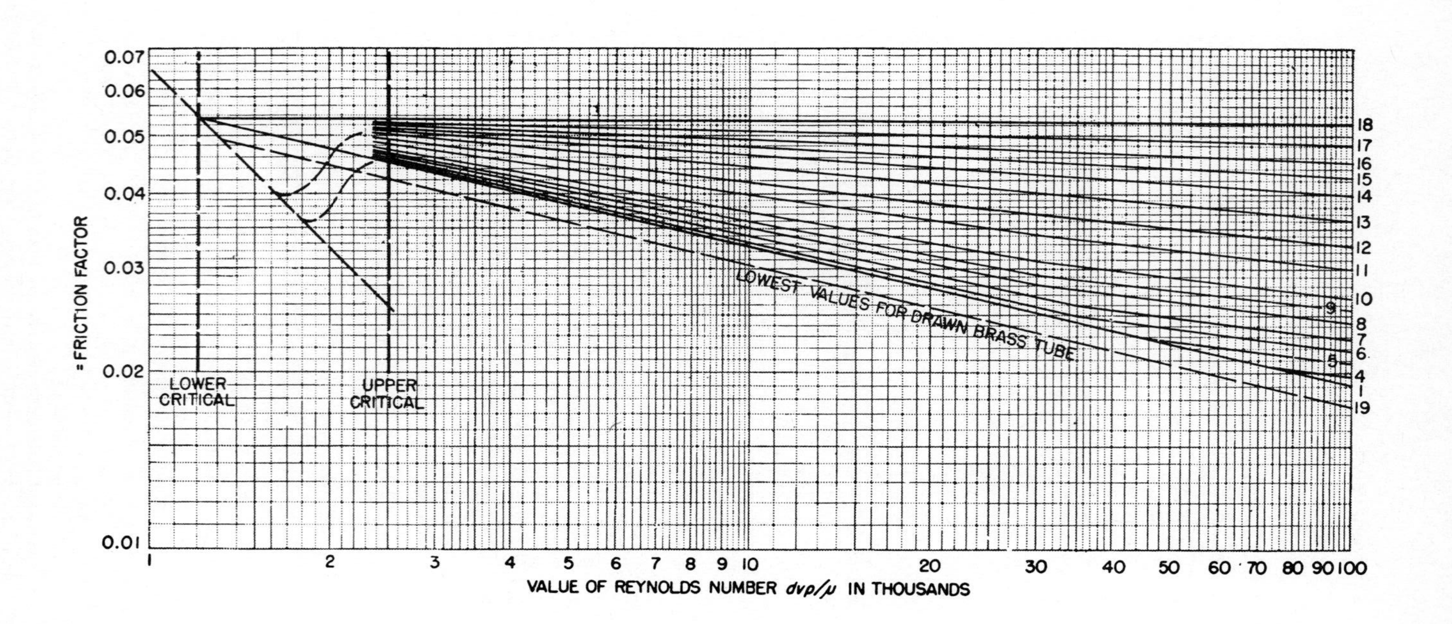

= FRICTION FACTOR
0.07
0.06
0.05
0.04
0.03
0.02
0.01
LOWER CRITICAL
UPPER CRITICAL
LOWEST VALUES FOR DRAWN BRASS TUBE
18
17
16
15
14
13
12
11
10
9
8
7
6
5
4
1
19
1
2
3
4
5
6
7
8
9
10
20
30
40
50
60
70
80
90
100
VALUE OF REYNOLDS NUMBER $dv\rho/\mu$ IN THOUSANDS

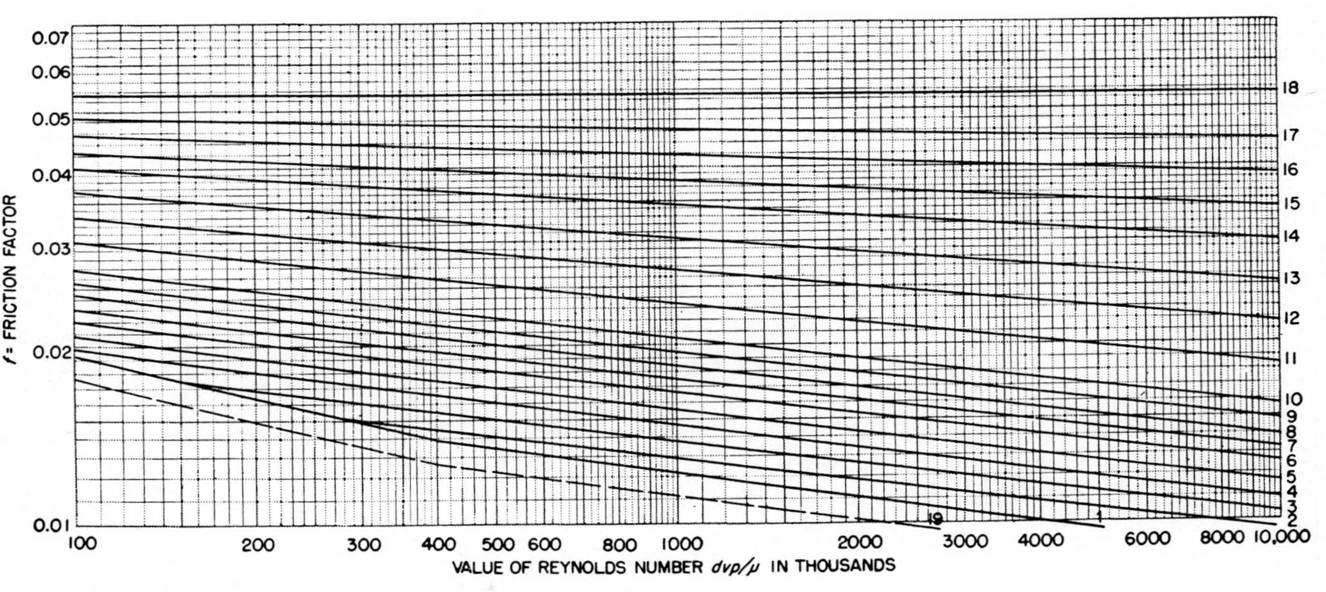

FIG. 13 Friction-factor curves. *(Mechanical Engineering.)*

TABLE 21 Viscosities of Liquids
Coordinates for use with Fig. 12

No.	Liquid	X	Y
1	Acetaldehyde	15.2	4.8
	Acetic acid:		
2	100%	12.1	14.2
3	70%	9.5	17.0
4	Acetic anhydride	12.7	12.8
	Acetone:		
5	100%	14.5	7.2
6	35%	7.9	15.0
7	Allyl alcohol	10.2	14.3
	Ammonia:		
8	100%	12.6	2.0
9	26%	10.1	13.9
10	Amyl acetate	11.8	12.5
11	Amyl alcohol	7.5	18.4
12	Aniline	8.1	18.7
13	Anisole	12.3	13.5
14	Arsenic trichloride	13.9	14.5
15	Benzene	12.5	10.9
	Brine:		
16	$CaCl_2$, 25%	6.6	15.9
17	NaCl, 25%	10.2	16.6
18	Bromine	14.2	13.2
19	Bromotoluene	20.0	15.9
20	Butyl acetate	12.3	11.0
21	Butyl alcohol	8.6	17.2
22	Butyric acid	12.1	15.3
23	Carbon dioxide	11.6	0.3
24	Carbon disulfide	16.1	7.5
25	Carbon tetrachloride	12.7	13.1
26	Chlorobenzene	12.3	12.4
27	Chloroform	14.4	10.2
28	Chlorosulfonic acid	11.2	18.1
	Chlorotoluene:		
29	Ortho	13.0	13.3
30	Meta	13.3	12.5
31	Para	13.3	12.5
32	Cresol, meta	2.5	20.8
33	Cyclohexanol	2.9	24.3
34	Dibromoethane	12.7	15.8
35	Dichloroethane	13.2	12.2
36	Dichloromethane	14.6	8.9
37	Diethyl oxalate	11.0	16.4
38	Dimethyl oxalate	12.3	15.8
39	Diphenyl	12.0	18.3
40	Dipropyl oxalate	10.3	17.7
41	Ethyl acetate	13.7	9.1
	Ethyl alcohol:		
42	100%	10.5	13.8
43	95%	9.8	14.3
44	40%	6.5	16.6
45	Ethyl benzene	13.2	11.5
46	Ethyl bromide	14.5	8.1
47	Ethyl chloride	14.8	6.0
48	Ethyl ether	14.5	5.3
49	Ethyl formate		
50	Ethyl iodide	14.7	10.3
51	Ethylene glycol	6.0	23.6
52	Formic acid	10.7	15.8
53	Freon-11	14.4	9.0
54	Freon-12	16.8	5.6
55	Freon-21	15.7	7.5
56	Freon-22	17.2	4.7
57	Freon-13	12.5	11.4
	Glycerol:		
58	100%	2.0	30.0
59	50%	6.9	19.6
60	Heptene	14.1	8.4
61	Hexane	14.7	7.0
62	Hydrochloric acid, 31.5%	13.0	16.6
63	Isobutyl alcohol	7.1	18.0
64	Isobutyric acid	12.2	14.4
65	Isopropyl alcohol	8.2	16.0
66	Kerosene	10.2	16.9
67	Linseed oil, raw	7.5	27.2
68	Mercury	18.4	16.4
	Methanol:		
69	100%	12.4	10.5
70	90%	12.3	11.8
71	40%	7.8	15.5
72	Methyl acetate	14.2	8.2
73	Methyl chloride	15.0	3.8
74	Methyl ethyl ketone	13.9	8.6
75	Naphthalene	7.9	18.1
	Nitric acid:		
76	95%	12.8	13.8
77	60%	10.8	17.0
78	Nitrobenzene	10.6	16.2
79	Nitrotoluene	11.0	17.0
80	Octane	13.7	10.0
81	Octyl alcohol	6.6	21.1
82	Pentachloroethane	10.9	17.3
83	Pentane	14.9	5.2
84	Phenol	6.9	20.8
85	Phosphorus tribromide	13.8	16.7
86	Phosphorus trichloride	16.2	10.9
87	Propionic acid	12.8	13.8
88	Propyl alcohol	9.1	16.5
89	Propyl bromide	14.5	9.6
90	Propyl chloride	14.4	7.5
91	Propyl iodide	14.1	11.6
92	Sodium	16.4	13.9
93	Sodium hydroxide, 50%	3.2	25.8
94	Stannic chloride	13.5	12.8
95	Sulfur dioxide	15.2	7.1
	Sulfuric acid:		
96	110%	7.2	27.4
97	98%	7.0	24.8
98	60%	10.2	21.3
99	Sulfuryl chloride	15.2	12.4
100	Tetrachloroethane	11.9	15.7
101	Tetrachloroethylene	14.2	12.7
102	Titanium tetrachloride	14.4	12.3
103	Toluene	13.7	10.4
104	Trichloroethylene	14.8	10.5
105	Turpentine	11.5	14.9
106	Vinyl acetate		
107	Water	10.2	13.0
	Xylene:		
108	Ortho	13.5	12.1
109	Meta	13.9	10.6
110	Para	13.9	10.9

TABLE 22 Data for Fig. 13

Percentage of roughness	For value of f see curve	Diameter (actual of drawn tubing, nominal of standard-weight pipe)											
		Drawn tubing, brass, tin, lead, glass		Clean steel, wrought iron		Clean, galvanized		Best cast iron		Average cast iron		Heavy riveted, spiral riveted	
		in	mm	in	mm	in	mm	in	mm	in	mm	in	mm
0.2	1	0.35 up	8.89 up	72	1829	—	—	—	—	—	—	—	—
1.35	4	—	—	6–12	152–305	10–24	254–610	20–48	508–1219	42–96	1067–2438	84–204	2134–5182
2.1	5	—	—	4–5	102–127	6–8	152–203	12–16	305–406	24–36	610–914	48–72	1219–1829
3.0	6	—	—	2–3	51–76	305	76–127	5–10	127–254	10–20	254–508	20–42	508–1067
3.8	7	—	—	1½	38	2½	64	3–4	76–102	6–8	152–203	16–18	406–457
4.8	8	—	—	1–1¼	25–32	1½–2	38–51	2–2½	51–64	4–5	102–127	10–14	254–356
6.0	9	—	—	¾	19	1¼	32	1½	38	3	76	8	203
7.2	10	—	—	½	13	1	25	1¼	32	—	—	5	127
10.5	11	—	—	⅜	9.5	¾	19	1	35	—	—	4	102
14.5	12	—	—	¼	6.4	½	13	—	—	—	—	3	76
24.0	14	0.125	3.18	—	—	⅜	9.5	—	—	—	—	—	—
31.5	16	—	—	—	—	¼	6.4	—	—	—	—	—	—
37.5	18	0.0625	1.588	—	—	⅛	3.2	—	—	—	—	—	—

Use this procedure for any petroleum product—crude oil, kerosene, benzene, gasoline, naphtha, fuel oil, Bunker C, diesel oil toluene, etc. The tables and charts presented here and in the *Pipe Friction Manual* save computation time.

FLOW RATE AND PRESSURE LOSS IN COMPRESSED-AIR AND GAS PIPING

Dry air at 80°F (26.7°C) and 150 lb/in^2 (abs) (1034 kPa) flows at the rate of 500 ft^3/min (14.2 m^3/min) through a 4-in (101.6-mm) schedule 40 pipe from the discharge of an air compressor. What are the flow rate in lb/h and the air velocity in ft/s? Using the Fanning formula, determine the pressure loss if the total equivalent length of the pipe is 500 ft (152.4 m).

Calculation Procedure:

1. *Determine the density of the air or gas in the pipe*

For air or a gas, $pV = MRT$, where p = absolute pressure of the gas, lb/ft^2 (abs); V = volume of M lb of gas, ft^3; M = weight of gas, lb; R = gas constant, ft·lb/(lb·°F); T = absolute temperature of the gas, °R. For this installation, using 1 ft^3 of air, $M = pV/(RT)$, $M = (150)(144)/[(53.33)(80 + 459.7)] = 0.754$ lb/ft^3 (12.1 kg/m^3). The value of R in this equation was obtained from Table 23.

2. *Compute the flow rate of the air or gas*

For air or a gas, the flow rate W_h lb/h = (60) (density, lb/ft^3)(flow rate, ft^3/min); or, $W_h = (60)(0.754)(500) = 22{,}620$ lb/h (10,179 kg/h).

3. *Compute the velocity of the air or gas in the pipe*

For any air or gas pipe, velocity of the moving fluid v ft/s $= 183.4\ W_h/3600\ d^2\rho$, where d = internal diameter of pipe, in; ρ = density of fluid, lb/ft^3. For this system, $v = (183.4)(22{,}620)/[(3600)(4.026)^2(0.754)] = 95.7$ ft/s (29.2 m/s).

4. *Compute the Reynolds number of the air or gas*

The viscosity of air at 80°F (26.7°C) is 0.0186 cP, obtained from Crocker and King—*Piping Handbook*, Perry et al.—*Chemical Engineers' Handbook*, or a similar reference. Then, by using the Reynolds number relation given in Table 20, $R = 6.32W/(dz) = (6.32)(22{,}620)/[(4.026)(0.0186)] = 3{,}560{,}000$.

TABLE 23 Gas Constants

Gas	R ft·lb/(lb·°F)	R J/(kg·K)	C for critical-velocity equation
Air	53.33	286.9	2870
Ammonia	89.42	481.1	2080
Carbon dioxide	34.87	187.6	3330
Carbon monoxide	55.14	296.7	2820
Ethane	50.82	273.4	
Ethylene	54.70	294.3	2480
Hydrogen	767.04	4126.9	750
Hydrogen sulfide	44.79	240.9	
Isobutane	25.79	138.8	
Methane	96.18	517.5	2030
Natural gas	—	—	2070–2670
Nitrogen	55.13	296.6	2800
n-butane	25.57	137.6	
Oxygen	48.24	259.5	2990
Propane	34.13	183.6	
Propylene	36.01	193.7	
Sulfur dioxide	23.53	126.6	3870

5. *Compute the pressure loss in the pipe*

Using Fig. 13 or the Hydraulic Institute *Pipe Friction Manual,* we get $f = 0.0142$ to 0.0162 for a 4-in (101.6-mm) schedule 40 pipe when the Reynolds number = 3,560,000. From the Fanning formula from an earlier calculation procedure and the higher value of f, $p_d = 3.36(10^{-6})fLW^2/d^5\rho$, or $p_d = 3.36(10^{-6})(0.0162)(500)(22{,}620)^2/[(4.026)^5(0.754)] = 17.52$ lb/in^2 (120.8 kPa).

Related Calculations: Use this procedure to compute the pressure loss, velocity, and flow rate in compressed-air and gas lines of any length. Gases for which this procedure can be used include ammonia, carbon dioxide, carbon monoxide, ethane, ethylene, hydrogen, hydrogen sulfide, isobutane, methane, nitrogen, n-butane, oxygen, propane, propylene, and sulfur dioxide

Alternate relations for computing the velocity of air or gas in a pipe are $v = 144W_s/a\rho$; $v = 183.4W_s/d^2\rho$; $v = 0.0509W_s v_g/d^2$, where W_s = flow rate, lb/s; a = cross-sectional area of pipe, in^2, v_g = specific volume of the air or gas at the operating pressure and temperature, ft^3/lb.

FLOW RATE AND PRESSURE LOSS IN GAS PIPELINES

Using the Weymouth formula, determine the flow rate in a 10-mi (16.1-km) long 4-in (101.6-mm) schedule 40 gas pipeline when the inlet pressure is 200 lb/in^2 (gage) (1378.8 kPa), the outlet pressure is 20 lb/in^2 (gage) (137.9 kPa), the gas has a specific gravity of 0.80, a temperature of 60°F (15.6°C), and the atmospheric pressure is 14.7 lb/in^2 (abs) (101.34 kPa).

Calculation Procedure:

1. *Compute the flow rate from the Weymouth formula*

The Weymouth formula for flow rate is $Q = 28.05[(p_i^2 - p_0^2)d^{5.33}/sL]^{0.5}$, where p_i = inlet pressure, lb/in^2 (abs); p_0 = outlet pressure, lb/in^2 (abs); d = inside diameter of pipe, in; s = specific gravity of gas; L = length of pipeline, mi. For this pipe, $Q = 28.05 \times [(214.7^2 - 34.7^2)4.026^{5.33}/0.8 \times 10]^{0.5} = 86{,}500$ lb/h (38,925 kg/h).

2. *Determine if the acoustic velocity limits flow*

If the outlet pressure of a pipe is less than the critical pressure p_c lb/in^2 (abs), the flow rate in the pipe cannot exceed that obtained with a velocity equal to the critical or acoustic velocity, i.e., the velocity of sound in the gas. For any gas, $p_c = Q(T_i)^{0.5}/d^2C$, where T_i = inlet temperature, °R; C = a constant for the gas being considered.

Using $C = 2070$ from Table 23, or Crocker and King—*Piping Handbook*, $p_c = (86{,}500)(60 + 460)^{0.5}/[(4.026)^2(2070)] = 58.8$ lb/in^2 (abs) (405.4 kPa). Since the outlet pressure $p_0 = 34.7$ lb/in^2 (abs) (239.2 kPa), the critical or acoustic velocity limits the flow in this pipe because $p_c > p_0$. When $p_c < p_0$, critical velocity does not limit the flow.

Related Calculations: Where a number of gas pipeline calculations must be made, use the tabulations in Crocker and King—*Piping Handbook* and Bell—*Petroleum Transportation Handbook*. These tabulations will save much time. Other useful formulas for gas flow include the Panhandle, Unwin, Fritsche, and rational. Results obtained with these formulas agree within satisfactory limits for normal engineering practice.

Where the outlet pressure is unknown, assume a value for it and compute the flow rate that will be obtained. If the computed flow is less than desired, check to see that the outlet pressure is less than the critical. If it is, increase the diameter of the pipe. Use this procedure for natural gas from any gas field, manufactured gas, or any other similar gas.

To find the volume of gas that can be stored per mile of pipe, solve $V_m = 1.955p_m d^2K$, where p_m = mean pressure in pipe, lb/in^2 (abs) $\approx (p_i + p_0)/2$; $K = (1/Z)^{0.5}$, where Z = supercompressibility factor of the gas, as given in Baumeister and Marks—*Standard Handbook for Mechanical Engineers* and Perry—*Chemical Engineer's Handbook*. For exact computation of p_m, use $p_m = (2/3)(p_i + p_0 - p_i p_0/p_i + p_0)$.

SELECTING HANGERS FOR PIPES AT ELEVATED TEMPERATURES

Select the number, capacity, and types of pipe hanger needed to support the 6-in (152.5-mm) schedule 80 pipe in Fig. 14 when the installation temperature is 60°F (15.6°C) and the operating temperature is 700°F (371.1°C). The pipe is insulated with 85 percent magnesia weighing 11.4

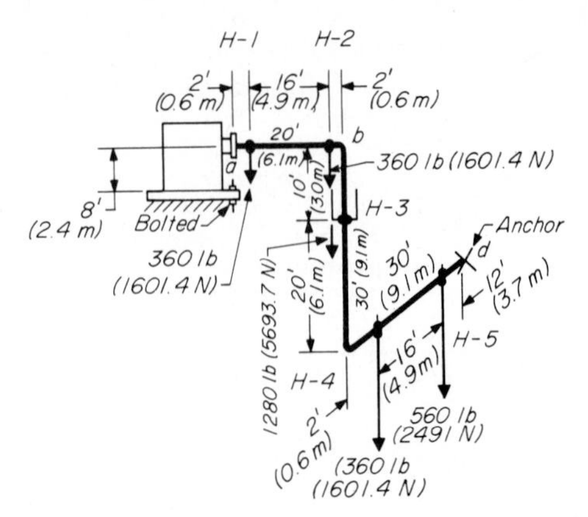

FIG. 14 Typical complex pipe operating at high temperature.

lb/ft (16.63 N/m). The pipe and unit served by the pipe have a coefficient of thermal expansion of 0.0575 in/ft (0.48 cm/m) between the 60°F (15.6°C) installation temperature and the 700°F (371.1°C) operating temperature.

Calculation Procedure:

1. *Draw a freehand sketch of the pipe expansion*

Use Fig. 15 as a guide and sketch the expanded pipe, using a dashed line. The sketch need not be exactly to scale; if the proportions are accurate, satisfactory results will be obtained. The shapes shown in Fig. 15 cover the 11 most common situations met in practice.

2. *Tentatively locate the required hangers*

Begin by locating hangers *H*-1 and *H*-5 close to the supply and using units, Fig. 14. Keeping a hanger close to each unit (boiler, turbine, pump, engine, etc.) prevents overloading the connection on the unit.

Space intermediate hangers *H*-2, *H*-3, and *H*-4 so that the recommended distances in Table 24 or hanger engineering data (e.g., Grinnell Corporation *Pipe Hanger Design and Engineering*) are not exceeded. Indicate the hangers on the piping drawing as shown in Fig. 14.

3. *Adjust the hanger locations to suit structural conditions*

Study the building structural steel in the vicinity of the hanger locations, and adjust these locations so that each hanger can be attached to a support having adequate strength.

4. *Compute the load each hanger must support*

From a table of pipe properties, such as in Crocker and King—*Piping Handbook*, find the weight of 6-in (152.4-mm) schedule 80 pipe as 28.6 lb/ft (41.7 N/m). The insulation weighs 11.4 lb/ft (16.6 N/m), giving a total weight of insulated pipe of 28.6 + 11.4 = 40.0 lb/ft (58.4 N/m).

Compute the load on the hangers supporting horizontal pipes by taking half the length of the pipe on each side of the hanger. Thus, for hanger *H*-1, there is (2 ft)(½) + (16 ft) × (½) = 9 ft (2.7 m) of horizontal pipe, Fig. 14, which it supports. Since this pipe weighs 40 lb/ft (58.4 N/m), the total load on hanger *H*-1 = (9 ft)(40 lb/ft) = 360 lb (1601.4 N). A similar analysis for hanger *H*-2 shows that it supports (8 + 1)(40) = 360 lb (1601.4 N).

Hanger *H*-3 supports the entire weight of the vertical pipe, 30 ft (9.14 m), plus 1 ft (0.3 m) at the top bend and 1 ft (0.3 m) at the bottom bend, or a total of 1 + 30 + 1 = 32 ft (9.75 m). The total load on hanger *H*-3 is therefore (32)(40) = 1280 lb (5693.7 N).

Hanger *H*-4 supports (1 + 8)(40) = 360 lb (1601.4 N), and hanger *H*-5 supports (8 + 6)(40) = 560 lb (2491 N).

As a check, compute the total weight of the pipe and compare it with the sum of the endpoint

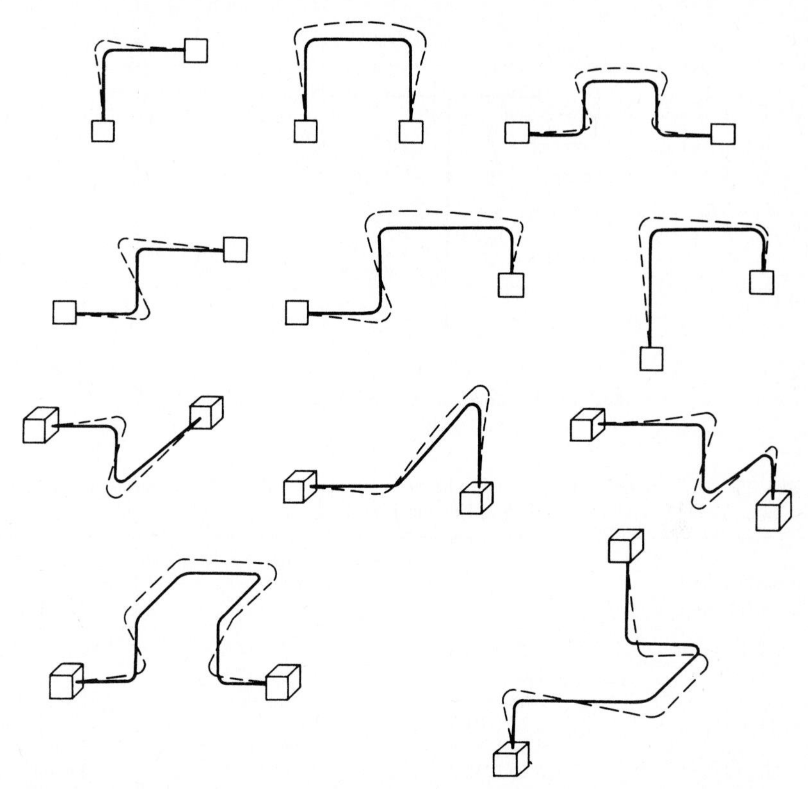

FIG. 15 Pipe shapes commonly used in power and process plants assume the approximate forms shown by the dotted lines when the pipe temperature rises. *(Power.)*

and hanger loads. Thus, there is 100 ft (30.5 m) of pipe weighing (80)(40) = 3200 lb (14.2 kN). The total load the hangers will support is 360 + 360 + 1280 + 360 + 560 = 2920 lb (12.9 kN). The first endpoint will support (1)(40) = 40 lb (177.9 N), and the anchor will support (6)(40) = 240 lb (1067 N). The total hanger and endpoint support = 2920 + 40 + 240 = 3200 lb (14.2 kN); therefore, the pipe weight = the hanger load.

5. *Sketch the shape of the hot pipe*

Use Fig. 15 as a guide, and draw a dotted outline of the approximate shape the pipe will take when hot. Start with the first corner point nearest the unit on the left, Fig. 16. This point will move away from the unit, as in Fig. 16. Do the same for the first corner point near the other unit served by the pipe and for intermediate corner points. Use arrows to indicate the probable direction of pipe movement at each corner. When sketching the shape of the hot pipe, remember that a straight pipe expanding against a piece of pipe at right angles to itself will bend the latter. The

TABLE 24 Maximum Recommended Spacing between Pipe Hangers

Nominal pipe size, in (mm)	4 (101.6)	5 (127)	6 (152.4)	8 (203.2)	10 (254)	12 (304.8)
Maximum span, ft (m)	14 (4.3)	16 (4.9)	17 (5.2)	19 (5.8)	22 (6.7)	23 (7.0)

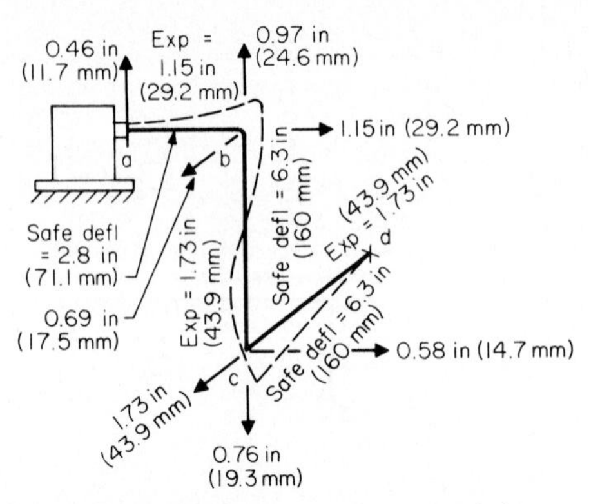

FIG. 16 Expansion of the various parts of the pipe shown in Fig. 14. *(Power.)*

TABLE 25 Deflection, in (mm), that Produces 14,000-lb/in² (96,530-kPa) Tensile Stress in Pipe Legs Acting as a Cantilever Beam, Load at Free End

Cantilever length, ft (m)	Nominal pipe size, in (mm)		
	4 (101.6)	6 (152.4)	8 (203.2)
5 (1.5)	0.26 (6.6)	0.17 (4.3)	0.13 (3.3)
10 (3.0)	1.03 (26.2)	0.70 (17.8)	0.54 (13.7)
15 (4.6)	2.32 (58.9)	1.58 (40.1)	1.21 (30.7)
20 (6.1)	4.12 (104.6)	2.80 (71.1)	2.15 (54.6)
25 (7.6)	6.44 (163.6)	4.38 (111.3)	3.35 (85.1)
30 (9.1)	9.26 (235.2)	6.30 (160.0)	4.83 (122.7)

distance that various lengths of pipe will bend while producing a tensile stress of 14,000 lb/in² (96.5 MPa) is given in Table 25. This stress is a typical allowable value for pipes in industrial systems.

6. *Determine the thermal movement of units served by the pipe*

If either or both fixed units (boiler, turbine, etc.) operate at a temperature above or below atmospheric, determine the amount of movement at the flange of the unit to which the piping connects, using the thermal data in Table 26. Do this by applying the thermal expansion coefficient for the metal of which the unit is made. Determine the vertical and horizontal distance of the flange face from the point of no movement of the unit.

The point of no movement is the point or surface where the unit is fastened to *cold* structural steel or concrete.

The flange, point *a*, Fig. 16, is 8 ft (2.4 m) above the bolted end of the unit and directly in line with the bolt, Fig. 14. Since the bolt and flange are on a common vertical line, there will not be any *horizontal* movement of the flange because the bolt is the no-movement point of the unit.

Since the flange is 8 ft (2.4 m) away from the point of no movement, the amount that the flange will move = (distance away from the point of no movement, ft)(coefficient of ther-

TABLE 26 Thermal Expansion of Pipe, in/ft (mm/m) (Carbon and Carbon-Moly Steel and WI)

Operating temperature, °F (°C)	Installation temperature	
	32°F (0°C)	60°F (15.6°C)
600 (316)	0.050 (4.17)	0.0475 (3.96)
650 (343)	0.055 (4.58)	0.0525 (4.38)
700 (371)	0.060 (5.0)	0.0575 (4.79)
750 (399)	0.065 (5.42)	0.0624 (5.2)
800 (427)	0.070 (5.83)	0.0674 (5.62)

mal expansion, in/ft) = (8)(0.0575) = 0.46 in (11.7 mm) *away* (up) from the point of no movement. If the unit were operating at a temperature *less than* atmospheric, it would contract and the flange would move *toward* (down) the point of no movement. Mark the flange movement on the piping sketch, Fig. 16.

Anchor *d*, Fig. 16, does not move because it is attached to either cold structural steel or concrete.

7. *Compute the amount of expansion in each pipe leg*

Expansion of the pipe, in = (pipe length, ft)(coefficient of linear expansion, in/ft). For length *ab*, Fig. 14, the expansion = (20)(0.0575) = 1.15 in (29.2 mm); for *bc*, (30)(0.0575) = 1.73 in (43.9 mm); for *cd*, (30)(0.0575) = 1.73 in (43.9 mm). Mark the amount and direction of expansion on Fig. 16.

8. *Determine the allowable deflection for each pipe leg*

Enter Table 25 at the nominal pipe size and find the allowable deflection for a 14,000-lb/in^2 (96.5-MPa) tensile stress for each pipe leg. Thus, for *ab*, the allowable deflection = 2.80 in (71.1 mm) for a 20-ft (6.1-m) long leg; for *bc*, 6.30 in (160 mm) for a 30-ft (9.1-m) long leg; for *cd*, 6.30 in (160 mm) for a 30-ft (9.1-m) long leg. Mark these allowable deflections on Fig. 16, using dashed arrows.

9. *Compute the actual vertical and horizontal deflections*

Sketch the vertical deflection diagram, Fig. 17*a*, by drawing a triangle showing the total expansion in each direction in proportion to the length of the parts at right angles to the expansion. Thus,

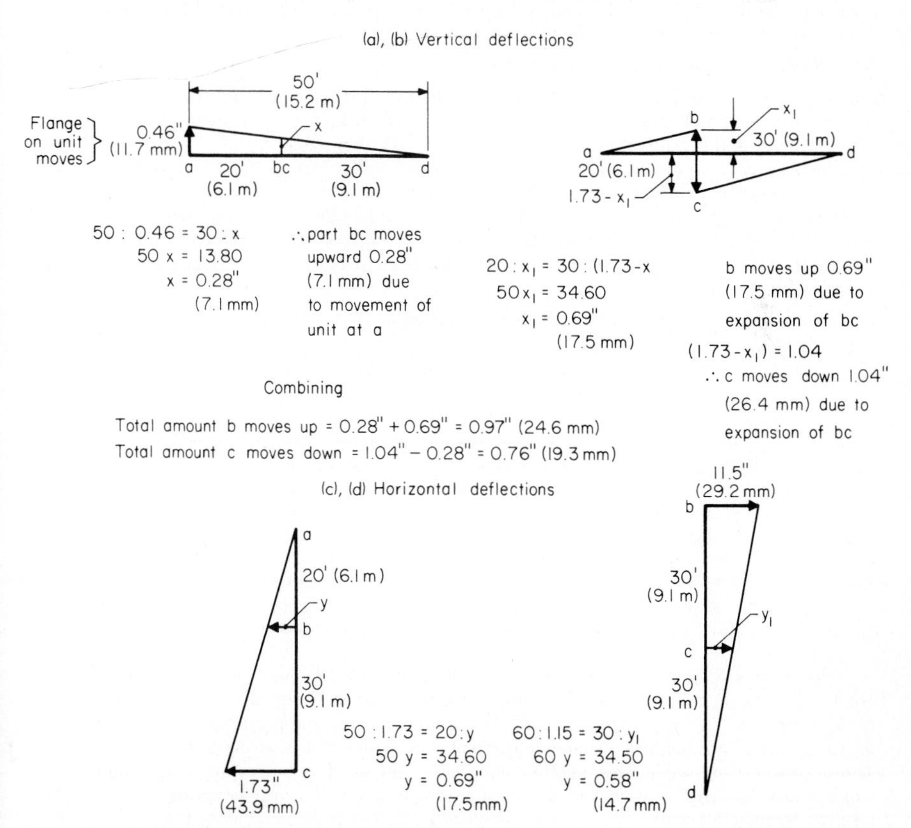

FIG. 17 (*a*), (*b*) Vertical deflection diagrams for the pipe in Fig. 14; (*c*), (*d*) horizontal deflection diagrams for the pipe in Fig. 14. *(Power.)*

the 0.46-in (11.7-mm) upward expansion at the flange, *a*, is at right angles to leg *ab* and is drawn as the altitude of the right triangle. Lay off 20 ft (6.1 m), *ab*, on the base of the triangle. Since *bc* is parallel to the direction of the flange movement, it is shown as a point, *bc*, on the base of this triangle. From point *bc*, lay off *cd* on the base of the triangle, Fig. 17*a*, since it is at right angles to the expansion of point *a*. Then, by similar triangles, $50:46 = 30:x$; $x = 0.28$ in (7.1 mm). Therefore, leg *bc* moves upward 0.28 in (7.1 mm) because of the flange movement at *a*.

Now draw the deflection diagram, Fig. 17*b*, showing the upward movement of leg *ab* and the downward movement of leg *cd* along the length of each leg, or 20 and 30 ft (6.1 and 9.1 m), respectively. Solve the similar triangles, or $20:x_1 = 30:(1.73 - x_1)$; $x_1 = 0.69$ in (17.5 mm). Therefore, point *b* moves *up* 0.69 in (17.5 mm) as a result of the expansion of leg *bc*. Then $1.73 - x_1 = 1.73 - 0.69 = 1.04$ in (26.4 mm). Thus, point *c* moves *down* 1.04 in (26.4 mm) as a result of the expansion of *bc*. The total distance *b* moves up $= 0.28 + 0.69 = 0.97$ in (24.6 mm), whereas the total distance *c* moves down $= 1.04 - 0.28 = 0.76$ in (19.3 mm). Mark these actual deflections on Fig. 16.

Find the actual horizontal deflections in a similar fashion by constructing the triangle, Fig. 17*c*, formed by the vertical pipe *bc* and the horizontal pipe *ab*. Since point *a* does not move horizontally but point *b* does, lay off leg *ab* at right angles to the direction of movement, as shown. From point *b* lay off leg *bc*. Then, since leg *bc* expands 1.73 in (43.9 mm), lay this distance off perpendicular to *ac*, Fig. 17*c*. By similar triangles, $20 + 30:1.73 = 20:y$; $y = 0.69$ in (17.5 mm). Hence, point *b* deflects 0.69 in (17.5 mm) in the direction shown in Fig. 16.

Follow the same procedure for leg *cd*, constructing the triangle in Fig. 17*d*. Beginning with point *b*, lay off legs *bc* and *cd*. The altitude of this right triangle is then the distance point *c* moves when leg *ab* expands, or 1.15 in (29.2 mm). By similar triangles, $30 + 30:1.15 = 30:y_1$; $y_1 =$ deflection of point $c = 0.58$ in (14.7 mm).

10. *Select the type of pipe hanger to use*

Figure 18 shows several popular types of pipe hangers, together with the movements that they are designed to absorb. For hangers *H*-1 and *H*-2, use type *E*, Fig. 18, because the pipe moves both vertically and horizontally at these points, as Fig. 17 shows. Use type *F*, Fig. 18, for hanger *H*-3, because riser *bc* moves both vertically and horizontally. Hangers *H*-4 and *H*-5 should be type *E*, because they must absorb both horizontal and vertical movements.

Once the hangers are selected from Fig. 18, refer to hanger engineering data for the exact design details of the hangers that will be selected. During the study of the data, look for other hangers that absorb the same movement or movements but may be more adaptable to the existing structural steel conditions.

11. *Select the hanger-rod diameter for each hanger*

Use Table 27 to find the required hanger-rod diameter. Since the pipe operates at 700°F (371°C), select the maximum safe load from the 750°F (399°C) column. Tabulate the loads and diameters as follows:

Hanger	Load, lb (kN)	Rod diameter, in (mm)
H-1	360 (1.6)	⅜ (9.5)
H-2	360 (1.6)	⅜ (9.5)
H-3	1280 (5.7)	2–½ each (2–12.7)
H-4	360 (1.6)	⅜ (9.5)
H-5	560 (2.5)	½ (12.7)

Select standard springs for spring-loaded hangers from pipe-hanger engineering data. Springs are listed in the data on the basis of loading per inch of travel. For small movements [less than 1 in (25.4 mm)], it is generally desirable to select a lighter spring and precompress it at installation so that it has a light loading. Hanger movement will then load the spring to the desired value. This approach is desirable from another standpoint: any error in estimating hanger movement will not cause as large an unbalanced load on the pipe as would a heavier spring with a greater loading per inch of travel.

Related Calculations: Use this procedure for any type of pipe operating at elevated temperature—steam, oil, water, gas, etc.—serving a load in a power plant, process plant, ship, barge,

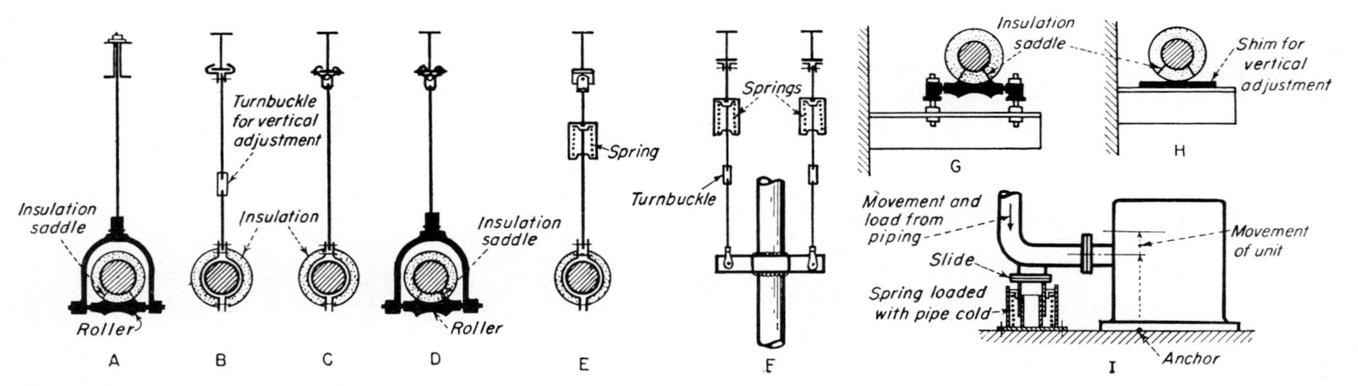

FIG. 18 Pipe hangers chosen depend on the movement expected. Hangers *A* and *B* are suitable for pipe movement in one horizontal direction. Hangers *C* and *D* permit pipe movement in two horizontal directions. Vertical and horizontal movement requires use of hangers such as *E* for horizontal pipes and *F* for vertical pipes. (*G*) Cantilever support; (*H*) sliding movement in two horizontal directions; (*I*) base elbow support.

TABLE 27 Hanger-Rod Load-Carrying Capacity (Hot-Rolled Steel Rod)

Nominal diameter of rod, in (mm)	Thread root area, in² (mm²)	Maximum safe load on rod, lb (kN), at rod temperature of: 450°F (232°C)	750°F (399°C)
⅜ (9.5)	0.068 (43.9)	610 (2.7)	510 (2.3)
½ (12.7)	0.126 (81.3)	1130 (5.0)	940 (4.2)
⅝ (15.9)	0.202 (130.3)	1810 (8.1)	1510 (6.7)
¾ (19.1)	0.302 (194.8)	2710 (12.1)	2260 (10.1)
⅞ (22.2)	0.419 (270.3)	3770 (16.8)	3150 (14.0)
1 (25.4)	0.552 (356.1)	4960 (22.1)	4150 (18.5)

aircraft, or other type of installation. In piping systems having very little or no increase in temperature during operation, the steps for computing the expansion can be eliminated. In this type of installation, the weight of the piping is the primary consideration in the choice of the hangers.

If desired, hanger loads can also be determined by taking moments about an arbitrarily selected axis on either side of the hanger. This method gives the same results as the procedure used above. The weight of bends is assumed to be concentrated at the center of gravity of each bend, whereas the weight of valves is assumed to be concentrated at the vertical centerline of the valve. Figure 19 shows typical moment arms, *a* and *c*, for valves and other fittings. The moment

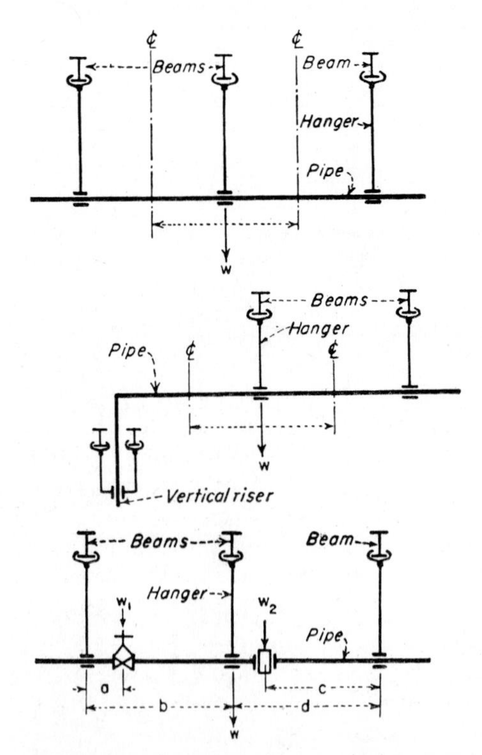

FIG. 19 Compute hanger loads of uniformly loaded pipes as shown. Use beam relations for concentrated loads. *(Power.)*

for W_1 about the hanger to the left of it is W_1a, and the moment for W_2 is W_2c about the hanger to the right of it. The weight of the pipe is assumed to be concentrated at a point midway between the hangers, and the moment is (weight of pipe, lb)(distance between hangers, ft/2). The method given here was developed by Frank Kamarck, Mechanical Engineer, and reported in *Power* magazine.

HANGER SPACING AND PIPE SLOPE FOR AN ALLOWABLE STRESS

An 8-in (203.2-mm) schedule 40 water pipe has an allowable bending stress of 10,000 lb/in² (68,950 kPa). What is the maximum allowable distance between hangers for this pipe? What slope will the allowable hanger span require to prevent pocketing of water in the pipe? Describe the method for computing hanger span and pipe slope for empty pipe. How are hanger distances computed when the pipe contains concentrated loads?

Calculation Procedure:

1. *Compute the allowable span between hangers*

For a pipe filled with water, $S = WL^2/8m$, where S = bending stress in pipe, lb/in²; W = weight of pipe and water lb/lin in; L = maximum allowable distance between hangers, in; m = section modulus of pipe, in³. By using a table of pipe properties, as in Crocker and King—*Piping Handbook*, $L = (8mS/W)^{0.5} = (8 \times 16.81 \times 10{,}000/4.18)^{0.5} = 568$ in, or $568/12 = 47.4$ ft (14.5 m).

2. *Compute the pipe slope required by the span*

To prevent pocketing of water or condensate at the low point in the pipe, the pipe must be pitched so that the outlet is lower than the lowest point in the span. When the pipe has no concentrated loads—such as valves, cross connections, or meters—the deflection of the pipe is y in $= 22.5wl^4/(EI)$, where w = weight of the pipe and its contents, lb/ft; l = distance between hangers, ft; E = modulus of elasticity of pipe, lb/in² $= 30 \times 10^6$ for steel; I = moment of inertia of the pipe, in⁴. Substituting values gives $y = (22.5)(50.24)(47.4)^4/[(30 \times 10^6)(72.5)] = 2.61$ in (66.3 mm).

With the deflection y known, the pipe slope, expressed as 1 in (25.4 mm) per G ft of pipe length, is 1 in (25.4 mm) per G ft $= \frac{1}{4}y$, or $G = (47.4)/[(4)(2.61)] = 4.53$. Thus, a pipe slope of 1 in (25.4 mm) in 4.53 ft (1.38 m) is necessary to prevent pocketing of the water when the hanger span is 47.4 ft (14.5 m). With this slope, the outlet of the pipe would be $47.4/4.53 = 10.45$ in (265.4 mm) below the inlet.

3. *Compute the empty-pipe hanger span and pipe slope*

Use the same procedure as in steps 1 and 2, except that the empty weight of the pipe is substituted in the equations instead of the weight of the pipe when full of water. For pipes containing steam, gas, or vapor, compute the flowing-fluid weight and add it to the pipe weight. Follow the same procedure for insulated pipes, adding the insulation weight to the pipe weight.

4. *Determine the hanger span and slope with concentrated loads*

Hanger span and pipe slope can be computed from standard beam relations. However, most piping designers use the deflection chart and deflection factors for concentrated loads in Crocker and King—*Piping Handbook*. The chart and correction factors simplify the calculations considerably. The computation involves only simple multiplication and division.

Related Calculations: Use this procedure for piping in any type of installation—power, process, marine, industrial, or utility—for any type of liquid, vapor, or gas.

EFFECT OF COLD SPRING ON PIPE ANCHOR FORCES AND STRESSES

A carbon molybdenum pipe operates at 800°F (427°C) and has an anchor force of 5000 lb (22.2 kN) and a maximum bending stress s_b of 15,000 lb/in² (103.4 MPa) without cold spring. Compute the anchor force and bending stress in the hot and cold condition when the pipe is cold-sprung an amount equal to the expansion e and $0.5e$. The total expansion of the pipe is 24 in (609.6 mm).

Calculation Procedure:

1. *Compute the hot-condition force and stress*

The allowable cold-spring adjustment is expressed as a ratio $(e - 2S/3)/e$, where e = the total expansion of the pipe, in; S = cold-spring distance, in. This ratio is multiplied by the original anchor force and bending stress at the maximum operating temperature *without* cold spring to find the anchor force and bending stress *with* cold spring in the hot condition. If the ratio is less than 2/3, the value of 2/3 is used where maximum credit for cold spring is desired.

For this pipe, with maximum cold spring, the ratio = $(24 - 2 \times 24/3)/24 = 1/3$. Since this is less than 2/3, use 2/3. Then, the anchor force $F = (2/3)(5000) = 3333$ lb (14.8 kN), and the bending stress $s_b = (2/3)(15{,}000) = 10{,}000$ lb/in^2 (68.9 MPa).

With $S = 0.5e = (0.5)(24) = 12$ in (304.8 mm), the ratio = $(24 - 2 \times 12/3)/24 = 2/3$. Hence, $F = (2/3)(5000) = 3333$ lb (14.8 kN); $s_b = (2/3)(15{,}000) = 10{,}000$ lb/in^2 (68.9 MPa).

2. *Compute the cold-condition force and stress*

For the cold condition, the adjustment ratio = $-S/eM_R$, where M_R = modulus ratio for the pipe material = modulus of elasticity, lb/in^2, of the pipe material at the operating temperature, °F/ modulus of elasticity of the pipe material, lb/in^2, at 70°F (21.1°C). For this pipe, $M_R = 0.865$, from a table of pipe properties. The minus sign in the ratio indicates that the anchor force and stress are reversed in the cold condition as compared with the hot condition.

For this pipe, with maximum cold spring, the ratio = $-24/[(24)(0.865)] = -1.156$. Then the anchor force in the cold condition = $(-1.156)(15{,}000) = -5790$ lb (25.7 kN), and the bending stress = $(-1.156)(15{,}000) = -17{,}350$ lb/in^2 (119.6 MPa).

With $S = 0.5e = (0.5)(24) = 12$ in (304.8 mm), the ratio = $-12/[(24)(0.865)] = -0.578$. Then the anchor force in the cold condition $(-0.578)(5000) = -2895$ lb (12.9 kN), and the bending stress = $(-0.578)(15{,}000) = -\ 8670$ lb/in^2 (59,771 kPa).

These calculations show that cold spring reduces the anchor force and bending stress when the pipe is in the hot condition, step 1. With a cold spring of one-half the pipe expansion, the anchor force and bending stress are reduced and reversed when in the cold condition, step 2. When the cold spring equals the expansion, the anchor force and bending stress increase in the cold condition, step 2.

Related Calculations: Use this procedure for a pipe conveying steam, oil, gas, water, and similar vapors, liquids, and gases.

REACTING FORCES AND BENDING STRESS IN SINGLE-PLANE PIPE BEND

Determine the horizontal and vertical reacting forces in the single-plane pipe bend of Fig. 20 if the pipe is 6-in (152.4-mm) schedule 40 carbon steel A106 seamless operating at 500°F (260°C). What is the maximum bending stress in the pipe and the resultant reacting or anchor force? Determine the maximum bending stress if a long-radius welded elbow is used at point C, Fig. 20. Use the tabular method of solution.

Calculation Procedure:

1. *Compute the horizontal reacting or anchor force*

Several methods are available for determining the reacting or anchor forces and maximum bending stress in a single-plane pipe bend. Crocker and King—*Piping Handbook* presents simplified, analytical, and graphical methods for computing forces and stresses in single- and multiplane piping systems. Another useful reference, *Design of Piping Systems*, written by members of the engineering departments of the M. W. Kellogg Company, presents both simplified and analytical methods and an excellent history and discussion of piping flexibility analysis. Probably the simplest method for routine piping flexibility analyses is that developed by the Grinnell Company, Inc., and S. W. Spielvogel. This method uses tabulated constants for specific pipe shapes in one, two, and three planes. It is satisfactory for the majority of piping problems met in normal engineering practice. To assist the practicing engineer, a number of Grinnell-Spielvogel tabulations for common pipe shapes are included here. For uncommon pipe shapes, refer to Grinnell Company—

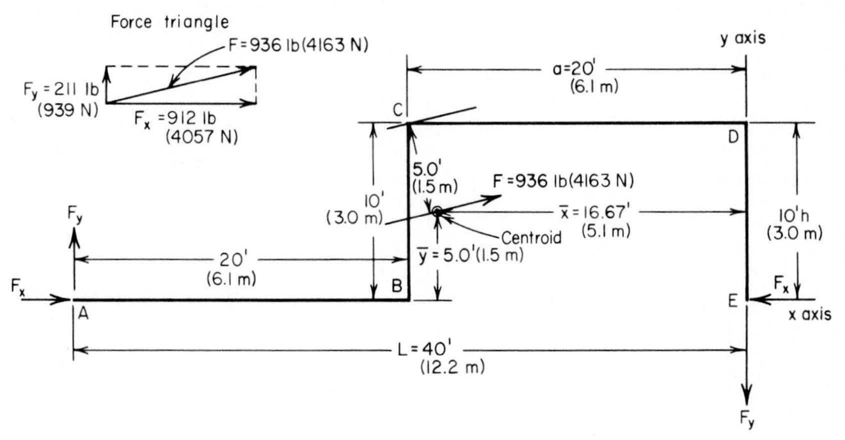

FIG. 20 U-shaped pipe with single tangent.

Piping Design and Engineering or to Spielvogel—*Piping Stress Calculations Simplified*. Both these references contain complete tabulations for a variety of pipe shapes.

To apply the Grinnell-Spielvogel solution procedure, compute the horizontal reaction force F_x lb from $F_x = k_x cI_p/L^2$, where K_x = a constant from Table 28 for the bend shape shown in Fig. 20; c = expansion factor = (pipe expansion, in/100 ft)(EM_R/172,800), where E = modulus of elasticity of the pipe material being used, lb/in²; M_R = modulus ratio = E at the operating temperature, F/E at 70°F (21.1°C); I_p = moment of inertia of pipe cross section, in⁴; L = length of bend, ft, as shown in Fig. 20.

To enter Table 28 for the shape in Fig. 20, the values of L/a and L/h must be known, or $L/a = 40/20 = 2$; $L/h = 40/10 = 4$. Entering Table 28 at these values, read $k_x = 91$; $k_y = 21$; $K_b = 120$. From the Spielvogel c table or by computation, $c = 570$ for carbon-steel pipe operating at 500°F (260°C). From a table of pipe properties, $I_p = 28.14$ in⁴ (1171.3 cm⁴) for 6-in (152.4-mm) schedule 40 pipe. Then $F_x = (91)(570)(28.14)/(40)^2 = 912$ lb (4057 N).

2. *Compute the vertical reacting or anchor force*

Use the same procedure as in step 1, except that the vertical reacting force F_y lb $= k_y cI_p/L^2$, or $F_y = (21)(570)(28.14)/(40)^2 = 211$ lb (939 N), by using the appropriate value from Table 28.

3. *Compute the resultant reacting or anchor force*

The resultant reacting or anchor force F lb is found by drawing and solving the force triangle in Fig. 20. From the pythagorean theorem $F = (912^2 + 211^2)^{0.5} = 936$ lb (4163 N). Draw the force triangle to scale, as shown in Fig. 20.

4. *Compute the maximum bending stress in the pipe*

The pipe bending stress s_b lb/in² is found in a similar manner from $s_b = k_b cD/L$, where k_b = bending-stress factor from Table 28; D = outside diameter of pipe, in. For 6-in (152.4-mm) schedule 40 pipe having an outside diameter of 6.625 in (168.3 mm), $s_b = (120)(570)(6.625)/40 = 11{,}330$ lb/in² (78.1 MPa).

5. *Determine the bending stress in the welded elbow*

The tables presented here are accurate when all the turns in the piping system analyzed are miters or rigid fittings. When all the turns are welded elbows or bends, the anchor forces derived from Table 28 are accurate for practical systems. The actual forces will be somewhat smaller than the values obtained from Table 28. Stresses in the elbows or bends may, however, exceed the values computed from Table 28 if the stress intensification factor β for these curved sections is >1. If the proportion of the straight to curved pipe is large, use the following procedure to obtain a close approximation of the stress in the curved section:

Determine the value of β from a table of pipe properties. For a 6-in (152.4-mm) schedule 40

TABLE 28 U Shape with Single Tangent

Reacting Force $F_x = k_x \cdot c \cdot \frac{I_p}{L^2}$ lb (N)

Reacting Force $F_y = k_y \cdot c \cdot \frac{I_p}{L^2}$ lb (N)

Maximum Bending Stress $s_B = k_b \cdot c \cdot \frac{D}{L}$ lb/in² (Pa)

I_p in in⁴ (cm⁴) L in ft (m) D in in (cm)

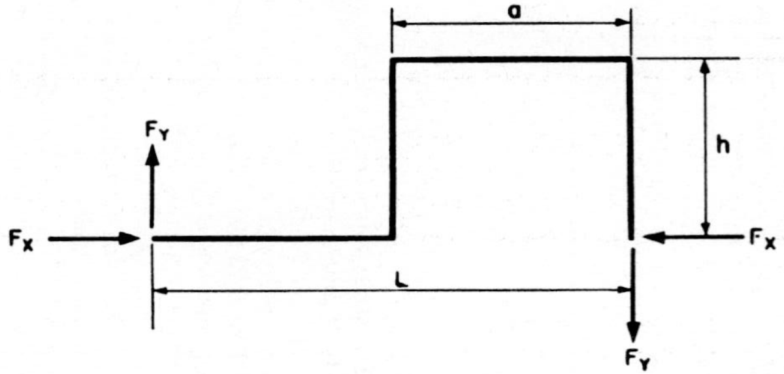

	L/a								
	1.5			2			3		
L/h	k_x	k_y	k_b	k_x	k_y	k_b	k_x	k_y	k_b
1.0	2.63 (0.0261)	0.75 (0.0074)	10.5 (8,690)	2.8 (0.0278)	1.41 (0.0140)	11.3 (9,350)	3.3 (0.0327)	2.3 (0.0228)	12.5 (10,300)
2.0	14.5 (0.1439)	3.4 (0.0337)	33.6 (27,800)	16 (0.1588)	5.8 (0.0575)	38 (31,400)	20 (0.1984)	8.4 (0.0833)	42 (34,700)
3.0	39 (0.3870)	7.7 (0.0764)	67 (55,400)	45 (0.4465)	12.4 (0.1230)	75 (62,100)	53 (0.5259)	16 (0.1588)	77 (63,700)
4.0	79 (0.7838)	13.5 (0.1339)	108 (89,400)	91 (0.9029)	21 (0.2084)	120 (99,300)	108 (1.072)	26 (0.2580)	124 (103,000)
5.0	139 (1.3792)	21.8 (0.2163)	159 (131,600)	156 (1.548)	31 (0.3076)	173 (143,000)	185 (1.836)	37 (0.3671)	174 (144,000)

long-radius welded elbow, β = 2.22. Therefore, the actual stress may exceed the table-computed stress, because $\beta > 1$.

Lay out the pipe bend to scale and compute the centroid of the bend by taking line moments about the x and y axes, Fig. 20.

	x axis			y axis	
AB	(20′)(0′) =	0	AB	(20′)(30′) =	600
BC	(10′)(5′) =	50	BC	(10′)(20′) =	200
CD	(20′)(10′) =	200	CD	(20′)(10′) =	200
DE	(10′)(5′) =	50	DE	(10′)(0′) =	0
	60	300		60	1000

$\bar{y}$ = 300/60 = 5 ft (1.5 m) $\bar{x}$ = 1000/60 = 16.67 ft (5.1 m)

In this calculation, the first value is the length of the pipe segment, and the second value is the distance of the center of gravity of the segment from the axis. For a straight section of pipe, the center of gravity is taken as the midpoint of the pipe section. The welded elbows are ignored in stress calculations based on table values.

Lay off $\bar{x}$ and $\bar{y}$ to scale, Fig. 20. Scale the distance to the tangent to the centerline of the long-radius elbow at C. This distance d = 5.0 ft (1.52 m). The moment at point C, $M_c = Fd$ lb·ft; or $m_c = 12Fd$ lb·in (1.36 N·m), and m_c = 12(936)(5) = 56,160 lb·in (6.34 kN·m), after force F is transposed from the force triangle to the centroid, Fig. 20.

The bending stress at any point in a pipe is $s_b = m\beta/S_m$, where S_m = section modulus of the pipe cross section, in^3. For 6-in (152.4-mm) schedule 40 pipe, S_m = 8.50 in^3 (139.3 cm^3), from a table of pipe properties. Then s_b = (56,160)(2.22)/8.50 = 14,700 lb/in^2 (101.3 MPa). This is somewhat greater than the 11,330 lb/in^2 (78.1 MPa) computed in step 4 but within the allowable stress of 15,000 lb/in^2 (103.4 MPa) for seamless carbon steel A106 pipe at 500°F (260°C).

By inspection of the scale drawing, Fig. 20, the stress in the long-radius elbows at B and D is less than at C because the moment arm at each of these points is less than at C.

Related Calculations: Tables 29, 30, and 31 present Grinnell-Spielvogel reaction and stress factors for three other single-plane bends—90° turn, U shape with equal tangents, and U shape with unequal legs. Use these tables and the factors in them in the same way as described above. Correct for curved elbows in the same manner. The tables can be used for piping conveying steam, water, gas, oil, and similar liquids, vapors, or gases. For bends of different shape, the analytical method must be used.

REACTING FORCES AND BENDING STRESS IN A TWO-PLANE PIPE BEND

Determine the horizontal reacting forces and bending and torsional stresses in the two-plane pipe bend shown in Table 32 if the dimensions of the bend are L = 20 ft (6.1 m); h = 5 ft (1.5 m); a = 5 ft (1.5 m); b = 5 ft (1.5 m). Use the tabular method of solution. The pipe is a 10-in (254-mm) carbon steel schedule 80 line operating at 750 lb/in^2 (gage) (5170.5 kPa) and 750°F (398.9°C). Determine the combined stress in the pipe.

Calculation Procedure:

1. *Compute the tabular factors for the pipe bend*

To apply the Grinnell-Spielvogel method to two-plane pipe bends, three tabular factors are required: L/a, a/b, and L/h. From the given values, L/a = 20/5 = 4; a/b = 5/5 = 1; L/h = 20/5 = 4.

TABLE 29 90° Turn

Reacting Force $F_x = k_x \cdot c \cdot \dfrac{I_p}{L^2}$ lb (N)

Reacting Force $F_y = k_y \cdot c \cdot \dfrac{I_p}{L^2}$ lb (N)

Maximum Bending Stress $s_B = k_b \cdot c \cdot \dfrac{D}{L}$ lb/in² (Pa)

I_p in in⁴ (cm⁴) L in ft (m) D in in (cm)

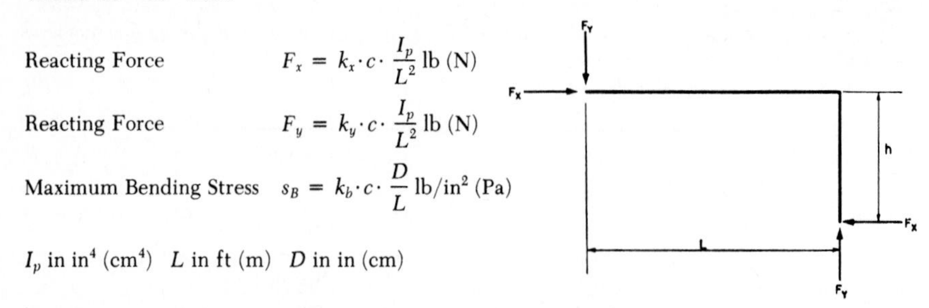

L/h	k_x	k_y	k_b
1.0	12.0 (0.1191)	12.0 (0.1191)	36 (29,800)
2.0	54.0 (0.5358)	16.6 (0.1647)	102 (84,400)
3.0	150 (1.488)	23.5 (0.2332)	209 (173,000)
4.0	315 (3.125)	31.5 (0.3125)	349 (289,000)
5.0	570 (5.656)	39.5 (0.3919)	528 (437,000)

2. *Determine the force and stress factors for the pipe*

From Table 32, for the factors in step 1, $k_x = 21.3$; $k_b = 24.5$; $k_t = 7.40$.

3. *Compute the horizontal reacting force of the bend*

The horizontal reacting force $F_x = k_x cI_p/L^2$, where the symbols are the same as in the preceding calculation procedure, except for L. Substituting the values for 10-in (254-mm) carbon-steel schedule 80 pipe operating at 750°F (399°C), we get $F_x = (21.3)(874)(244.9)/(20)^2 = 11{,}380$ lb (52.6 kN).

4. *Compute the bending stress in the pipe*

The bending stress in the pipe is found from $s_b = k\,cD/L$, where the symbols are the same as in the previous calculation procedure. Substituting values gives $s_b = (24.5)(874)(10.75)/(20) =$ 11,510 lb/in² (79.4 MPa). Table 32 shows that the maximum combined stress in the pipe occurs at the two upper bends, D.

5. *Compute the torsional stress in the pipe*

The torsional stress in the pipe is found from $s_t = k_t\,cD/L$, where the symbols are the same as in the previous calculation procedure. Substituting values yields $s_t = (7.40)(874)(10.75)/20 =$ 3475 lb/in² (24 MPa). Table 32 shows that the maximum combined stress in the pipe occurs at the two upper bends, D.

6. *Determine the combined stress in the pipe*

For any multiplane piping system, the combined stress s_{co} lb/in² $= 0.5\{s_1 + s_c + [4s_t^2 + (s_1 - s_c)^2]^{0.5}\}$. In this equation, $s_1 = s_b + s_p$, where s_p = pressure due to internal pressure, lb/in² (kPa); s_c = circumferential or hoop stress, lb/in² (kPa); other symbols are as given earlier. Also, $s_p = pA_i/A_m$, where p = operating pressure, lb/in² (gage) (kPa); A_i = inside area of pipe cross section, in² (cm²); A_m = metal area of pipe cross section, in² (cm²). Likewise, $s_c = p(D - t)/2t$, where D = outside diameter of pipe, in; t = pipe-wall thickness, in (cm).

TABLE 30 U Shape with Equal Tangents

Reacting Force $F_x = k_x \cdot c \cdot \frac{I_P}{L^2}$ lb (N)

Maximum Bending Stress $s_B = k_b \cdot c \cdot \frac{D}{L}$ lb/in² (Pa)

I_P in in⁴ (cm⁴) L in ft (m) D in in (cm)

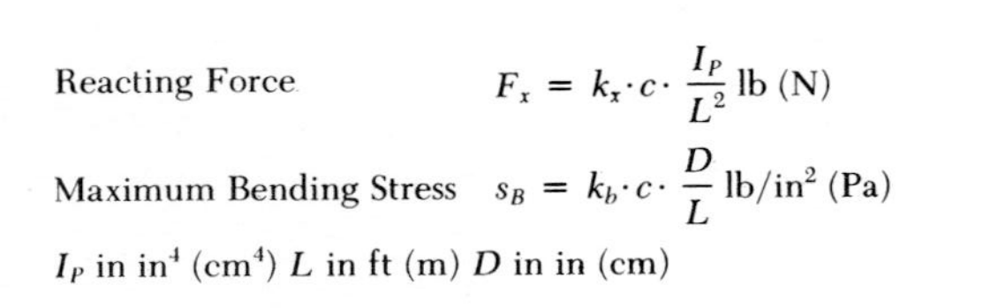

	L/a									
	2		3		4		5		6	
L/h	k_x	k_b	k_x	k_b	k_x	k_b	k_x	k_b	k_x	k_b
1.0	2.40 (0.0238)	7.20 (5,960)	2.46 (0.0244)	8.2 (6,780)	2.52 (0.0250)	8.82 (7,300)	2.58 (0.0256)	9.29 (7,690)	2.64 (0.0262)	9.69 (8,020)
2.0	12.00 (0.1191)	18.00 (14,900)	12.5 (0.1240)	21.8 (18,000)	13.24 (0.1314)	24.8 (20,500)	13.87 (0.1376)	27.1 (22,400)	14.4 (0.1429)	28.8 (23,800)
3.0	29.45 (0.2922)	29.45 (24,400)	31.2 (0.3096)	37.4 (30,900)	33.6 (0.3334)	43.7 (36,200)	35.8 (0.3552)	48.7 (40,300)	37.7 (0.3741)	52.7 (43,600)
4.0	54.9 (0.5447)	41.1 (34,000)	58.5 (0.5804)	53.6 (44,300)	64.0 (0.6350)	64.0 (53,000)	69.1 (0.6856)	72.5 (60,000)	73.6 (0.7303)	79.7 (65,900)
5.0	88.2 (0.8751)	52.9 (43,800)	95.3 (0.9456)	70.8 (58,600)	104.6 (1.0378)	85.2 (70,500)	114.7 (1.1380)	97.8 (80,900)	122.5 (1.2154)	107.5 (88,900)

TABLE 31 U Shape with Unequal Legs

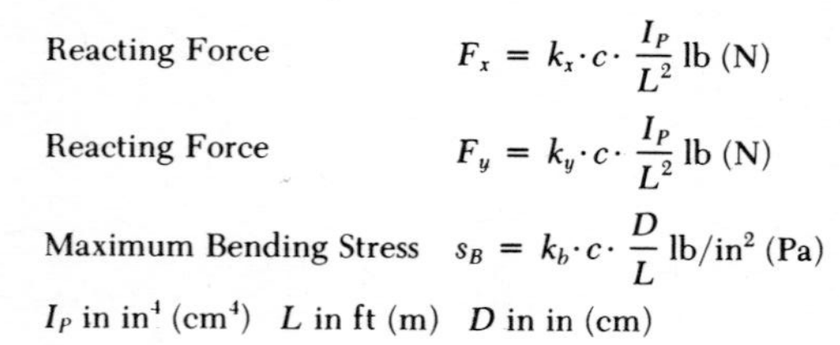

Reacting Force $F_x = k_x \cdot c \cdot \dfrac{I_P}{L^2}$ lb (N)

Reacting Force $F_y = k_y \cdot c \cdot \dfrac{I_P}{L^2}$ lb (N)

Maximum Bending Stress $s_B = k_b \cdot c \cdot \dfrac{D}{L}$ lb/in² (Pa)

I_P in in⁴ (cm⁴) L in ft (m) D in in (cm)

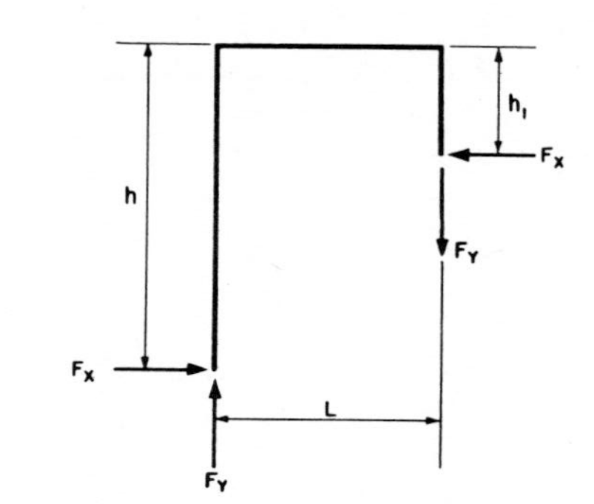

	h/h_1								
	4/3			2			3		
L/h	k_x	k_y	k_b	k_x	k_y	k_b	k_x	k_y	k_b
0.2	0.07 (0.0007)	0.6 (0.0059)	1.5 (1,240)	0.29 (0.0029)	1.8 (0.0179)	7 (5,790)	0.53 (0.0053)	3.4 (0.0337)	11 (9,100)
0.8	2.4 (0.0238)	0.9 (0.0089)	9.5 (7,860)	3.6 (0.0357)	2.5 (0.0248)	15 (12,400)	4.8 (0.0476)	4.4 (0.0436)	20 (16,500)
1.0	4.3 (0.0127)	1.2 (0.0119)	16 (13,200)	6.2 (0.0615)	3.0 (0.0298)	21 (17,400)	8 (0.0794)	4.9 (0.0486)	26 (21,500)
2.0	27 (0.2679)	2.3 (0.0228)	58 (48,000)	37 (0.3671)	6.0 (0.0595)	75 (62,100)	44 (0.4366)	8.0 (0.0883)	88 (72,800)
3.0	81 (0.8037)	3.8 (0.0377)	124 (102,600)	110 (1.0914)	10.0 (0.0992)	162 (134,000)	128 (1.2700)	15 (0.1488)	185 (153,000)

TABLE 32 Two-Plane U with Tangents

Reacting Force $F_x = k_x \cdot c \cdot \frac{I_P}{L^2}$ lb (N)

Bending Stress $s_B = k_B \cdot c \cdot \frac{D}{L}$ lb/in² (Pa)

Torsional Stress $s_T = k_T \cdot c \cdot \frac{D}{L}$ lb/in² (Pa)

I_P in in⁴ (cm⁴) L in ft (m) D in in (cm)

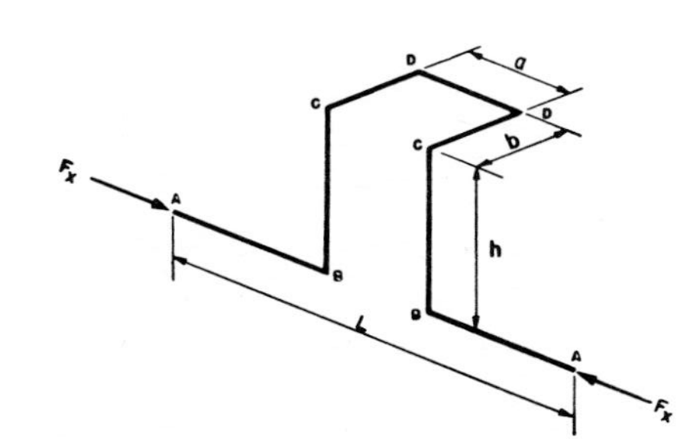

	a/b						L/a = 4					
	0.25			0.5			1			2		
L/h	k_x	k_b	k_t	k_x	k_b	k_t	k_x	k_b	k_t	k_x	k_b	k_t
1	0.67 (0.0066)	*D* 3.20 (2,650)	. . .	1.22 (0.0121)	*A* 4.35 (3,600)	*A* 0.30 (250)	1.67 (0.0166)	*A* 5.2 (4,300)	*A* 0.15 (120)	2.0 (0.0198)	*C* 6.3 (5,200)	
2	1.35 (0.0134)	*D* 5.80 (6,800)	. . .	4.30 (0.0427)	*D* 9.96 (8,240)	*D* 2.45 (2,030)	6.96 (0.0691)	*A* 11.0 (9,100)	. . .	9.3 (0.0923)	*C* 15.0 (12,400)	
3	1.70 (0.0169)	*D* 7.00 (5,790)	. . .	6.23 (0.0618)	*D* 13.8 (11,400)	*D* 2.28 (1,890)	14.0 (0.1389)	*D* 16.5 (13,700)	6.55 (5,420)	21.2 (0.2103)	*C* 24.0 (19,900)	
4	1.88 (0.0187)	*D* 7.44 (6,150)	. . .	7.84 (0.0778)	*D* 16.9 (14,000)	*D* 2.09 (1,730)	21.3 (0.2113)	*D* 24.5 (20,300)	*D* 7.40 (6,120)	36.2 (0.3592)	*C* 30.0 (24,800)	
5	2.01 (0.0199)	*D* 7.75 (6,410)	. . .	8.94 (0.0887)	*D* 18.8 (15,600)	*D* 1.89 (1,560)	27.8 (0.2758)	31.4 (26,000)	*D* 7.75 (6,410)	52.6 (0.5219)	*D* 31.0 (25,600)	*D* 17.3 (14,300)

Computing stress values for this 10-in (254-mm) schedule 80 carbon-steel pipe operating at 750 lb/in^2 (gage) (5.2 MPa) and 750°F (399°C), and using values from a table of pipe properties, we get s_p = (750)(71.8/18.92) = 2845 lb/in^2 (19.6 MPa); s_c = (750)(10.75 − 0.593)/2(0.593) = 6420 lb/in^2 (44.3 MPa). Then $s_1 = s_b + s_p$ = 11,510 + 2845 = 14,355 lb/in^2 (99 MPa), where s_b is from step 4. By substituting in the combined-stress equation, $s_{co} = 0.5\{14{,}355 + 6420 + [4 \times 3475^2 + (14{,}355 - 6420)^2]^{0.5}\}$ = 16,648 lb/in^2 (114.8 MPa). This is higher than the stress allowed in carbon-steel pipe by the ANSA *Piping Code*, unless the pipe conforms to the special conditions of certain paragraphs of the *Code*.

Related Calculations: Use this procedure for piping conveying steam, water, gas, oil, and similar liquids, vapors, or gases. For bends of different shape, the analytical method is generally used.

REACTING FORCES AND BENDING STRESS IN A THREE-PLANE PIPE BEND

Determine the three reacting forces and moments and bending and torsional stresses in the three-plane pipe bend shown in Table 33 if the dimensions of the bend are L_1 = 20 ft (6.1 m), L_2 = 10 ft (3.0 m), L_3 = 5 ft (1.5 m). The pipe is 10-in (254-mm) carbon-steel schedule 80 operating at 750 lb/in^2 (gage) (5.2 MPa) and 750°F (399°C).

Calculation Procedure:

1. *Compute the tabular factors for the pipe bend*

From the Grinnell-Spielvogel method, the two tabular factors required are $m = L_1/L_3$ and $n = L_2/L_3$; or m = 20/5 = 4 and n = 10/5 = 2.

2. *Determine the force and stress factors for the pipe*

From Table 33, for the factors in step 1, k_b = 8.0; k_t = 3.6; k_x = 1.48; k_y = 0.13; k_z = 0.80; k_{xy} = 1.3; k_{xz} = 1.2; k_{yz} = 0.51.

3. *Compute the longitudinal reaction force of the bend*

The horizontal reacting force $F_x = k_x C i_p/L_3^2$, where the symbols are the same as in the preceding calculation procedure, except for L_3. By substituting values for 10-in (254-mm) carbon-steel schedule 80 pipe operating at 750°F (398.9°C), F_x = (1.48)(874)(244.9)/(5)^2 = 12,680 lb (56.4 kN).

4. *Compute the vertical reacting force of the bend*

The vertical reacting force $F_y = k_y c I_p L_3^2$, where the symbols are the same as in the preceding calculation procedure, except for L_3. Substituting values for this pipe, we find F_y = (0.13)(874)(244.9)/(5)^2 = 1115 lb (4.96 kN).

5. *Compute the horizontal reacting force of the bend*

The horizontal reacting force $F_z = k_z c I_p/L_3^2$, where the symbols are the same as in the preceding calculation procedure, except for K_z and L_3. Substituting values for this pipe gives F_z = (0.80)(874)(244.9)/(5)^2 = 6850 lb (30.5 kN).

6. *Compute the bending and torsional stresses in the pipe*

The bending stress $s_b = k_b cD/L_3$, where the symbols are the same as in the preceding calculation procedure, except for L_3. Substituting values for this pipe gives s_b = (8.0)(874)(10.75)/5 = 15,020 lb/in^2 (103.6 MPa).

The torsional stress $s_t = k_t cD/L_3$, where the symbols are the same as in the preceding calculation procedure, except for L_3. By substituting values for this pipe, s_t = (3.6)(874)(10.75)/5 = 6760 lb/in^2 (46.6 MPa).

7. *Compute the three reacting moments at the pipe end*

For each bending moment M ft·lb = kcI_p/L_3, where the symbols are the same as given in the previous steps in this calculation procedure, except that k is the appropriate bending-moment factor.

For the xy moment M_{xy} = (1.3)(874)(244.9)/5 = 55,700 ft·lb (75.5 kN·m). For the xz moment M_{xz} = (1.2)(874)(244.9)/5 = 51,400 ft·lb (69.7 kN·m). For the yz moment M_{yz} = (0.51)(874)(244.9)/5 = 22,250 ft·lb (30.2 kN·m).

$L_1 \geq L_3 \quad L_1 = m \quad \dfrac{L_2}{L_3} = n$

Bending Stress $\quad s_B = k_b \cdot c \cdot \dfrac{D}{L}$ lb/in² (Pa)

Torsional Stress $\quad s_T = k_t \cdot c \cdot \dfrac{D}{L_3}$ lb/in² (Pa)

Reacting Force $\quad F_x = k_x \cdot c \cdot \dfrac{I_P}{L_3^2}$ lb (N)

Reacting Force $\quad F_y = k_y \cdot c \cdot \dfrac{I_P}{L_3^2}$ lb (N)

Reacting Force $\quad F_z = k_z \cdot c \cdot \dfrac{I_P}{L_3^2}$ lb (N)

Reacting Moment $\quad M_{xy} = k_{xy} \cdot x \cdot \dfrac{I_P}{L_3}$ ft·lb (N·m)

Reacting Moment $\quad M_{xx} = k_{xx} \cdot c \cdot \dfrac{I_P}{L_3}$ ft·lb (N·m)

Reacting Moment $\quad M_{yz} = k_{yz} \cdot c \cdot \dfrac{I_P}{L_3}$ ft·lb (N·m)

I_p in in⁴ (cm⁴) $\quad L$ in ft (m) $\quad D$ in in (cm)

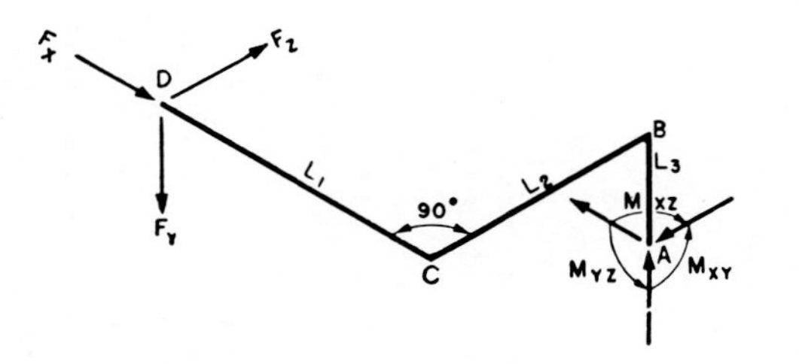

	m = 3								m = 4								
n	k_b	k_t	k_x	k_y	k_z	k_{xy}	k_{xz}	k_{yz}	k_b	k_t	k_x	k_y	k_z	k_{xy}	k_{xz}	k_{yz}	n
1	A 22.3 (18,500)	A 4.86 (4,020)	4.5 (0.045)	0.54 (0.0054)	1.50 (0.0148)	3.6 (0.0357)	1.62 (0.0161)	0.74 (0.0073)	A 24.0 (19,900)	A 6.0 (4,960)	4.80 (0.0476)	0.37 (0.0037)	1.10 (0.0110)	3.9 (0.0387)	2.0 (0.0199)	0.61 (0.0061)	1
2	D 9.3 (7,690)	D 0.15 (120)	1.4 (0.014)	0.22 (0.0022)	1.10 (0.0109)	1.1 (0.0109)	1.00 (0.0099)	0.71 (0.0070)	A 8.0 (6,620)	A 3.6 (2,980)	1.48 (0.0147)	0.13 (0.0013)	0.80 (0.0079)	1.3 (0.0129)	1.2 (0.0119)	0.51 (0.0051)	2
3	D 10.0 (8,270)	D 0.24 (199)	0.76 (0.007)	0.13 (0.0013)	1.08 (0.0107)	0.60 (0.0059)	0.74 (0.0073)	0.70 (0.0069)	D 7.26 (6,000)	D 0.10 (83)	0.76 (0.0075)	0.09 (0.0009)	0.65 (0.0064)	0.6 (0.0059)	0.88 (0.0088)	0.42 (0.0042)	3

Related Calculations: Use this procedure for piping conveying steam, water, gas, oil, and similar liquids, vapors, or gases. For bends of different shape, the analytical method must be used. Compute the combined stress in the same way as in step 6 of the previous calculation procedure. Table 33 shows that the maximum combined stress occurs at point A in this piping system.

ANCHOR FORCE, STRESS, AND DEFLECTION OF EXPANSION BENDS

Determine the deflection and anchor force in an 8-in (203.2-mm) schedule 40 double-offset expansion U bend having a radius of 64 in (1626 mm) if the bending stress is 10,000 lb/in^2 (68.9 MPa). What would the deflection and anchor force be with a bending stress of 15,000 lb/in^2 (103.4 MPa) if the bend tangents are guided and the pipe is carbon steel operating at 500°F (260°C)? With a bending stress of 8000 lb/in^2 (55.2 MPa)? Tabulate the deflection and anchor-force equations for the popular types of expansion bends when the expanding pipe is guided axially.

Calculation Procedure:

1. *Compute the deflection of the pipe bend*

For a double-offset expansion U bend, the deflection d in $= 0.728R^2K/D\beta$, where R = bend radius, ft; K = flexibility factor for curved pipe, from a table of pipe properties or from $K = (12\lambda^2 + 10)/(12\lambda^2 + 1)$, where $\lambda = tR/r^2$, where t = pipe thickness, in, $r = (D - t)/2$; D = outside diameter of pipe, in; β = stress coefficient for curved pipe from a table of pipe properties or from $\beta = (2K/3)[6\lambda^2 + 5)/18]^{0.5}$ when $\lambda \leq 1.47$; $\beta = (12\lambda^2 - 2)/(12\lambda^2 + 1)$ when $\lambda > 1.47$.

For this bend, $R = 64/12 = 5.33$ ft (1.62 m); $K = 1.49$ from a table of pipe properties or by computation; $D = 8.625$ in (219.1 mm) from a table of pipe properties; $\beta = 0.86$ from a table of pipe properties, or by computation. Then $d = (0.728)(5.33)^2(1.49)/[(8.625)(0.86)] = 4.15$ in (105.4 mm).

2. *Compute the anchor force of the pipe bend*

For a double-offset expansion U bend, the anchor force F_x lb $= 976\ I_p/(RD\beta)$, where I_p = moment of inertia of pipe cross section, in^4. For this pipe, use values from a table of pipe properties; or computing the values, $F_x = (976)(72.5)/[(5.33)(8.625)(0.86)] = 1790$ lb (7.96 kN).

3. *Compute the deflection and anchor force for a larger bending stress*

With a larger bending stress—15,000 lb/in^2 (103.4 MPa) in this instance—and a greater deflection at the higher stress d_h, solve $d_h = (d)$(allowable stress, lb/in^2)/$(10{,}000M_R)$, or $d_h = (4.15)(15{,}000)/[10{,}000(0.932)] = 6.68$ in (169.7 mm).

The anchor force at the larger bending stress is $F_h = F_x d_h M_R/d$, or $F_x = (1790)(6.68)(0.932)/4.15 = 2680$ lb (11,921 N).

4. *Compute the deflection and anchor force for a smaller bending stress*

Use the same equation as in step 3, except that the lower bending stress is substituted for the higher one. Or, $d_l = (4.15)(8000)/[10{,}000(0.932)] = 3.56$ in (90.4 mm), and $F_l = (1790)(3.56)(0.932)/4.15 = 1432$ lb (6370 N).

5. *Tabulate the deflection and anchor-force equations*

Do this as shown in the table on page 3.442.

Related Calculations: Use the procedures given here for piping conveying steam, water, oil, gas, air, and similar vapors, liquids, and gases. The value of E in step 5, 29×10^6 lb/in^2 (199.9 MPa), is satisfactory for pipes made of carbon steel, carbon moly steel, chromium moly steel, nickel steel, and chromium nickel steel. These materials are commonly used in piping systems requiring expansion bends.

Note that the equations in step 5 apply to pipe bends having guides to direct the axial expansion of the pipe. This is the usual arrangement used today because unguided bends require too much space. For design of unrestrained bends, multiply d by 1.5 to find the deflection at the higher stress, as in step 3. This factor, 1.5, is an approximation, but it is on the safe side in almost every case. The equations given in step 5 are presented in great detail in Grinnell—*Piping Design and Engineering* and Crocker and King—*Piping Handbook*.

Bend type	Deflection for 10,000 lb/in² (68.9-MPa) s_b		Anchor force for 10,000 lb/in² (68.9-MPa) s_b	
	$E = 29 \times 10^6$ lb/in²	$E = 199.9$ MPa	$E = 29 \times 10^6$ lb/in²	$E = 199.9$ MPa
Double-offset U	$d = 0.728R^2K/D\beta$	$d = 5.056 \times 10^{-3}R^2K/D\beta$	$F_x = 976\ I_pRD\beta$	$F_x = 8070 \times I_p/RD\beta$
Expansion U bend (no tangents)	$d = 0.312R^2K/D\beta$	$d = 2.167 \times 10^{-3}R^2K/D\beta$	$F_x = 1667I_p/RD\beta$	$F_x = 13{,}780 \times I_p/RD\beta$
Expansion U bend [tangents = 2 ft (0.6 m)]	$d = [(0.312R^3 + 0.795R^2 + 0.624R)K + 0.132]/(R + 1)D\beta$	$d = [(2.167R^3 + 165R^2 + 3895R) \times 10^{-3}K + 24.7]/(R + 30)D\beta$	$F_x = 1667I_p/(R + 1)D\beta$	$F_x = 13{,}780I_p/(R + 30)D\beta$
Expansion U bend (tangents = R)	$d = (0.577 + 0.011)R^2/D\beta$	$d = (4.007K + 0.076)R^2/D\beta$	$F_x = 1111I_p/RD\beta$	$F_x = 9190I_p/RD\beta$
Expansion U bend (tangents = $2R$)	$d = (0.865K + 0.0662)R^2/D\beta$	$d = (6.00 \times 10^{-3}K + 0.459\ 10^{-3})R^2/D\beta$	$F_x = 833I_p/RD\beta$	$F_x = 6890I_p/RD\beta$
Expansion U bend (tangents = $4R$)	$d = (1.465K + 0.353)R^2/D\beta$	$d = (10.17K + 2.45) \times 10^{-3} \times R^2/D\beta$	$F_x = 556I_p/RD\beta$	$F_x = 4600I_p/RD\beta$
Double-offset U bend	$d = 0.260R^2K/D\beta$	$d = 1.806 \times 10^{-3}R^2K/D\beta$	$F_x = 1209I_p/RD\beta$	$F_x = 9997I_p/RD\beta$
Single-offset quarter bend	$d = 0.0366R^2K/D\beta$	$d = 2.5 \times 10^{-4}R^2K/D\beta$	$F_x = 2763I_p/RD\beta$	$F_x = 22{,}850I_p/RD\beta$
			$F_y = 0.066F_x$	$F_y = 0.066F_x$
Circle bend	$d = 0.312R^2K/D\beta$	$d = 2.167 \times 10^{-3}R^2K/D\beta$	$F_x = 1667I_p/RD\beta$	$F_x = 13{,}780 \times I_p/RD\beta$

SLIP-TYPE EXPANSION JOINT SELECTION AND APPLICATION

Select and size slip-type expansion joints for the 20-in (508-mm) carbon-steel schedule 40 pipeline in Fig. 21 if the pipe conveys 125-lb/in^2 (gage) (861.6-kPa) steam having a temperature of 380°F (193°C). The minimum temperature expected in the area where the pipe is installed is 0°F (−17.8°C). Determine the anchor loads that can be expected. The steam inlet to the pipe is at *A*; the outlet is at *F*.

Calculation Procedure:

1. *Determine the expansion of each section of pipe*

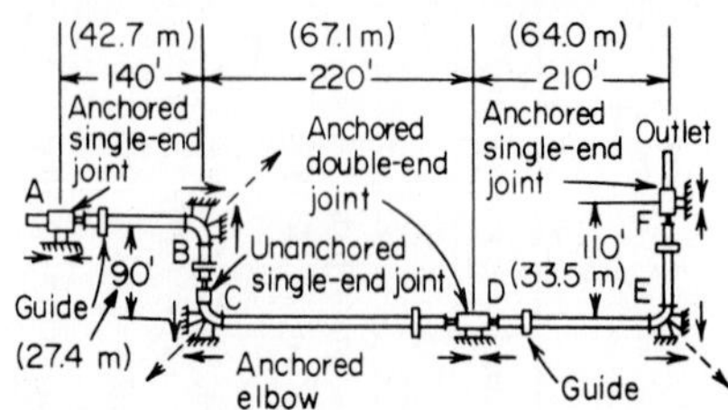

FIG. 21 Slip-type expansion joints in a piping system. *(Yarway Corporation.)*

From Fig. 22, the expansion of steel pipe at 380°F (193°C) with a 0°F (−17.8°C) minimum temperature is 3.4 in (88.9 mm) per 100 ft (30.5 m) of pipe. Expansion of each section of pipe is then *e* in = (3.4)(pipe length, ft/100). For *AB*, *e* = (3.4)(140/100) = 4.76 in (120.9 mm); for *BC*, *e* = (3.4)(90/100) = 3.06 in (77.7 mm); for *CD*, *e* = (3.4)(220/100) = 7.48 in (190 mm); for *DE*, *e* = (3.4)(210/100) = 71.4 in (1813.6 mm); for *EF*, *e* = (3.4)(110/100) = 3.74 in (95 mm).

2. *Select the type and the traverse of each expansion joint*

The slip-type expansion joint at *A* will absorb expansion from only one direction—the right-hand side. This expansion will occur in pipe section *AB* and is 4.76 in (120.9 mm) from step 1. Therefore, a single-end slip-type expansion joint (one that absorbs expansion on only one side) can be used. The traverse—the amount of expansion a slip joint will absorb—is usually given in multiples of 4 in (101.6 mm), that is, 4, 8, and 12 in (101.6, 203.2, and 304.8 mm). Hence, an 8-in (203.2-mm) traverse slip-type single-end joint will be suitable at *A* because the expansion is 4.76 in (120.9 mm). A 4-in (101.6-mm) traverse joint would be unsatisfactory because it could not absorb at 4.76-in (120.9-mm) expansion.

The next joint, at *C*, must absorb the expansion in the vertical pipe *BC*. Since the elbow beneath the joint is anchored, an unanchored joint can be used. With pipe expansion in only one direction—from *B* to *C*—a single-end joint can be used. Since the expansion of section *BC* is 3.06

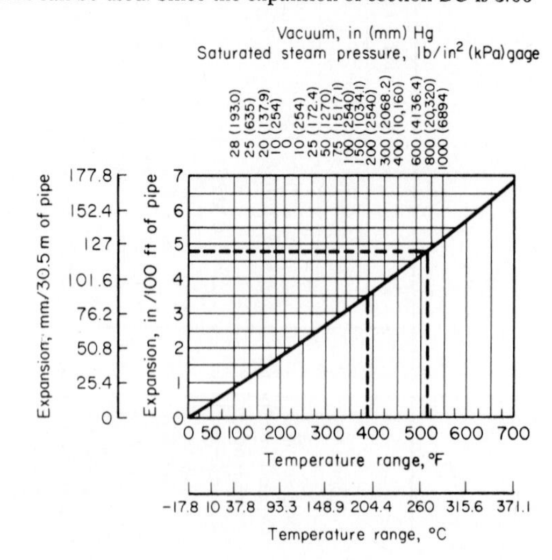

FIG. 22 Expansion of steel pipe. *(Yarway Corporation.)*

in (77.7 mm), use a single-end 4-in (101.6-mm) traverse slip-type expansion joint, unanchored at *C*.

The expansion joint at *D* must absorb expansion from two directions—from *C* to *D* and from *E* to *D*. Therefore, a double-end joint (one that can absorb expansion on each end) must be used. The double-end joint must be anchored because the pipe expands *away* from the anchored elbow *C* in section *CD* and *away* from the anchored elbow *E* in section *DE*. In both instances the pipe expands *toward* the expansion joint at *D*.

The expansion in section *CD* is, from step 1, 7.48 in (190 mm), whereas the expansion in *DE* is 7.14 in (181.4 mm). Therefore, a double-end anchored joint with an 8-in (203.2-mm) traverse at *each* end will be suitable.

Since the pipe outlet is at *F* and there is no anchor in the pipe at *F*, the expansion joint at this point must be anchored. The pipe section between *E* and *F* will expand vertically upward into the joint for a distance of 3.74 in (95 mm), as computed in step 1. Therefore, a single-end anchored joint with a 4-in (101.6-mm) traverse will be suitable.

3. *Compute the anchor loads in the pipeline*

Use Fig. 23 to determine the anchor loads on intermediate and end anchors (those where the pipe makes a sharp change in direction). Enter Fig. 23 at the bottom at a pipe size of 20-in (508-mm) diameter, and project vertically upward to the dashed curve labeled *intermediate anchor—all pressures*. At the left read the anchor load at each intermediate anchor, *A*, *D*, and *F*, as 20,000 lb (88.9 kN). Note that the joint expansion load = joint contraction load = 20,000 lb (88.9 kN).

The end anchors, *B*, *C*, and *E*, have, from Fig. 23, a possible maximum load of 58,000 lb (258 kN), found by projecting vertically upward from the 20-in (508-mm) pipe size to 125-lb/in^2 (gage) (862-kPa) steam pressure, which lies midway between the 100- and 150-lb/in^2 (gage) (689.5- and 1034-kPa) curves. Indicate the possible maximum end-anchor loads by the solid arrows at each elbow, as shown in Fig. 21. The resultant *R* of the loads at any end anchor is found by the pythagorean theorem to be $R = (58{,}000^2 + 58{,}000^2)^{0.5} = 82{,}200$ lb (365.6 kN). Indicate the resultant by a dotted arrow, as shown in Fig. 21.

Contraction loads on the end anchors are in the reverse direction and consist only of friction. This friction load equals the joint expansion load, or 20,000 lb (88.9-kN). The resultant of the joint expansion loads is $(20{,}000^2 + 20{,}000^2)^{0.5} = 28{,}350$ lb (126.1 kN).

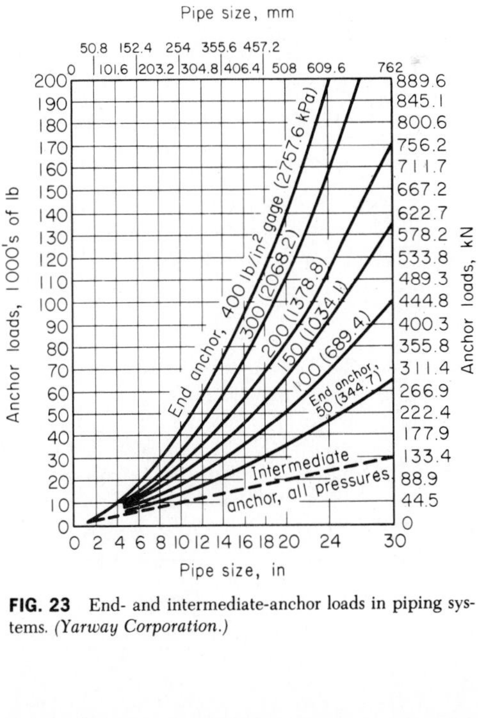

FIG. 23 End- and intermediate-anchor loads in piping systems. *(Yarway Corporation.)*

TABLE 34 Guide and Support Spacing

Nominal pipe size, in (mm)	Distance between guide and joint, ft (m) Packing type		Distance between guides, ft (m)
	Gun	Gland	
18 (457)	24 (7.3)	11 (3.4)	100 (30.5)
20 (508)	25 (7.6)	12 (3.7)	105 (32)
24 (610)	26 (7.9)	12 (3.7)	110 (33.5)

Locate guides within 25 or 12 ft (7.62 or 3.66 m) of the expansion joint, depending on the type of packing used, Table 34. These guides should allow free axial movement of the pipe into and out of the joint with minimum friction.

Related Calculations: Use this procedure to choose slip-type expansion joints for pipes conveying steam, water, air, oil, gas, and similar vapors, liquids, and gases. In some instances, the gland friction and pressure thrust is used instead of Fig. 23 to determine anchor loads. With either method, the results are about the same.

CORRUGATED EXPANSION JOINT SELECTION AND APPLICATION

Select corrugated expansion joints for the 8-, 6-, and 4-in (203.2-, 152.4-, and 101.6-mm) carbon-steel pipeline in Fig. 24 if the steam pressure in the pipe is 75 lb/in² (gage) (517.1 kPa), the steam temperature is 340°F (171°C), and the installation temperature is 60°F (15.6°C).

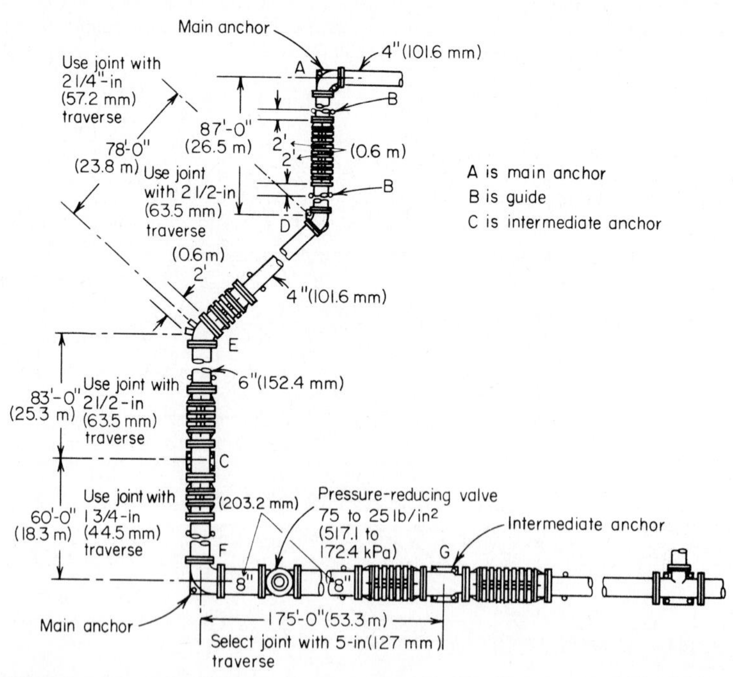

FIG. 24 Piping system fitted with expansion joints. *(Flexonics Division, Universal Oil Products Company.)*

Calculation Procedure:

1. *Determine the expansion of each section of pipe*

From a table of thermal expansion of pipe, the expansion of carbon-steel pipe at 340°F (171°C) is 2.717 in/100 ft (2.26 mm/30.5 m) from 0 to 340°F (−17.8 to 171°C). Between 0 and 60°F (−17.8 and 15.6°C) the expansion is 0.448 in/100 ft (50 mm/30.5 m). Hence, the expansion between 60 and 340°F (15.6 and 171°C) is 2.717 − 0.448 = 2.269 in/100 ft (1.89 mm/m). This factor can now be applied to each length of pipe by finding the product of (pipe-section length, ft/100)(expansion, in/100 ft) = expansion of section, in = e.

For section AD, e = (87/100)(2.269) = 1.97 in (50 mm); for DE, e = (78/100)(2.269) = 1.77 in (45 mm); for EC, e = (83/100)(2.269) = 1.88 in (47.8 mm); for CF, e = (60/100)(2.269) = 1.36 in (34.5 mm); for FG, e = (175/100)(2.269) = 3.97 in (100.8 mm).

In selecting corrugated expansion joints, the usual practice is to increase the computed expansion by a suitable safety factor to allow for any inaccuracies in temperature measurement. By applying a 25 percent safety[1] factor: for AD, e = (1.97)(1.25) = 2.46 in (62.5 mm); for DE, e = (1.77)(1.25) = 2.13 in (54.1 mm); for EC, e = (1.88)(1.25) = 2.35 in (59.7 mm); for CF, e = (1.36)(1.25) = 1.70 (43.2 mm); for FG, e = (3.97)(1.25) = 4.96 in (126 mm).

2. *Select the traverse for, and type of, each expansion joint*

Obtain corrugated-expansion joint engineering data, and select a joint with the next largest traverse for each section of pipe. Thus, traverse $AD \geq$ 2½ in (63.5 mm); traverse $DE \geq$ 2¼ in (57.2 mm); traverse $EC \geq$ 2½ in (63.5 mm); traverse $CF \geq$ 1¾ in (44.5 mm); traverse $FG \geq$ 5.0 in (127 mm).

Two types of expansion joints are commonly used: free-flexing and controlled-flexing. Free-flexing joints are generally used where the pressures in the pipeline are relatively low and the required motion is relatively small. Controlled-flexing expansion joints are generally used for higher pressures and larger motions. Both types of expansion joints are available in stainless steel in both single and dual units. For precise data on a given joint being considered, consult the expansion-joint manufacturer. Corrugated expansion joints are characterized by their freedom from any maintenance needs.

3. *Compute the anchor loads in the pipeline*

Main anchors are used between expansion joints, as at F and A, Fig. 24, and at turns such as at F and A. The force[2] a main anchor must absorb is given by F_i lb = $F_p + F_e$, where F_p = pressure thrust in the pipe, lb = pA, where p = pressure in pipe, lb/in² (gage); A = effective internal cross-sectional area of expansion joint, in² (see Table 35 for cross-sectional areas of typical corrugated joints); F_e = force required to compress the expansion joint, lb = [300 lb/in (52.5 N/mm)] (joint inside diameter, in) for stainless-steel self-equalizing joints, and [200 lb/in (35 N/mm)] (joint inside diameter, in) for copper nonequalizing joints. Determining the main anchor force for the 8-in (203.2-mm) pipeline gives F_i = (75)(85) + (300)(8) = 8775 lb (39.0 kN). In this equation, the area of 85 in² (548.3 cm²) in the first term is obtained from Table 35.

TABLE 35 Effective Area of Corrugated Expansion Joints

Joint inside diameter		Joint effective area	
in	mm	in²	cm²
6	152.4	51.0	329.0
8	203.2	85.0	548.4
10	254.0	120.0	774.2
12	304.8	174.0	1122.6
14	355.6	215.0	1387.1
16	406.4	270.0	1741.9
18	457.2	310.0	1999.9
20	508.0	390.0	2516.1
24	609.6	540.0	3483.9

The total force at a main anchor, as at A and F, Fig. 24, is the vector sum of the forces in each line leading to the anchor. Thus, at F, there is a force of 8775 lb (39.0 kN) in the 8-in

[1]This value is for illustration purposes only. Contact the expansion-joint manufacturer for the exact value of the safety factor to use.

[2]This is an approximate method for finding the anchor force. For a specific make of expansion joint, consult the joint manufacturer.

(203.2-mm) line and a force of F_i = (75)(51) + (300)(6) = 5625 lb (25 kN) in the 6-in (152.4-mm) line connected to the elbow outlet. Since the elbow at *F* is a right angle, use the pythagorean theorem, or *R* = resultant anchor force, lb = $(8775^2 + 5625^2)^{0.5}$ = 10,400 lb (46.3 kN).

Where two lines containing corrugated expansion joints are connected by a bend of other than 90°, as at *D* and *E*, use a force triangle to determine the anchor force after computing F_i for each pipe. Thus, at *E*, F_i for the 6-in (152.4-mm) pipe = 5625 lb (25 kN), and F_i = (75) × (23.5) + (300)(4) = 2963 lb (13.2 kN) for the 4-in (101.6-m) pipe. Draw the force triangle in Fig. 25 with the 6-in (152.4-mm) pipe F_i and the 4-in (101.6-mm) pipe F_i as two sides and the bend angle, 45°, as the included angle. Connect the third side, or resultant, to the ends of the force vectors, and scale the resultant as 4125 lb (18.4 kN), or compute the resultant from the law of cosines. Find the resultant force at *D* in a similar manner as 2963 lb (13.2 kN).

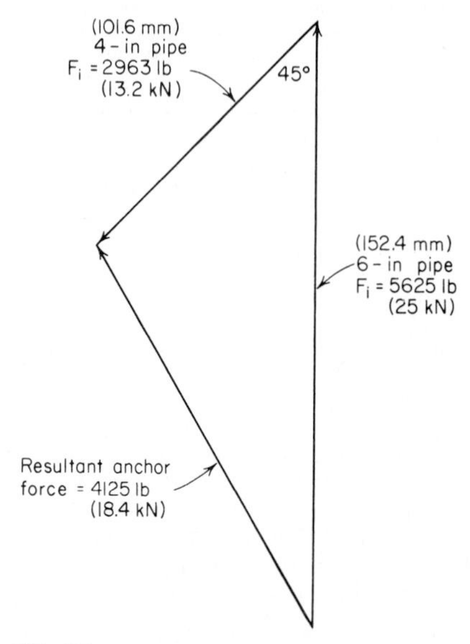

FIG. 25 Force triangle for determining piping anchor force.

Intermediate anchors, as at *C* and *G*, must withstand only one force—the unbalanced (differential) spring force. With approximate force calculations,[1] starting at *C*, for a 6-in (152.4-mm) expansion joint, F_e = (300)(6) = 1800 lb (8 kN). At *G*, for an 8-in (203.2-mm) expansion joint, F_e = (300)(8) = 2400 lb (10.7 kN). Thus, the loads the intermediate anchors must withstand are considerably less than the main-anchor loads.

Provide the pipe guides at suitable locations in accordance with the joint manufacturer's recommendations and at suitable intervals on the pipeline to prevent any lateral and buckling forces on the joint and adjacent piping. Intermediate anchors between two joints in a straight run of pipe ensure that each joint will absorb its share of the total pipe motion. Slope the pipe in the direction of fluid flow to prevent condensate accumulation. Use enough pipe hangers to prevent sagging of the pipe.

Related Calculations: Use this procedure to choose corrugated-type expansion joints for pipes conveying steam, water, air, oil, gas, and similar vapors, liquids, and gases. When choosing a specific make of corrugated expansion joint, use the manufacturer's engineering data, where available, to determine the maximum allowable traverse. One popular make has a maximum traverse of 7.5 in (190.5 mm) or a maximum allowable lateral motion of 1.104 in (28.0 mm) in its various joint sizes. The larger the lateral motion, the greater the number of corrugations required in the joint.

In some pipelines there is an appreciable pressure thrust caused by a change in direction of the pipe. This pressure or centrifugal thrust F_c is usually negligible, but the wise designer makes a practice of computing this thrust from $F_c = (2A\rho v^2/32.2) \times (\sin \theta/2)$ lb, where *A* = inside area of pipe, ft²; ρ = density of fluid or vapor, lb/ft³; v = fluid or vapor velocity, ft/s; θ = change in direction of the pipeline.

The number of corrugations required in a joint varies with the expansion and lateral motion to be absorbed. A typical free-flexing joint can absorb 6.25 in (158.8 mm) of expansion and a variable amount of lateral motion, depending on joint size and operating condition. Free-flexing joints are commonly built in diameters up to 48 in (1219 mm), while controlled-flexing joints are commonly built in diameters up to 24 in (609.6 mm). For a more precise calculation procedure, consult the Flexonics Division, Universal Oil Products Company.

[1]Consult the expansion-joint manufacturer for an exact procedure for computing the anchor forces.

DESIGN OF STEAM TRANSMISSION PIPING

Design a steam transmission pipe to supply a load that is 1700 ft (518.2 m) from the power plant. The terrain permits a horizontal run between the power plant and the load. Maximum steam flow required by the load is 300,000 lb/h (135,000 kg/h), whereas the average steam flow required is estimated as 150,000 lb/h (67,500 kg/h). The maximum steam pressure at the load must not exceed 150 lb/in^2 (abs) (1034.1 kPa) saturated. Superheated steam at 450 lb/in^2 (abs) (3102.7 kPa) and 600°F (316°C) is available at the power plant. Two schemes are proposed for the line: (1) Reduce the steam pressure to 180 lb/in^2 (abs) (1240.9 kPa) at the line inlet, thus allowing a 180 − 150 = 30-lb/in^2 (206.8-kPa) loss in the 1700-ft (518.2-m) long line. This scheme is called the *nominal pressure-loss line*. (2) Admit high-pressure steam to the line and thereby allow the steam pressure to fall to a level slightly greater than 150 lb/in^2 (abs) (1034.1 kPa). Since 600°F (316°C) steam would probably cause expansion and heat-loss difficulties in the pipe, assume that the inlet temperature of the steam is reduced to 455°F (235°C) in a desuperheater in the power plant. There is a 10-lb/in^2 (68.9-kPa) pressure loss between the power plant and the line, reducing the line inlet pressure to 440 lb/in^2 (abs) (3033.4 kPa). Since the pressure can fall about 440 − 150 = 290 lb/in^2 (1999.3 kPa), this will be called the *maximum pressure-loss line*. During design, determine which line is the most economical.

Calculation Procedure:

1. *Determine the required pipe diameter for each condition*

The average steam pressure in the nominal pressure-loss line is (inlet pressure + outlet pressure)/2 = (180 + 150)/2 = 165 lb/in^2 (abs) (1138 kPa). Use this average pressure to determine the pipe size, because the average pressure is more representative of actual conditions in the pipe. Assume that there will be a 5-lb/in^2 (34.5-kPa) pressure drop through any expansion bends and other fittings in the pipe. Then, the allowable friction-pressure drop = 30 − 5 = 25 lb/in^2 (172.4 kPa).

Use the Thomas saturated-steam formula to determine the required pipe diameter, or $d = (80{,}000W/Pv)^{0.5}$, where d = inside pipe diameter, in; W = weight of steam flowing, lb/min; P = average steam pressure, lb/in^2 (abs); v = steam velocity, ft/min. Assuming a steam velocity of 10,000 ft/min (3048 m/min), which is typical for a long steam transmission line, we get $d = [(80{,}000 \times 300{,}000/60)/(165 \times 10{,}000)]^{0.5}$ = 15.32 in (389.1 mm).

The inside diameter of a schedule 40 16-in (406-mm) outside-diameter pipe is, from a table of pipe properties, 15.00 in (381 mm). Assume that a 16-in (406-mm) pipe will be used if schedule 40 wall thickness is satisfactory for the nominal pressure-loss line. Note that the larger flow was used in computing the size of this line because a pipe satisfactory for the larger flow will be acceptable for the smaller flow.

The maximum pressure-loss line will have an average pressure that is a function of the inlet pressure at the pressure-reducing valve at the line outlet. Assume that there is a 10-lb/in^2 (68.9-kPa) drop through this reducing valve. Then steam will enter the valve at 150 + 10 = 160 lb/in^2 (abs) (1103 kPa), and the average line pressure = (440 + 160)/2 = 300 lb/in^2 (abs) (2068 kPa). Using a higher steam velocity [15,000 ft/min (4572 m/min)] for this maximum pressure-loss line than for the nominal pressure-loss line [10,000 ft/min (3048 m/min)], because there is a larger allowable pressure drop, compute the required inside diameter from the Thomas saturated-steam formula because the steam has a superheat of only 456.28 − 455.00 = 1.28°F (2.3°C). Or, $d = [(80{,}000 \times 300{,}000/60)/(300 \times 15{,}000)]^{0.5}$ = 9.44 in (239.8 mm). Since a 10-in (254-mm) schedule 40 pipe has an inside diameter of 10,020 in (254.5 mm), use this size for the maximum pressure-loss line.

2. *Compute the required pipe-wall thickness*

As shown in an earlier calculation procedure, the schedule number SN = $1000P_i/S$. Assuming that seamless carbon-steel ASTM A53 grade A pipe is used for both lines, the *Piping Code* allows a stress of 12,000 lb/in^2 (82.7 MPa) for this material at 600°F (316°C). Then SN = (1000) × (435)/12,000 = 36.2; use schedule 40 pipe, the next largest schedule number for both lines. This computation verifies the assumption in step 1 of the suitability of schedule 40 for each line.

3. *Check the pipeline for critical velocity*

In a steam line, $p_c = W'/Cd^2$, where p_c = critical pressure in pipe, lb/in² (abs), W' = steam flow rate, lb/h; C = constant from Crocker and King—*Piping Handbook; d* = inside diameter of pipe, in.

When the pressure loss in a pipe exceeds 50 to 58 percent of the initial pressure, flow may be limited by the fluid velocity. The limiting velocity that occurs under these conditions is called the *critical velocity,* and the coexisting pipeline pressure, the *critical pressure.*

Critical velocity may limit flow in the 10-in (254-mm) maximum pressure-loss line because the terminal pressure of 150 lb/in² (abs) (1034 kPa) is less than 58 percent of 440 lb/in² (abs) (3033.4 kPa), the inlet pressure. Use the above equation to find the critical pressure. Or, $p_c = (300{,}000)/[(75.15)(10.02)^2]$ = 39.7 lb/in² (abs) (273.7 kPa), using the constant from the *Piping Handbook* after interpolating for the initial enthalpy of 1205.4 Btu/lb (2804 kJ/kg), which is obtained from steam-table values.

Critical velocity would limit flow if the pipeline terminal pressure were equal to, or less than, 39.7 lb/in² (abs) (273.7 kPa). Since the terminal pressure of 150 lb/in² (abs) (1034.1 kPa) is greater than 39.7 lb/in² (abs) (237.7 kPa), critical velocity does not limit the steam flow. With smaller flow rates, the critical pressure will be lower because the denominator in the equation remains constant for a given pipe. Hence, the 10-in (254-mm) line will readily transmit 300,000-lb/h (135,000-kg/h) and smaller flows.

If critical pressure existed in the pipeline, the diameter of the pipe might have to be increased to transmit the desired flow. The 16-in (406.4-mm) line does not have to be checked for critical pressure because its final pressure is more than 58 percent of the initial pressure.

4. *Compute the heat loss for each line*

Assume that 2-in (50.8-mm) thick 85 percent magnesia insulation is used on each line and that the lines will run above the ground in an area having a minimum temperature of 40°F (4.4°C). Set up a computation form as follows:

Pipe size, in (mm)	16 (406.5)	10 (254.0)
Steam temperature, °F (°C)	373 (189)	455 (235)
Air temperature, °F (°C)	40 (4.4)	40 (4.4)
Temperature difference, °F (°C)	333 (184.6)	415 (184.6)
Insulation heat loss, Btu/(h·ft²·°F)* [W/(m²·°C)]	1.11 (6.3)	0.704 (3.99)
Heat loss, Btu/(h·lin ft) (W/m)	370 (356)	292 (281)
Heat loss, Btu/h (kW), for 1700 ft (518 m)	629,000 (184)	496,400 (145.6)
Total heat loss, Btu/h (kW), with a 25% safety factor	786,250 (230)	620,500 (182.0)
Heat loss, Btu/lb (W/kg) of steam, for 300,000-lb/h (135,000-kg/h) flow	2.62 (1.7)	2.07 (1.35)
Heat loss, Btu/lb (W/kg), for the average flow of 150,000 lb/h (67,500 kg/h)	5.24 (3.4)	4.14 (2.69)

*From table of pipe insulation, Ehret Magnesia Manufacturing Company.

In this form, the following computations were made for both pipes: heat loss, Btu/(h·lin ft) = [insulation heat loss, Btu/(h·ft·°F)] (temperature difference,°F); heat loss, Btu/h for 1700 ft (518.2 m) = [heat loss, Btu/(h·lin ft)] (1700); total heat loss, Btu/h, 25 percent safety factor = (heat loss, Btu/h) (1700 ft)(1.25); heat loss, Btu/lb steam = (total heat loss, Btu/h, with a 25 percent safety factor)/(300,000-lb steam).

5. *Compute the leaving enthalpy of the steam in each line*

Acceleration of steam in each line results from an enthalpy decrease of $h_a = (v_2^2 - v_1^2)/2g(778)$, where h_a = enthalpy decrease, Btu/lb; v_2 and v_1 = final and initial velocity of the steam, respectively, ft/s; g = 32.2 ft/s². The velocity at any point x in the pipe is found from the continuity equation $v_x = (W'v_g)/3600\ A_x$, where v_x = steam velocity, ft/s, when the steam volume is v_g ft^3/lb, and A_x is the cross-sectional area of pipe, ft^2, at the point being considered.

For the 16-in (406.4-mm) nominal pressure-loss line with a flow of 300,000 lb/h (135,000 kg/h) at 180-lb/in^2 (abs) (1241-kPa) entering and 150-lb/in^2 (abs) (1034.1-kPa) leaving pressure, using steam and piping table values, v_1 = (300,000)(2.53)/[(3600)(1.23)] = 171.5 ft/s (52.3 m/s); v_2 = 300,000(3.015)/[(3600)(1.23)] = 205 ft/s (62.5 m/s). Then h_a = [(204.5)^2 − (171.5)^2]/[(64.4)(778)] = 0.2504 Btu/lb (0.58 kJ/kg), say 0.25 Btu/lb (0.58 kJ/kg).

By an identical calculation, h_a = 3.7 Btu/lb (8.6 kJ/kg) for the 10-in (254-mm) maximum pressure-loss line when the leaving steam is assumed to be 150 lb/in^2 (abs) (1034.1 kPa), saturated.

Enthalpy of the 180-lb/in^2 (abs) (1241-kPa) saturated steam entering the 16-in (406.4-mm) line is 1196.9 Btu/lb (2784 kJ/kg). Heat loss during 300,000-lb/h (135,000-kg/h) flow is 2.62 Btu/lb (6.1 kJ/kg), as computed in step 4. The enthalpy drop of 0.25 Btu/lb (0.58 kJ/kg) accelerates the steam. Hence, the calculated leaving enthalpy is 1196.9 − (2.62 + 0.25) = 1194.03 Btu/lb (2777.3 kJ/kg). The enthalpy of the leaving steam at 150 lb/in^2 (abs) (1034.1 kPa) saturated is 1194.1 Btu/lb (2777.5 kJ/kg). To have saturated steam leave the line, 1194.10 − 1194.03, or 0.07 Btu/lb (0.16 kJ/kg), must be supplied to the steam. This heat will be obtained from the enthalpy of vaporization given off by condensation of some of the steam in the line.

Make a group of identical calculations for the 10-in (254-mm) maximum pressure-loss line. The enthalpy of 440-lb/in (abs) (3033.4-kPa) 455°F (235°C) entering steam is 1205.4 Btu/lb (2803.8 kJ/kg), found by interpolation in the steam tables. Heat loss during 300,000-lb/h (135,000-kg/h) flow is 2.07 Btu/lb (4.81 kJ/kg). An enthalpy drop of 3.7 Btu/lb (8.6 kJ/kg) accelerates the steam. Hence, the calculated leaving enthalpy = 1205.4 − (2.07 + 3.7) = 1199.63 Btu/lb (2790.3 kJ/kg).

The enthalpy of the leaving steam at 150-lb/in^2 (abs) (1034-kPa) saturated is 1194.1 Btu/lb (2777.5 kJ/kg). As a result, under maximum flow conditions, the steam will be superheated from the entering point to the leaving point of the line. The enthalpy difference of 5.53 Btu/lb = 1199.63 − 1194.10) (12.9 kJ/kg) produces this superheat. Because the steam is superheated throughout the line length, condensation of the steam will not occur during maximum flow conditions.

For most industrial applications, the steam leaving the line may be considered as saturated at the desired pressure. But for precise temperature regulation, some form of pressure-temperature control must be used at the end of long lines.

During average flow conditions of 150,000 lb/h (67,500 kg/h), the line heat loss is 4.14 Btu/lb (9.6 kJ/kg), as computed in step 4. The enthalpy drop to accelerate the steam is 0.925 Btu/lb (2.2 kJ/kg). As in the case of maximum flow, the steam is superheated throughout the length of the 10-in (254-mm) maximum pressure-loss line because the calculated leaving enthalpy is 1205.40 − 5.07 = 1200.33 Btu/lb (2791.9 kJ/kg).

6. *Compute the quantity of condensate formed in each line*

For either line, the quantity of condensate formed, lb/h = C = W'(h_g at leaving pressure − calculated leaving h_g)/outlet pressure h_{fg}.

Using computed values from step 5 and steam-table values, we see the 16-in (406.4-mm) line with 300,000 lb/h (135,000 kg/h) flowing forms C = (300,000)(0.07)/863.6 = 24.35, say 24.4 lb/h (10.9 kg/h) of condensate.

Condensation during an average flow of 150,000 lb/h (67,500 kg/h) is found in the same way. The enthalpy drop to accelerate the steam is neglected for average flow in normal pressure-loss lines because the value is generally small. For the 150,000-lb/h (67,500-kg/h) flow, the calculated leaving enthalpy = 1196.90 − 5.24 = 1191.66 Btu/lb (536.3 kg/h). Hence, C = (150,000)(1194.10 − 1191.66)/863.6 = 424 lb/h (190.8 kg/h) say 425 lb/h (191.3 kg/h).

The largest amount of condensate is formed during line warm-up. Condensate-removal equipment—traps and related piping—must be sized up on the basis of the warm-up not the average steam flow. Using a warm-up time of 30 min and the method of an earlier calculation procedure, we see the condensate formed in 16-in (406.4-mm) schedule 40 pipe weighing 83 lb/ft (122.8 kg/m) is, with a 25 percent safety factor to account for radiation, C = 1.25 × (60)(83)(1700)(373 − 40)(0.12)/[(30)(850.8)] = 16,550 lb/h (7448 kg/h). Thus, the trap or traps should have a capacity of about 17,000 lb/h (7650 kg/h) to remove the condensate during the 30-min warm-up period.

Condensate does not form in the 10-in (254-mm) maximum pressure-loss line during either maximum or average flow. Warm-up condensate for a 30-min warm-up period and a 25 percent safety factor is $C = 1.25(60)(40.5)(1700)(455 - 40)(0.12)/[(30)(770.0)] = 11{,}120$ lb/h (5004 kg/h). Thus, the trap or traps should have a capacity of about 11,500 lb/h (5175 kg/h) to remove the condensate during the 30-min warm-up period.

In general, traps sized on a warm-up basis have adequate capacity for the condensate formed during the maximum and average flows. However, the condensate formed under all three conditions must be computed to determine the maximum rate of formation for trap and drain-line sizing.

7. *Determine the number of plain U bends needed*

A 1700-ft (518-m) long steel steam line operating at a temperature in the 400°F (204°C) range will expand nearly 50 in (1270 mm) during operation. This expansion must be absorbed in some way without damaging the pipe. There are four popular methods for absorbing expansion in long transmission lines: plain U bends, double-offset expansion U bends, slip or corrugated expansion joints, and welded-elbow expansion bends. Each of these will be investigated to determine which is the most economical.

Assume that the governing code for piping design in the locality in which the line will be installed requires that the combined stress resulting from bending and pressure S_{bp} not exceed three-fourths the sum of the allowable stress for the piping material at atmospheric temperature S_a and the allowable stress at the operating temperature S_o of the pipe. This is a common requirement. In equation form, $S_{bp} = 0.75(S_a + S_o)$, where each stress is in lb/in^2.

By using allowable stress values from the *Piping Code* or the local code for 16-in (406.4-mm) seamless carbon-steel ASTM A53 grade A pipe operating at 373°F (189°C), $S_{bp} = 0.75(12{,}000 + 12{,}000) = 18{,}000$ lb/in^2 (124.1 MPa).

Determine the longitudinal pressure stress P_L by dividing the end force due to internal pressure F_e lb by the cross-sectional area of the pipe wall a_m in^2, or $P_L = F_e/a_m$. In this equation, $F_e = pa$, where p = pipe operating pressure, lb/in^2 (gage); a = cross-sectional area of the pipe, in^2. Since the 16-in (406.4-mm) line operates at $180 - 14.7 = 165.3$ lb/in^2 (gage) (1139.6 kPa) and, from a table of pipe properties, $a = 176.7$ in^2 (1140 cm^2) and $a_m = 24.35$ in^2 (157.1 cm^2), $P_L = (165.3)(176.7)/24.35 = 1197$, say 1200 lb/in^2 (8.3 MPa). The allowable bending stress at 373°F (189°C), the pipe operating temperature, is then $S_{np} - P_L = 18{,}000 - 1200 = 16{,}800$ lb/in^2 (115.8 MPa).

Assume that the expansion U bend will have a radius of seven times the nominal pipe diameter, or (7)(16 in) = 112 in (284.5 cm). The allowable bending stress is 16,800 lb/in^2 (115.8 MPa). Full *Piping Code* allowable credit will be taken for cold spring; i.e., the pipe will be cut short by 50 percent or more of the computed expansion and sprung into position.

Referring to Crocker and King—*Piping Handbook*, or a similar tabulation of allowable U-bend overall lengths for various operating temperatures, and choosing the length for 400°F (204°C), we see that the next higher tabulated temperature greater than the 373°F (189.4°C) operating temperature, an allowable length of 157.0 ft (47.9 m) is obtained for the bend. Plot a curve of the allowable bend length vs. temperature at 200, 300, 400, and 500°F (93.3, 148.9, 204.4, and 260°C). From this curve, the allowable bend length at 373°F (189.4°C) is found to be 175 ft (53.3 m). This length is based on an allowable pipe stress of 12,000 lb/in^2 (82.7 MPa) and no cold spring. Since the allowable stress is 16,800 lb/in^2 (115.8 MPa) and maximum cold spring is used, permitting a length 1.5 times the tabulated length, the total allowable length per bend = (175.0)(16,800/12,000)(1.5) = 367.5 ft (112 m). With a total length of pipe between the power plant and a load of 1700 ft (518.2 m), the number of bends required = 1700/367.5 = 4.64 bends. Since only a whole number of bends can be used, the next larger whole number, or five bends, would be satisfactory for this 16-in (406.4-mm) line. Each bend would have an overall length, Fig. 26, of 1700/5 = 340 ft (103.6 m).

Find the actual stress S_a in the pipe when five 340-ft (103.6-m) bends are used by setting up a proportion between the tabulated stress and bend length. Thus, the *Piping Handbook* chart is based on a stress of 12,000 lb/in^2 (82.7 MPa) without cold spring. For this stress, the maximum allowable bend length is 175 ft (53.3 m) found by graphical interpolation of the tabular values, as discussed above. When a 340-ft (103.6-m) bend with maximum cold spring is used, the pipe stress is such that the allowable bend length is 340/1.5 = 226.5 ft (69 m). The actual stress in the pipe is therefore $S_a/12{,}000 = 226.5/175$, or $S_a = 15{,}520$ lb/in^2 (107 MPa). This compares favorably with the allowable stress of 16,800 lb/in^2 (115.8 MPa). The actual stress is less because the overall bend length was reduced.

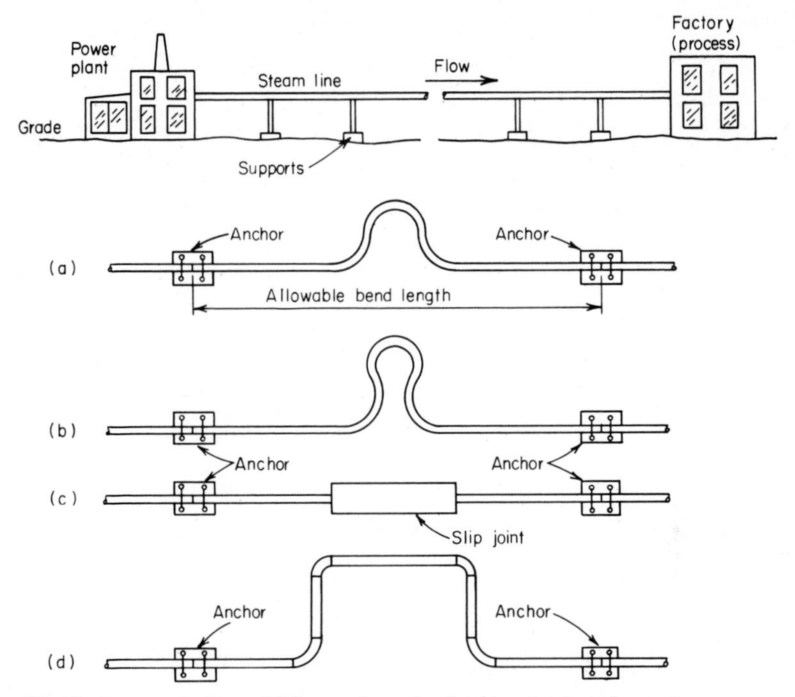

FIG. 26 Process steam line and different schemes for absorbing pipe thermal expansion.

Use the *Piping Handbook* or the method of an earlier calculation procedure to find the anchor reaction forces for these bends. Using the *Piping Handbook* method with graphical interpolation, the anchor reacting force for a 16-in (406.4-mm) schedule 80 bend having a radius of seven times the pipe diameter is 10,550 lb (46.9 kN) at 373°F (189.4°C), based on a 12,000-lb/in^2 (82.7-MPa) stress in the pipe. This tabular reaction must be corrected for the actual pipe stress and for schedule 40 pipe instead of schedule 80 pipe. Thus, the actual anchor reaction, lb = (tabular reaction, lb) [(actual stress, lb/in^2) (tabular stress, lb/in^2)] (moment of inertia, schedule 40 pipe, in^4/moment of inertia of schedule 80 pipe, in^4) = (10,550)(15,520/12,000)(731.9/1156.6) = 8650 lb (38.5 kN). With a reaction of this magnitude, each anchor would be designed to withstand a force of 10,000 lb (44.5 kN). Good design would locate the bends midway between the anchor points; that is, there would be an anchor at each end of each bend. Adjustment for cold spring is not necessary, because it has negligible effect on anchor forces.

Use the same procedure for the 10-in (254-mm) maximum pressure-loss line. If 100-in (254-cm) radius bends are used, seven are required. The bending stress is 14,700 lb/in^2 (65.4 kN), and the anchor force is 2935 lb (13.1 kN). Anchors designed to withstand 3000 lb (13.3 kN) would be used.

8. *Determine the number of double-offset U bends needed*

By the same procedure and the *Piping Handbook* tabulation similar to that in step 7, the 16-in (406.4-mm) nominal pressure-loss line requires two 850-ft (259.1-m) long 112-in (284.5-cm) radius bends. Stress in the pipe is 15,610 lb/in^2 (107.6 MPa), and the anchor reaction is 4780 lb (21.3 kN).

The 10-in (254-mm) maximum pressure-loss line requires five 340-ft (103.6-m) long 70-in (177.8-cm) radius bends. Stress in the pipe is 12,980 lb/in^2 (89.5 MPa), and the anchor reaction is 2090 lb (9.3 kN).

Note that a smaller number of double-offset U bends are required—two rather than five for the 16-in (406.4-mm) pipe and five rather than seven for the 10-in (254-mm) pipe. This shows that double-offset U bends can absorb more expansion than plain U bends.

9. *Determine the number of expansion joints needed*

For any pipe, the total linear expansion e_t in at an elevated temperature above 32°F (0°C) is $e_t = (c_e)(\Delta t)(l)$, where c_e = coefficient of linear expansion, in/(ft·°F); Δt = operating temperature, °F − installation temperature, °F; l = length of straight pipe, ft. Using Crocker and King—*Piping Handbook* as the source for c_e for both lines, we see the expansion of the 373°F (189°C) 16-in (406.4-mm) line with a 40°F (4.4°C) installation temperature is e_t = (12)(0.0000069)(373 − 40)(1700) = 46.8 in (1189 mm). For the 10-in (254-mm) 455°F (235°C) line, e_t = (12)(0.0000072)(455 − 40)(1700) = 61 in (1549 mm). The factor 12 is used in each of these computations because Crocker and King give c_e in in/in; therefore, the pipe total length must be converted to inches by multiplying by 12.

Double-ended slip-type expansion joints that can absorb up to 24 in (609.6 mm) of expansion are available. Hence, the number of joints N needed for each line is: 16-in (406.4-mm) line, N = 46.8/24, or 2; 10-in (254-mm) line, N = 61/24, or 3.

The joints for each line would be installed midway between anchors, Fig. 26. Joints in both lines would be anchored to the ground or a supporting structure. Between the joints, the pipe must be adequately supported and free to move. Roller supports that guide and permit longitudinal movement are usually best for this service. Whereas roller-support friction varies, it is usually assumed to be about 100 lb (444.8 N) per support. At least six supports per 100 ft (30.5 m) are needed for the 16-in (406.4-mm) line and sever per 100 ft (30.5 m) for the 10-in (254-mm) line. Support friction and the number of rollers required are obtained from Crocker and King—*Piping Handbook* or piping engineering data.

The required anchor size and strength depend on the pipe diameter, steam pressure, slip-joint construction, and type of supports used. During expansion of the pipe, friction at the supports and in the joint packing sets up a force that must be absorbed by the anchor. Also, steam pressure in the joint tends to force it apart. The magnitude of these forces is easily computed. With the total force known, a satisfactory anchor can be designed. Slip-joint packing-gland friction varies with different manufacturers, type of joint, and packing used. Gland friction in one popular type of slip joint is about 2200 lb/in (385.3 N/mm) of pipe diameter. Assuming use of these joints in both lines, compute the anchor forces as follows:

	lb	kN
16-in (406.4-mm) nominal pressure-loss line		
Support friction = [(1700 ft)(6 supports per 100 ft)(100 lb per support)]/100	= 10,200	45.3
Gland friction = (2200 lb/in diameter)(16 in)	= 35,200	156.6
Pressure force = [165.3 lb/in² (gage)](176.7-in² pipe area)	= 29,200	129.9
Total force to be absorbed by anchor	= 74,600	331.8
10-in (254-mm) maximum pressure-loss line		
Support friction = [(1700)(7)(100)]/100	= 11,900	52.8
Gland friction = (2200)(10)	= 22,000	97.9
Pressure force = (425.3)(78.9)	= 33,600	149.5
Total force to be absorbed by anchor	= 67,500	300.2

Comparing these results shows that the 10-in (254-mm) line requires smaller anchors than does the 16-in (406.4-mm) line. However, the 16-in (406.4-mm) line requires only three anchors whereas the 10-in (254-mm) line needs four anchors. The total cost of anchors for both lines will be about equal because of the difference in size of the anchors.

The advantages of slip joints become apparent when the piping layout is studied. Only a minimum of pipe is needed because the pipe runs in a straight line between the point of supply and point of use. The amount of insulation is likewise a minimum.

Corrugated expansion joints could be used in place of slip-type joints. These would reduce the required anchor size somewhat because there would be no gland friction. The selection procedure resembles that given for slip-type joints.

10. *Select welded-elbow expansion bends*

Use the graphical analysis in Crocker and King—*Piping Handbook* or in any welding fittings engineering data. Using either method shows that three bends of the most economical shape are suitable for the 16-in (406.4-mm) line and four for the 10-in (254-mm) line. The most economical bend is obtained when the bend width, divided by the distance between the anchor points, is 0.50. With these proportions, the longitudinal stress at the top and bottom of the bend is the same. Use of such bends, although desirable, is not always feasible, because existing piping or structures interfere.

When bend dimensions other than the most economical must be used, the maximum longitudinal stress occurs at the top of the bend when the width/anchor distance < 0.5. When this ratio is > 0.5, the maximum stress occurs at the bottom of the bend. Regardless of the bend type—plain U, double-offset U, or welded—the actual stress in the pipe should not exceed 40 percent of the tensile strength of the pipe material.

11. *Determine the materials, quantities, and costs*

Set up tabulations showing the materials needed and their cost. Table 36 shows the materials required. Piping length is computed by using standard bend tables available in the cited references.

Table 37 shows the approximate material costs for each pipeline. The costs used in preparing this table were the most accurate available at the time of writing. However, the actual numerical values given in the table should not be used for similar design work because price changes may cause them to be incorrect. The important findings in such a tabulation are the differences in total cost. These differences will remain substantially constant even though prices change. Hence, if an $8000 difference exists between two sizes of pipe, this difference will not change appreciably with a moderate rise or fall in unit prices of materials.

Study of Table 37 shows that, in general, lines using double-offset U bends or welding elbows have the lowest material first cost. However, higher first costs do not rule out slip joints or plain U bends. Frequently, use of slip joints will eliminate offsets to clear existing buildings or piping because the pipe path is a straight line. Plain U bends have smaller overall heights than double-offset U bends. For this reason, the plain bend is often preferable where the pipe is run through congested areas of factories.

In some cases, past piping practice will govern line selection. For instance, in a factory that has made wide use of slip joints, the slightly higher cost of such a line might be overlooked. Preference might also be shown for plain U bends, double-offset U bends, or welded bends.

The values given in Table 37 do not include installation, annual operating costs, or depreciation. These have been omitted because accurate estimates are difficult to make unless actual conditions are known. Thus, installation costs may vary considerably according to who does the work. Annual costs are a function of the allowable depreciation, nature of process served, and location of the line. For a given transmission line of the type considered here, annual costs will usually be less for the smaller line.

The economic analysis, as made by the pipeline designer, should include all costs relative to the installation and operation of the line. The allowable cost of money and recommended depreciation period can be obtained from the accounting department.

TABLE 36 Summary of Material Requirements for Various Lines

Means used to absorb expansion	Number of anchors required		Approximate number of supports required		Approximate feet (meters) of pipe and insulation required	
	Pipe size, in (mm)					
	10 (254)	16 (406.4)	10 (254)	16 (406.4)	10 (254)	16 (406.4)
Plain U bends	9	5	127	120	2120 (646.2)	1970 (600.5)
Double-offset U bends	6	3	119	114	1985 (605.0)	1820 (554.7)
Slip joints	4	3	102	102	1700 (518.2)	1700 (518.2)
Welding elbows	5	4	106	106	1760 (536.4)	1760 (536.4)

TABLE 37 Approximate Material Costs for Various Lines

Means used to absorb expansion	Total material cost, $		Condensate removal equipment		Cost of anchors, $		Cost of supports, $		Cost of insulation, $		Cost of pipe and bends or joints, $	
	Pipe size, in (mm)											
	10 (254)	16 (406.4)	10 (254)	16 (406.4)	10 (254)	16 (406.4)	10 (254)	16 (406.4)	10 (254)	16 (406.4)	10 (254)	16 (406.4)
Plain U bends	26,500	51,350	2,000	3,000	1,000	1,800	1,800	2,400	3,700	5,650	18,000	38,500
Double-offset U bends	23,800	44,000	2,000	3,000	600	800	1,700	2,300	3,500	5,400	16,000	32,500
Slip joints	29,650	51,775	2,000	3,000	400	600	1,500	2,000	3,000	4,675	22,750	41,500
Welding elbows	23,800	43,975	2,000	3,000	400	800	1,800	2,300	3,600	5,375	16,000	32,500

12. *Select the most economical pipe size*

Table 37 shows that from the standpoint of first costs, the smaller line is more economical. This lower first cost is not, however, obtained without losing some large-line advantages.

Thus, steam leaves the 16-in (406.4-mm) line at 150 lb/in^2 (abs) (1034.1 kPa) saturated, the desired outlet condition. Special controls are unnecessary. With the 10-in (254-mm) line, the desired leaving conditions are not obtained. Slightly superheated steam leaves the line unless special controls are used. Where an exact leaving temperature is needed by the process served, a desuperheater at the end of the 10-in (254-mm) line will be needed. Neglecting this disadvantage, the 10-in (254-mm) line is more economical than is the 16-in (406.4-mm) line.

Besides lower first cost, the small line loses less heat to the atmosphere, has smaller anchor forces, and does not cause steam condensation during average flows. Lower heat losses and condensation reduce operating costs. Therefore, if special temperature controls are acceptable, the 10-in (254-mm) maximum pressure-loss line will be a more economical investment.

Such a conclusion neglects the possibility of future plant expansion. Where expansion is anticipated, installation of a small line now and another line later to handle increased steam requirements is uneconomical. Instead, installation of a large nominal pressure-loss line now that can later be operated as a maximum pressure-loss line will be found more economical. Besides the advantage of a single line in crowded spaces, there is a reduction in installation and maintenance costs.

13. *Provide for condensate removal*

Fit a condensate drip line for every 100 ft (30.5 m) of pipe, regardless of size. Attach a trap of suitable capacity (see step 6) to each drip line. Pitch the steam-transmission pipe toward the trap, if possible. Where the condensate must flow *against* the steam, the steam-transmission pipe *must* be sloped in the direction of condensate flow. Every vertical rise of the main line must also be dripped. Where water is scarce, return the condensate to the boiler.

Related Calculations: Use this method to design long steam, gas, liquid, or vapor lines for factories, refineries, power plants, ships, process plants, steam heating systems, and similar installation. Follow the applicable piping code when designing the pipeline.

STEAM DESUPERHEATER ANALYSIS

A spray- or direct-contact-type desuperheater is to remove the superheat from 100,000 lb/h (45,000 kg/h) of 300-lb/in^2 (abs) (2068-kPa) 700°F (371°C) steam. Water at 200°F (93.3°C) is available for desuperheating. How much water must be furnished per hour to produce 30-lb/in^2 (abs) (206.8-kPa) saturated steam? How much steam leaves the desuperheater? If a shell-and-tube type of noncontact desuperheater is used, determine the required water flow rate if the overall coefficient of heat transfer U = 500 Btu/(h·ft^2·°F) [2.8 kW/(m^3·°C)]. How much tube area A is required? How much steam leaves the desuperheater? Assume that the desuperheating water is not allowed to vaporize in the desuperheater.

Calculation Procedure:

1. *Compute the heat absorbed by the water*

Water entering the desuperheater must be heated from the entering temperature, 200°F (93.3°C), to the saturation temperature of 300-lb/in^2 (abs) (2068-kPa) steam, or 417.3°F (214°C). Using the steam tables, we see the sensible heat that must be absorbed by the water = h_f at 417.3°F (214°C) − h_f at 200°F (93.3°C) = 393.81 − 167.99 = 255.81 Btu/lb (525.2 kJ/kg) of water used.

Once the desuperheating water is at 417.3°F (214°C), the saturation temperature of 300°F (148.9°C) steam, the water must be vaporized if additional heat is to be absorbed. From the steam tables, the enthalpy of vaporization at 300 lb/in^2 (abs) (2068 kPa) is h_{fg} = 809.0 Btu/lb (1881.7 kJ/kg). This is the amount of heat the water will absorb when vaporized from 417.3°F (214°C).

Superheated steam at 300 lb/in^2 (abs) (2068 kPa) and 700°F (371°C) has an enthalpy of h_g = 1368.3 Btu/lb (3182.7 kJ/kg), and the enthalpy of 300-lb/in^2 (abs) (2068-kPa) saturated steam is h_g = 1202.8 Btu/lb (2797.7 kJ/kg). Thus 1368.3 − 1202.8 = 165.5 Btu/lb (384.9 kJ/kg) must be absorbed by the water to desuperheat the steam from 700°F (371°C) to saturation at 300 lb/in^2 (abs) (2068 kPa).

2. *Compute the weight of water required for the spray*

The weight of water evaporated by 1 lb (0.45 kg) of steam while it is being desuperheated = heat absorbed by water, Btu/lb of steam/heat required to evaporate 1 lb (0.45 kg) of water entering the desuperheater at 200°F (93.3°C), Btu = 165.5/(225.81 + 809.0) = 0.16 lb (0.07 kg) of water. Since 100,000-lb/h (45,000 kg/h) of steam is being desuperheated, the water flow rate required = (0.16)(100,000) = 16,000 lb/h (7200 kg/h). Water for direct-contact desuperheating can be taken from the feedwater piping or from the boiler.

Note that 16,000 lb/h (7200 kg/h) of additional steam will leave the desuperheater because the superheated steam is not condensed while being desuperheated. Thus, the total flow from the desuperheater = 100,000 + 16,000 = 116,000 lb/h (52,200 kg/h).

3. *Compute the tube area required in the desuperheater*

The total heat transferred in the desuperheater, Btu/h = UAt_m, where t_m = logarithmic mean temperature difference across the heater. Using the method for computing the logarithmic temperature difference given elsewhere in this handbook, or a graphical solution as in Perry—*Chemical Engineers' Handbook*, we find t_m = 134°F (74.4°C) with desuperheating water entering at 200°F (93.3°C) and leaving at 430°F (221.1°C), a temperature about 13°F (7°C) higher than the leaving temperature of the saturated steam, 417.3°F (214°C). Steam enters the desuperheater at 700°F (371°C). Assumption of a leaving water temperature 10 to 15°F (5.6 to 8.3°C) higher than the steam temperature is usually made to ensure an adequate temperature difference so that the desired heat-transfer rate will be obtained. If the graphical solution is used, the greatest temperature difference then becomes 700 − 200 = 500°F (278°C), and the least temperature difference = 430 − 417.3 = 12.7°F (7°C).

Then the heat transferred = (500)(A)(134), whereas the heat given up by the steam is, from step 1, (100,000 lb/h)(165.5 Btu/lb) [(45,000 kg/h)(384.9 kJ/kg)]. Since the heat transferred = the heat absorbed, (500)(A)(134) = (100,000)(165.5); A = 247 ft^2 (22.9 m^2), say 250 ft^2 (23.2 m^2).

4. *Compute the required water flow*

Heat transferred to the water = (500)(247)(134) Btu/h (W). The temperature rise of the water during passage through the desuperheater = outlet temperature, °F − inlet temperature = outlet temperature, °F = 430 − 200 = 230°F (127.8°C). Since the specific heat of water = 1.0, closely, the heat absorbed by the water = (flow rate, lb/h)(230)(1.0). Then the heat transferred = heat absorbed, or (500)(247)(134) = (flow rate, lb/h)(230)(1.0); flow rate = 72,000 lb/h (32,400 kg/h). Since the water and steam do *not* mix, the steam output of the desuperheater = steam input = 100,000 lb/h (45,000 kg/h).

Only about 25 percent as much water, 16,000 lb/h (7200 kg/h), is required by the direct-contact desuperheater as compared with the indirect desuperheater. The indirect type of superheater requires more cooling water because the enthalpy of vaporization, nearly 1000 Btu/lb (2326 kJ/kg) of water, is not used to absorb heat. Some indirect-type desuperheaters are designed to permit the desuperheating water to vaporize. This steam is returned to the boiler. The water-consumption determination and the calculation procedure for this type are similar to the spray-type discussed earlier. Where the water does not vaporize, it must be kept at a high enough pressure to prevent vaporization.

Related Calculations: Use this method to analyze steam desuperheaters for any type of steam system—industrial, utility, heating, process, or commercial.

STEAM ACCUMULATOR SELECTION AND SIZING

Select and size a steam accumulator to deliver 10,000 lb/h (4500 kg/h) of 25-lb/in^2 (abs) (172.4-kPa) steam for peak loads in a steam system. Charging steam is available at 75 lb/in^2 (abs) (517.1 kPa). Room is available for an accumulator not more than 30 ft (9.1 m) long, 20 ft (6.1 m) wide, and 20 ft (6.1 m) high. How much steam is required for startup?

Calculation Procedure:

1. *Determine the required water capacity of the accumulator*

One lb (0.45 kg) of water stored in this accumulator at 75 lb/in^2 (abs) (517.1 kPa) has a saturated liquid enthalpy h_f = 277.43 Btu/lb (645.3 kJ/kg) from the steam tables; whereas for 1 lb (0.45

kg) of water at 25 lb/in^2 (abs) (172.4 kPa), h_f = 208.42 Btu/lb (484.8 kJ/kg). In an accumulator, the stored water flashes to steam when the pressure on the outlet is reduced. For this accumulator, when the pressure on the 75-lb/in^2 (abs) (517.1-kPa) water is reduced to 25 lb/in^2 (abs) (172.4 kPa) by a demand for steam, each pound of stored 75-lb/in^2 (517.1-kPa) water flashes to steam, releasing 277.43 − 208.42 = 69.01 Btu/lb (160.5 kJ/kg).

The enthalpy of vaporization of 25 lb/in^2 (abs) (172.4-kPa) steam is h_{fg} = 952.1 Btu/lb (2215 kJ/kg). Thus, 1 lb (0.45 kg) of 75-lb/in^2 (abs) (517.1-kPa) water will form 69.01/952.1 = 0.0725 lb (0.03 kg) of steam. To supply 10,000 lb/h (4500 kg/h) of steam, the accumulator must store 10,000/0.0725 = 138,000 lb/h (62,100 kg/h) of 75-lb/in^2 (abs) (517.1-kPa) water.

Saturated water at 75 lb/in^2 (abs) (517.1 kPa) has a specific volume of 0.01753 ft^3/lb (0.001 m^3/kg) from the steam tables. Since density = 1/specific volume, the density of 75-lb/in^2 (abs) (517.1-kPa) saturated water = 1/0.01753 = 57 lb/ft^3 (912.6 kg/m^3). The volume required in the accumulator to store 138,000 lb (62,100 kg) of 75-lb/in^2 (abs) (517.1-kPa) water = total weight, lb/density of water = 138,000/57 = 2420 ft^3 (68.5 m^3).

2. *Select the accumulator dimensions*

Many steam accumulators are cylindrical because this shape permits convenient manufacture. Other shapes—rectangular, cubic, etc.—may also be used. However, a cylindrical shape is assumed here because it is the most common.

The usual accumulator that serves as a reserve steam supply between a boiler and a load (often called a Ruths-type accumulator) can safely release steam at the rate of 0.3 [accumulator storage pressure, lb/in^2 (abs)] lb/ft^2 of water surface per hour [kg/(m^2·h)]. Thus, this accumulator can release (0.3)(75) = 22.5 lb/(ft^2·h) [112.5 kg/(m^2·h)]. Since a release rate of 10,000 lb/h (4500 kg/h) is desired, the surface area required = 10,000/225 = 445 ft^2 (41.3 m^2).

Space is available for a 30-ft (9.1-m) long accumulator. A cylindrical accumulator of this length would require a diameter of 445/30 = 14.82 ft (4.5 m), say 15 ft (4.6 m). When half full of water, the accumulator would have a surface area (30)(15) = 450 ft^2 (41.8 m^2).

Once the accumulator dimensions are known, its storage capacity must be checked. The volume of a horizontal cylinder of d-ft diameter and l-ft length = $(\pi d^2/4)(l)$ = $(\pi \times 15^2/4)(3)$ = 5300 ft^3 (150 m^3). When half full, this accumulator could store 5300/2 = 2650 ft^3 (75 m^3). Since, from step 1, a capacity of 2420 ft^3 (68.5 m^3) is required, a 15 × 30 ft (4.6 × 9.1 m) accumulator is satisfactory. A water-level controller must be fitted to the accumulator to prevent filling beyond about the midpoint. In this accumulator, the water level could rise to about 60 percent, or (0.60)(15) = 9 ft (2.7 m), without seriously reducing the steam capacity. When an accumulator delivers steam from a more-than-half-full condition, its releasing capacity increases as the water level falls to the midpoint, where the release area is a maximum. Since most accumulators function for only short periods, say 5 or 10 min, it is more important that the vessel be capable of delivering the desired rate of flow than that it deliver the last pound of steam in its lb/h rating.

If the size of the accumulator computed as shown above is unsatisfactory from the standpoint of space, alter the dimensions and recompute the size.

3. *Compute the quantity of charging steam required*

To start an accumulator, it must first be partially filled with water and then charged with steam at the charging pressure. The usual procedure is to fill the accumulator from the plant feedwater system. Assume that the water used for this accumulator is at 14.7 lb/in^2 (abs) and 212°F (101.3 kPa and 100°C) and that the accumulator vessel is half-full at the start.

For any accumulator, the weight of charging steam required is found by solving the following heat-balance equation: (weight of starting water, lb)(h_f of starting water, Btu/lb) + (weight of charging steam, lb)(charging steam h_g, Btu/lb) = (weight of charging steam, lb + weight of starting water, lb)(h_f at charging pressure, Btu/lb). For this accumulator with a 75-lb/in^2 (abs) (517-kPa) charging pressure and 212°F (100°C) starting water, the first step is to compute the weight of water in the half-full accumulator. Since, from step 2, the accumulator must contain 2420 ft^3 (68.5 m^3) of water, this water has a total weight of (volume of water, ft^3)/(specific volume of water, ft^3/lb) = 2420/0.01672 = 144,600 lb (65,070 kg). However, the accumulator can actually store 2650 ft^3 (75 m^3) of water. Hence, the actual weight of water = 2650/0.1672 = 158,300 lb (71,235 kg). Then, with C = weight of charging steam, lb, (158,300)(180.07) + $(C)(1181.9) = (C + 158{,}300)(277.43)$; C = 17,080 lb (7686 kg) of steam.

Once the accumulator is started up, less steam will be required. The exact amount is computed in the same manner, by using the steam and water conditions existing in the accumulator.

Related Calculations: Use this method to size an accumulator for any type of steam service—heating, industrial, process, utility. The operating pressure of the accumulator may be greater or less than atmospheric.

SELECTING PLASTIC PIPING FOR INDUSTRIAL USE

Select the material, schedule number, and support spacing for a 1-in (25.4-mm) nominal-diameter plastic pipe conveying ethyl alcohol liquid having a temperature of 75°F (23.9°C) and a pressure of 400 lb/in^2 (2758 kPa). What expansion must be anticipated if a 1000-ft (304.8-m) length of the pipe is installed at a temperature of 50°F (10°C)? How does the cost of this plastic pipe compare with galvanized-steel pipe of the same size and length?

Calculation Procedure:

1. *Determine the required schedule number*

Refer to Baumeister and Marks—*Standard Handbook for Mechanical Engineers* or a plastic pipe manufacturer's engineering data for the required schedule number. Table 38 shows typical pressure ratings for various sizes and schedule number polyvinyl chloride (PVC) (plastic) piping.

Table 38 shows that schedule 40 normal-impact grade 1-in (25.4-mm) pipe is unsuitable because its maximum operating pressure with fluid at 75°F (24°C) is 310 lb/in^2 (2.13 MPa). Plain-end 1-in (25.4-mm) schedule 80 pipe is, however, satisfactory because it can withstand pressures up to 435 lb/in^2 (2.99 MPa). Note that threaded schedule 80 pipe can withstand pressures only to 255 lb/in^2 (1757 kPa). Therefore, plain-end normal-impact grade pipe must be used for this installation. High-impact grade pipe, in general, has lower allowable pressure ratings at 75°F (24°C) because the additive used to increase the impact resistance lowers the tensile strength, temperature, and chemical resistance. Data shown in Table 38 are also presented in graphical form in some engineering data.

2. *Select a suitable piping material*

Refer to piping engineering data to determine the corrosion resistance of PVC to ethyl alcohol. A Grinnell Company data sheet rates PVC normal-impact and high-impact pipe as having excellent corrosion resistance to ethyl alcohol at 72 and 140°F (22.2 and 60°C). Therefore, PVC is a suitable piping material for this liquid at its operating temperature of 75°F (24°C).

3. *Find the required support spacing*

Use a tabulation or chart in the plastic-pipe engineering data to find the required support spacing for the pipe. Be sure to read the spacing under the correct schedule number. Thus, a Grinnell Company plastic-piping tabulation recommends a 5-ft 4-in (162.6-cm) spacing for schedule 80 1-in (25.4-mm) PVC pipe that weighs 0.382 lb/ft (0.57 kg/m) when empty. The pipe hangers should not clamp the pipe tightly; instead, free axial movement should be allowed.

4. *Compute the expansion of the pipe*

The temperature of the pipe rises from 50 to 75°F (10 to 24°C) when it is put in operation. This is a rise of 75 − 50 = 25°F (14°C). Table 39 shows the thermal expansion of various types of plastic piping.

TABLE 38 Maximum Operating Pressure, PVC Pipe [normal-impact grade, fluid temperature 75°F (23.9°C) or less]

Pipe size		Schedule 40, plain end		Schedule 80			
				Plain end		Threaded	
in	mm	lb/in^2	MPa	lb/in^2	MPa	lb/in^2	MPa
½	12.7	410	2.83	575	3.96	330	2.28
¾	19.1	335	2.31	470	3.24	285	1.97
1	25.4	310	2.14	435	2.99	255	1.76
1½	38.1	230	1.59	325	2.24	205	1.41

TABLE 39 Thermal Expansion of Plastic Pipe

Piping material	Expansion	
	in/(ft·°F)	cm/(m·°C)
Butyrate	0.00118	0.018
Kralastic	0.00067	0.010
Polyethylene	0.00108	0.016
Polyvinyl chloride	0.00054	0.008
Saran	0.00126	0.019

The thermal expansion of any plastic pipe is found from $E_t = LC\ \Delta t$, where E_t = total expansion, in; L = pipe length, ft; C = coefficient of thermal expansion, in/(ft·°F), from Table 39, Δt = temperature change of the pipe, °F. For this pipe, $E_t = (1000)(0.00054)(25) = 13.5$ in (342.9 mm) when the temperature rises from 50 to 75°F (10 to 24°C).

5. ***Determine the relative cost of the pipe***

Check the prices of galvanized-steel and PVC pipe as quoted by various suppliers. These quotations will permit easy comparison. In this case, the two materials will be approximately equal in per-foot cost.

Related Calculations: Use the method given here for selecting plastic pipe for any service—process, domestic, or commercial—conveying any fluid or gas. Note that the maximum operating pressure of plastic piping is normally taken as about 20 percent of the bursting pressure. The allowable operating pressure decreases with an increase in temperature. The maximum allowable operating temperature is usually 150°F (65.5°C). The pressure loss caused by pipe friction in plastic pipe is usually about one-half the pressure loss in galvanized-steel pipe of the same diameter. Pressure loss for plastic piping is computed in the same way as for steel piping.

FRICTION LOSS IN PIPES HANDLING SOLIDS IN SUSPENSION

What is the friction loss in 800 ft (243.8 m) of 6-in (152.4-mm) schedule 40 pipe when 400 gal/min (25.2 L/s) of sulfate paper stock is flowing? The consistency of the sulfate stock is 6 percent.

Calculation Procedure:

1. ***Determine the friction loss in the pipe***

There are few general equations for friction loss in pipes conveying liquids having solids in suspension. Therefore, most practicing engineers use plots of friction loss available in engineering handbooks, *Cameron Hydraulic Data, Standards of the Hydraulic Institute,* and from pump engineering data. Figure 27 shows one set of typical friction-loss curves based on work done at the University of Maine on the data of Brecht and Heller of the Technical College, Darmstadt, Germany, and published by Goulds Pumps, Inc. There is a similar series of curves for commonly used pipe sizes from 2 through 36 in (50.8 through 914.4 mm).

Enter Fig. 27 at the pipe flow rate, 400 gal/min (25.2 L/s), and project vertically upward to the 6 percent consistency curve. From the intersection, project horizontally to the left to read the friction loss as 60 ft (18.3 m) of liquid per 100 ft (30.5 m) of pipe. Since this pipe is 800 ft (243.8 m) long the total friction-head loss in the pipe = (800/100)(60) = 480 ft (146.3 m) of liquid flowing.

2. ***Correct the friction loss for the liquid consistency***

Friction-loss factors are usually plotted for one type of liquid, and correction factors are applied to determine the loss for similar, but different, liquids. Thus, with the Goulds charts, a factor of 0.9 is used for soda, sulfate, bleached sulfite, and reclaimed paper stocks. For ground wood, the factor is 1.40.

When the stock consistency is less than 1.5 percent, water-friction values are used. Below a

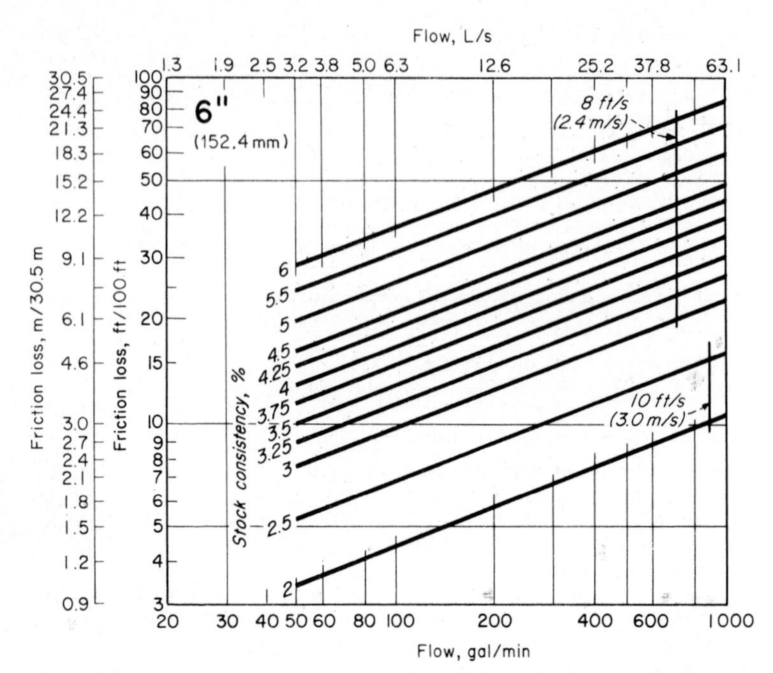

FIG. 27 Friction loss of paper stock in 4-in (101.6-mm) steel pipe. *(Goulds Pumps, Inc.)*

consistency of 3 percent, the velocity of flow should not exceed 10 ft/s (3.05 m/s). For suspensions of 3 percent and above, limit the maximum velocity in the pipe to 8 ft/s (2.4 m/s).

Since the liquid flowing in this pipe is sulfate stock, use the 0.9 correction factor, or the actual total friction head = (0.9)(480) = 432 ft (131.7 m) of sulfate liquid. Note that Fig. 27 shows that the liquid velocity is less than 8 ft/s (2.4 m/s).

Related Calculations: Use this procedure for soda, sulfate, bleached sulfite, and reclaimed and ground-wood paper stock. The values obtained are valid for both suction and discharge piping. The same general procedure can be used for sand mixtures, sewage, slurries, trash, sludge, and foods in suspension in a liquid.

Heat Transfer and Heat Exchangers

REFERENCES: Goldstein—*Heat Transfer in Energy Conservation*, ASME; Isachenko—*Heat Transfer*, Mir (Moscow); Karlekar and Desmond—*Engineering Heat Transfer*, West; Kays and Crawford—*Convective Heat and Mass Transfer*, McGraw-Hill; Butterworth and Hewitt—*Two Phase Flow and Heat Transfer*, Oxford University Press; French—*Heat Transfer and Fluid Flow in Nuclear Systems*, Pergamon Press; Frost—*Heat Transfer at Low Temperatures*, Plenum Press; McAdams—*Heat Transmission*, McGraw-Hill; Kern—*Process Heat Transfer*, McGraw-Hill; General Electric Company—*Electric Heaters and Heating Devices*; Jakob—*Heat Transfer*, Wiley; Bosworth—*Heat Transfer Phenomena*, Wiley; Kays and London—*Compact Heat Exchangers*, McGraw-Hill; Kraus—*Cooling Electronic Equipment*, Prentice-Hall; Fraas and Ozisik—*Heat Exchanger Design*, Wiley; Heat Transfer Research, Inc.—*Design Manual*; API Standards—*Heat Exchangers for General Refinery Service*; Giedt—*Principles of Engineering Heat Transfer*, Van Nostrand; Eckert and Drake—*Heat and Mass Transfer*, McGraw-Hill; Schnieder—*Conduction Heat Transfer*, Addison-Wesley; Kreith—*Principles of Heat Transfer*, International Textbook; Perry—*Chemical Engineers' Handbook*, McGraw-Hill; Carslaw and Jaeger—*Conduction of Heat in Solids*, Oxford; Wilkes—*Heat Insulation*, Wiley.

SELECTING TYPE OF HEAT EXCHANGER FOR A SPECIFIC APPLICATION

Determine the type of heat exchanger to use for each of the following applications: (1) heating oil with steam; (2) cooling internal-combustion engine liquid coolant; (3) evaporating a hot liquid. For each heater chosen, specify the typical pressure range for which the heater is usually built and the typical range of the overall coefficient of heat transfer U.

Calculation Procedure:

1. *Determine the heat-transfer process involved*

In a heat exchanger, one or more of four processes may occur: heating, cooling, boiling, or condensing. Table 1 lists each of these four processes and shows the usual heat-transfer fluids involved. Thus, the heat exchangers being considered here involve (*a*) oil heater—heating—vapor-liquid; (*b*) internal-combustion engine coolant—cooling—gas-liquid; (*c*) hot-liquid evaporation—boiling—liquid-liquid.

2. *Specify the heater action and the usual type selected*

Using the same identifying letters for the heaters being selected, Table 1 shows the action and usual type of heater chosen. Thus,

Action	Type
a. Steam condensed; oil heated	Shell-and-tube
b. Air heated; water cooled	Tubes in open air
c. Waste liquid cooled; water boiled	Shell-and-tube

3. *Specify the usual pressure range and typical U*

Using the same identifying letters for the heaters being selected, Table 1 shows the action and usual type of heater chosen. Thus,

	Typical U range	
Usual pressure range	Btu/(h·°F·ft²)	W/(m²·°C)
a. 0–500 lb/in² (abs) (0 to 3447 kPa)	20–60	113.6–340.7
b. 0–100 lb/in² (abs) (0 to 689.4 kPa)	2–10	11.4–56.8
c. 0–500 lb/in² (abs) (0 to 3447 kPa)	40–150	227.1–851.7

4. *Select the heater for each service*

Where the heat-transfer conditions are normal for the type of service met, the type of heater listed in step 2 can be safely used. When the heat-transfer conditions are unusual, a special type of heater may be needed. To select such a heater, study the data in Table 1 and make a tentative selection. Check the selection by using the methods given in the following calculation procedures in this section.

Related Calculations: Use Table 1 as a general guide to heat-exchanger selection in any industry—petroleum, chemical, power, marine, textile, lumber, etc. Once the general type of heater and its typical U value are known, compute the required size, using the procedures given later in this section.

SHELL-AND-TUBE HEAT EXCHANGER SIZE

What is the required heat-transfer area for a parallel-flow shell-and-tube heat exchanger used to heat oil if the entering oil temperature is 60°F (15.6°C), the leaving oil temperature is 120°F (48.9°C), and the heating medium is steam at 200 lb/in² (abs) (1378.8 kPa)? There is no subcool-

TABLE 1 Heat-Exchanger Selection Guide°

	Heat-transfer fluids	Equipment	Action	Type†	Pressure range‡	Typical range of U§	
Heating	Liquid-liquid	Boiler-water blowdown exchanger	Blowdown cooled, feedwater heated	S	M, H	50–300	(0.28–1.7)
		Laundry-water heat reclaimer	Waste water cooled, feed heated	S	L	30–200	(0.17–1.1)
		Service-water heater	Waste liquid cooled, water heated	S	L, H	50–300	(0.28–1.7)
	Vapor-liquid	Bleeder heater	Steam condensed, feedwater heated	S	L, H	200–800	(1.1–4/5)
		Deaerating feed heater	Steam condensed, feedwater heated	M	L, M	DC	
		Jet heater	Steam condensed, water heated	M	L	DC	
		Process kettle	Steam condensed, liquid heated	S	L, M	100–500	(0.57–2.8)
		Oil heater	Steam condensed, oil heated	S	L, M	20–60	(0.11–0.34)
		Service-water heater	Steam condensed, water heated	S	L, M	200–800	(1.1–4.5)
		Open flow-through heater	Steam condensed, water heated	M	L	DC	
		Liquid-sodium steam superheater	Sodium cooled, steam superheated	S	M, H	50–200	(0.28–1.1)
	Gas-liquid	Waste-heat water heater	Waste gas cooled, water heated	T	L	2–10	(0.011–0.057)
		Boiler economizer	Flue gas cooled, feedwater heated	T	M, H	2–10	(0.011–0.057)
		Hot-water radiator	Water cooled, air heated	T	L	1–10	(0.0057–0.057)
	Gas-gas	Boiler air heater	Flue gas cooled, combustion air heated	T, R	L	2–10	(0.011–0.057)
		Gas-turbine regenerator	Flue gas cooled, combustion air heated	T	L	2–10	(0.011–0.057)
	Vapor-gas	Boiler superheater	Combustion gas cooled, steam superheated	T	M, H	2–20	(0.011–0.11)
		Steam pipe coils	Steam condensed, air heated	T	L, M	2–10	(0.011–0.057)
		Steam radiator	Steam condensed, air heated	T	L	2–10	(0.011–0.057)
Cooling	Liquid-liquid	Oil cooler	Water heated, oil cooled	S, D	L, M	20–200	(0.11–1.1)
		Water chiller	Refrigerant boiled, water cooled	S	L, M	30–151	(0.17–0.86)
		Brine cooler	Refrigerant boiled, brine cooled	S	L, M	30–150	(0.17–0.86)
		Transformer-oil cooler	Water heated, oil cooled	S	L, M	20–50	(0.11–0.88)
	Vapor-liquid	Boiler desuperheater	Boiler water heated, steam desuperheated	S, M	M, H	150–800	(0.85–4.5)

Cooling	Gas-liquid	Compressor intercoolers and aftercoolers	Water heated, compressed air cooled	S	L, H	10–20	(0.057–0.11)
		Internal-combustion-engine radiator	Air heated, water cooled	T	L	2–10	(0.011–0.057)
		Generator hydrogen, air coolers	Water heated, hydrogen or air cooled	S	L	2–10	(0.011–0.057)
		Air-conditioning cooler	Water heated, air cooled	T	L	2–10	(0.011–0.057)
		Refrigeration heat exchanger	Brine heated, air cooled	T	L, M	2–10	(0.011–0.057)
		Refrigeration evaporator	Refrigerant boiled, air cooled	T	L, M	2–10	(0.011–0.057)
	Vapor-gas	Boiler desuperheater	Flue gas heated, steam desuperheated	T	M, H	2–8	(0.011–0.045)
Boiling	Liquid-liquid	Hot-liquid evaporator	Waste liquid cooled, water boiled	S	L, H	40–150	(0.23–0.85)
		Liquid-sodium steam generator	Sodium cooled, water boiled	S	M, H	500–1000	(2.8–5.7)
	Vapor-liquid	Evaporator (vacuum)	Steam condensed, water boiled	S	L	400–600	(2.3–3.4)
		Evaporator (high pressure)	Steam condensed, water boiled	S	L, M	400–600	(2.3–3.4)
		Mercury condenser-boiler	Mercury condensed, water boiled	S	M, H	500–700	(2.8–4.0)
	Gas-liquid	Waste-heat steam boiler	Flue gas cooled, water boiled	T	L, H	2–10	(0.011–0.057)
		Direct-fired steam boiler	Combustion gas cooled, water boiled	T	L, H	2–10	(0.011–0.057)
Condensing	Vapor-liquid	Refrigeration condenser	Water heated, refrigerant condensed	S, D	L, M	80–250	(0.45–1.4)
		Steam surface condenser	Water heated, steam condensed	S	L	300–800	(1.7–4.5)
		Steam mixing condenser	Water heated, steam condensed	M	L	DC	
		Intercondenser and aftercondenser	Condensate heated, steam condensed	S	L	15–300	(0.085–1.7)
	Vapor-gas	Air-cooled surface condenser	Air heated, steam condensed	T	L	2–16	(0.011–0.091)

°*Power*.

†*S*—shell-and-tube exchanger; *M*—direct contact mixing exchanger; *T*—tubes in path of moving fluid, or exchanger open to surrounding air; *R*—regenerative plate-type or simple plate-type exchanger; *D*—double-tube exchanger.

‡*L*—highest pressure ranges from 0 to 100 lb/in^2 (abs) (0 to 689.4 kPa); *M*—highest pressure from 100 to 500 lb/in^2 (abs) (689.4 to 3447 kPa); *H*—500 lb/in^2 (abs) (3447 kPa) up.

§Values of *U* represent range of overall heat-transfer coefficients that might be expected in various exchangers. Coefficients are stated in Btu/(h·°F·ft^2) [W/(m^2·°C)] of heating surface. Total heat transferred in exchanger, in Btu/h, is obtained by multiplying a specific value of *U* for that type of exchanger by the surface and the log mean temperature difference. *DC* indicates direct exchange of heat.

ing of condensate in the heat exchanger. The overall coefficient of heat transfer $U = 25$ Btu/(h·°F·ft^2) [141.9 W/(m^2·°C)]. How much heating steam is required if the oil flow rate through the heater is 100 gal/min (6.3 L/s), the specific gravity of the oil is 0.9, and the specific heat of the oil is 0.5 Btu/(lb·°F) [2.84 W/(m^2·°C)]?

Calculation Procedure:

1. *Compute the heat-transfer rate of the heater*

With a flow rate of 100 gal/min (6.3 L/s) or (100 gal/min)(60 min/h) = 6000 gal/h (22,710 L/h), the weight flow rate of the oil, using the weight of water of specific gravity 1.0 as 8.33 lb/gal, is (6000 gal/h) (0.9 specific gravity)(8.33 lb/gal) = 45,000 lb/h (20,250 kg/h), closely.

Since the temperature of the oil rises 120 − 60 = 60°F (33.3°C) during passage through the heat exchanger and the oil has a specific heat of 0.50, find the heat-transfer rate of the heater from the general relation $Q = wc\,\Delta t$, where Q = heat-transfer rate, Btu/h; w = oil flow rate, lb/h; c = specific heat of the oil, Btu/(lb·°F); Δt = temperature rise of the oil during passage through the heater. Thus, $Q = (45{,}000)(0.5)(60) = 1{,}350{,}000$ Btu/h (0.4 MW).

2. *Compute the heater logarithmic mean temperature difference*

The logarithmic mean temperature difference (LMTD) is found from LMTD = $(G - L)/\ln(G/L)$, where G = greater terminal temperature difference of the heater, °F; L = lower terminal temperature difference of the heater, °F; ln = logarithm to the base e. This relation is valid for heat exchangers in which the number of shell passes equals the number of tube passes.

In general, for parallel flow of the fluid streams, $G = T_1 - t_1$ and $L = T_2 - t_2$, where T_1 = heating fluid inlet temperature, °F; T_2 = heating fluid outlet temperature, °F; t_1 = heated fluid inlet temperature, °F; t_2 = heated fluid outlet temperature, °F. Figure 1 shows the maximum and minimum terminal temperature differences for various fluid flow paths.

For this parallel-flow exchanger, $G = T_1 - t_1 = 382 - 60 = 322$°F (179°C), where 382°F (194°C) = the temperature of 200-lb/in^2 (abs) (1379-kPa) saturated steam, from a table of steam properties. Also, $L = T_2 - t_2 = 382 - 120 = 262$°F (145.6°C), where the condensate temperature = the saturated steam temperature because there is no subcooling of the condensate. Then LMTD = $G - L/\ln(G/L) = (322 - 262)/\ln(322/262) = 290$°F (161°C).

3. *Compute the required heat-transfer area*

Use the relation $A = Q/U \times$ LMTD, where A = required heat-transfer area, ft^2; U = overall coefficient of heat transfer, Btu/(ft^2·h·°F). Thus, $A = 1{,}350{,}000/[(25)(290)] = 186.4$ ft^2 (17.3 m^2), say 200 ft^2 (18.6 m^2),

4. *Compute the required quantity of heating system*

The heat added to the oil = Q = 1,350,000 Btu/h, from step 1. The enthalpy of vaporization of 200-lb/in^2 (abs) (1379-kPa) saturated steam is, from the steam tables, 843.0 Btu/lb (1960.8 kJ/kg). Use the relation $W = Q/h_{fg}$, where W = flow rate of heating steam, lb/h; h_{fg} = enthalpy of vaporization of the heating steam, Btu/lb. Hence, $W = 1{,}350{,}000/843.0 = 1600$ lb/h (720 kg/h).

Related Calculations: Use this general procedure to find the heat-transfer area, fluid outlet temperature, and required heating-fluid flow rate when true parallel flow or counterflow of the fluids occurs in the heat exchanger. When such a true flow does *not* exist, use a suitable correction factor, as shown in the next calculation procedure.

The procedure described here can be used for heat exchangers in power plants, heating systems, marine propulsion, air-conditioning systems, etc. Any heating or cooling fluid—steam, gas, chilled water, etc.—can be used.

To select a heat exchanger by using the results of this calculation procedure, enter the engineering data tables available from manufacturers at the computed heat-transfer area. Read the heater dimensions directly from the table. Be sure to use the next *larger* heat-transfer area when the exact required area is not available.

When there is little movement of the fluid on either side of the heat-transfer area, such as occurs during heat transmission through a building wall, the arithmetic mean (average) temperature difference can be used instead of the LMTD. Use the LMTD when there is rapid movement of the fluids on either side of the heat-transfer area and a rapid change in temperature in one, or both, fluids. When one of the two fluids is partially, but not totally, evaporated or condensed, the

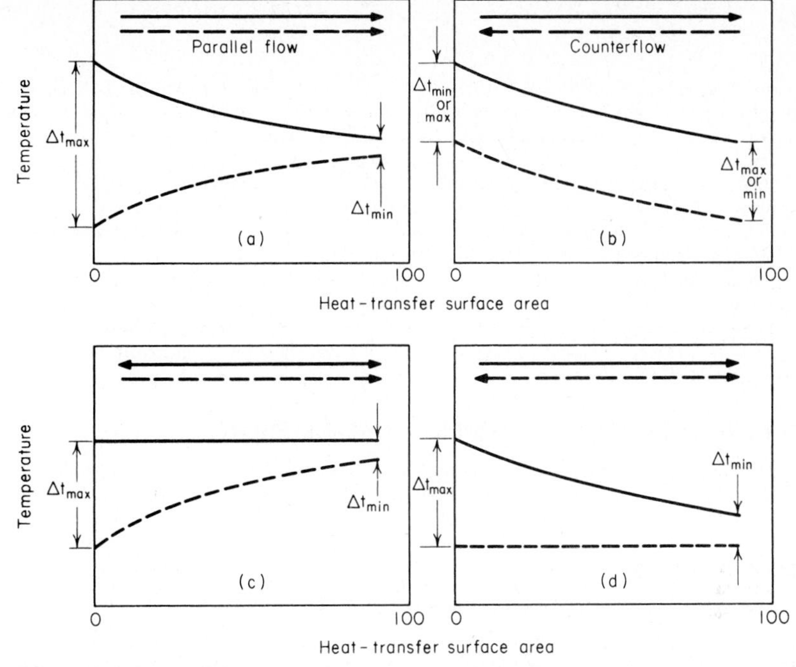

FIG. 1 Temperature relations in typical parallel-flow and counterflow heat exchangers.

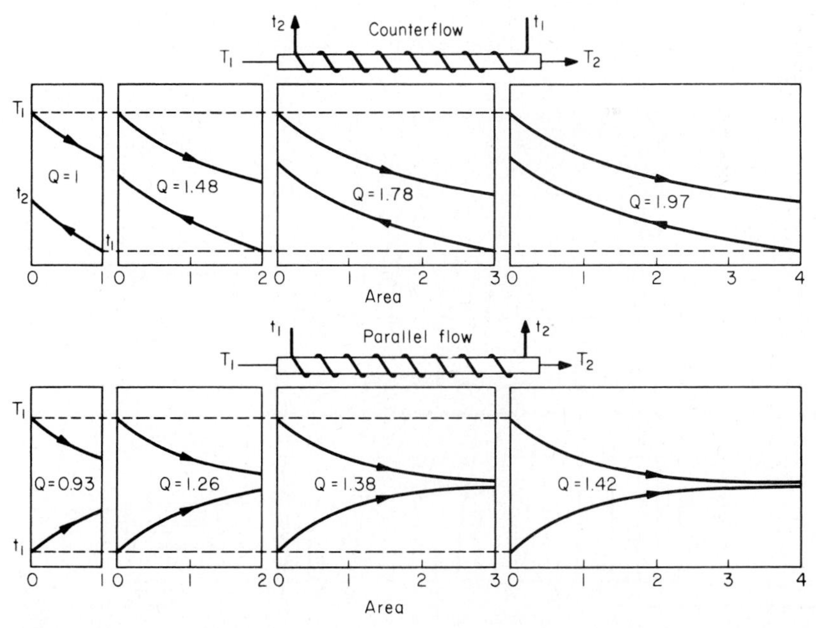

FIG. 2 For certain conditions, the area between the temperature curves measures the amount of heat being transferred.

true mean temperature difference is different from the arithmetic mean and the LMTD. Special methods, such as those presented in Perry—*Chemical Engineers' Handbook*, must be used to compute the actual temperature difference under these conditions.

When two liquids or gases with constant specific heats are exchanging heat in a heat exchanger, the area between their temperature curves, Fig. 2, is a measure of the total heat being transferred. Figure 2 shows how the temperature curves vary with the amount of heat-transfer area for counterflow and parallel-flow exchangers when the fluid inlet temperatures are kept constant. As Fig. 2 shows, the counterflow arrangement is superior.

If enough heating surface is provided, in a counterflow exchanger, the leaving cold-fluid temperature can be raised above the leaving hot-fluid temperature. This cannot be done in a parallel-flow exchanger, where the temperatures can only approach each other regardless of how much surface is used. The counterflow arrangement transfers more heat for given conditions and usually proves more economical to use.

HEAT-EXCHANGER ACTUAL TEMPERATURE DIFFERENCE

A counterflow shell-and-tube heat exchanger has one shell pass for the heating fluid and two shell passes for the fluid being heated. What is the actual LMTD for this exchanger if $T_1 = 300°F$ (148.9°C), $T_2 = 250°F$ (121°C), $t_1 = 100°F$ (37.8°C), and $t_2 = 230°F$ (110°C)?

Calculation Procedure:

1. *Determine how the LMTD should be computed*

When the numbers of shell and tube passes are unequal, true counterflow does not exist in the heat exchanger. To allow for this deviation from true counterflow, a correction factor must be applied to the logarithmic mean temperature difference (LMTD). Figure 3 gives the correction factor to use.

2. *Compute the variables for the correction factor*

The two variables that determine the correction factor are shown in Fig. 3 as $P = (t_2 - t_1)/(T_1 - t_1)$ and $R = (T_1 - T_2)/(t_2 - t_1)$. Thus, $P = (230 - 100)/(300 - 100) = 0.65$, and $R = (300 - 250)/(230 - 100) = 0.385$. From Fig. 3, the correction factor is $F = 0.90$ for these values of P and R.

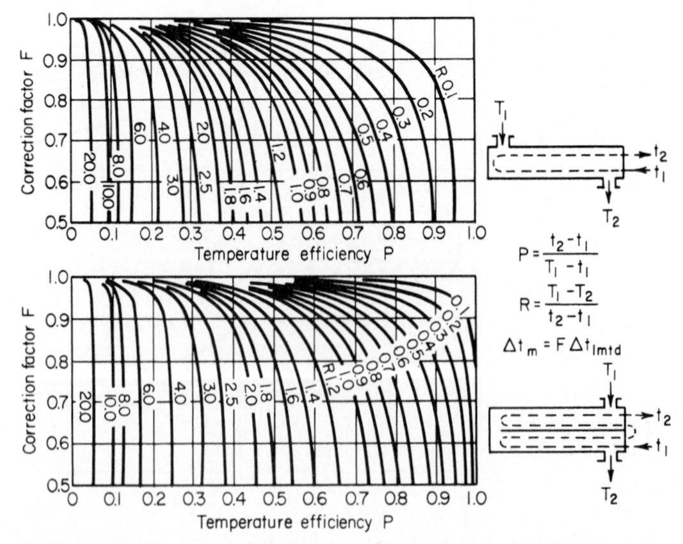

FIG. 3 Correction factors for LMTD when the heater flow path differs from true counterflow. *(Power.)*

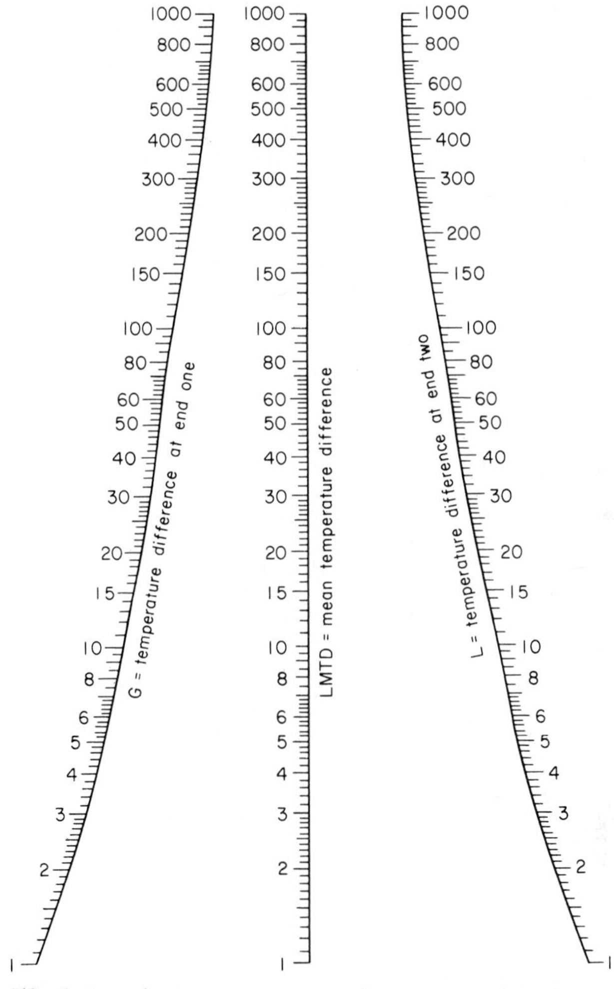

FIG. 4 Logarithmic mean temperature for a variety of heat-transfer applications.

3. *Compute the theoretical LMTD*

Use the relation LMTD = $(G - L)/\ln(G/L)$, where the symbols for counterflow heat exchange are $G = T_2 - t_1$; $L = T_1 - t_2$; ln = logarithm to the base e. All temperatures in this equation are expressed in °F. Thus, $G = 250 - 100 = 150$°F (83.3°C); $L = 300 - 230 = 70$°F (38.9°C). Then LMTD = $(150 - 70)/\ln(150/70) = 105$°F (58.3°C).

4. *Compute the actual LMTD for this exchanger*

The actual LMTD for this or any other heat exchanger is $\text{LMTD}_{\text{actual}} = F(\text{LMTD}_{\text{computed}}) = 0.9(105) = 94.5$°F (52.5°C). Use the actual LMTD to compute the required exchanger heat-transfer area.

Related Calculations: Once the corrected LMTD is known, compute the required heat-exchanger size in the manner shown in the previous calculation procedure. The method given here is valid for both two- and four-pass shell-and-tube heat exchangers. Figure 4 simplifies the computation of the uncorrected LMTD for temperature differences ranging from 1 to 1000°F (−17 to 537.8°C). It gives LMTD with sufficient accuracy for all normal industrial and commercial heat-exchanger applications. Correction-factor charts for three shell passes, six or more tube passes, four shell passes, and eight or more tube passes are published in the *Standards of the Tubular Exchanger Manufacturers Association*.

FOULING FACTORS IN HEAT-EXCHANGER SIZING AND SELECTION

A heat exchanger having an overall coefficient of heat transfer of U = 100 Btu/(ft²·h·°F) [567.8 W/(m²·°C)] is used to cool lean oil. What effect will the tube fouling have on the value of U for this exchanger?

Calculation Procedure:

1. *Determine the heat exchanger fouling factor*

Use Table 2 to determine the fouling factor for this exchanger. Thus, the fouling factor for lean oil = 0.0020.

2. *Determine the actual U for the heat exchanger*

Enter Fig. 5 at the bottom with the clean heat-transfer coefficient of U = 100 Btu/(h·ft²·°F) [567.8 W/(m²·°C)] and project vertically upward to the 0.002 fouling-factor curve. From the intersection with this curve, project horizontally to the left to read the design or actual heat-transfer coefficient as U_a = 78 Btu/(h·ft²·°F) [442.9 W/(m²·°C)]. Thus, the fouling of the tubes causes a reduction of the U value of 100 − 78 = 22 Btu/(h·ft²·°F) [124.9 W/(m²·°C)]. This means that the required heat transfer area must be increased by nearly 25 percent to compensate for the reduction in heat transfer caused by fouling.

Related Calculations: Table 2 gives fouling factors for a wide variety of service conditions in applications of many types. Use these factors as described above; or add the fouling factor to the film resistance for the heat exchanger to obtain the total resistance to heat transfer. Then U = the reciprocal of the total resistance. Use the actual value U_a of the heat-transfer coefficient when sizing a heat exchanger. The method given here is that used by Condenser Service and Engineering Company, Inc.

TABLE 2 Heat-Exchanger Fouling Factors*

Fluid heated or cooled	Fouling factor
Fuel oil	0.0055
Lean oil	0.0020
Clean recirculated oil	0.0010
Quench oils	0.0042
Refrigerants (liquid)	0.0011
Gasoline	0.0006
Steam-clean and oil-free	0.0001
Refrigerant vapors	0.0023
Diesel exhaust	0.013
Compressed air	0.0022
Clean air	0.0011
Seawater under 130°F (54°C)	0.0006
Seawater over 130°F (54°C)	0.0011
City or well water under 130°F (54°C)	0.0011
City or well water over 130°F (54°C)	0.0021
Treated boiler feedwater under 130°F, 3 ft/s (54°C, 0.9 m/s)	0.0008
Treated boiler feedwater over 130°F, 3 ft/s (54°C, 0.9 m/s)	0.0009
Boiler blowdown	0.0022

*Condenser Service and Engineering Company, Inc.

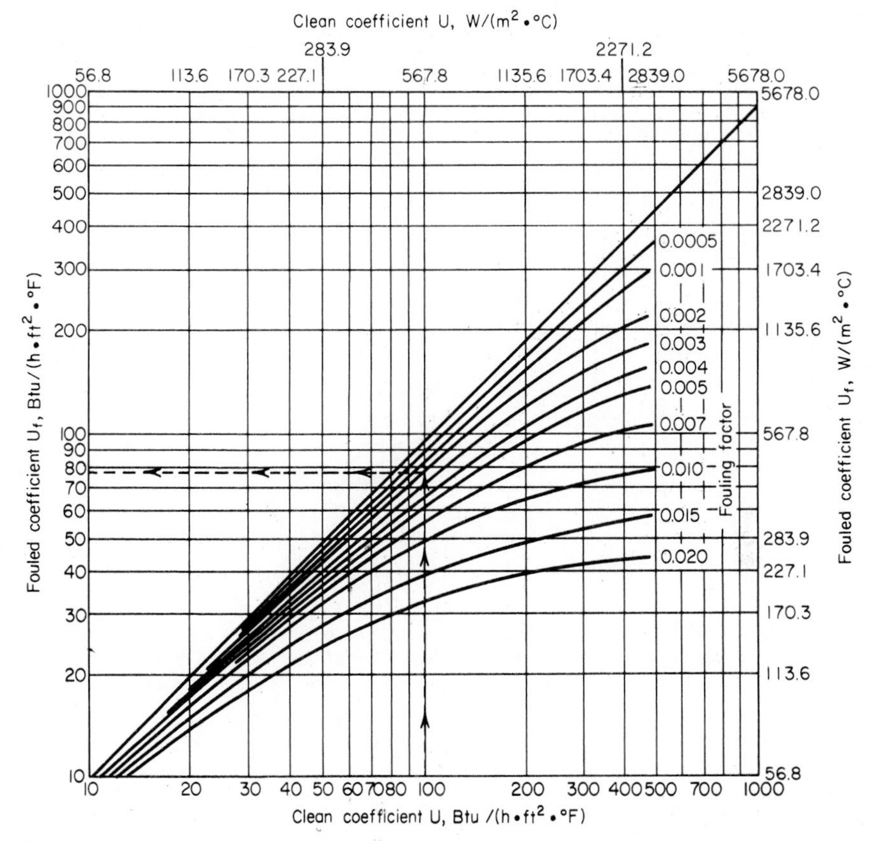

FIG. 5 Effect of heat-exchanger fouling on the overall coefficient of heat transfer. *(Condenser Service and Engineering Co., Inc.)*

HEAT TRANSFER IN BAROMETRIC AND JET CONDENSERS

A counterflow barometric condenser must maintain an exhaust pressure of 2 lb/in^2 (abs) (13.8 kPa) for an industrial process. What condensing-water flow rate is required with a cooling-water inlet temperature of 60°F (15.6°C); of 80°F (26.7°C)? How much air must be removed from this barometric condenser if the steam flow rate is 25,000 lb/h (11,250 kg/h); 250,000 lb/h (112,500 kg/h)?

Calculation Procedure:

1. *Compute the required unit cooling-water flow rate*

Use Fig. 6 as a quick guide to the required cooling-water flow rate for counterflow barometric condensers. Thus, entering the bottom of Fig. 6 at 2-lb/in^2 (abs) (13.8-kPa) exhaust pressure and projecting vertically upward to the 60°F (15.6°C) and 80°F (26.7°C) cooling-water inlet temperature curves show that the required flow rate is 52 gal/min (3.2 L/s) and 120 gal/min (7.6 L/s), respectively, per 1000 lb/h (450 kg/h) of steam condensed.

2. *Compute the total cooling-water flow rate required*

Use this relation: total cooling water required, gal/min = (unit cooling-water flow rate, gal/min per 1000 lb/h of steam condensed) (steam flow, lb/h)/1000. Or, total gpm = (52)(250,000/1000)

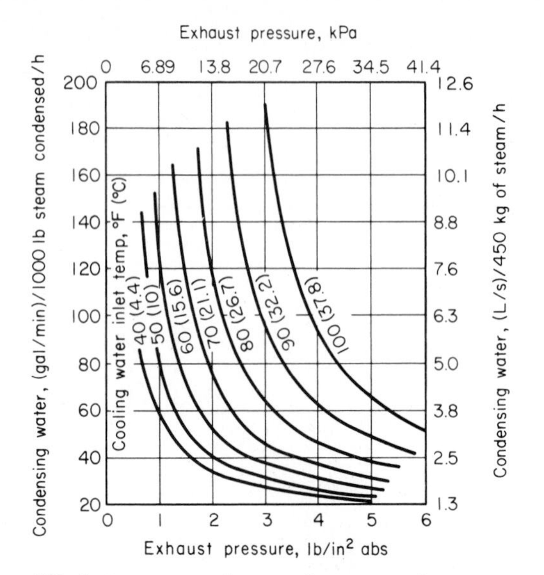

FIG. 6 Barometric condenser condensing-water flow rate.

= 13,000 gal/min (820.2 L/s) of 60°F (15.6°C) cooling water. For 80°F (26.7°C) cooling water, total gpm = (120)(250,000/1000) = 30,000 gal/min (1892.7 L/s). Thus, a 20°F (11.1°C) rise in the cooling-water temperature raises the flow rate required by 30,000 − 13,000 = 17,000 gal/min (1072.5 L/s).

3. *Compute the quantity of air that must be handled*

With a steam flow of 25,000 lb/h (11,250 kg/h) to a barometric condenser, manufacturers' engineering data show that the quantity of air entering with the steam is 3 ft^3/min (0.08 m^3/min); with a steam flow of 250,000 lb/h (112,500 kg/h), air enters at the rate of 10 ft^3/min (0.28 m^3/min). Hence, the quantity of air in the steam that must be handled by this condenser is 10 ft^3/min (0.28 m^3/min).

Air entering with the cooling water varies from about 2 ft^3/min per 1000 gal/min of 100°F (0.06 m^3/min per 3785 L/min of 37.8°C) water to 4 ft^3/min per 1000 gal/min at 35°F (0.11 m^3/min per 3785 L/min at 1.7°C). Using a value of 3 ft^3/min (0.08 m^3/min) for this condenser, we see the quantity of air that must be handled is (ft^3/min per 1000 gal/min)(cooling-water flow rate, gal/min)/1000, or cfm of air = (3)(13,000/1000) = 39 ft^3/min at 60°F (1.1 m^3/min at 15.6°C). At 80°F (26.7°C) cfm = (3)(30,000/1000) = 90 ft^3/min (2.6 m^3/min).

Hence, the total air quantity that must be handled is 39 + 10 = 49 ft^3/min (1.4 m^3/min) with 60°F (15.6°C) cooling water, and 90 + 10 = 100 ft^3/min (2.8 m^3/min) with 80°F (26.7°C) cooling water. The air is usually removed from the barometric condenser by a two-stage air ejector.

Related Calculations: For help in specifying conditions for parallel-flow and counterflow barometric condensers, refer to *Standards of Heat Exchange Institute—Barometric and Low-Level Jet Condensers*. Whereas Fig. 6 can be used for a first approximation of the cooling water required for parallel-flow barometric condensers, the results obtained will not be as accurate as for counterflow condensers.

SELECTION OF A FINNED-TUBE HEAT EXCHANGER

Choose a finned-tube heat exchanger for a 1000-hp (746-kW) four-cycle turbocharged diesel engine having oil-cooled pistons and a cooled exhaust manifold. The heat exchanger will be used only for jacket-water cooling.

TABLE 3 Approximate Rates of Heat Rejection to Cooling Systems*

Four-cycle engines				
Engine type	Normally aspirated, dry pistons, water-jacketed exhaust manifold, Btu/(bhp·hr) (kJ/kWh)	Normally aspirated, oil-cooled pistons, water-jacketed manifold, Btu/(bhp·h) (kJ/kWh)	Turbocharged, oil-cooled pistons, dry manifold, Btu/(bhp·h) (kJ/kWh)	Turbocharged, oil-cooled pistons, cooled manifold, Btu/(bhp·h) (kJ/kWh)
Jacket water	2200–2600 (12.5–14.8)	2000–2500 (11.3–14.2)	1450–1750 (8.2–9.9)	1800–2200 (10.2–12.5)
Lubricating oil	175–350 (1.0–2.0)	300–600 (1.7–3.4)	300–500 (1.7–2.8)	300–500 (1.7–2.8)
Raw water	2375–2950 (13.5–16.7)	2300–3100 (13.1–17.6)	1750–2250 (9.9–12.8)	2100–2700 (11.9–15.3)

Two-cycle engines			
Engine type	Loop scavenging oil-cooled pistons, Btu/(bhp·h) (kJ/kWh)	Uniflow scavenging oil-cooled pistons	
		Opposed piston, Btu/(bhp·h) (kJ/kWh)	Valve in head, Btu/(bhp·h) (kJ/kWh)
Jacket water	1300–1900 (7.4–10.8)	1200–1600 (6.8–9.1)	1700–2100 (9.6–11.9)
Lubricating oil	500–700 (2.8–4.0)	900–1100 (5.1–6.2)	400–750 (2.3–4.3)
Raw water	1800–2600 (10.2–14.8)	2100–2700 (11.9–15.3)	2100–2850 (11.9–16.2)

*Diesel Engine Manufacturers Association; SI values added by handbook editor.

Calculation Procedure:

1. *Determine the heat-exchanger cooling load*

The Diesel Engine Manufacturers Association (DEMA) tabulation, Table 3, lists the heat rejection to the cooling system by various types of diesel engines. Table 3 shows that the heat rejection from the jacket water of a four-cycle turbocharged engine having oil-cooled pistons and a cooled manifold is 1800 to 2200 Btu/(bhp·h) (0.71 to 0.86 kW/kW). Using the higher value, we see the jacket-water heat rejection by this engine is (1000 bhp)[2200 Btu/(bhp·h)] = 2,200,000 Btu/h (644.8 kW).

2. *Determine the jacket-water temperature rise*

DEMA reports that a water temperature rise of 15 to 20°F (8.3 to 11.1°C) is common during passage of the cooling water through the engine. The maximum water discharge temperature reported by DEMA ranges from 140 to 180°F (60 to 82.2°C). Assume a 20°F (11.1°C) water temperature rise and a 160°F (71.1°C) water discharge temperature for this engine.

3. *Determine the air inlet and outlet temperatures*

Refer to weather data for the locality of the engine installation. Assume that the weather data for the locality of this engine show that the maximum dry-bulb temperature met in summer is 90°F (32.2°C). Use this as the air inlet temperature.

Before the required surface area can be determined, the air outlet temperature from the radiator must be known. This outlet temperature cannot be computed directly. Hence, it must be assumed and a trial calculation made. If the area obtained is too large, a higher outlet air temperature must be assumed and the calculation redone. Assume an outlet air temperature of 150°F (65.6°C).

4. *Compute the LMTD for the radiator*

The largest temperature difference for this exchanger is 160 − 90 = 70°F (38.9°C), and the smallest temperature difference is 150 − 140 = 10°F (5.6°C). In the smallest temperature difference expression, 140°F (77.8°C) = water discharge temperature from the engine − cooling-water temperature rise during passage through the engine, or 160 − 20 = 140°F (77.8°C). Then LMTD = (70 − 10)/[ln(70/10)] = 30°F (16.7°C). (Figure 4 could also be used to compute the LMTD).

5. *Compute the required exchanger surface area*

Use the relation $A = Q/U \times \text{LMTD}$, where A = surface area required, ft²; Q = rate of heat transfer, Btu/h; U = overall coefficient of heat transfer, Btu/(h·ft²·°F). To solve this equation, U must be known.

Table 1 in the first calculation procedure in this section shows that U ranges from 2 to 10 Btu/(h·ft²·°F) [56.8 W/(m²·°C)] in the usual internal-combustion-engine finned-tube radiator. Using a value of 5 for U, we get A = 2,200,000/[(5)(30)] = 14,650 ft² (1361.0 m²).

6. *Determine the length of finned tubing required*

The total area of a finned tube is the sum of the tube and fin area per unit length. The tube area is a function of the tube diameter, whereas the finned area is a function of the number of fins per inch of tube length and the tube diameter.

Assume that 1-in (2.5-cm) tubes having 4 fins per inch (6.35 mm per fin) are used in this radiator. A tube manufacturer's engineering data show that a finned tube of these dimensions has 5.8 ft² of area per linear foot (1.8 m²/lin m) of tube.

To compute the linear feet L of finned tubing required, use the relation $L = A/(\text{ft}^2/\text{ft})$, or L = 14,650/5.8 = 2530 lin ft (771.1 m) of tubing.

7. *Compute the number of individual tubes required*

Assume a length for the radiator tubes. Typical lengths range between 4 and 20 ft (1.2 and 6.1 m), depending on the size of the radiator. With a length of 16 ft (4.9 m) per tube, the total number of tubes required = 2530/16 = 158 tubes. This number is typical for finned-tube heat exchangers having large heat-transfer rates [more than 10^6 Btu/h (100 kW)].

8. *Determine the fan horsepower required*

The fan horsepower required can be computed by determining the quantity of air that must be moved through the heat exchanger, after assuming a resistance—say 1.0 in of water (0.025 Pa)—for the exchanger. However, the more common way of determining the fan horsepower is by referring to the manufacturer's engineering data.

Thus, one manufacturer recommends three 5-hp (3.7-kW) fans for this cooling load, and another recommends two 8-hp (5.9-kW) fans. Hence, about 16 hp (11.9 kW) is required for the radiator.

Related Calculations: The steps given here are suitable for the initial sizing of finned-tube heat exchangers for a variety of applications. For exact sizing, it may be necessary to apply a correction factor to the LMTD. These correction factors are published in Kern—*Process Heat Transfer*, McGraw-Hill, and McAdams—*Heat Transfer*, McGraw-Hill.

The method presented here can be used for finned-tube heat exchangers used for air heating or cooling, gas heating or cooling, and similar industrial and commercial applications.

SPIRAL-TYPE HEATING COIL SELECTION

How many feet of heating coil are required to heat 1000 gal/h (1.1 L/s) of 0.85-specific-gravity oil if the specific heat of the oil is 0.50 Btu/(lb·°F) [2.1 kJ/(kg·°C)], the heating medium is 65-lb/in² (gage) (448.2-kPa) steam, and the oil enters at 60°F (15.6°C) and leaves at 125°F (51.7°C)? There is no subcooling of the condensate.

Calculation Procedure:

1. *Compute the LMTD for the heater*

Steam at 65 + 14.7 = 79.7 lb/in^2 (abs) (549.5 kPa) has a temperature of approximately 312°F (155.6°C), as given by the steam tables. Condensate at this pressure has the same approximate temperature. Hence, the entering and leaving temperatures of the heating fluid are approximately the same.

Oil enters the heater at 60°F (15.6°C) and leaves at 125°F (51.7°C). Therefore, the greater temperature G across the heater is G = 312 − 60 = 252°F (140.0°C), and the lesser temperature difference L is L = 312 − 125 = 187°F (103.9°C). Hence, the LMTD = $(G - L)/[\ln(G/L)]$, or (252 − 187)/[ln (252/187)] = 222°F (123.3°C). In this relation, ln = logarithm to the base e = 2.7183. (Figure 4 could also be used to determine the LMTD.)

2. *Compute the heat required to raise the oil temperature*

Water weighs 8.33 lb/gal (1.0 kg/L). Since this oil has a specific gravity of 0.85, it weighs (8.33)(0.85) = 7.08 lb/gal (0.85 kg/L). With 1000 gal/h (1.1 L/s) of oil to be heated, the weight of oil heated is (1000 gal/h)(7.08 lb/gal) = 7080 lb/h (0.89 kg/s). Since the oil has a specific heat of 0.5 Btu/(lb·°F) [2.1 kJ/(kg·°C)] and this oil is heated through a temperature range of 125 − 60 = 65°F (36.1°C), the quantity of heat Q required to raise the temperature of the oil is Q = (7080 lb/h) [0.5 Btu/(lb·°F) (65°F)] = 230,000 Btu/h (67.4 kW).

3. *Compute the heat-transfer area required*

Use the relation $A = Q/(U \times \text{LMTD})$, where Q = heat-transfer rate, Btu/h; U = overall coefficient of heat transfer, Btu/(h·ft^2·°F). For heating oil to 125°F (51.7°C), the U value given in Table 1 is 20 to 60 Btu/(h·ft^2·°F) [0.11 to 0.34 kW/(m^2·°C)]. Using a value of U = 30 Btu/(h·ft^2·°F) [0.17 kW/(m^2·°C)] to produce a conservatively sized heater, we find A = 230,000/[(30)(222)] = 33.4 ft^2 (3.1 m^2) of heating surface.

4. *Choose the coil material for the heater*

Spiral-type tank heating coils are usually made of steel because this material has a good corrosion resistance in oil. Hence, this coil will be assumed to be made of steel.

5. *Compute the heating steam flow required*

To determine the steam flow rate required, use the relation $S = Q/h_{fg}$, where S = steam flow, lb/h; h_{fg} = latent heat of vaporization of the heating steam, Btu/lb, from the steam tables; other symbols as before. Hence, S = 230,000/901.1 = 256 lb/h (0.03 kg/s), closely.

6. *Compute the heating coil pipe diameter*

Steam-heating coils submerged in the liquid being heated are usually chosen for a steam velocity of 4000 to 5000 ft/min (20.3 to 25.4 m/s). Compute the heating pipe cross-sectional area a in^2 from $a = 2.4Sv_g/V$, where v_g = specific volume of the steam at the coil operating pressure, ft^3/lb, from the steam tables; V = steam velocity in the heating coil, ft/min; other symbols as before. With a steam velocity of 4000 ft/min (20.3 m/s), a = 2.4(256)(5.47)/4000 = 0.838 in^2 (5.4 cm^2).

Refer to a tabulation of pipe properties. Such a tabulation shows that the internal transverse area of a schedule 40 1-in (2.5-cm) diameter nominal steel pipe is 0.863 in^2 (5.6 cm^2). Hence, a 1-in (2.5-cm) pipe will be suitable for this heating coil.

7. *Determine the length of coil required*

A pipe property tabulation shows that 2.9 lin ft (0.9 m) of 1-in (2.5-cm) schedule 40 pipe has 1.0 ft^2 (0.09 m^2) of external area. Hence, the total length of pipe required in this heating coil = (33.1 ft^2)(2.9 ft/ft^2) = 96 ft (29.3 m).

Related Calculations: Use this general procedure to find the area and length of spiral heating coil required to heat water, industrial solutions, oils, etc. This procedure also can be used to find the area and length of cooling coils used to cool brine, oils, alcohol, wine, etc. In every case, be certain to substitute the correct specific heat for the liquid being heated or cooled. For typical values of U, consult Perry—*Chemical Engineers' Handbook*, McGraw-Hill; McAdams—*Heat Transmission*, McGraw-Hill; or Kern—*Process Heat Transfer*, McGraw-Hill.

SIZING ELECTRIC HEATERS FOR INDUSTRIAL USE

Choose the heating capacity of an electric heater to heat a pot containing 600 lb (272.2 kg) of lead from the charging temperature of 70°F (21.1°C) to a temperature of 750°F (398.9°C) if 600 lb (272.2 kg) of the lead is to be melted and heated per hour. The pot is 30 in (76.2 cm) in diameter and 18 in (45.7 cm) deep.

Calculation Procedure:

1. *Compute the heat needed to reach the melting point*

When a solid is melted, first it must be raised from its ambient or room temperature to the melting temperature. The quantity of heat required is H = (weight of solid, lb)[specific heat of solid, Btu/(lb·°F)]($t_m - t_i$), where H = Btu required to raise the temperature of the solid, °F; t_i = room, charging, or initial temperature of the solid, °F; t_m = melting temperature of the solid, °F.

For this pot with lead having a melting temperature of 620°F (326.7°C) and an average specific heat of 0.031 Btu/(lb·°F) [0.13 kJ/(kg·°C)], $H = (600)(0.031)(620 - 70) =$ 10,240 Btu/h (3.0 kW), or (10,240 Btu/h)/(3412 Btu/kWh) = 2.98 kWh.

2. *Compute the heat required to melt the solid*

The heat H_m Btu required to melt a solid is H_m = (weight of solid melted, lb)(heat of fusion of the solid, Btu/lb). Since the heat of fusion of lead is 10 Btu/lb (23.3 kJ/kg), $H_m = (600)(10)$ = 6000 Btu/h, or 6000/3412 = 1.752 kWh.

3. *Compute the heat required to reach the working temperature*

Use the same relation as in step 1, except that the temperature range is expressed as $t_w - t_m$, where t_w = working temperature of the melted solid. Thus, for this pot, $H = (600)(0.031)(750 - 620)$ = 2420 Btu/h (709.3 W), or 2420/3412 = 0.709 kWh.

4. *Determine the heat loss from the pot*

Use Fig. 7 to determine the heat loss from the pot. Enter at the bottom of Fig. 7 at 750°F (398.9°C), and project vertically upward to the 10-in (25.4-cm) diameter pot curve. At the left, read the heat loss at 7.3 kWh/h.

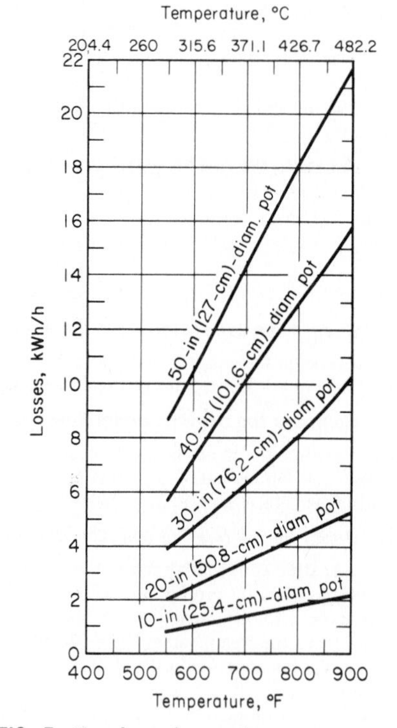

FIG. 7 Heat losses from melting pots. *(General Electric Co.)*

5. *Compute the total heating capacity required*

The total heating capacity required is the sum of the individual capacities, or 2.98 + 1.752 + 0.708 + 7.30 = 12.74 kWh. A 15-kW electric heater would be chosen because this is a standard size and it provides a moderate extra capacity for overloads.

Related Calculations: Use this general procedure to compute the capacity required for an electric heater used to melt a solid of any kind—lead, tin, type metal, solder, etc. When the substance being heated is a liquid—water, dye, paint, varnish, oil, etc.—use the relation H = (weight of liquid heated, lb) [specific heat of liquid, Btu/(lb·°F)] (temperature rise desired, °F), when the liquid is heated to approximately its boiling temperature, or a lower temperature.

For space heating of commercial and residential buildings, two methods used for computing

TABLE 4 Two Methods for Determining Wattage for Heating Buildings Electrically*

	W/ft^3 method	W/m^3 method
1. Interior rooms with no or little outside exposure	0.75 to 1.25	25.6 to 44.1
2. Average rooms with moderate windows and doors	1.25 to 1.75	44.1 to 61.8
3. Rooms with severe exposure and great window and door space	1.0 to 4.0	35.3 to 141.3
4. Isolated rooms, cabins, watchhouses, and similar buildings	3.0 to 6.0	105.9 to 211.9

	The "35" method
1. Volume in ft^3 for one air change × 0.35 =	0.01 W
2. Exposed net wall, roof, or ceiling and floor in ft^2 × 3.5 =	0.1 W
3. Area of exposed glass and doors in ft^2 × 35.0 =	1 W

*General Electric Company.

the approximate wattage required are the W/ft^3 and the "35" method. These are summarized in Table 4. In many cases, the results given by these methods agree closely with more involved calculations. When the desired room temperature is different from 70°F (21.1°C), increase or decrease the required kilowatt capacity proportionately, depending on whether the desired temperature is higher than or lower than 70°F (21.1°C).

For heating pipes with electric heaters, use a heater capacity of 0.8 W/ft^2 (8.6 W/m^2) of uninsulated exterior pipe surface per °F temperature difference between the pipe and the surrounding air. If the pipe is insulated with 1 in (2.5 cm) of insulation, use 30 percent of this value, or 0.24 (W/(ft^2·°F) [4.7 W/(m^2·°C)].

The types of electric heaters used today include immersion (for water, oil, plating, liquids, etc.), strip, cartridge, tubular, vane, fin, unit, and edgewound resistor heaters. These heaters are used in a wide variety of applications including liquid heating, gas and air heating, oven warming, deicing, humidifying, plastics heating, pipe heating, etc.

For pipe heating, a tubular heating element can be fastened to the bottom of the pipe and run parallel with it. For large-wattage applications, the heater can be spiraled around the pipe. For temperatures below 165°F (73.9°C), heating cable can be used. Electric heating is often used in place of steam tracing of outdoor pipes.

The procedure presented above is the work of General Electric Company.

Refrigeration

REFERENCES: Trott—*Refrigeration and Air Conditioning*, McGraw-Hill; Hallowell—*Cold and Freezer Storage Manual*, AVI; Munton and Stott—*Refrigeration at Sea*, Applied Science (England); International Institute of Refrigeration—*Low Temperature and Electric Power*, Pergamon; Betts—*Refrigeration and Thermometry below One Kelvin*, Crane-Russak; Emerick—*Heating Design and Practice*, McGraw-Hill; Carrier Air Conditioning Company—*Handbook of Air Conditioning System Design*, McGraw-Hill; Severns and Fellows—*Air Conditioning and Refrigeration*, Wiley; ASHRAE—*Guide and Data Book: Fundamentals and Equipment*; ASHRAE—*Guide and Data Book: Applications*; Strock and Koral—*Handbook of Heating, Air Conditioning, and Ventilation*, Industrial Press; American Blower Corporation—*Air Conditioning and Engineering*; MacIntire-Hutchinson—*Refrigeration Engineering*, Wiley.

REFRIGERATION SYSTEM SELECTION

Choose a refrigeration system for a given load. Show the steps that the designer should follow in choosing a suitable refrigeration system for various types of loads.

Calculation Procedure:

1. *Determine the refrigeration load*

Use the method given in the next calculation procedure. In any refrigeration plant, the total refrigeration load = heat gain from external sources, tons + product load, tons + sensible heat load, tons.

2. *Choose the type of refrigeration system to use*

Table 1 shows the usual compressor choices for various refrigeration loads. Thus, reciprocating compressors find wide use for refrigeration loads up to 400 tons (362.9 t). Up to loads of about 5 tons (4.5 t), *unit systems* that combine the compressor, drive, evaporator, and condenser in a compact unit are popular. In some instances, larger-capacity unit systems may be available from certain manufacturers. Some large unit systems, called *central-station systems,* are built with capacities of 100 to 150 tons (90.7 to 136.1 t).

From 5- to 400-ton (4.5- to 362.9-t) capacity, *built-up central systems* are popular. In these systems, the manufacturer supplies the compressor, evaporator, and condenser as separate units. These are connected by suitable piping. The refrigeration equipment manufacturer may or may not supply the compressor driving unit. This driver may be an electric motor, steam turbine, internal-combustion engine, or some other type of prime mover.

For loads greater than 400 tons (362.9 t), the centrifugal refrigeration compressor is often chosen. Whereas this may be a built-up system, more and more manufacturers today supply completely fabricated systems containing all the needed components, including the controls, driver, etc.

Steam-jet refrigeration units find some application for loads of 50 tons (45.4 t) or more. The steam-jet refrigeration system is used for a large number of applications where steam is available.

TABLE 1 Typical Refrigeration System Choices*

System load		System type		
tons	t	Often used	Occasionally used	Rarely used
0–5	0–4.5	Unit system with reciprocating compressor	Central-station built-up system; reciprocating compressor	Central-station built-up units
5–25	4.5–22.7	Central-station built-up systems; reciprocating compressor	Central station built-up systems; reciprocating compressor	Absorption or adsorption units
25–50	22.7–45.4	Central-station built-up systems; reciprocating compressor	Central-station built-up systems; centrifugal compressor	Absorption units
50–400	45.4–362.9	Central-station built up systems; reciprocating compressor	Central-station built-up systems; steam-jet and centrifugal compressors	
400 and up	362.9 and up	Central system; centrifugal and/or absorption unit	Central-station built-up steam-jet unit	

*Adapted from ASHRAE data.

Typical applications include comfort air conditioning, industrial process cooling, and similar service. In recent years, some large office buildings have used steam-jet systems mounted in the building penthouse. These units provide the cooling needed for the building air-conditioning system.

Absorption refrigeration systems were once popular for a variety of cooling tasks in industry, food storage, etc. In recent years, the absorption system has found renewed use in medium- and large-size air-conditioning systems. The usual absorbent used today is lithium bromide; the refrigerant is ordinary tap water. Absorption refrigeration systems are popular in areas where fuel costs are low, electric rates are high, waste steam is available, low-pressure heating boilers are unused during the cooling season, or steam or gas utility companies desire to promote summer loads. Absorption refrigeration systems can be installed in almost any location in a building where the floor is of adequate strength and reasonably level. Absence of heavy moving parts practically eliminates vibration and reduces the noise level to a minimum.

Combination absorption-centrifugal refrigeration systems are well suited for many large-tonnage air-conditioning and industrial loads. These systems are extremely economical where medium- or high-pressure steam is used as the energy source.

Using the expected refrigeration load from step 1 and the data above, make a preliminary choice of the type of refrigeration systems to use. Remember that the necessity for part-load operation might change the preliminary choice of the system type.

3. *Choose the system components*

Manufacturers' engineering data generally list compatible components for a given capacity compressor. These components include the condenser, expansion valve, evaporator, receiver, cooling tower, etc. Later calculation procedures in this section give specific instructions for selecting these and other important components of the system. When a unit system is chosen, the important components are preselected by the manufacturer.

4. *Have the system choice verified*

Have the manufacturer whose equipment will be used verify the selection for the given load. This ensures a correct choice.

Related Calculations: Use this general procedure to select the type of refrigeration system serving air-conditioning, product-cooling, liquid-cooling, ice-making, and similar applications in stationary (land) and marine service.

SELECTION OF A REFRIGERATION UNIT FOR PRODUCT COOLING

What capacity and type of refrigeration system are needed for a walk-in cooler having inside dimensions of 8 × 6 × 10 ft (2.4 × 1.8 × 3.1 m) if it is insulated with 4-in (10.2-cm) thick cork? The user estimates that a maximum of 400 lb (181.4 kg) of beef will be placed in the cooler daily, arriving at 70°F (21.1°C). The average hottest summer day in the cooler locality is, according to weather bureau records, 92°F (33.3°C). The meat is to be stored at 36°F (2.2°C). A ⅙-hp (0.12-kW) blower circulates air in the cooler. What refrigeration capacity is required for the same cooler, if the meat is stored at −10°F (−23.3°C) and the cork insulation is 8 in (20.3 cm) thick? Two ⅛-hp (0.09-kW) blowers will be used in the cooler.

Calculation Procedure:

1. *Compute the outside area of the cooler*

The outside dimensions of this cooler are 9 ft (2.7 m) high, 7 ft (2.1 m) wide, and 11 ft (3.4 m) long, including the cork insulation and the supporting structure. Hence, the total outside area of the cooler, including the floor and roof, is $2(9 \times 7) + 2(9 \times 11) + 2(7 \times 11) = 478$ ft^2 (44.4 m^2).

2. *Compute the heat gain and service load*

There is a heat gain into the cooler through the insulated surfaces caused by the difference between the inside and outside temperatures. Also, there is a service load, that is, a heat gain caused by the opening and shutting of the cooler door. Since meat will be loaded only once a day, it is safe to assume that the service load is a normal one—i.e., the door will be opened less than 5 times per hour.

TABLE 2 Heat Leakage Factors[*]

	Btu and kJ per degree temperature difference per ft^2 (m^2) of outside surface											
	Insulation thickness											
	in						cm					
	1	2	4	6	8	10	2.5	3.1	10.2	15.2	20.3	25.4
Heat leakage only	0.178	0.127	0.079	0.059	0.046	0.038	0.0010	0.0007	0.0005	0.0003	0.0003	0.0002
Heat leakage plus normal service load	0.216	0.163	0.110	0.090	0.077	0.069	0.0012	0.0009	0.0006	0.0005	0.0004	0.0004

[*]Brunner Manufacturing Company; SI values added by handbook editor.

Note: Light duty—multiply factor by 0.90; heavy duty—multiply factor by 1.10; single glass—multiply factor by 15; double glass—multiply factor by 6.5; triple glass—multiply factor by 5. If any wall or ceiling of a cooler is exposed to the sun, increase the temperature difference by 20°F (11.1°C) for that wall.

For product storage, cooling, heat, and service load, Btu/h = (total outside area of cooler, ft^2)(maximum outside temperature, °F − minimum inside temperature, °F)(factor from Table 2), or (478)(92 − 36)(0.110) = 2944 Btu/h (0.86 kW).

3. *Compute the product heat load*

Use this relation: product heat load, Btu/h = (lb/h of product cooled)(temperature of product entering cooler, °F − temperature of product leaving cooler, °F)[specific heat of product, Btu/(lb·°F)]. For this cooler, given the specific heat from Table 3, the product heat load = (400 lb/24 h)(70 − 36)(0.8) = 453 Btu/h (132.8 W).

4. *Compute the total heat load*

The total heat load = sum of heat gain and service load + product heat load + supplementary heat load, Btu/h, or 2994 + 453 + 424 = 3821 Btu/h (1.1 kW).

5. *Compute the refrigeration-system capacity required*

In cooler operation, it is essential to ensure defrosting of the evaporator during the off cycle. To permit this defrosting, select a condensing unit to operate 18 h per 24-h day. With an 18-h operating time, the required condensing-unit capacity to handle the 24-h load is (24 h/operating time, h)(total heat load, Btu/h) = (24/18)(3821) = 5082 Btu/h (1.5 kW).

6. *Select the refrigeration unit*

Since the required capacity of this refrigeration system is between 0 and 5 tons (0 and 4.5 t), the previous calculation procedure indicates that a unit system with a reciprocating compressor is the most common type used. Referring to a manufacturer's engineering data shows that a 5000-Btu/h (1.46-kW) 0.5-hp (0.37-kW) air-cooled unit having a 20°F (−6.7°C) suction temperature is available. This unit will operate about 18.5 h/day to carry the actual heat load of 5082 Btu/h (1.48 kW) if the evaporator is chosen on a 16°F (36°F − 20°F) (8.9°C) temperature difference between the room and refrigerant.

The exact size of a condensing unit cannot be selected until a choice is made of the evaporating temperature, or suction pressure, at which the compressor is to work. In general, a difference of between 10 and 20°F (5.6 and 11.1°C) should be maintained between the product or room temperature and the evaporator temperature. Thus, the 16°F (8.9°C) temperature difference used above is within the normal working range.

A better plan for this product cooler would be to select a larger evaporator on a 10°F (5.6°C) temperature-difference basis. The running time of the condensing unit would be decreased because of the higher operating suction temperature, that is, 26 instead of 20°F (−3.3 instead of −6.7°C).

A standard refrigeration unit having the same characteristics as the unit described above, except that the evaporating temperature is 25°F (−3.9°C), has a capacity of 5550 Btu/h (1.6 kW) with refrigerant 12. The evaporating pressure is 24.6 lb/in^2 (169.6 kPa). The compressor is a two-cylinder unit and is belt-driven by an electric motor. A finned-tube air-cooled condenser is used. The receiver is mounted below the compressor, on the same frame. Refrigerant 12 (formerly called Freon-12) is most satisfactory for low-temperature systems.

7. *Compute the required capacity at below-freezing temperature*

The six steps above are for product storage at temperatures above freezing, i.e., above 32°F (0°C). For temperatures below 32°F (0°C), the same procedure is followed except that the product load is computed in three steps—cooling to 32°F (0°C), freezing, and cooling to the final temperature.

Thus, heat and service load = 478[92 − (−10)](0.085) = 4140 Btu/h (1.2 kW), given the heavy-duty factor from Table 2.

For cooling to 32°F (0°C), product load = (400/24)(70 − 32)(0.8) = 504 Btu/h (0.15 kW). For the freezing process, heat removal = (lb/h of product cooled)(latent heat or enthalpy of freezing, Btu/lb, from Table 3), or (400 lb/24 h)(98) = 1635 Btu/h (0.48 kW). To cool below freezing, heat removal = (lb/h of product cooled)(32°F − temperature of storage room, °F)[specific heat of product at temperature below freezing, Btu/(lb·°F), from Table 3], or (400/24)[32 − (−10)](0.404) = 282 Btu/h (0.08 kW). Then the total product load = 504 + 1635 + 282 = 2421 Btu/h (0.71 kW).

The supplementary load with two ⅛-hp (0.09-kW) blowers is (2)(⅛)(2545) = 635 Btu/h (0.19 kW).

The total load is thus 4140 + 2421 + 635 = 7196 Btu/h (2.1 kW). Assuming a 16-h/day

TABLE 3 Typical Specific and Latent Heats°

Article	Specific heat				Latent heat of freezing		Cold-storage temperature	
	Above freezing		Below freezing					
	Btu/(lb·°F)	kJ/(kg·°C)	Btu/(lb·°F)	kJ/(kg·°C)	Btu/(lb·°F)	kJ/(kg·°C)	°F	°C
Canned goods:								
Fruits	As fresh	As fresh	As fresh	As fresh	...	...	35–40	1.7–4.4
Meats	As fresh	As fresh	As fresh	As fresh	...	...	35–40	1.7–4.4
Sardines	0.760	3.18	0.410	1.72	101.0	234.9	35–40	1.7–4.4
Butter, eggs, etc.:								
Butter	0.302	1.26	0.238	1.00	18.4	42.8	18–20	−7.8–6.7
Cheese	0.480	2.01	0.305	1.28	50.5	117.5	34	1.1
Eggs	0.760	3.18	0.410	1.72	100.0	232.6	31	−0.6
Milk, ice cream	0.900	3.77	0.462	1.93	124.0	288.4	35	1.7
Flour, meal (wheat)	0.26–0.38	1.1–1.6	0.21–0.28	0.9–1.2	14.4–28.8	34–67	36–40	2.2–4.4
Vegetables:								
Asparagus	0.952	3.99	0.482	2.02	134.0	311.7	34–35	1.1–1.7
Cabbage	0.928	3.88	0.473	1.98	131.0	304.7	34–35	1.1–1.7
Carrots	0.864	3.62	0.449	1.88	119.5	278.0	34–35	1.1–1.7
Celery (edible portion)	0.952	3.99	0.482	2.02	135.0	314.0	34–35	1.1–1.7
Dried beans	0.300	1.26	0.237	0.99	18.0	41.9	32–45	0.0–7.2
Dried corn	0.284	1.19	0.231	0.97	15.1	35.1	35–45	1.7–7.2

Dried peas	0.276	1.16	0.224	0.94	13.7	31.9	35–45	1.7–7.2
Onions	0.900	3.77	0.462	1.93	126.0	293.1	36	2.2
Parsnips	0.864	3.62	0.449	1.88	119.5	278.0	34–35	1.1–1.7
Potatoes	0.792	3.32	0.422	1.97	106.5	247.7	36–40	2.2–4.4
Sauerkraut	0.912	3.82	0.467	1.45	128.0	297.7	35	1.7
Miscellaneous:								
Cigars, tobacco	. . .	. . .	. . .	. . .	. . .	. . .	35–42	1.7–5.6
Furs, woolens, etc.	. . .	. . .	. . .	. . .	. . .	. . .	35	1.7
Honey	0.344	1.44	0.254	1.06	25.9	60.2	36–40	2.2–4.4
Hops	. . .	. . .	. . .	. . .	. . .	. . .	32–40	0.0–4.4
Maple syrup	0.488	2.08	0.308	1.29	51.8	120.5	40–45	4.4–7.2
Maple sugar	0.240	1.00	0.215	0.98	7.2	16.7	40–45	4.4–7.2
Poultry, dressed and iced	0.790	3.31	0.421	1.76	105.0	244.7	28–30	−2.2–1.1
Poultry, dry-packed	0.720	3.01	0.395	1.65	93.5	217.5	26–28	−2.2–1.1
Poultry, scalded	0.800	3.35	0.425	1.65	108.0	251.2	20	−6.7
Game, frozen	0.680	2.85	0.380	1.59	86.5	201.2	15–28	−9.4–2.2
Poultry, frozen	0.680	2.85	0.380	1.59	86.5	201.2	15–28	−9.4–2.2
Nuts (dried)	0.21–0.29	0.9–1.2	0.20–0.24	0.8–1.0	4.3–14.4	10–34	35–40	1.7–4.4
Water	1.000	4.20	0.500	2.09	144.0	334.9	. . .	. . .
Meats:								
Fresh (typical only)	0.800	3.35	0.404	1.69	. . .	. . .	20–40	−6.7–4.4
Fruits:								
Fresh (typical only)	0.700	2.93	0.387	1.62	. . .	. . .	32–55	0.0–12.2

°Brunner Manufacturing Company; SI values added by handbook editor.

operating time for the refrigeration unit, the condensing capacity required is 7196(24/16) = 10,800 Btu/h (3.2 kW).

Choose an evaporator for a 10°F (5.6°C) temperature difference with a capacity of 10,800 Btu/h (3.2 kW) at a suction temperature of −20°F (−28.9°C). Checking a manufacturer's engineering data shows that a 3-hp (2.2-kW) air-cooled two-cylinder unit will be suitable.

Related Calculations: Use this general procedure to choose refrigeration units for stationary, mobile (truck), and marine applications of walk-in coolers, display cases, milk and bottle coolers, ice cream freezers and hardeners, air conditioning, etc. Note that one procedure is used for applications above 32°F (0°C) and another for applications below 32°F (0°C).

In general, choose a unit for a 10 to 20°F (5.6 to 11.1°C) difference between the product and evaporator temperatures. Thus, where a room is maintained at 40°F (4.4°C), choose a condensing unit capacity corresponding to above a 25°F (−3.9°C) evaporator temperature. If brine is to be cooled to 5°F (−15.0°C), select a condensing unit for about −10°F (−23.3°C). Where a high relative humidity is desired in a cold room, select cooling coils with a large surface area, so that the minimum operating differential temperature can be maintained between the room and the coil. The procedure and data given here were published by the Brunner Manufacturing Company based on ASHRAE data.

ENERGY REQUIRED FOR STEAM-JET REFRIGERATION

A steam-jet refrigeration system operates with an evaporator temperature of 45°F (7.2°C) and a chilled-water inlet temperature of 60°F (15.6°C). The condenser operating pressure is 1.135 lb/in^2 (abs) (7.8 kPa), and the steam-jet ejectors use 3.1 lb of boiler steam per pound (1.4 kg/kg) of vapor removed from the evaporator. How many pounds of boiler steam are required per ton of refrigeration produced? How much steam is required per hour for a 100-ton (90.6-t) capacity steam-jet refrigeration unit?

Calculation Procedure:

1. *Determine the system pressures and enthalpies*

Using the steam tables, find the following values. At 45°F (7.2°C), P = 0.1475 lb/in^2 (abs) (1.0 kPa); h_f = 13.06 Btu/lb (30.6 kJ/kg); h_{fg} = 1068.4 Btu/lb (2485.1 kJ/kg). At 60°F (15.6°C), h_f = 28.06 Btu/lb (65.3 kJ/kg). At 1.135 lb/in^2 (abs) (7.8 kPa), h_f = 73.95 Btu/lb (172.0 kJ/kg), where P = absolute pressure, lb/in^2 (abs); h_f = enthalpy of liquid, Btu/lb; h_{fg} = enthalpy of vaporization, Btu/lb.

2. *Compute the chilled-water heat pickup*

The chilled-water inlet temperature is 60°F (15.6°C), and the chilled-water outlet temperature is the same as the evaporator temperature, or 45°F (7.2°C), as shown in Fig. 1. Hence, the chilled-water heat pickup = enthalpy at 60°F (15.6°C) − enthalpy at 45°F (7.2°C), both expressed in Btu/lb. Or, heat pickup = 28.06 − 13.06 = 15.0 Btu/lb (34.9 kJ/kg).

3. *Compute the required chilled-water flow rate*

Since a ton of refrigeration corresponds to a heat removal rate of 12,000 Btu/h (3.5 kW), the chilled-water flow rate = (12,000 Btu/h)/(chilled-water heat pickup, Btu/lb) = 12,000/15 = 8000 lb/(h·ton) [1.0 kg/(s·t)].

4. *Compute the quantity of chilled water that vaporizes*

Figure 1 shows the three fluid cycles involved: (*a*) chilled-water flow from the evaporator to the cooling coils and back, (*b*) chilled-water vapor flow from the evaporator through the ejector to the condenser and back as makeup, and (*c*) boiler steam flow from the boiler to the ejector to the condenser and back to the boiler as condensate.

Base the calculations on 1 lb (0.5 kg) of chilled water flowing through the cooling coils. For the throttling process from 3 to 4 in the evapoator, Fig. 1, the enthalpy remains constant, but part of the chilled water vaporizes at the lower, or evaporator, pressure. Hence, $H_3 = H_4 = h_f + xh_{fg}$, where x = lb of vapor formed per lb of chilled water entering, or 28.06 = 13.06 + x(1068.4); x = 0.01405 lb of vapor per lb (0.0063 kg/kg) of chilled water entering. The quantity of chilled water remaining at 1 in the evaporator is 1.0 − 0.01405 = 0.98595 lb/lb (0.4436 kg/kg) of chilled water recirculating.

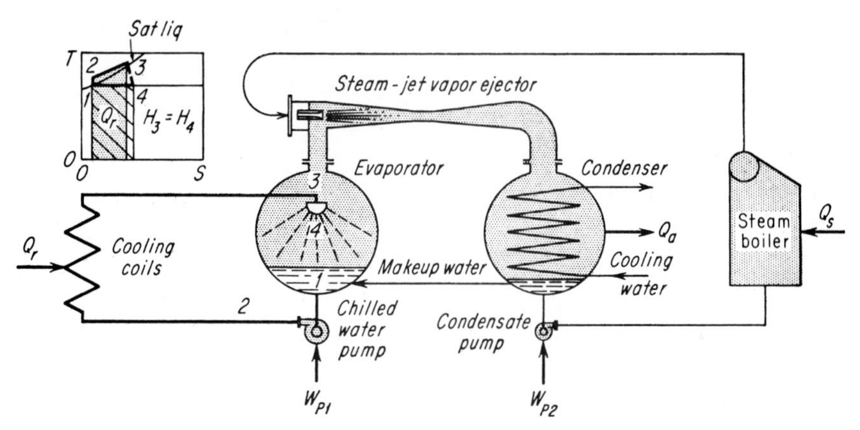

FIG. 1 Steam-jet refrigeration unit and *T-S* diagram of its operating cycle.

5. *Compute the quantity of makeup vaporized*

Some of the condensate in the condenser returns to the evaporator as makeup, Fig. 1. This makeup throttles into the evaporator and part of it evaporates. Hence, $H_m = h_f + x_m h_{fg}$, where H_m = enthalpy of condensate, Btu/lb; x_m = quantity of makeup vaporized, lb/lb of makeup water. Since the enthalpy of the condensate at the condenser pressure of 1.135 lb/in^2 (abs) (7.8 kPa) is 73.95 Btu/lb (172.0 kJ/kg), 73.95 = 13.06 + x_m(1068.4); x_m = 0.057 lb of makeup vaporized per lb (0.025 kg/kg) of makeup water entering the evaporator.

Makeup vapor simply recirculates between the evaporator and the condenser. So the total makeup water entering the evaporator must replace both the chilled-water vapor and the makeup vapor formed by the two throttling processes.

6. *Compute the makeup vapor and water quantities*

The lb of makeup vapor per lb of makeup water remaining in the evaporator = $x_m/(1.0 - x_m)$ = 0.0570/(1.0 − 0.0570) = 0.0604.

The total makeup water to the evaporator needed to replace the vapor = x(1 + lb of makeup vapor per lb of makeup water) = 0.01405(1 + 0.0604) = 0.01491 lb/lb (0.0067 kg/kg) of chilled water circulating. This is also the vapor removed from the evaporator by the ejector.

7. *Compute the total vapor removed from the evaporator*

The total vapor removed from the evaporator = [lb/(h·ton) chilled water] × (makeup water per lb of chilled water circulated) = (8000)(0.01491) = 119.3 lb/ton (54.1 kg/t) of refrigeration.

8. *Compute the boiler steam required*

The boiler steam required = (vapor removed from the evaporator, lb/ton of refrigeration)(steam-jet steam, lb/lb of vapor removed from the evaporator) = (119.3)(3.1) = 370 lb of boiler steam per ton of refrigeration (167.8 kg/t). For a 100-ton (90.6-t) machine, the boiler steam required = (100)(370) = 37,000 lb/h (4.7 kg/s).

Related Calculations: Use this general method for any steam-jet refrigeration system using water and steam to produce a low temperature for air conditioning, product cooling, manufacturing processes, or other applications. Note that any of the eight items computed can be found when the other variables are known.

REFRIGERATION COMPRESSOR CYCLE ANALYSIS

An ammonia refrigeration compressor takes its suction from the evaporator, Fig. 2*a*, at a temperature of −20°F (−28.9°C) and a quality of 95 percent. The compressor discharges at a pressure of 100 lb/in^2 (abs) (689.5 kPa). Liquid ammonia leaves the condenser at 50°F (10.0°C). Find the heat absorbed by the evaporator, the work input to the compressor, the heat rejected to the condenser, the coefficient of performance (COP) of the cycle, horsepower per ton of refrigeration, the quality of the refrigerant at state 2, quantity of refrigerant circulated per ton of refrigeration, required rate of condensing-water flow for a 100-ton (90.6-t) load, compressor displacement for a 100-ton (90.6-t) capacity. What cylinder dimensions are required for a 100-ton (90.6-t) capacity if the stroke = 1.3(cylinder bore) and the compressor makes 200 r/min?

Calculation Procedure:

1. *Compute the enthalpy and entropy at cycle points*

Assume a constant-entropy compression process for this cycle. This is the usual procedure in analyzing a refrigeration compressor whose actual performance is not known.

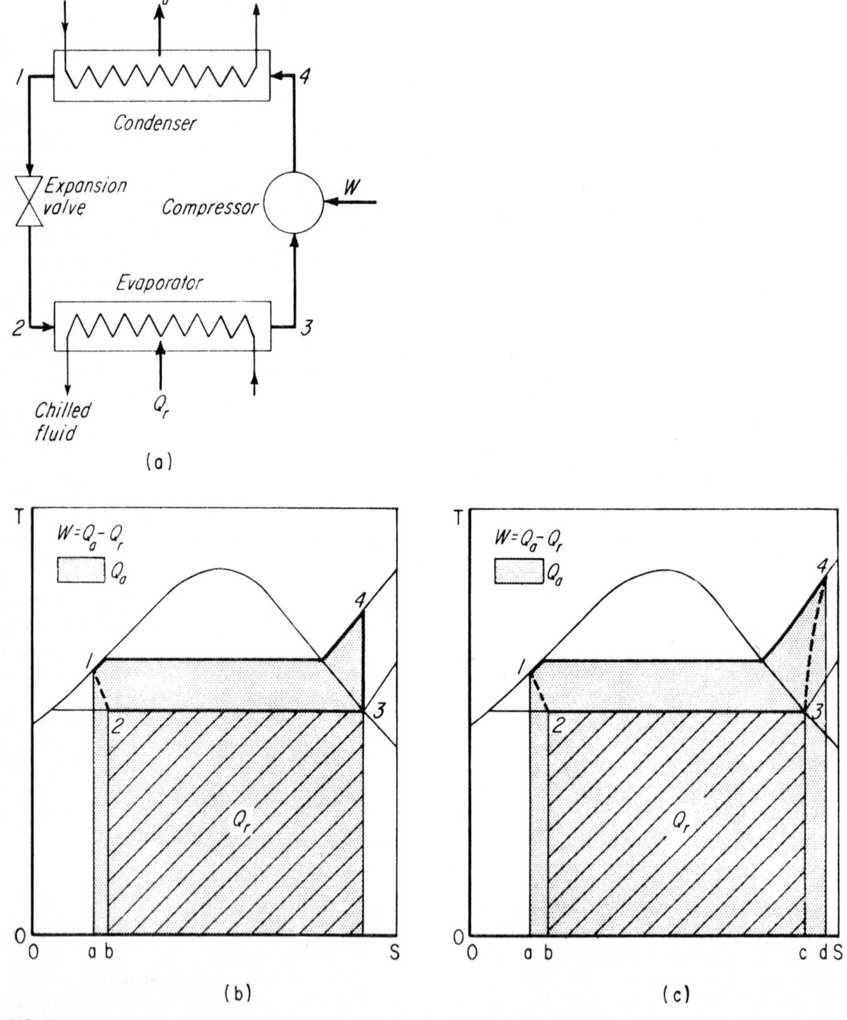

FIG. 2 (*a*) Components of a vapor refrigeration system; (*b*) ideal refrigeration cycle *T-S* diagram; (*c*) actual refrigeration cycle *T-S* diagram.

TABLE 4 Thermodynamic Properties of Ammonia°

Saturated ammonia

Temperature t, °F (°C)	Pressure p, lb/in² (abs) (kPa)	Volume, ft³/lb (m³/kg)		Enthalpy, Btu/lb (kJ/kg)			Entropy, Btu/(lb·°F) [kJ/(kg·°C)]	
		Liquid v_f	Vapor v_g	Liquid h_f	Evaporation h_{fg}	Vapor h_g	Liquid s_f	Vapor s_g
0	30.42	0.0242	9.116	42.9	568.9	611.8	0.0975	1.3352
(−17.8)	(209.7)	(0.00151)	(0.569)	(99.8)	(1323.3)	(1423.0)	(0.408)	(5.590)
20	48.21	0.0247	5.910	64.7	553.1	617.8	0.1437	1.2969
(−6.7)	(332.4)	(0.00154)	(0.369)	(150.5)	(1286.5)	(1437.0)	(0.564)	(5.430)
100	211.9	0.0272	1.419	155.2	477.8	633.0	0.3166	1.1705
(37.8)	(1461.1)	(0.00170)	(0.089)	(361.0)	(1111.4)	(1472.4)	(1.326)	(4.901)
120	286.4	0.0284	1.047	179.0	455.0	634.0	0.3576	1.1427
(48.9)	(1974.7)	(0.00177)	(0.065)	(416.4)	(1058.3)	(1474.7)	(1.497)	(4.787)

Superheated ammonia

Temperature, °F (°C)	50 lb/in² (abs) (344.8 kPa) [21.67°F (−5.7°C) saturation]			100 lb/in² (abs) (689.5 kPa) [56.05°F (13.4°C) saturation]			150 lb/in² (abs) (1034.3 kPa) [78.81°F (26.0°C) saturation]		
	v	h	s	v	h	s	v	h	s
100	6.843	663.7	1.3816	3.304	655.2	1.2891	2.118	645.9	1.2289
(37.8)	(0.427)	(1543.8)	(5.783)	(0.206)	(1524.0)	(5.397)	(0.132)	(1502.4)	(5.145)
120	7.117	674.7	1.4009	3.454)	667.3	1.3104	2.228	659.4	1.2526
(48.9)	(0.444)	(1569.4)	(5.865)	(0.216)	(1552.1)	(5.486)	(0.139)	(1533.98)	(5.244)
140	7.387	685.7	1.4195	3.600	679.2	1.3305	2.334	672.3	1.2745
(60.0)	(0.461)	(1594.9)	(5.943)	(0.225)	(1579.8)	(5.571)	(0.146)	(1563.8)	(5.336)

Using Fig. 2*b* as a guide, we see that $H_3 = h_f + xh_{fg}$, where H_3 = enthalpy at point 3, Btu/lb; h_f = enthalpy of liquid ammonia, Btu/lb from Table 4; h_{fg} = enthalpy of evaporation, Btu/lb, from the same table; x = vapor quality, expressed as a decimal. Since point 3 represents the suction conditions of the compressor, $H_3 = 21.4 + 0.95(583.6) = 575.8$ Btu/lb (1339.3 kJ/kg).

The entropy at point 3 is $S_3 = s_f + xs_{fg}$, where the subscripts refer to the same fluid states as above and the S and s values are the entropy. Or, $S_3 = 0.0497 + 0.95(1.3277) = 1.3110$ Btu/(lb·°F) [5.465 kJ/(kg·°C)].

2. Compute the final cycle temperature and enthalpy

The compressor discharges at 100 lb/in² (abs) (689.5 kPa) at an entropy of $S_4 = 1.3110$ Btu/(lb·°F) [5.465 kJ/(kg·°C)]. Inspection of the saturated ammonia properties, Table 4, shows that at 100 lb/in² (abs) (689.5 kPa) the entropy of saturated vapor is less than that computed. Hence, the vapor discharged by the compressor must be superheated.

Enter Table 4 at $S_4 = 1.3110$ Btu/(lb·°F) [5.465 kJ/(kg·°C)]. Inspection shows that the final cycle temperature T_4 lies between 120 and 130°F (48.9 and 54.4°C) because the actual entropy value lies between the entropy values for these two temperatures. Interpolating gives $T_4 = 130 - [(S_{130} - S_4)/(S_{130} - S_{120})] \times (130 - 120)$, where the subscripts refer to the respective temperatures. Or, $T_4 = 130 - [(1.3206 - 1.3110)/(1.3206 - 1.3104)](130 - 120) = 120.6$°F (49.2°C).

Interpolating in a similar fashion for the final enthalpy, using the enthalpy at 130°F (54.4°C) as the base, we find $H_4 = 673.3 - [(1.3206 - 1.3110)/(1.3206 - 1.3104)](673.3 - 667.3) = 667.7$ Btu/lb (1553.1 kJ/kg).

3. Compute the heat absorbed by the evaporator

The heat absorbed by the evaporator is $Q_r = H_3 - H_2$, where Q_r = Btu/lb of refrigerant. Or, for this system, $Q_r = 575.8 - 97.9 = 477.9$ Btu/lb (1111.6 kJ/kg).

4. Compute the work input to the compressor

Find the work input to the compressor from $W = H_4 - H_3$, where W = work input, Btu/lb of refrigerant. Or, $W = 667.7 - 575.8 = 91.9$ Btu/lb (213.8 kJ/kg) of refrigerant circulated.

5. Compute heat rejected to the condenser

The heat rejected to the condenser is $Q_a = H_4 - H_1$, where Q_a = heat rejection, Btu/lb of refrigerant. Or, $Q_a = 667.7 - 97.9 = 569.8$ Btu/lb (1325.4 kJ/kg) of refrigerant circulated.

6. Compute the coefficient of performance of the machine

For any refrigerating machine, the coefficient of performance (COP) = Q_r/W, where the symbols are as defined earlier. Or COP = 477.9/91.9 = 5.20.

7. Compute the horsepower per ton for this system

For any refrigerating system, the horsepower per ton hp_t = 4.72/COP. Or, for this system, hp_t = 4.72/5.20 = 0.908 hp/ton (0.68 kW).

8. Compute the refrigerant quality at the evaporator inlet

At the evaporator inlet, or point 2, the quality of the refrigerant $x = (H_2 - h_f)/h_{fg}$, where the enthalpies are those at −20°F (−28.9°C), the evaporator operating temperature. Or, $x = (97.9 - 21.4)/583.6 = 0.1311$, or 13.11 percent quality.

9. Compute the quantity of refrigerant circulated per ton capacity

Find the quantity of refrigerant circulated, lb/(min·ton) of refrigeration produced from $q_t = 200/Q_r$, or $q_t = 200/477.9 = 0.419$ lb/(min·ton) [(0.0035 kg/(s·t)] of refrigeration.

10. Compute the required rate of condensing-water flow

The heat rejected to the condenser Q_a must be absorbed by the condenser cooling water. The quantity of water that must be circulated is $q_w = Q_a/\Delta t$, where q_w = weight of water circulated per lb of refrigerant; Δt = temperature rise of the cooling water during passage through the condenser, °F. Assuming a 20°F (11.1°C) temperature rise of the cooling water, we find $q_w = 569.8/20 = 28.49$, say 28.5 lb of water per lb (12.8 kg/kg) of refrigerant circulated.

Since 0.419 lb/(min·ton) [0.0035 kg/(t·s)] of ammonia must be circulated, step 9, at a load of 100 tons (90.7 t), the quantity of refrigerant circulated will be 100(0.419) = 41.9 lb/min (0.32 kg/s). The condenser cooling water required is then (28.5)(41.9) = 1191 lb/min (9.0 kg/s), or 1191/8.33 = 143.4 gal/min (9.1 L/s).

11. *Compute the compressor displacement*

Use the relation $V_d = q_t v_g T$, where V_d = required compressor displacement, ft^3/min; q_t = quantity of refrigerant circulated, lb/(ton·min); v_g = specific volume of suction gas, ft^3/lb; T = refrigeration capacity, tons. For a 100-ton (90.7-t) capacity with the suction gas at −20°F (−28.9°C), V_d = (0.419)(14.68)(100) = 614 ft^3/min (0.29 m^3/s), given the specific volume for −20°F (−28.9°C) suction gas from Table 4.

12. *Compute the compressor cylinder dimensions*

For any reciprocating refrigeration compressor, V_d = (shaft rpm)(piston displacement, ft^3/stroke) = v_d, or 614 = 200v_d; v_d = 3.07 ft^3 (0.087 m^3).

Also, $D = (V_d/0.785)^{1/3}$, where D = piston diameter, ft; r = ratio of stroke length to cylinder bore. Or, $D = [3.07/(0.785 \times 1.3)]^{1/3}$ = 1.447 ft (0.44 m). Then $L = 1.3D$ = 1.3(1.447) = 1.88 ft (0.57 m).

Related Calculations: Employ the method given here for any reciprocating compressor using any refrigerant. Note that where the volumetric efficiency E_V of a compressor is given, the actual volume of gas drawn into the cylinder, ft^3 = E_V × piston displacement, ft^3. When analyzing an actual compressor, be sure you use the enthalpies which actually prevail. Thus, the gas entering the compressor suction may be superheated instead of saturated, as assumed here.

RECIPROCATING REFRIGERATION COMPRESSOR SELECTION

Choose the compressor capacity and hp, and determine the heat rejection rate for a 36-ton (32.7-t) load, a 30°F (−1.1°C) evaporator temperature, a 20°F (−6.7°C) evaporator coil superheat, a suction-line pressure drop of 2 lb/in² (13.8 kPa), a condensing temperature of 105°F (40.6°C), a compressor speed of 1750 r/min, a subcooling of the refrigerant of 5°F (2.8°C) in the water-cooled condenser, and use of refrigerant 12. Determine the required condensing-water flow rate when the entering water temperature is 70°F (21.1°C). How many gal/min of chilled water can be handled if the water temperature is reduced 10°F (5.6°C) by the evaporator chiller?

Calculation Procedure:

1. *Compute the compressor suction temperature*

With refrigerant 12, a pressure change of 1 lb/in² (6.9 kPa) at 0°F (−17.8°C) is equivalent to a temperature change of 2°F (1.1°C); at 50°F (10.0°C), a 1-lb/in² (6.9-kPa) pressure change is equivalent to 1°F (0.6°C) temperature change. At the evaporator temperature of 30°F (−1.1°C), the temperature change is about 1.4°F·in²/lb (0.11°C/kPa), obtained by interpolation between the ranges given above. Then, suction temperature, °F = evaporator temperature, °F − (suction-line loss, °F·in²/lb)(suction-line pressure drop, lb/in²), or 30 − 1.4 × 2 = 27.2, say 27°F (−2.8°C).

2. *Compute the compressor equivalent capacity*

To compute the compressor equivalent capacity, two correction factors must be applied: the superheat correction factor and the subcooling correction factor. Both are given in the engineering data available from compressor manufacturers.

To apply correction-factor listings, such as those in Table 5, use the following as guides: (*a*)

TABLE 5 Open Compressor Ratings

Suction temperature		Condensing temperature, 105°F (40.6°C)					
		Capacity		Power input		Heat rejection	
°F	°C	tons	t	bhp	kW	tons	t
10	−12.2	26.2	23.8	41.3	30.8	34.9	31.7
20	−6.7	34.0	30.8	45.3	33.8	43.6	39.6
30	−1.1	43.0	39.0	48.6	36.2	53.2	48.3

TABLE 6 Rating Basis and Capacity Multipliers—Refrigerant 12 and Refrigerant 500*

Saturated suction temperature		Actual suction gas temperature to compressor			
°F	°C	30°F (−1.1°C)	40°F (4.4°C)	50°F (10.0°C)	60°F (15.6°C)
20	−6.7	0.969	0.978	0.987	0.996
30	−1.1	0.970	0.979	0.987	0.996
40	4.4	. . .	0.987	0.992	0.997

*Carrier Air Conditioning Company; SI values added by handbook editor.

Superheating of the suction gas can result from heat pickup by the gas outside the cooled space. Superheating increases the refrigeration compressor capacity 0.3 to 1.0 percent per 10°F (5.6°C) with refrigerant 12 or 500 if the heat absorbed represents useful refrigeration, such as coil superheat, and not superheating from a liquid suction heat exchanger. (*b*) Subcooling increases the potential refrigeration effect by reducing the percentage of liquid flashed during expansion. For each °F of subcooling, the compressor capacity is increased about 0.5 percent owing to the increased refrigeration effect per pound of refrigerant flow.

Applying guide (*a*) to a 27°F (−2.8°C) suction, 20°F (−6.7°C) superheat, interpolate in Table 6 between the 40 and 50°F (4.4 and 10.0°C) actual suction-gas temperatures for a 30°F (−1.1°C) saturated suction temperature, because the actual suction temperature is 27 + 20 = 47°F (8.3°C) and the saturated suction temperature is given as 30°F (−1.1°C). Or, (0.987 − 0.979)[(47 − 40)/(50 − 40)] + 0.979 = 0.9846, say 0.985.

Applying guide (*b*), we see that subcooling = 5°F (2.8°C), as given. Then subcooling correction = 1 − 0.0005(15 − 5) = 0.95, where 0.005 = 0.5 percent, expressed as a decimal; 15°F (−9.4°C) = the liquid subcooling on which the compressor capacity is based. This value is given in the compressor rating, Table 6.

With the superheat and subcooling correction factors known, compute the compressor equivalent capacity from (load, tons)/[(superheat correction factor)(subcooling correction factor)], or 36/[(0.985)(0.95)] = 38.5 tons (34.9 t).

3. *Select the compressor unit*

Use Table 5. Choose an eight-cylinder compressor. Interpolate for a 27°F (−2.8°C) suction and 105°F (40.6°C) condensing temperature to find compressor capacity = 40.3 tons (36.6 t); power input = 47.6 bhp (35.5 kW); heat rejection = 50.3 tons (45.6 t).

4. *Compute the required condensing-water flow rate*

From step 3, the condensing temperature of the compressor chosen is 105°F (40.6°C). Assume a condenser-water outlet temperature of 95°F (35.0°C), a typical value. Then the required condenser-water flow rate, gal/min = 24 × condenser load/(condensing-water outlet temperature, °F − entering condenser-water temperature, °F). Or 24(50.3)/(95 − 70) = 48.4 gal/min (3.1 L/s). This is within the normal flow for water-cooled condensing units. Thus, city-water quantities range from 1 to 2 gal/(min·ton) [0.07 to 0.14 L/(s·t)]; cooling-tower quantities are usually chosen for 3 gal/(min·ton) [0.21 L/s·t)].

5. *Compute the quantity of chilled water that can be handled*

Use this relation: chilled water, gal/min = 24 × capacity, tons/chilled-water temperature range, or inlet − outlet temperature, °F. Since, from step 3, the compressor capacity is 40.3 tons (36.6 t) and the chilled-water temperature range is 10°F (5.6°C), *gpm* = 24(40.3)/10 = 96.7 gal/min (6.1 L/s).

The temperature of the chilled water leaving the evaporator chiller is selected so that it equals the inlet temperature required at the heat-load source. The required inlet temperature is a function of the type of heat exchanger, type of load, and similar factors.

Related Calculations: The standard operating conditions for an air-conditioning refrigeration system, as usually published by the manufacturer, are based on an entering saturated refrigerant vapor temperature of 40°F (4.4°C), an actual entering refrigerant vapor temperature of

55°F (12.8°C), a leaving saturated refrigerant vapor temperature of 105°F (40.6°C), and an ambient of 90°F (32.2°C) and no liquid subcooling.

The Air Conditioning and Refrigeration Institute (ARI) standards for a reciprocating compressor liquid-chilling package establish a standard rating condition for a water-cooled model of a leaving chilled-water temperature of 44°F (6.7°C), a chilled-water range of 10°F (5.6°C), a 0.0005 fouling factor in the cooler and the condenser, a leaving condenser-water temperature of 95°F (35.0°C), and a condenser-water temperature rise of 10°F (5.6°C). The standard rating conditions for a condenserless model are a leaving chilled-water temperature of 44°F (6.7°C), a chilled-water temperature range of 10°F (5.6°C), a 0.0005 fouling factor in the cooler, and a condensing temperature of 105 or 120°F (40.6 or 48.9°C).

Use these standard rating conditions to make comparisons between compressors. When catalog ratings of compressors of different manufacturers are compared, the rating conditions must be known, particularly the amount of subcooling and superheating needed to produce the capacities shown.

General guides for reciprocating compressors using refrigerants 12, 22, and 500 are as follows:

1. Lowering the evaporator temperature 10°F (5.6°C) from a base of 40 and 105°F (4.4 and 40.6°C) reduces the system (evaporator) capacity about 24 percent and at the same time increases the compressor hp/ton by about 18 percent.
2. Increasing the condensing temperature 15°F (8.3 °C) from a base of 40 and 105°F (4.4 and 40.6°C) reduces the capacity about 13 percent and at the same time increases the compressor hp/ton by about 27 percent.
3. In air-conditioning service at normal loads, a piping loss equivalent to approximately 2°F (1.1°C) is allowed in the suction piping and to 2°F (1.1°C) in the hot-gas discharge piping. Thus when an evaporator requires a refrigerant temperature of 42°F (5.6 °C) to handle a load, the compressor must be selected for a 40°F (4.4°C) suction temperature. Correspondingly, if the condenser requires 103°F (39.4°C) to reject the proper amount of heat, the compressor must be selected for a 103 + 2 = 105°F (40.6°C) condensing temperature.
4. Compressor manufacturers generally state the operating limits for each compressor in the capacity table describing it. These limits should not be exceeded.
5. To select a condenser to match a compressor, the heat rejection of the compressor must be known. For an open-type compressor, heat rejection, tons = 0.212(compressor power input, bhp) + tons refrigeration capacity of the compressor. For a gas-cooled hermetic-type compressor, heat rejection, tons = 0.285 (kW input to the compressor) + refrigeration capacity, tons. The selection procedure and other data given here were developed by the Carrier Air Conditioning Company.

CENTRIFUGAL REFRIGERATION MACHINE LOAD ANALYSIS

Select a centrifugal refrigeration machine to cool 720 gal/min (45.4 L/s) of chilled water from an entering temperature of 60°F (15.6°C) to a leaving temperature of 45°F (7.2°C).

Calculation Procedure:

1. *Compute the load on the machine*

Use this relation: load, tons = $gpm \times \Delta t/24$, where gpm = quantity of chilled water cooled, gal/min; Δt = temperature reduction of the chilled water during passage through the evaporator chiller, °F. For this machine, load = 720(50 − 45)/24 = 450 tons (408.2 t).

2. *Choose the compressor to use*

Table 7 shows typical hermetic centrifugal refrigeration machine ratings. In a hermetic machine, the driver is built into the housing, completely isolating the refrigerant space from the atmosphere. An open machine has a shaft that projects outside the compressor housing. The shaft must be fitted with a suitable seal to prevent refrigerant leakage. Open machines are available in capacities up to approximately 4500 tons (4085 t) at air-conditioning load temperatures. Hermetic machines are available in capacities up to approximately 2000-ton (1814-t) capacity.

TABLE 7 Typical Hermetic Centrifugal Machine Ratings° [Refrigeration Capacity, tons (t)]

Leaving chilled-water temperature		Leaving condenser-water temperature					
°F	°C	85°F	29.4°C	90°F	32.2°C	95°F	35.0°C
44	6.7	442†	401.0†	435	394.6	424	384.6
45	7.2	450	408.2	441	400.1	430	390.1
46	7.8	457†	414.6†	447	405.5	435	394.6

°Carrier Air Conditioning Company.
†These ratings require less than 330-kW input. All ratings shown are based on a two-pass cooler using 380 to 1260 gal/min (24.0 to 79.5 L/s) and on a two-pass condenser using 430 to 1430 gal/min (27.1 to 90.2 L/s).

Study of Table 7 shows that a 450-ton (408.5-t) unit is available with a leaving chilled-water temperature of 45°F (7.2°C) and a leaving condenser-water temperature of 85°F (29.4°C). If the condenser water were available at temperatures of 75°F (23.9°C) or lower, this machine would probably be chosen.

Related Calculations: The factors involved in the selection of a centrifugal machine are load; chilled-water, or brine quantity; temperature of the chilled water or brine; condensing medium (usually water) to be used; quantity of the condensing medium and its temperature; type and quantity of power available; fouling-factor allowance; amount of usable space available; and the nature of the load, whether variable or constant. The final selection is usually based on the least expensive combination of machine and heat rejection device, as well as a reasonable machine operating cost.

Brine cooling normally requires special selection of the machine by the manufacturer. As a general rule, multiple-machine applications are seldom made on normal air-conditioning loads less than about 400 tons (362.9 t).

The optimum machine selection involves matching the correct machine and cooling tower as well as the correct entering chilled-water temperature and temperature reduction. A selection of several machines and cooling towers often results in finding one combination having a minimum first cost. In many instances, it is possible to reduce the condenser-water quantity and increase the leaving condenser-water temperature, resulting in a smaller tower.

Centrifugal refrigeration machines are used for air-conditioning, process, marine, manufacturing, and many other cooling applications throughout industry.

HEAT PUMP CYCLE ANALYSIS AND COMPARISON

Determine the quantity of water required to supply heat to a heat pump that must deliver 70,000 Btu/h (20.5 kW) to a building. Refrigerant 12 is used; the temperature of the water in the heat sink is 50°F (10.0°C). Air must be delivered to the heating system at a temperature of 118°F (47.8°C).

Calculation Procedure:

1. *Determine the compressor suction temperature to use*

To produce sufficient heat transfer between the water and the evaporator, a temperature difference of at least 10°F (5.6°C) must exist. With a water temperature of 50°F (10.0°C), this means that a suction temperature of 40°F (4.4°C) might be satisfactory. A suction temperature of 40°F (4.4°C) corresponds to a suction pressure of 51.68 lb/in^2 (abs) (356.3 kPa), as a table of thermodynamic properties of refrigerant 12 shows.

Since water entering the evaporator heat exchanger cannot be reduced to 40°F (4.4°C), the refrigerant temperature, the actual outlet temperature must be either assumed or computed. Assume that the water leaves the evaporator heat exchanger at 44°F (6.7°C). Then, each pound of water passing through the evaporator yields 50°F − 44°F = 6 Btu (6.3 kJ). Since 1 gal of water weighs 8.33 lb (1 kg/L), the quantity of heat released by the water is (6 Btu/lb)(8.33 lb/gal) = 49.98 Btu/gal, say 50 Btu/gal (13.9 kJ/L).

As an alternative solution, assume a suction temperature of 35°F (1.7°C) and an evaporator exit temperature of 39°F (3.9°C). Then each pound of water will yield 50 − 39 = 11 Btu (11.6 kJ). This is equal to (11)(8.33) = 91.6 Btu/gal (25.5 kJ/L). This comparison indicates that for every °F the cooling range of the water heat sink is extended, an additional 8.33 Btu/gal (2.3 kJ/L) of water is obtained.

2. *Evaluate the effect of suction-temperature decrease*

As the compressor suction temperature is reduced, the specific volume of the suction gas increases. Thus the compressor must handle more gas to evaporate the same quantity of refrigerant. However, the displacement of the usual reciprocating compressor used in a heat-pump system cannot be varied easily, if at all, in some designs. Also, at the lower suction temperature, the enthalpy of vaporization of the refrigerant increases only slightly.

Study of a table of thermodynamic properties of refrigerant 12 shows that reducing the suction temperature from 40 to 35°F (4.4 to 1.7°C) increases the specific volume from 0.792 to 0.862 ft^3/lb (0.0224 to 0.0244 m^3/kg). The enthalpy of vaporization increases from 65.71 to 66.28 Btu/lb (152.8 to 154.2 kJ/kg), but the total enthalpy decreases from 82.71 to 82.16 Btu/lb (192.4 to 191.1 kJ/kg). Hence, the advisability of reducing the suction temperature must be carefully investigated before a final decision is made.

3. *Determine the required compressor discharge temperature*

Air must be delivered to the heating system at 118°F (47.8°C), according to the design requirements. To produce a satisfactory transfer of heat between the condenser and the air, a 10°F (5.6°C) temperature difference is necessary. Hence, the compressor discharge temperature must be at least 118 + 10 = 128°F (53.3°C).

Checking a table of thermodynamic properties of refrigerant 12 shows that a temperature of 128°F (53.3°C) corresponds to a discharge pressure of 190.1 lb/in^2 (abs) (1310.7 kPa). The table also shows that the enthalpy of the vapor at the 118°F (47.8°C) condensing temperature is 90.01 Btu/lb (209.4 kJ/kg), whereas the enthalpy of the liquid is 35.65 Btu/lb (82.9 kJ/kg).

With a suction temperature of 40°F (4.4°C), the enthalpy of the vapor is 82.71 Btu/lb (192.4 kJ/kg). Hence, the heat supplied by the evaporator is: enthalpy of vapor at 40°F (4.4°C) − enthalpy of liquid at 118°F = 82.71 − 35.65 = 47.06 Btu/lb (109.5 kJ/kg). This heat is abstracted from the water that is drawn from the heat sink.

The gas leaving the evaporator contains 82.71 Btu/lb (192.4 kJ/kg). When this gas enters the condenser, it contains 90.64 Btu/lb (210.8 kJ/kg). The difference, or 90.64 − 82.71 = 7.93 Btu/lb (18.4 kJ/kg), is added to the gas by the compressor and represents a portion of the work input to the compressor.

4. *Compute the evaporator and compressor heat contribution*

The total heat delivered to the air = evaporator heat + compressor heat = 47.06 + 7.93 = 54.99 Btu/lb (127.9 kJ/kg). Then the evaporator supplies 47.06/54.99 = 0.856, or 85.6 percent of the total heat, and the compressor supplies 7.93/54.99 = 0.144, or 14.4 percent of the total heat.

5. *Determine the actual evaporator and compressor heat contribution*

Since this heat pump is rated at 70,000 Btu/h (20.5 kW), the evaporator contributes 0.856 × 70,000 = 59,920 Btu/h (17.5 kW), and the compressor supplies 0.144(70,000) = 10,080 Btu/h (3.0 kW). As a check, 59,920 + 10,080 = 70,000 Btu/h (20.5 kW).

6. *Compute the sink-water flow rate required*

The evaporator obtains its heat, or 59,920 Btu/h (17.5 kW), from the sink water. Since, from step 1, each gallon of water delivers 50 Btu (52.8 kJ) at a 40°F (4.4°C) suction temperature, the flow rate required to contribute the evaporator heat is 59,920/50 = 1198.4 gal/h, or 1198.4/60 = 19.9 gal/min (1.3 L/s).

7. *Evaluate the lower suction temperature*

At 35°F (1.7°C), the evaporator will supply 82.16 − 35.65 = 46.51 Btu/lb (108.2 kJ/kg), by the same reasoning as in step 3. The balance, or 90.64 − 82.16 = 8.48 Btu/lb (19.7 kJ/kg), must be supplied by the compressor.

8. *Compute the required refrigerant gas flow*

At a 40°F (4.4°C) suction temperature, a table of thermodynamic properties of refrigerant 12 shows that the specific volume of the gas is 0.792 ft^3/lb (0.050 m^3/kg). Step 4 shows that the heat pump must deliver 54.99 Btu/lb (127.9 kJ/kg) of refrigerant to the air, or 54.99/0.792 = 69.4 Btu/ft^3 (2585.8 kJ/m^3) of gas. With a total heat requirement of 70,000 Btu/h (20.5 kW), the compressor must handle 70,000/69.4 = 1010 ft^3/h (0.0079 m^3/s).

As noted earlier, the cubic capacity of a reciprocating compressor is a fixed value at a given speed. Hence, a compressor chosen to handle this quantity of gas cannot handle a larger heat load.

At a 35°F (1.7°C) suction temperature, using the same procedure as above, we see that the required heat content of the gas is 54.99/0.862 = 63.7 Btu/ft^3 (2373.4 kJ/m^3). The compressor capacity must be 70,000/63.7 = 1099 ft^3/h (0.0086 m^3/s).

If the compressor were selected to handle 1010 ft^3/h (0.0079 m^3/s), then reducing the suction temperature to 35°F (1.7°C) would give a heat capacity of only (1010)(63.7) = 64,400 Btu/h (18.9 kW). This is inadequate because the system requires 70,000 Btu/h (20.5 kW).

9. *Compute the water flow rate at the lower suction temperature*

Step 6 shows the procedure for finding the sink-water flow rate required at a 40°F (4.4°C) suction temperature. Suppose, however, that the heat output of 64,400 Btu/h (18.9 kW) at the 35°F (1.7°C) suction temperature was acceptable. The evaporator portion of this load, by the method of step 4, is (82.16 − 35.65)/54.99 = 0.847, or 84.7 percent. Hence, the quantity of water required is (64,400)(0.847)/[(91.6)(60)] = 9.92 gal/min (0.63 L/s). In this equation, the value of 91.6 Btu/gal is obtained from step 1. The factor 60 converts hours to minutes. Thus, reducing the suction temperature from 40 to 35°F (4.4 to 1.7°C) just about halves the water quantity—from 19.9 to 9.92 gal/min (1.26 to 0.63 L/s).

10. *Compare the pumping power requirements*

The power input to the pump is hp = 8.33(*gpm*)(head, ft)/33,000(pump efficiency)(motor efficiency). If the total head on the pump is computed as being 40 ft (12.2 m), the efficiency of the pump is 60 percent, and the efficiency of the motor is 85 percent, then with a 40°F (4.4°C) suction temperature, the pump horsepower is 8.33(19.9)(40)/33,000(0.60)(0.85) = 0.394, say 0.40 hp (0.30 kW).

At a 35°F (1.7°C) suction temperature with a flow rate of 9.92 gal/min (0.03 L/s) and all the other factors the same, hp = 8.33(9.92)(40)/33,000(0.60)(0.85) = 0.1965, say 0.20 hp (0.15 kW). Thus, the 35°F (1.7°C) suction temperature requires only half the pump hp that the 40°F (4.4°C) suction temperature requires.

11. *Compute the compressor power input and power cost*

At the 40°F (4.4°C) suction temperature, the compressor delivers 7.93 Btu/lb (18.4 kJ/kg) of refrigerant gas, step 3. Since the total weight of gas delivered by the compressor per hour is (70,000 Btu/h)/(54.99 Btu/lb) = 1272 lb/h (0.16 kg/s), the compressor's total heat contribution is (1272 lb/h)(7.93 Btu/lb) = 10,100 Btu/h (3.0 kW).

With a compressor-driving motor having an efficiency of 85 percent, the hourly motor input is equivalent to 10,100/0.85 = 11,880 Btu/h (3.5 kW), or 11,880/2545 Btu/(hp·h) = 4.66 hp·h = (4.66)[746 Wh/(hp·h)] = 3480 Wh = 3.48 kWh. Also, the pump requires 0.4 hp·h, step 10, or (0.4)(746) = 299 Wh = 0.299 kWh. Hence, the total power consumption at a 40°F (4.4°C) suction temperature is 3.48 + 0.299 = 3.779 kWh. At a power cost of 5 cents per kilowatthour, the energy cost is 3.779(5.0) = 18.9 cents per hour.

With a 35°F (1.7°C) suction temperature and using the lower heating capacity obtained with the smaller, fixed-capacity compressor, step 9, we see the weight of gas handled by the compressor will be (64,400 Btu/h)/(54.99 Btu/lb) = 1172 lb/h (0.15 kg/s). From step 7, the compressor must supply 8.48 Btu/lb (19.7 kJ/kg). Therefore, the compressor's total heat contribution is (1172 lb/h)(8.48 Btu/lb) = 9950 Btu/h (2.9 kW). With a motor efficiency of 85 percent, the hourly motor input is equivalent to 9950/0.85 = 11,700 Btu/h (3.4 kW).

Using the same procedure as above, we see that the electric power input to the compressor will be 3.43 kWh, while the pump electric power input is 0.150 kWh. The total electric power input is 3.58 kWh at a cost of (3.58)(5.0) = 17.9 cents per hour. Hence, the hourly savings with the 35°F (1.7°C) suction temperature is 18.9 − 17.9 = 1.0 cent. Note, however, that the heat output at the 35°F (1.7°C) suction temperature, 64,400 Btu/h (18.9 kW), is 5600 Btu/h (1.6 kW) less than at the 40°F (4.4°C) suction temperature. If the lower heat output were unacceptable, the higher suction temperature or a larger compressor would have to be used. Either alternative would increase the power cost.

Related Calculations: With a water sink as the heat source, the usual water consumption of a heat pump ranges from 1.1 to more than 4 gal/(min·ton) [0.06 to 0.23 L/(s·t)]. A consumption range this broad requires that the actual flow rate be computed because a guess could be considerably in error.

Either air or the earth may be used as a heat source instead of water. When the cooling load rather than the heating load establishes the basic equipment size, an ideal situation exists for the use of the heat pump with air as the heat source. This occurs in localities where the minimum outdoor temperature in the winter is 20°F (−6.7°C) or higher.

Ground coils can be bulky, costly, and troublesome. One study shows that the temperature difference between the evaporating refrigerant in a ground coil and the surrounding earth is about equal to the number of Btu/h that may be drawn from each linear foot of coil. Thus, with a temperature difference of 15°F (8.3°C) and a 70,000-Btu (73,853.9-kJ) load of which 85 percent is supplied by the coil, the length of coil needed is (70,000)(0.85)/15 = 3970 ft (1210.1 m).

The coefficient of performance of a heat pump = heat rejected by condenser, Btu/heat equivalent of the net work of compression, Btu. The usual single-stage air source heat pump has a coefficient of performance ranging between 2.25 and 3.0. The procedure and data presented were developed by Robert Henderson Emerick, P.E., Consulting Mechanical Engineer.

Energy Conservation

REFERENCES: Hunt—*Windpower,* Van Nostrand Reinhold; Burberry—*Building for Energy Conservation,* Halsted Press; Chiogioji—*Industrial Energy Conservation,* Dekker; Courtney—*Energy Conservation in the Built Environment,* Longman; Culp—*Principles of Energy Conversion,* McGraw-Hill; Dorf—*The Energy Factbook,* McGraw-Hill; Dubin—*Energy Conservation Standards,* McGraw-Hill; *Energy Conservation in the International Energy Agency,* OECD; Grant—*Energy Conservation in the Chemical & Process Industries,* Institute of Chemical Engineers, England; Helcke—*The Energy Saving Guide,* Commission of the European Communities; Jarmul—*The Architect's Guide to Energy Conservation,* McGraw-Hill; Kovah—*Thermal Energy Storage,* Pergamon; Meckler—*Energy Conservation in Buildings & Industrial Plants,* McGraw-Hill; Payne—*Energy Managers' Handbook,* Butterworth; Pindyck—*The Structure of World Energy Demands,* M.I.T. Press; Reay—*Industrial Energy Conservation,* Pergamon; Smith—*Industrial Energy Management for Cost Reduction,* Ann Arbor Science; Yaverbaum—*Energy Saving by Increasing Boiler Efficiency,* Noyes; Considine—*Energy Technology Handbook,* McGraw-Hill.

CHOICE OF WIND-ENERGY CONVERSION SYSTEM

Select a wind-energy conversion system to generate electric power at constant speed and constant frequency in a sea-level area where winds average 18 m/h (29 km/h), a cut-in speed of 8 mi/h (13 km/h) is sought, blades will be fully feathered (cut out) at wind speeds greater than 60 mi/h (100 km/h), and the system must withstand maximum wind velocities of 150 mi/h (240 km/h). Determine typical costs which might be expected. The maximum rotor diameter allowable for the site is 125 ft (38 m).

Calculation Procedure:

1. *Determine the total available wind power*

Figure 1 shows the total available power in a freely flowing windstream at sea level for various wind speeds and cross-sectional areas of windstream. Since the maximum blade diameter, given that a blade-type conversion device will be used, is 125 ft (38 m), the area of the windstream will be $A = \pi d^2/4 = \pi(125)^2/4 = 12{,}271.9\ \text{ft}^2\ (1140.1\ \text{m}^2)$. Entering Fig. 1 at this area and projecting vertically to a wind speed of 18 mi/h (29 km/h), we see that the total available power is 200 kW.

2. *Select a suitable wind machine*

Typical modern wind machines are shown in Fig. 2. In any wind-energy conversion system there are three basic subsystems: the aerodynamic system, the mechanical transmission system (gears, shafts, bearings, etc.), and the electrical generating system. Figure 2 gives the taxonomy of the

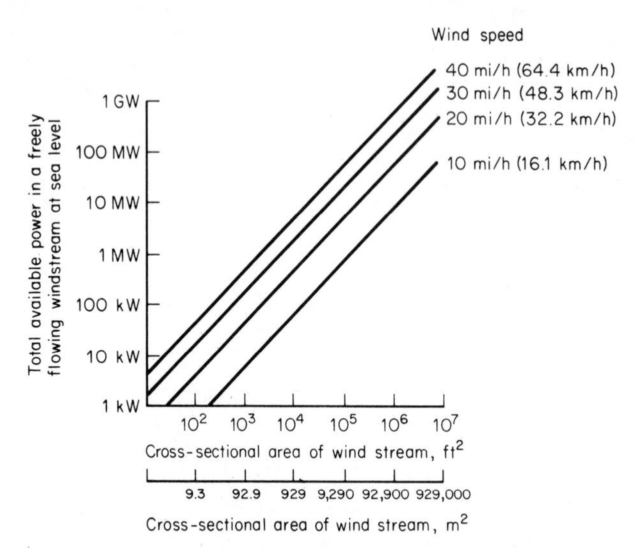

FIG. 1 The total available power in a freely flowing windstream at sea level versus the cross-sectional area of the windstream and the wind speed. *(Mechanical Engineering.)*

more practical versions of wind machines (the aerodynamic system) available today. "Almost any physical configuration which produces an asymmetric force in the wind can be made to rotate, translate, or oscillate—thereby generating power. The governing consideration is economic—how much power for how much size and cost," according to Fritz Hirschfeld, Member, ASME.

Continuing, Hirschfeld notes, "The power coefficient of an ideal wind machine rotor varies with the ratio of blade tip speed to free-flow windstream speed, and approaches the maximum of 0.59 when this ratio reaches a value of 5 or 6. Experimental evidence indicates that two-bladed rotors of good aerodynamic design—running at high rotational speeds where the ratio of the blade-tip-speed-to-free-flow-speed of the windstream is 5 or 6—will have power coefficients as high as 0.47. Figure 3 outlines the maximum power coefficients obtainable for several rotor designs. Figure 4 plots the typical performance curves of a number of different wind machines."

Choose a horizontal-axis double-bladed rotor wind machine for this application with a power coefficient C_p of 0.375. This type of wind machine is being chosen because (1) the power coefficient is relatively high (0.375), providing efficient conversion of the energy of the wind; (2) the allowable blade diameter, 125 ft (38 m), is suitable for the double-bladed design; (3) a double-bladed rotor will operate well in the average wind speed, 18 mi/h (29 km/h), prevailing in the installation area; and (4) the double-bladed rotor is well suited for the constant-speed constant-frequency (CSCF) system desired for this installation.

3. *Compute the maximum electric power output of the wind machine*

The power of the wind P_w is converted to mechanical power P_m by the wind machine. In any wind machine, $P_m = C_p P_w$, where C_p = power coefficient. The mechanical power is then converted to electric power by the generator. Since there is an applicable efficiency for each of the systems, that is, C_p for the aerodynamic system, η_m for the mechanical system (gears, usually), or η_g for the generator, the electric power generated is $P_e = P_w \eta_m \eta_g$.

In actual practice, the maximum electric power output in kilowatts of horizontal-axis bladed wind machines geared to a 70 percent efficiency electric generator can be quickly computed from $P_e = 0.38d^2V^3/10^6$, where d = blade diameter, ft (m); V = maximum wind velocity, ft/s (m/s). For this wind machine, $P_e = (0.38)(125)^2(26.4)^3/10^6 = 109.2$ kW. This result agrees closely with the actual machine on which the calculation procedure is based, which has a rated output of 100 kW.

FIG. 2 Wind machines come in all shapes and sizes. Some of the more practical design categories are illustrated in this taxonomy. *(Mechanical Engineering.)*

4. *Determine the typical capital cost of this machine*

Figures 5 and 6 show typical capital costs for small conventional wind machines. Larger wind machines, such as the one being considered here, are estimated to have a cost of $150,000 for a 100-kW unit, or $1500 per kilowatt. Such costs may be safely used in first approximations with the base-year cost given in the illustration being suitably adjusted by a factor for inflation.

Related Calculations: Use this general procedure to choose wind machines for other duties—pumping, battery charging, supplying power to utility lines, etc. Be certain to check with manufacturers to determine whether the calculated results agree with actual practice in the field. In general, good agreement will be found to exist.

The illustrations and much of the data in this procedure are the work of Fritz Hirschfeld, as reported in *Mechanical Engineering* magazine. Also reported in the magazine is a proposal by J. S. Goela of Physical Sciences, Inc., to use kites to extract energy from the wind.

Kites avoid the use of high-capital-cost components such as windmill towers and large rotors.

FIG. 2 (*Continued*)

Further, a kite can utilize the full available potential of the wind. As Fig. 7 shows, the earth's boundary layer extends up to 5000 ft (1500 m) above sea level. In this boundary layer, the average wind velocity increases while the air density decreases with altitude. Consequently, the total available wind power per unit area ($= \frac{1}{2}\rho V^3$) increases with altitude until at an altitude of 5000 ft (1500 m) a maximum is reached. The ratio of available wind power in New England at 5000 ft (1500 m) and 150 ft [many wind systems operate at an altitude of 150 ft (50 m) or less] is 25. This is a large factor which makes it very attractive to employ systems that use an energy extraction device located at an altitude of 5000 ft (1500 m). Even at an altitude of 1000 ft (300 m), this ratio is large, approximately equal to 10.

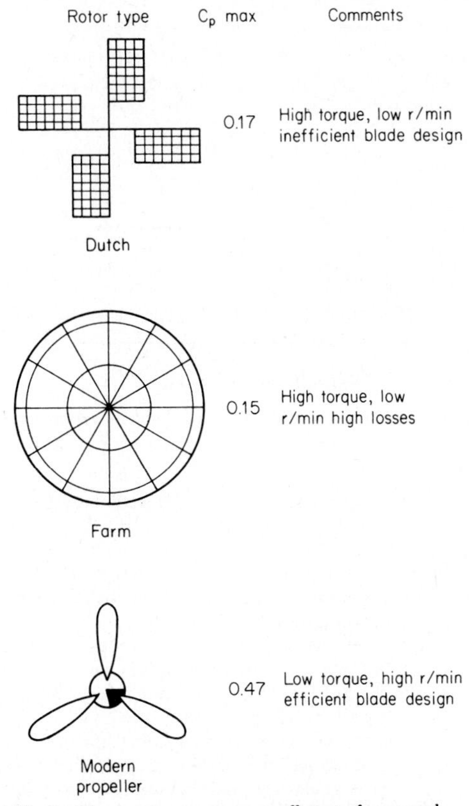

FIG. 3 The maximum power coefficients for several types of rotor designs. *(Mechanical Engineering.)*

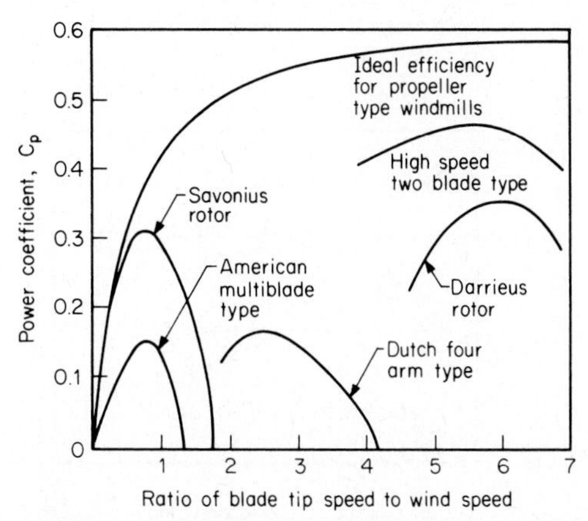

FIG. 4 Typical performance curves for different types of wind machines. *(Mechanical Engineering.)*

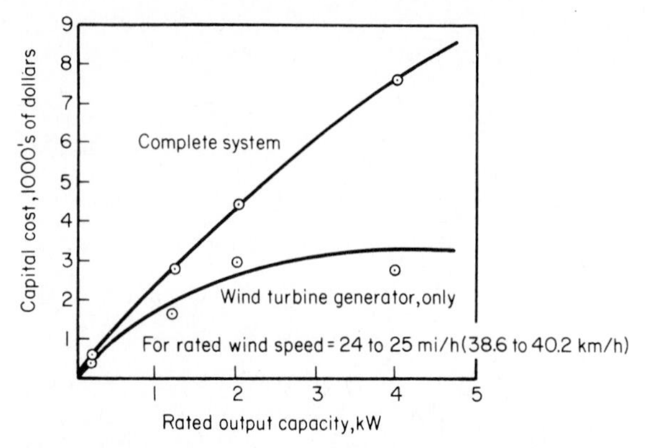

FIG. 5 Capital cost of small conventional wind machine. *(Mechanical Engineering.)*

To understand how the proposed scheme will extract energy from the wind, consider the following: The motion of air generates a pull in the rope that holds the kite. This pull is a function of both the angle of attack of the kite and the kite area normal to the wind direction, and by varying any of these we can vary the pull on the rope. On the surface of the earth, this rope will be suitably connected to an energy system which will convert the variation in developed force on the rope to the rotational energy of a rotor.

Whenever a period of calm occurs, the kite will tend to lose its altitude. One solution to this difficulty is to fill the empty spaces in the kite with helium gas such that the upward pull from the helium gas will balance the downward gravity force due to the weight of the kite and its string. Another possibility is to tie a ballon to the kite.

A detailed theoretical analysis of the proposed scheme has been carried out. This analysis indicates that the proposed scheme is scalable, that the drag on the kite string is small in comparison to the pull in the string for large devices, and that approximately 0.38 kW of power theoretically can be obtained from a kite 1.2 yd^2 (1 m^2) in area. In addition, there are no material or systems constraints which will prevent the kite from achieving an elevation of 5000 ft (1500 m).

Even though it is difficult to estimate the cost of wind power from the proposed scheme, a rough estimate indicates that for a 100-kW system, the capital cost per unit of energy from the proposed scheme will be less—approximately by a factor of 3—in comparison with the capital cost of one unit of energy produced from other 100-kW wind-energy systems.

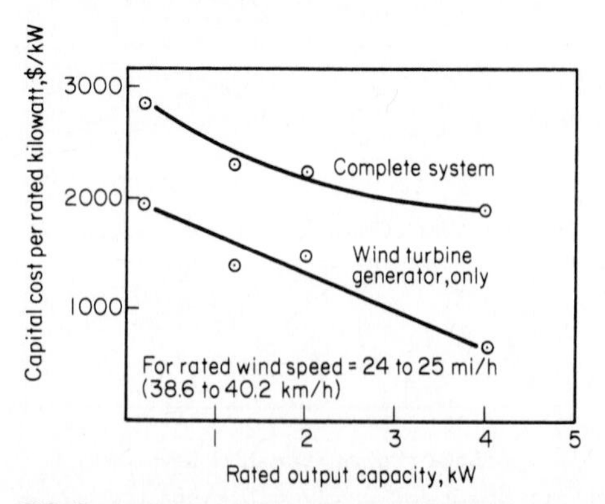

FIG. 6 Capital cost, per rated kilowatt, for small conventional wind machines. *(Mechanical Engineering.)*

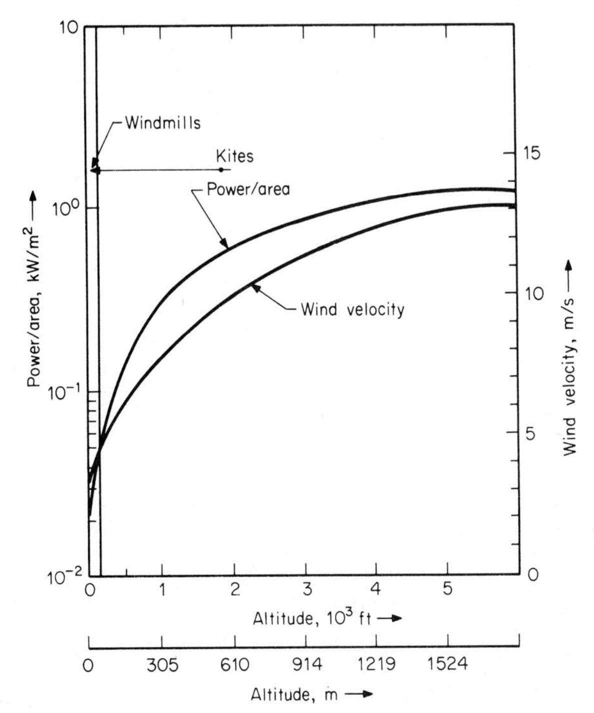

FIG. 7 Variation of mean annual free-air wind velocity and total available wind power per unit area ($=\frac{1}{2}\rho V^3$) with altitude in New England. *(Mechanical Engineering.)*

The most important application of kite-based energy systems is in developing countries where these systems can be used to pump water from wells, grind grain, and generate electricity. A majority of developing countries do not have an adequate supply of indigenous oil and gas, nor can they afford to buy substantial quantities of fossil fuel at international prices. What these countries prefer is a system that could generate useful energy by using as inputs resources that are available within the country. The simple scheme proposed here is ideally suited for the needs of developing countries.

The kite-based system may also be economically attractive in comparison to a small windmill of less than 1 hp (750 W) which has been used in rural and farm areas in the western United States to pump water, generate electricity, and irrigate land. Another application of the proposed scheme is to generate auxiliary power in large sailboats, motorboats, and ships where conventional wind-energy schemes cannot be employed. The wind-energy system employing kites can also be used as a fuel saver in conjunction with already existing transmission lines.

A kite flying at an altitude of 5000 ft (1500 m) may present a hazard to low-flying airplanes. One way to avoid a collision with the kite or an entanglement with the kite lines is to enhance their visibility by providing flashing lights around the kite structure and along its retaining line. Another approach is to fly the kites at lower altitudes. For instance, even at an altitude of 1000 ft (300 m), the total available wind power is larger by a factor of 10 in comparison with that at 150 ft (50 m), Fig. 7.

FUEL SAVINGS USING HIGH-TEMPERATURE HOT-WATER HEATING

Determine the fuel savings possible by using high-temperature-water (HTW) heating instead of steam if 50,000 lb/h (6.3 kg/s) of steam at 150 lb/in² (gage) (1034 kPa) is to be produced for delivering heat to equipment 1000 ft (305 m) from the boiler. The saturation temperature of the steam is 360°F (182°C); specific volume = 2.75 ft^3/lb (0.017 m^3/kg); enthalpy of evaporation = 857 Btu/lb (1996.8 kJ/kg); enthalpy of saturated vapor = 1195.6 Btu/lb (2785.7 kJ/kg); ambient temperature = 70°F (21.1°C); steam velocity = 5000 ft/min (1524 m/s); density of water at 240°F (171.1°C) = 56 lb/ft^3 (896.6 kg/m^3).

Calculation Procedure:

1. *Compute the required pipe cross-sectional area*

The required pipe cross-sectional area is A ft^2 (m^2) = Wv/V, where W = steam flow rate, lb/h (kg/h); V = specific volume of the steam, ft^3/lb (m^3/kg); V = steam velocity, ft/h (m/h). Or, A = 50,000(2.75)/[5000(60)] = 0.46 ft^2 or 66 in^2 (429 cm^2).

2. *Choose the size of the steam pipe*

For a 5-lb/in^2 (34.4-kPa) pressure drop in the 1000-ft (305-m) pipeline, standard pressure-loss calculations given elsewhere in this handbook show that a 10-in (25.4-cm) diameter pipe would be suitable when used in conjunction with a 5-in (12.7-cm) condensate-return line. The 10-in (25.4-cm) line would have 2-in (5.1-cm) thick calcium silicate insulation, while the 5-in (12.7-cm) line would have 1-in (2.5-cm) thick insulation of the same material.

3. *Compute the heat losses in the two lines*

Using the insulation heat-loss calculation methods given elsewhere in this handbook, we find the heat loss in the 10-in (25.4-cm) line is 183,200 Btu/h (53,678 W), while the heat loss in the 5-in (12.7-cm) condensate line is 78,100 Btu/h (22,883 W). Summing these, we see the total heat loss for the steam system is 261,300 Btu/h (76.6 kW).

4. *Determine the amount of condensate formed*

The amount of condensate formed w_c = (steam-line heat loss, Btu/h)/(enthalpy of vaporization, Btu/lb), or w_c = 183,200/857 = 214 lb/h (97.3 kg/h).

5. *Compute the amount of heat delivered to the load*

The amount of heat delivered to the load is H Btu/h = (steam flow rate, lb/h − condensate formation rate for heat loss, lb/h) (enthalpy of vaporization of the steam, Btu/lb). Or, for this steam system, H = (50,000 − 214)857 = 42,666,602 Btu/h (12,501 kW).

6. *Determine the condensate flash-out losses*

If the flash vapor is produced when the condensate is flashed out to atmospheric pressure in the return line and condensate receiver, the losses from flash-out will equal the enthalpy of the saturated water at 365°F (185°C) minus the enthalpy of the saturated water at 212°F (100°C), or 338.5 − 180 = 158.5 Btu/lb (369 kJ/kg).

To produce 857 Btu (904 J) of latent heat per pound of steam, the boiler must supply 1195.6 − 180 = 1015.6 Btu/lb (2366.3 kJ/kg), assuming that the condensate is returned to the boiler at 212°F (100°C). Hence, condensate losses from flash-out = (158.5/1015.6)100 = 15.6 percent. In addition, there is an approximate 5 percent loss due to leakage of steam and condensate, plus blowdown losses, which brings the total losses to 15.6 + 5.0 = 20.6, say 20 percent.

7. *Compute the total boiler heat input required*

With a condensate loss of 20 percent, as computed above, the amount of condensate returned to the boiler = 0.80(50,000 lb/h of steam) = 40,000 lb/h (5.04 kg/s). Hence, the enthalpy of the feedwater to the boiler, including makeup water, is 40,000 lb (18,181.8 kg) of condensate at 212°F (100°C) = 40,000(180 Btu/lb) = 7,200,000 Btu (7596 kJ); 10,000 lb (4545.4 kg) of makeup water at 50°F (10°C), 18 Btu/lb (41.9 kJ/kg), is 10,000(18) = 180,000 Btu (189,900 J); the sum = 7,380,000 Btu (7785 kJ). The boiler must therefore produce 50,000 × 1195.6 − 7,380,000 = 52,400,000 Btu/h (15,353.2 kW).

Assuming 75 percent boiler efficiency for this unit (a valid assumption for the usual steam heating boiler), we find the adjusted total amount of energy needed for steam heat = 52,400,000/0.75 = 69,867,000 Btu (73,710 kJ).

8. *Compute the hourly water flow rate*

To deliver 42,666,600 Btu/h (45,013.2 kJ) to the equipment, assume a 40°F (4.4°C) temperature drop between the supply and the return. Then the hourly flow rate = Btu/h heat required/temperature drop of the water, °F = 42,666,600/40 = 1,066,700 lb/h (134.3 kg/s).

9. *Choose the size pipe to use for the supply and return*

Assume a water flow velocity of 10 ft/s (3.05 m/s). Then the pipe area needed, from the relation in step 1 of this procedure, is 1,066,700/(3600)(10)(56) = 0.529 ft^2 (0.049 m^2). This area requires a 10-in (25.4-cm) pipe.

10. *Compute the heat loss in the piping*

The supply and return lines would require 2000 ft (609.6 m) of 10-in (25.4-cm) pipe with 2-in (5.1-cm) thick calcium silicate insulation. If the supply temperature is 360°F (182.2°C) and the return temperature is 320°F (160°C), the mean temperature would be (360 + 320)/2 = 340°F (171.1°C). Using the insulation heat-loss calculation methods given elsewhere in this handbook, we see that the heat loss in the supply and return lines is 326,800 Btu/h (96.8 kW). Hence, the total amount of heat which must be supplied to the water is 326,800 + 42,666,600 = 42,993,400 Btu/h (12,597 kW) before allowance is made for the efficiency of the boiler.

11. *Compare the steam and hot-water systems*

A typical hot-water heating boiler for a system such as this will have an operating efficiency of 77 percent. Using this value, we find the heat which must be supplied by the fuel = 42,993,400/0.77 = 55,835,600 Btu/h (16,360 kW).

As computed earlier, the heat required by the steam system exceeds that required by the hot-water system by 69,867,000 − 55,835,600 = 14,031,400 Btu/h (4111 kW), or 20 percent. This means that the high-temperature hot-water system will use 20 percent less fuel than the steam system for this installation.

Related Calculations: Use this approach when comparing or designing HTW systems for airports, military installations, hospitals, shopping centers, multifamily dwellings, garden apartments, industrial plants, central heating for large districts, university campuses, chemical-process plants, and similar installations. High-temperature-water systems are those using water in the 250 to 420°F (121.1 to 215.5°C) range, corresponding to a steam pressure of 300 lb/in^2 (gage) (2068.5 kPa). Mechanical problems caused by high water pressures above 420°F (215.5°C) make this temperature the practical upper limit. HTW systems can produce fuel savings 20 percent greater than systems using steam.

Studies show that conversion from steam to HTW is attractive—particularly for systems rated at 20,000,000 Btu/h (5860 kW) or higher. At this rating the conversion cost can usually be paid off in about 2 years. Smaller HTW systems, from 5,000,000 to 15,000,000 Btu/h (1470 to 4395 kW), are only marginally more economical to operate than steam, but they are still favored because they provide much more accurate and uniform temperature control.

HTW systems can give fuel savings of 20 to 50 percent, compared to an equivalent steam heating system. For new installations, the total capital investment is about the same for both steam and HTW systems. However, the savings in fuel costs and maintenance make the payout period for a new HTW system shorter than for conversion of an existing steam system.

Many plants use their steam boilers for both process and space heating. Cascade (direct-contact) heaters can generate up to 350°F (176.7°C) water from 150-lb/in^2 (1034-kPa) steam [or 400°F (204.4°C) from a 250-lb/in^2 (1724-kPa) boiler]. This water temperature is adequate for the rolls, presses, extruders, evaporators, conveyors, and reactors used in many industrial plants. Steam-pressure reducing valves are not needed to maintain the different temperature levels required by each machine.

Plants having steam boilers can convert to HTW heating simply and quickly by installing direct-contact water heaters in, or adjacent to, the boiler room. Such heaters can also serve as heat reservoirs, absorbing sudden peak loads and allowing the boilers to operate at fairly constant loads. HTW systems can easily supply water at elevated temperatures for process loads and water at lower temperatures for process loads and space-heating loads. Distribution efficiency of such systems approaches 95 percent overall.

For process applications requiring extremely close temperature control, the water circulating rate through a secondary loop can be designed to limit the difference between inlet and outlet temperatures to ±2°F (±1.11°C). The greater heat capacity of hot water over steam and the narrower pipelines required are other advantages. The usual HTW line need be only one or two sizes larger than the condensate line required in a steam system. The ratio of absolute heat-storing capacity is 42 to 1 in favor of HTW over steam. Where steam is needed in an all-HTW system, it can be obtained easily by flashing some of the water to steam. This calculation procedure is the work of William M. Teller, William Diskant, and Louis Malfitani, all of American Hydrotherm Corporation, as reported in *Chemical Engineering* magazine.

FUEL SAVINGS PRODUCED BY HEAT RECOVERY

Determine the primary-fuel saving which can be produced by heat recovery if 150 M Btu/h (158.3 MJ/h) in the form of 650-lb/in^2 (gage) (4481.1-kPa) steam superheated to 750°F (198.9°C) is recovered. The projected average primary-fuel cost (such as coal, gas, oil, etc.) over a 12-year evaluation period for this proposed heat recovery scheme is \$0.75 per 10^6 Btu (\$0.71 per million joules) lower heating value (LHV). Expected thermal efficiency of a conventional power boiler to produce steam at the equivalent pressure and temperature is 86 percent, based on the LHV of the fuel.

Calculation Procedure:

1. *Determine the value of the heat recovered during 1 year*

Enter Fig. 8 at the bottom at 1 year and project vertically to the curve marked \$0.75 per 10^6 LHV. From the intersection with the curve, project to the left to read the value of the heat recovered as \$5400 per year per MBtu/h recovered (\$5094 per MJ).

2. *Find the total value of the recovered heat*

The total value of the recovered heat = (hourly value of the heat recovered, \$/10^6 Btu)(heat recovered, 10^6 Btu/h)(life of scheme, years). For this scheme, total value of recovered heat = (\$5400)(150 $\times$ 10^6 Btu/h)(12 years) = \$9,720,000.

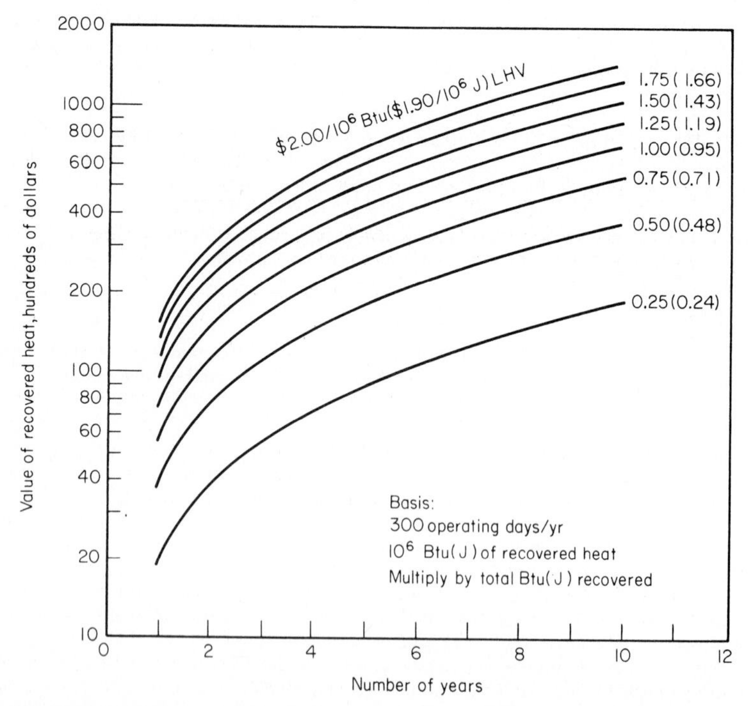

FIG. 8 Chart yields value of 1 million Btu of recovered heat. This value is based on the projected average costs for primary fuel. *(Chemical Engineering.)*

3. ***Compute the total value of the recovered heat, taking the boiler efficiency into consideration***

Since the power boiler has an efficiency of 86 percent, the equivalent cost of the primary fuel would be $0.75/0.86 = $0.872 per 10^6 Btu ($0.823 per million joules). The total value of the recovered heat if bought as primary fuel would be $9,720,000($0.872/$0.75) = $11,301,119. This is nearly $1 million a year for the 12-year evaluation period—a significant amount of money in almost any business. Thus, for a plant producing 1000 tons/day (900 t/day) of a product, the heat recovery noted above will reduce the cost of the product by about $3.14 per ton, based on 258 working days per year.

Related Calculations: This general procedure can be used for any engineered installation where heat is available for recovery, such as power-generating plants, chemical-process plants, petroleum refineries, marine steam-propulsion plants, nuclear generating facilities, air-conditioning and refrigeration plants, building heating systems, etc. Further, the procedure can be used for these and any other heat-recovery projects where the cost of the primary fuel can be determined. Offsetting the value of any heat saving will be the cost of the equipment needed to effect this saving. Typical equipment used for heat savings include waste-heat boilers, insulation, heat pipes, incinerators, etc.

With the almost certain continuing rise in fuel costs, designers are seeking new and proven ways to recover heat. Ways which are both popular and effective include the following:

1. Converting recovered heat to high-pressure steam in the 600- to 1500-lb/in^2 (gage) (4137- to 10,343-kPa) range where the economic value of the steam is significantly higher than at lower pressures.
2. Superheating steam using elevated-temperature streams to both recover heat and add to the economic value of the steam.
3. Using waste heat to raise the temperature of incoming streams of water, air, raw materials, etc.
4. Recovering heat from circulating streams of liquids which might otherwise be wasted.

In evaluating any heat-recovery system, the following facts should be included in the calculation of the potential savings:

1. The economic value of the recovered heat should exceed the value of the primary energy required to produce the equivalent heat at the same temperature and/or pressure level. An efficiency factor must be applied to the primary fuel in determining its value compared to that obtained from heat recovery. This was done in the above calculation.
2. An economic evaluation of a heat-recovery system must be based on a projection of fuel costs over the average life of the heat-recovery equipment.
3. Environmental pollution restrictions must be kept in mind at all times because they may force the use of a more costly fuel.
4. Many elevated-temperature process streams require cooling over a long temperature range. In such instances, the economic analysis should credit the heat-recovery installation with the savings that result from eliminating non-heat-recovery equipment that normally would have been provided. Also, if the heat-recovery equipment permits faster cooling of a stream and this time saving has an economic value, this value must be included in the study.
5. Where heat-recovery equipment reduces primary-fuel consumption, it is possible that plant operations can be continued with the use of such equipment whereas without the equipment the continued operation of a plant might not be possible.

The above calculations and comments on heat recovery are the work of J. P. Fanaritis and H. J. Streich, both of Struthers Wells Corp., as reported in *Chemical Engineering* magazine.

Where the primary-fuel cost exceeds or is different from the values plotted in Fig. 8, use the value of $1.00 per 10^6 Btu (J) (LHV) and multiply the result by the ratio of (actual cost, dollars per 10^6 Btu/$1) ($/J/$1). Thus, if the actual cost is $3 per 10^6 Btu (J), solve for $1 per 10^6 Btu (J) and multiply the result by 3. And if the actual cost were $0.80, the result would be multiplied by 0.8.

COST OF HEAT LOSS FOR UNINSULATED PIPES

What is the annual cost of the heat loss from 10 ft (3 m) of uninsulated 6-in (15-cm) diameter pipe conveying steam at 375°F (190.6°C) if the average ambient temperature is 75°F (23.9°C), the fuel cost is $1.44 per 10^6 Btu ($1.36 per MJ), and the net energy conversion efficiency is 72 percent? What effect would a wind velocity of 10 mi/h (16.1 km/h) have on the heat loss from the pipe?

Calculation Procedure:

1. *Determine the heat loss in still air per unit length of pipe*

The heat loss from any vessel or pipe depends on the temperature difference between the pipe or vessel and the medium in contact with the warmer surface, or ΔT. For this installation, $\Delta T = 375 - 75 = 300$°F (166.7°C).

To determine the cost of the heat loss per unit length of pipe, enter Fig. 9 at the bottom at 300°F (166.7°C) and project vertically upward to the intersection with the curve; from the intersection project horizontally to the left to read the cost as $7.60 per year per square foot ($84.44

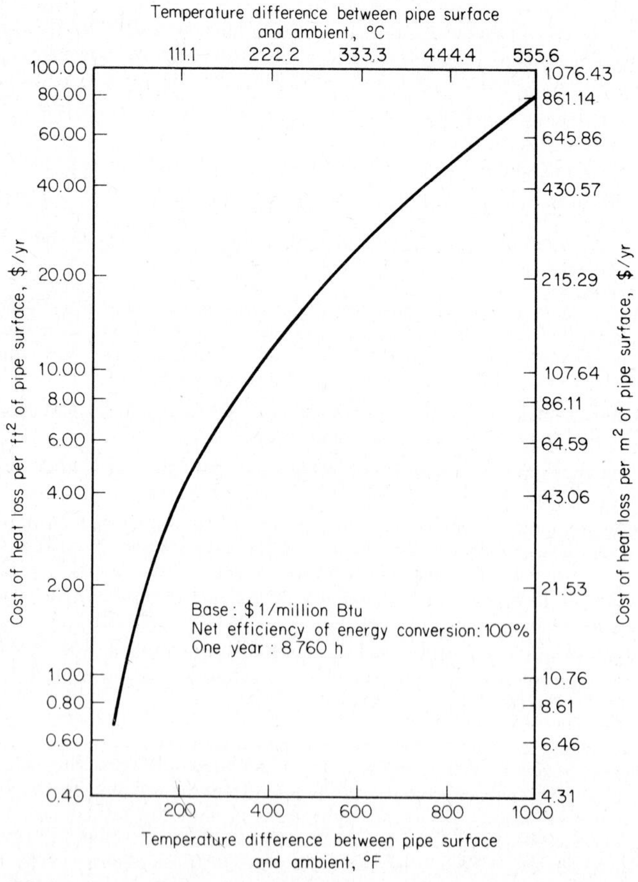

FIG. 9 Cost of heat loss when insulation is missing. *(Chemical Engineering.)*

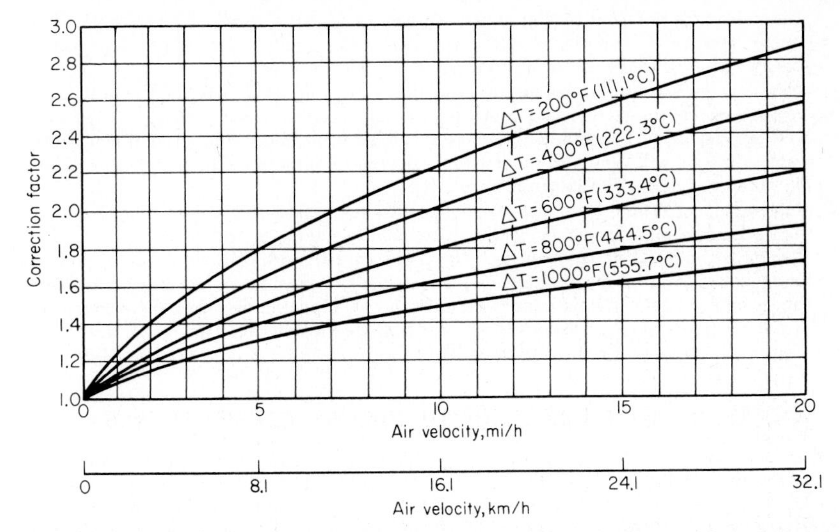

FIG. 10 Correction for wind velocity; ΔT is difference between pipe and ambient temperature. *(Chemical Engineering.)*

per year per square meter) of pipe surface. Convert this cost to the cost per linear foot (meter) of pipe by multiplying by 1.734 ft^2/ft, the surface area per foot of length of 6-in (15.2-cm) nominal pipe. Hence, (\$7.60)(1.734) = \$13.18 per foot per year (\$43.24 per meter per year) for this pipe.

To compute the annual cost for this pipe, multiply by the total pipe length, or 10(\$13.18) = \$131.80 per year.

2. *Correct the computed annual cost of the loss for net efficiency and fuel cost*

The net efficiency of energy conversion is 72 percent, and the annual fuel cost = (cost, \$ per foot per year)(ft of pipe)(fuel cost, \$ per 10^6 Btu per year)/(net energy conversion efficiency, %) = (\$131.80)(1.44/0.72) = \$263.60.

3. *Determine the annual cost of the heat loss when the pipe is exposed to the wind*

Enter Fig. 10 at the air velocity, 10 mi/h (16.1 km/h) and project upward to a ΔT of 300°F (166.7°C). At the left, read the wind correction factor as 2.14. Then the annual cost in a location having a prevailing wind = (annual cost for wind-free location)(air velocity correction factor) = (\$263.60)(2.14) = \$564.10.

Related Calculations: For piping whose surface has been left uninsulated or whose insulation has been so severely damaged as to be useless, the yearly (8760-h) cost of the heat loss per square foot can be found from Fig. 9. Costs so determined will generally be within the accuracy of data that can be obtained in the field, and are suitable for most applications without further refinement.

Figure 9 costs are based on a net efficiency of energy conversion of 100 percent, a fuel price of \$1.00 per 10^6 Btu, and heat transfer to still air. Corrections are necessary to arrive at annual costs for other conditions:

Net efficiency of less than 100 percent: To correct for a different net efficiency, divide the cost derived from Fig. 9 by the actual net efficiency of energy conversion. (For intermediate-size boilers most common in usual plants, the net efficiency is typically 72 percent.)

Fuel cost other than \$1.00 per 10^6 Btu: To correct for a different cost, multiply the cost derived from Fig. 9 by the actual cost of the fuel.

Outdoor installation exposed to moving air: To correct for variant wind conditions, multiply the cost obtained from Fig. 9 by a correction factor from Fig. 10 for the actual wind velocity. For most of the continental United States, 10 mi/h (16.1 km/h) represents a reasonable annual average.

The unit costs provided by Fig. 9 are average values for a range of pipe sizes in still air. Although there is some variation for different pipe sizes and ambient temperatures, the variations are small compared to the cost of heat loss caused by low air velocities.

The heat loss resulting from missing insulation is the difference between the heat loss from bare pipe and that through normal insulation. Insulation standards have changed in recent years because of varying fuel and insulation costs and other economic factors. And although the loss through insulation will vary with the standard adopted for a project, such variations will be negligible compared to the heat loss from bare pipe.

In general, corrections for precise ambient conditions, pipe size, and insulation thickness will not be justified because they go beyond the reasonable accuracy of field data.

The procedure and illustrations presented here are the work of Rene Cordero, piping and process mechanical equipment design engineer, Allied Chemical Corp., as reported in *Chemical Engineering* magazine. The data provided are valid for piping in chemical, petrochemical, factory, marine, power, and similar plants where it is desired to determine the cost of heat loss from uninsulated or partially insulated pipes.

HEAT-RATE IMPROVEMENT USING TURBINE-DRIVEN BOILER FANS

What is the net heat-rate improvement and net kilowatt gain in a steam power plant having a main generating unit rated at 870,000 kW at 2.5 in (6.35 cm) HgA, 0 percent makeup with motor-driven fans if turbine-driven fans are substituted? Plant data are as follows: (*a*) tandem-compound turbine, four-flow, 3600-r/min 33.5 in (85.1 cm) last-stage buckets with 264-ft^2 (24.5-m^2) total last-stage annulus area; (*b*) steam conditions 3500 lb/in^2 (gage) (24,133 kPa), 1000°F/1000°F (537.8°C/537.8°C); (*c*) with main-unit valves wide open, overpressure with motor-driven fans, generator output = 952,000 kW at 2.5 in (6.35 cm) HgA and 0 percent makeup; net heat rate = 7770 Btu/kWh (8197.4 kJ/kWh); (*d*) actual fan horsepower = 14,000(10,444 W) at valves wide open, overpressure with no flow or head margins; (*e*) motor efficiency = 93 percent; transmission efficiency = 98 percent; inlet-valve efficiency = 88 percent; total drive efficiency = 80 percent; difference between the example drive efficiency and base drive efficiency = 80 − 76.7 = 3.3 percent.

Calculation Procedure:

1. *Determine the percentage increase in net kilowatt output when turbine-driven fans are used*

Enter Fig. 11 at 264-ft^2 (24.5-m^2) annulus area and 14,000 required fan horsepower, and read the increase as 3.6 percent. Hence, the net plant output increase = 34,272 kW (= 0.036 × 952,000).

2. *Compute the net heat improvement*

From Fig. 12, the net heat rate improvement = 0.31 percent. Or, 0.0031(7770) = 24 Btu (25.3 J).

3. *Determine the increase in the throttle and reheater steam flow*

From Fig. 13, the increase in the throttle and reheater flow is 3.1 percent. This is the additional boiler steam flow required for the turbine-driven fan cycle.

4. *Compute the net kilowatt gain and the net heat-rate improvement*

From Fig. 14 the multipliers for the 2.5 in (6.35 cm) HgA backpressure are 0.98 for net kilowatt gain and 0.91 for net heat rate. Hence, net kW gain = 34,272(0.98) = 33.587 kW, and net heat-rate improvement = 24 × 0.91 = 22.0 Btu (23.0 J).

5. *Determine the overall cycle benefits*

From Fig. 15 the correction for a drive efficiency of 80 percent compared to the base case of 76.7 percent is obtained. Enter the curve with 3.3 percent (= 80 − 76.7) and read −6.6-Btu (−6.96-J) correction on the net heat rate and −0.08 percent of generated kilowatts.

To determine the overall cycle benefits, add algebraically to the values obtained from step 4, or net kW gain = 33,587 + (−0.0008 × 952,000) = 32,825 kW; net heat-rate improvement = 22.1 + (−6.6) = 15.5 Btu (16.4 J).

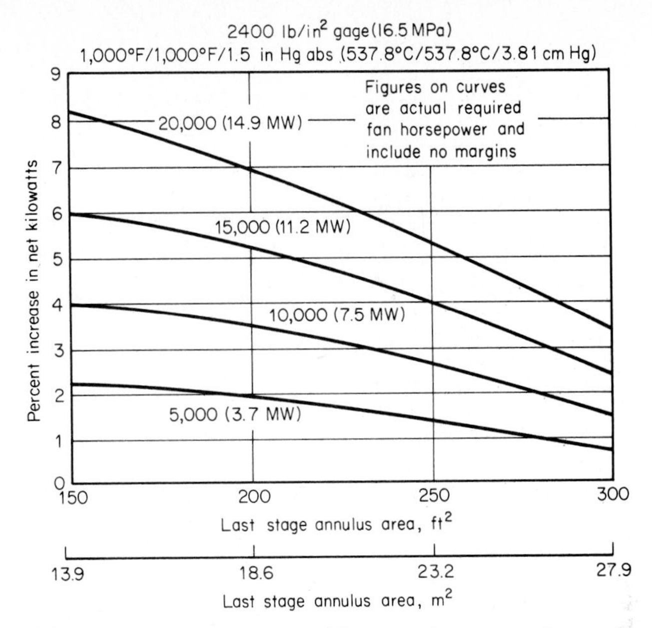

FIG. 11 Percentage increase in net kilowatts vs. last stage annulus area for 2400-lb/in^2 (gage) when turbine-driven fans are used as compared to motors. *(Combustion.)*

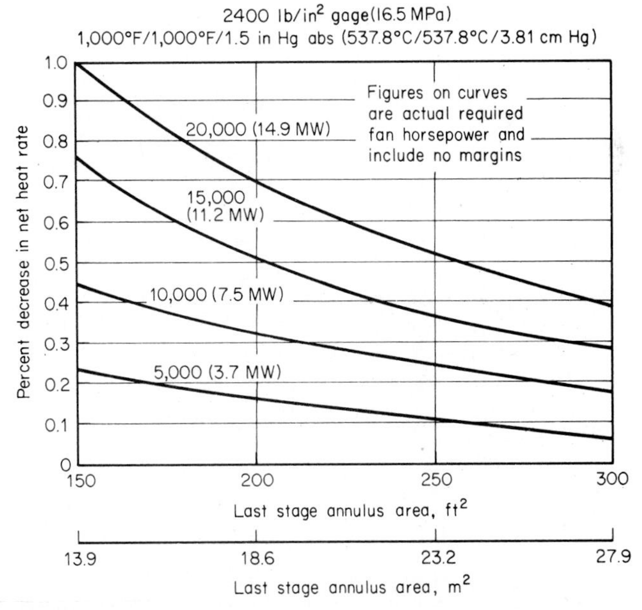

FIG. 12 Percentage decrease in net heat rate vs. last-stage annulus area for 2400-lb/in^2 (gage) when turbine-driven fans are used as compared to motors. *(Combustion.)*

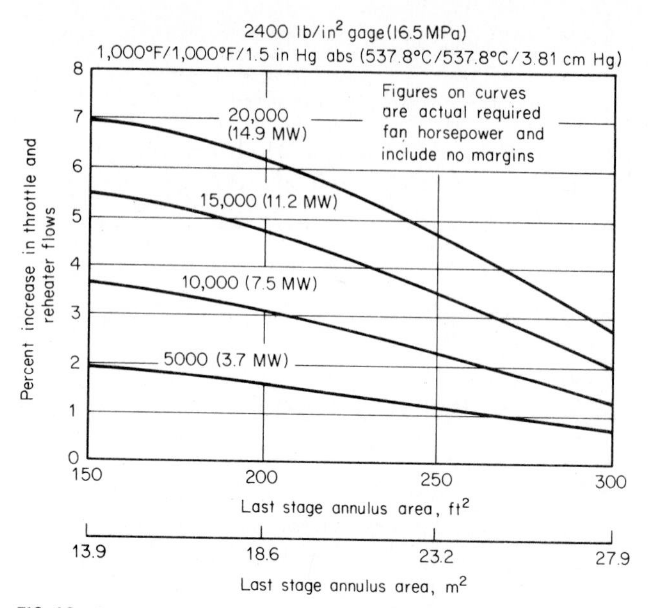

FIG. 13 Percentage increase in throttle and reheater flows vs. last-stage annulus area for 2400-lb/in^2 (gage) when turbine-driven fans are used as compared to motors. *(Combustion.)*

Related Calculations: This calculation procedure can be used for any maximum-loaded main turbine in utility stations serving electric loads in metropolitan or rural areas. A maximum-loaded main turbine is one designed and sized for the maximum allowable steam flow through its last-stage annulus area.

Turbine-driven fans have been in operation in some plants for more than 10 years. Next to feed pumps, the boiler fans are the second largest consumer of auxiliary power in utility stations.

Current studies indicate that turbine-driven fans can be economic at 700 MW and above, and possibly as low as 500 MW. Although the turbine-driven fan system will have a higher initial

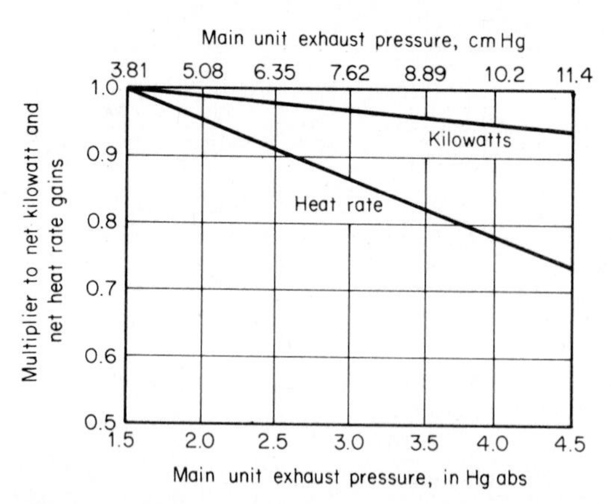

FIG. 14 Multiplier to net kilowatt and net heat-rate gains to correct for main-unit exhaust pressure higher than 1.5 inHg (38.1 mmHg). *(Combustion.)*

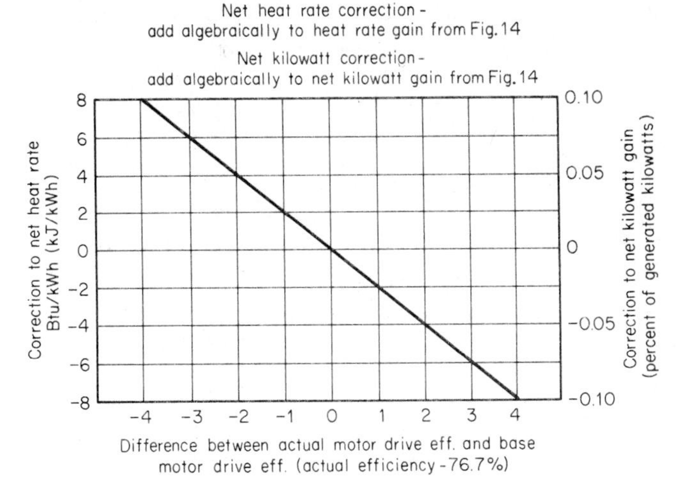

FIG. 15 Corrections for differences in motor-drive system efficiency. *(Combustion.)*

capital cost when compared to a motor-driven fan system, the additional cost will be more than offset by the additional net output in kilowatts. In certain cases, economic studies may show that turbine drives for fans may be advantageous in constant-throttle-flow evaluations.

As power plants for utility use get larger, fan power required for boilers is increasing. Environmental factors such as use of SO_2 removal equipment are also increasing the required fan power. With these increased fan-power requirements, turbine drive will be the more economic arrangement for many large fossil plants. Further, these drives enable the plant designer to obtain a greater output from each unit of fuel input.

This calculation procedure is based on the work of E. L. Williamson, J. C. Black, A. F. Destribats, and W. N. Iuliano, all of Southern Services, Inc., and F. A. Reed, General Electric Company, as reported in *Combustion* magazine and in a paper presented before the American Power Conference, Chicago.

COST SEPARATION OF STEAM AND ELECTRICITY IN A COGENERATION POWER PLANT USING THE ENERGY EQUIVALENCE METHOD

Allocate—using the energy equivalence method—the steam and electricity costs in a power plant having a double automatic-extraction, noncondensing steam turbine for process steam and electric generation. Turbine throttle steam flow is 800,000 lb/h (100.7 kg/s) at 865 lb/in^2 (abs) (5964.1 kPa). Process steam is extracted from the turbine in the amounts of 100,000 lb/h (12.6 kg/s) at 335 lb/in^2 (abs) (2309.8 kPa) and 200,000 lb/h (25.2 kg/s) at 150 lb/in^2 (abs) (1034.3 kPa) and is delivered to process plants. A total of 500,000 lb/h (62.9 kg/s) is exhausted at 35 lb/in^2 (abs) (241.3 kPa) with 100,000 lb/h (12.6 kg/s) of this exhaust steam for deaerator heating in the cycle and 400,000 lb/h (50.4 kg/s) sent to process plants. The turbine has a gross electric output of 51,743 kW, and the heat balance for the dual-purpose turbine cycle is shown in Fig. 16. Efficiency of the steam boiler is 85.4 percent, while the fuel is priced at \$0.50 per 10^6 Btu (\$0.47 per MJ). If a condensing turbine is used, an attainable backpressure is 1.75 inHg (abs) [43.75 mmHg (abs)], while the assumed turbine efficiency is 82 percent and the exhaust enthalpy is $h_{f'}$ = 1032 Btu/lb (2400.4 MJ/kg). Figure 17 shows the expansion-state curve of the turbine on a Mollier diagram. Final feedwater enthalpy is 228 Btu/lb (530.3 mJ/kg). Allocate the fuel cost to each energy use by using the energy equivalence method.

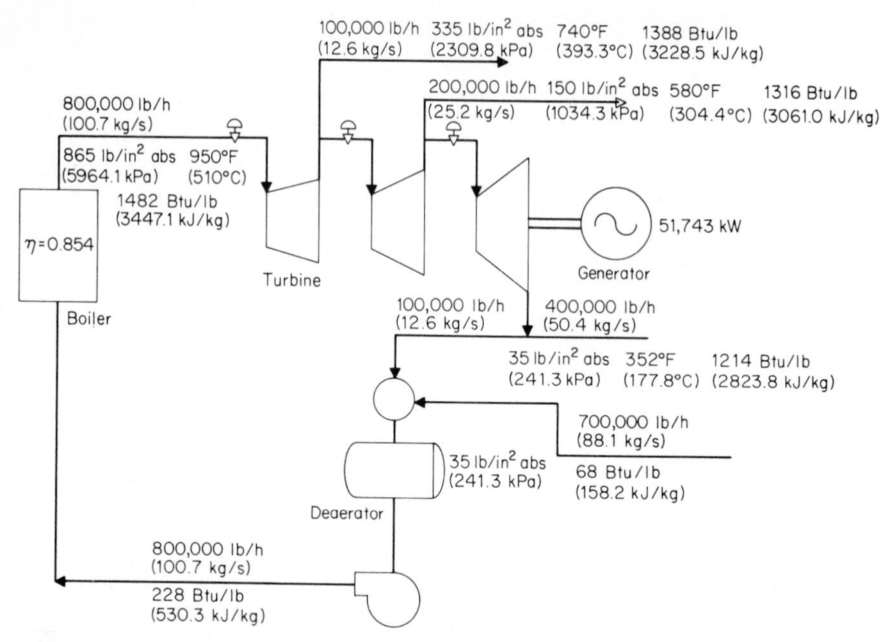

FIG. 16 Dual-purpose turbine heat balance. *(Combustion.)*

Calculation Procedure:

1. *Compute the hourly total fuel cost*

The total fuel cost for this plant per hour is C_f = (1/boiler efficiency)(0.50/10^6)(throttle steam flow rate m_t, lb/h)($h_t - h_{fw}$), where h_t = throttle enthalpy, Btu/lb, and h_{fw} = feedwater enthalpy, Btu/lb. Substituting gives C_f = (1/0.854)(0.50/10^6)(800,000)(1482 − 228) = \$587.35 per hour.

2. *Compute the nonextraction ultimate electric output*

The ultimate electric output is $E_u = m_t(h_i - h_{f'})/3413$, where m_t = total turbine inlet steam flow, lb/h; h_i = turbine initial enthalpy, Btu/lb; other symbols as before. Substituting, we find E_u = (800,00)(1482 − 1032)/3413 = 105,480 kW.

3. *Determine the actual electric output of the dual-purpose turbine*

Use the relation E_a = W(actual)/3413 = the work done by the extraction steam between the throttle inlet and the extraction point, plus the work done by the nonextraction steam between the throttle and the exhaust. Or, from the turbine expansion curve in Fig. 17, E_a = [100,000(1482 − 1388) + 200,000(1482 − 1316) + 500,000(1482 − 1214)]/3413 = 51,743 kW.

4. *Compute the extraction steam kilowatt equivalence*

Again from Fig. 17, $E_{x1} = (h_{x1} - h_{f'})/3413$ = 100,000(1388 − 1032)/3413 = 10,432 kW; E_{x2} = 200,000(1316 − 1032)/3413 = 16,642 kW; E_{x3} = 500,000(1214 − 1032)/3413 = 26,663 kW. Hence, the nonextraction turbine ultimate electric output = $E_a + E_{x1} + E_{x2} + E_{x3}$ = 51,743 + 10,432 + 16,642 + 26,663 = 105,480 kW.

5. *Determine the base fuel cost of electricity and steam*

The base fuel cost of electricity = C_f/E_u, or \$587.35/105,480 = \$0.005568 per kilowatthour, or 5.568 mil/kWh.

Now the base fuel cost of the steam at the different pressures can be found from (kW equiv-

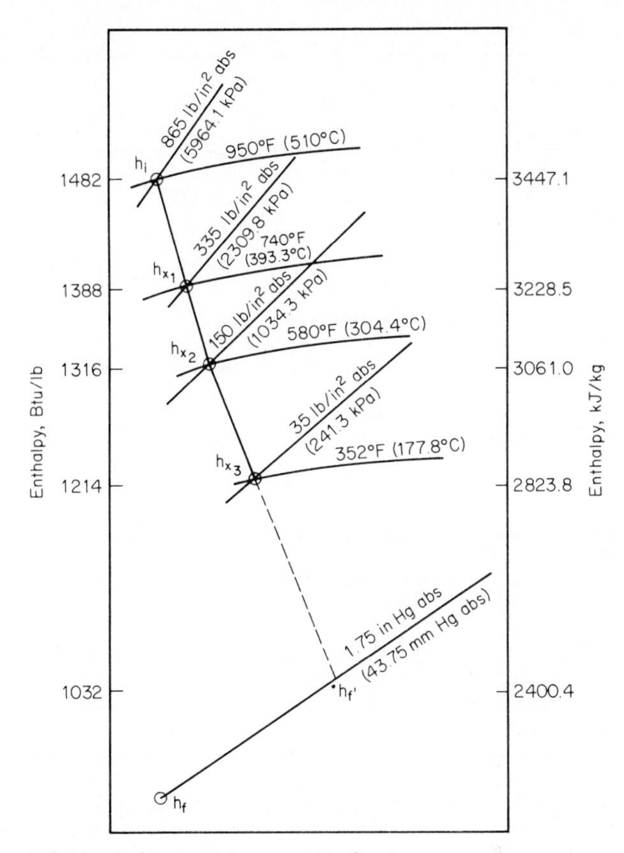

FIG. 17 Turbine expansion curve. *(Combustion.)*

alence)(base cost of electricity, mil)/(rate of steam use, lb/h). Thus, for the 335-lb/in² (abs) (2309.8-kPa) extraction steam used at the rate of 100,000 lb/h (12.6 kg/s), base fuel cost = 10,432(5.568)/100,000 = \$0.5808 per 1000 lb (\$0.2640 per 1000 kg). For the 150-lb/in² (abs) (1034.3-kPa) steam, base fuel cost = 16,642(5.568)/200,000 = \$0.4633 per 1000 lb (\$0.21059 per 1000 kg). And for the 35-lb/in² (abs) (241.3-kPa) steam, base fuel cost = 26,663(5.568)/500,000 = \$0.2969 per 1000 lb (\$0.13495 per 1000 kg). Since 100,000 lb/h (12.6 kg/s) of the 500,000-lb/h (62.9 kg/s) is used for deaerator heating, the cost of this heating steam = (100,000/1000)(\$0.2969) = \$29.69 per hour.

Since 100,000 lb (45,000 kg) of steam utilizes its energy for deaerator heating within the cycle, its equivalent electric output of 26,663/5 = 5333 kW should be deducted from the 26,663-kW electric energy equivalency of the 35-lb/in² (abs) (241.3-kPa) steam. The remaining equivalent energy of 21,330 kW (= 26,663 − 5333) represents 35-lb/in² (abs) (241.3-kPa) extraction steam to be delivered to process plants.

6. *Determine the added unit fuel cost*

The deaerator-steam fuel cost of \$29.69 per hour would be shared by both process steam and electricity in terms of energy equivalency as 105,480 − 5333 = 100,147 kW. Using this output as the denominator, we see that the added unit fuel cost for electricity based on sharing the cost of this heat energy input to the deaerator is \$26.69/100,147 = \$0.000296 per kilowatthour, or 0.296 mil/kWh.

Likewise, added fuel cost of 335-lb/in² (abs) (2309-kPa) steam = 10,432(100,000)/0.296 =

TABLE 1 Energy Equivalence Method of Fuel Cost Allocation*

Utility	Base	+	Unit cost heating steam	=	Total	Total fuel cost	Percent of total
Electricity 51,743 kW†	5.568		0.296		5.864 mi/kWh	$303.40	51.65
Steam @ 335 lb/in² (abs) (2309 kPa) 100,000 lb/h (45,000 kg/h)	0.5808		0.031		$0.6118 per 1000 lb (450 kg)	$ 61.18	10.41
Steam @ 150 lb/in² (abs) (1034 kPa) 200,000 lb/h (90,000 kg/h)	0.4633		0.0246		$0.4879 per 1000 lb (450 kg)	$ 97.57	16.62
Steam @ 35 lb/in² (abs) (241 kPa) 400,000 lb/h (180,000 kg/h)	0.2969		0.0158		0.3127 per 1000 lb (450 kg)	$125.20	21.32
					Total:	$587.35	100.00

**Combustion* magazine.

†Net kW delivered to process plants should be delivered after deducting fixed mechanical and electrical losses of the alternator. Electricity unit cost charged to production would be slightly higher after this adjustment.

$0.031 per 1000 lb (450 kg); added fuel cost of 150-lb/in² (abs) (1034-kPa) steam = $0.0246 per 1000 lb (450 kg); added fuel cost of 35-lb/in² (abs) (241-kPa) steam = $0.0158 per 1000 lb (450 kg). The fuel-cost allocation of steam and electricity is summarized in Table 1.

Related Calculations: The energy equivalence method is based on the fact that the basic energy source for process steam and electricity is the heat from fuel (combustion or fission). The cost of the fuel must be charged to the process steam and electricity. Since the analysis does not distinguish between types of fuels or methods of heat release, this procedure can be used for coal, oil, gas, wood, peat, bagasse, etc. Also, the procedure can be used for steam generated by nuclear fission.

Cogeneration is suitable for a multitude of industries such as steel, textile, shipbuilding, aircraft, food, chemical, petrochemical, city and town district heating, etc. With the increasing cost of all types of fuel, cogeneration will become more popular than in the past. This calculation procedure is the work of Paul Leung of Bechtel Corporation, as reported at the 34th Annual Meeting of the American Power Conference and published in *Combustion* magazine. Since the procedure is based on thermodynamic and economic principles, it has wide applicability in a variety of industries. For a complete view of the allocation of costs in cogeneration plants, the reader should carefully study the Related Calculations in the next calculation procedure.

COGENERATION FUEL COST ALLOCATION BASED ON AN ESTABLISHED ELECTRICITY COST

A turbine of the single-purpose type, operating at initial steam conditions identical to those in the previous calculation procedure, and a condenser backpressure of 1.75 in (43.75 mm) Hg (abs), would have a turbine heat rate of 9000 Btu/kWh (9495 kJ/kWh). Compute the fuel cost allocation to that of steam by using the established-electricity-cost method.

Calculation Procedure:

1. *Compute the unit cost of the electricity*

The unit cost of the electricity is F_e = (fuel price, $)(turbine heat rate, Btu/kWh)/(boiler efficiency). For this plant, $F_e = (0.5/10^6)(9000)/(0.854)$ = $0.00527 per kilowatthour, or 5.27 mil/kWh.

2. *Determine where the deaerator heating steam should be charged*

The turbine heat rate of 9000 Btu/kWh is a reasonable and economically justifiable heat rate of a regenerative cycle with a certain degree of feedwater heating. Hence, in this case, the deaerator heating steam should not be charged to the electricity. Instead, this portion of the deaerator-heating-steam cost should be charged to the process steam.

3. *Allocate the fuel cost to steam*

The total fuel cost from the previous calculation procedure is $587.35 per hour. The electricity cost allocation = (kW generated)(cost $/kWh) = (51,743)(0.00527) = $273. Hence, the fuel cost to the steam is $587.35 − $273.00 = $314.35.

4. *Compute the power equivalence of the steam*

From the previous calculation procedure, $E_x = E_{x1} + E_{x2} + E_{x3}$, where E_x = equivalent electric output of the extraction steam, kW; $E_{x1}, \ldots$ = equivalent electric output of the various extraction steam flows, kW. Hence, ΣE_x = 10,432 + 16,642 + 26,663 = 53,737 kW.

5. *Determine the ratio of each extraction steam flow to the total extraction steam flow*

The ratio for any flow is $E_x/\Sigma E_x$. Thus, $E_{x1}/\Sigma E_x$ = 10,432/53,737 = 0.194; $E_{x2}/\Sigma E_x$ = 16,663/53,737 = 0.310; and $E_{x3}/\Sigma E_x$ = 26,663/53,737 = 0.496.

6. *Compute the base unit fuel cost of steam*

Use the relation $(E_x/\Sigma E_x)$(fuel cost to steam)/m, where m = (steam flow rate, lb/h)/1000. Hence, for 335-lb/in² (abs) (2309.8-kPa) steam, base unit fuel cost = (0.194)($314.35)/100 = $0.610 per 1000 lb ($0.277 per 1000 kg); for 150-lb/in² (abs) (1034.3-kPa) steam, base unit fuel cost = (0.310)($314.35)/200 = $0.487 per 1000 lb ($0.2213 per 1000 kg); for 35-lb/in² (abs) (241.3-kPa) steam, base unit fuel cost = (0.496)($314.35)/500 = $0.312 per 1000 lb ($0.1418 per 1000 kg). Since the deaeration steam is at 35 lb/in² (abs) (241 kPa), the cost of this steam = (100,000/1000)($0.312) = $31.20 per hour.

7. *Determine the unit fuel cost from sharing the cost of the deaerator heating steam*

If the 5333-kW power equivalence of the deaerator heating steam is deducted from the electric power equivalence of the extraction steam, the kilowatt equivalence of all steam to production centers becomes 53,737 − 5333 = 48,404 kW. The unit fuel cost from sharing the cost of the deaerator heating steam is then ($31.20/h)/(48,404) = $0.000644 per kilowatthour, or 0.644 mil/kWh.

8. *Compute the added fuel cost of steam at each pressure*

The added fuel cost at each pressure is (kW output at that pressure/steam flow rate, lb/h)(0.644). Thus, added fuel cost of 335-lb/in² (abs) (2309.8-kPa) steam = (10,431/100,000)(0.644) = $0.067 per 1000 lb ($0.03045 per 1000 kg); added fuel cost for 150-lb/in² (abs) (1034.3-kPa) steam = $0.053 per 1000 lb ($0.02409 per 1000 kg); added fuel cost for 35-lb/in² (abs) (241.3-kPa) steam = $0.034 per 1000 lb ($0.01545 per 1000 kg). Table 2 summarizes the fuel-cost allocation of steam and electricity by using this approach.

Related Calculations: The established-electricity-cost method is based on the assumption (or existence) of a reasonable and economically justifiable heat rate of the cycle being considered or used. The cost of the fuel must be charged to the process steam and electricity. Since the analysis does not distinguish between types of fuels or methods of heat release, this procedure can be used for coal, oil, gas, wood, peat, bagasse, etc. Also, the procedure can be used for steam generated by nuclear fission.

Cogeneration is suitable for a multitude of industries such as steel, textile, shipbuilding, aircraft, food, chemical, petrochemical, city and town district heating, etc. With the increasing cost of all types of fuels, cogeneration will become more popular than in the past.

Other approaches to cost allocations for cogeneration include: (1) capital cost segregation, (2) capital cost allocation by cost separation of major functions, (3) cost separation of joint components, (4) capital cost allocation based on single-purpose electric generating plant capital cost, (5) unit cost based on fixed annual capacity factor, and (6) unit cost based on fixed peak demand. Each method has its advantages, depending on the particular design situation.

In the two examples given here (the present and previous calculation procedures), water return to the dual-purpose turbine cycle is assumed to be of condensate quality. Hence, no capital and

TABLE 2 Established-Electricity-Cost Method of Fuel-Cost Allocation*

Utility	Base	+	Unit cost heating steam	=	Total	Total fuel cost	Percent of total
Electricity 51,743 kW†	5.27		0		5.27 mil/kWh	$273.00	46.48
Steam @ 335 lb/in^2 (abs) (2309 kPa) 100,000 lb/h (45,000 kg/h)	0.610		0.067		$0.677 per 1000 lb (450 kg)	$ 67.70	11.53
Steam @ 150 lb/in^2 (abs) (1034 kPa) 200,000 lb/h (90,000 kg/h)	0.487		0.053		$0.541 per 1000 lb (450 kg)	$108.25	18.43
Steam @ 35 lb/in^2 (abs) (241 kPa) 400,000 lb/h (180,000 kg/h)	0.312		0.034		$0.346 per 1000 lb (450 kg)	$138.40	23.56
					Total:	$587.35	100.00

**Combustion* magazine.
†Net kw delivered to process plants should be delivered after deducting fixed mechanical loss and electrical loss of the alternator. Electricity unit cost charged to production would be slightly higher after this adjustment.

operating costs of water have been included. In actual cases, a cost account should be set up based on the quantity of the returned condensate. Special charges would be necessary for the unreturned portion of the water. Although the examples presented are for a fossil-fueled cycle, the methods are equally valid for a nuclear steam-turbine cycle. For a contrasting approach and for more data on where this procedure can be used, review the Related Calculations portion of the previous calculation procedure.

This calculation procedure is the work of Paul Leung of Bechtel Corporation, as reported at the 34th Annual Meeting of the American Power Conference and published in *Combustion* magazine. Since the procedure is based on thermodynamic and economic principles, it has wide applicability in a variety of industries.

BOILER FUEL CONVERSION FROM OIL OR GAS TO COAL

An industrial plant uses three 400,000-lb/h (50.4-kg/s) boilers fired by oil, a 600-MW generating unit, and two 400-MW units fired by oil. The high cost of oil, and the predictions that its cost will continue to rise in future years, led the plant owners to seek conversion of the boilers to coal firing. Outline the numerical and engineering design factors which must be considered in any such conversion.

Calculation Procedure:

1. *Evaluate the furnace size considerations*

The most important design consideration for a steam-generating unit is the fuel to be burned. Furnace size, fuel-burning and preparation equipment, heating-surface quantity and placement, heat-recovery equipment, and air-quality control devices are all fuel-dependent. Further, these items vary considerably among units, depending on the kind of fuel being used.

Figure 18 shows the difference in furnace size required between a coal-fired design boiler and an oil- or gas-fired design for the same steaming capacity in lb/h (kg/s). The major differences between coal firing and oil or natural-gas firing result from the solid form of coal prior to burning

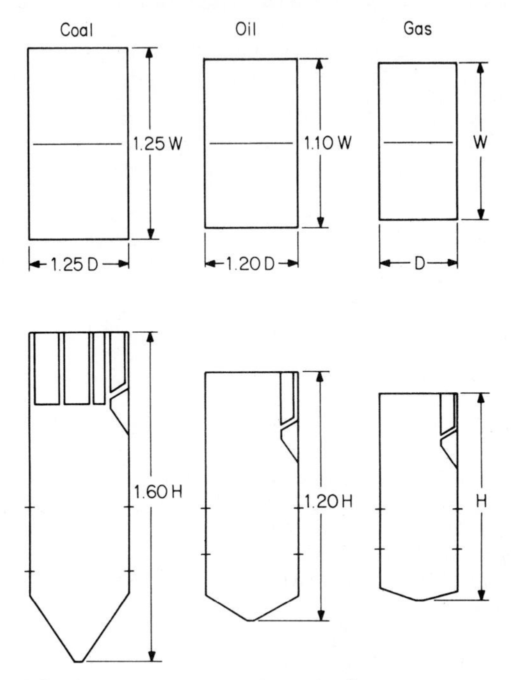

FIG. 18 Furnace size comparisons. *(Combustion.)*

and the ash in the products of combustion. Oil produces only small amounts of ash; natural gas produces no ash. Coal must be stored, conveyed, and pulverized before being introduced into a furnace. Oil and gas require little preparation. For these reasons, a boiler designed to burn oil as its primary fuel makes a poor conversion candidate for coal firing.

2. *Evaluate the coal properties from various sources*

Table 3 shows coal properties from many parts of the United States. Note that the heating values range from 12,000 Btu/lb (27,960 kJ/kg) to 6800 Btu/lb (15,844 kJ/kg). For a 600-MW unit, the

TABLE 3 Coal Properties—Nominal 600-MW Unit*

Type of coal	Eastern bituminous	Midwestern bituminous	Subbituminous C	Texas lignite	Northern plains lignite
HHV, Btu/lb (kJ/kg)	12,000 (27,912)	10,000 (23,260)	8,400 (19,538)	7,300 (16,980)	6,800 (15,817)
Moisture, %	6	12	27	32	37
lb $H_2O/10^6$ Btu (kg $H_2O/10^6$ kJ)	5 (0.00002)	12 (0.00005)	32 (0.000014)	44 (0.000019)	54 (0.000023)
Fuel fired, lb/h (kJ/h)	450,000 (202,500)	540,000 (243,000)	643,000 (289,350)	740,000 (333,000)	794,000 (357,300)

**Combustion* magazine.

TABLE 4 Pulverizer Requirements—Nominal 600-MW Unit*

	Eastern bituminous	Midwestern bituminous	Subbituminous	Texas lignite	Northern plains lignite
Hardgrove grindability	55	56	43	48	35
No. required	6	6	6	6	7
Nominal capacity†	50 tons/h (50.8 t/h)	63 tons/h (64 t/h)	85 tons/h (86.4 t/h)	92 tons/h (93.5 t/h)	100 tons/h (101.6 t/h)
Primary air temperature for drying coal	525°F (274°C)	640°F (338°C)	725°F (385°C)	750°F (399°C)	750°F (399°C)

***Combustion* magazine.
†Mill selection based on one full spare with remaining mills at 0.9 × new capacity.

coal firing rates [450,000 to 794,000 lb/h (56.7 or 99.9 kg/s)] to yield comparable heat inputs provide an appreciation of the coal storage yard and handling requirements for the various coals. On an hourly usage ratio alone, the lower-heating-value coal requires 1.76 times more fuel to be handled.

Pulverizer requirements are shown in Table 4 while furnace sizes needed for the various coals are shown in Fig. 19.

3. *Evaluate conversion to coal fuel*

Most gas-fired boilers can readily be converted to oil at reasonable cost. Little or no derating (reduction of steam or electricity output) is normally required.

From an industrial or utility view, conversion of oil- or gas-fired boilers not initially designed to fire coal is totally impractical from an economic viewpoint. Further, the output of the boiler would be severely reduced.

For example, the overall plant site requirements for a typical station having a pair of 400-MW units designed to fire natural gas could be an area of 624,000 ft^2 (57,970 m^2). This area would be

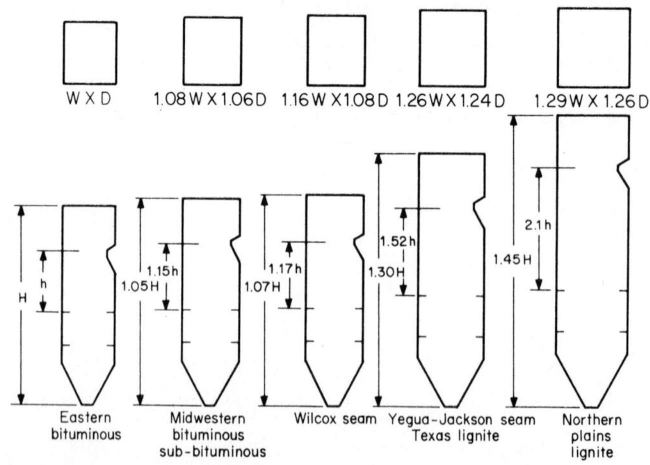

FIG. 19 Furnace sizes needed for various coals for efficient operation. *(Combustion.)*

for turbine bays, steam generators, and cooling towers. (With a condenser, the area required would be less.)

To accommodate the same facilities for a coal-fired plant with two 400-MW units, the ground area required would be 20 times greater. The additional facilities required include coal storage yard, ash disposal area, gas-cleaning equipment (scrubbers and precipitators), railroad siding, etc.

A coal-fired furnace is nominally twice the size of a gas-fired furnace. For some units the coal-fired boiler requires 4 times the volume of a gas-fired unit. Severe deratings of 40 to 70 percent are usually required for oil- and/or gas-fired boilers not originally designed for coal firing when they are switched to coal fuel. Further, such boilers cannot be economically converted to coal unless they were originally designed to be.

As an example of the derating required, the 400,000-lb/h (50.4-kg/s) units considered here would have to be derated to 265,000 lb/h (33.4 kg/s) if converted to pulverized-coal firing. This is 66 percent of the original rating. If a spreader stoker were used to fire the boiler, the maximum capacity obtainable would be 200,000 lb/h (25.2 kg/s) of steam. Extensive physical alteration of the boiler would also be required. Thus, a spreader stoker would provide only 50 percent of the original steaming capacity.

Related Calculations: Conversion of boilers from oil and/or gas firing to coal firing requires substantial capital investment, lengthy outage of the unit while alterations are being made, and derating of the boiler to about half the designed capacity. For these reasons, most engineers do not believe that conversion of oil- and/or gas-fired boilers to coal firing is economically feasible.

The types of boilers which are most readily convertible from oil or gas to coal are those which were originally designed to burn coal (termed *reconversion*). These are units which were mandated to convert to oil in the late 1960s because of environmental legislation.

Where the land originally used for coal storage was not sold or used for other purposes, the conversion problem is relatively minor. But if the land was sold or converted to other uses, there could be a difficult problem finding storage space for the coal.

Most of these units were designed to burn low-ash, low-moisture, high-heating-value, and high-ash-fusion coals. Fuels of this quality may no longer be available. Hence reconversion to coal firing may require significant downrating of the boiler.

Another important aspect of reconversion is the restoration of the coal storage, handling, and pulverizing equipment. This work will probably require considerable attention. Further, pulverizer capacity may not be sufficient, given the lower grade of fuel that would probably have to be burned.

This procedure is based on the work of C. L. Richards, Vice-President, Fossil Power Systems Engineering Research & Development, C-E Power Systems, Combustion Engineering, Inc., as reported in *Combustion* magazine.

RETURN ON INVESTMENT FOR ENERGY-SAVING PROJECTS

An industrial plant is considering an energy-saving installation of insulation which is projected to save 22.5 million Btu/h (65.9 MW) at a cost of $100,000. What is the return on investment (ROI) for this project if fuel is estimated to cost $1.25 per million Btu ($1.25 per 1,055,000 J), the plant is in continuous use, and 20 percent of the capital cost is for depreciation, tax, insurance, overhead, and maintenance?

Calculation Procedure:

1. *Compute the savings-to-capital ratio*

The savings-to-capital ratio is S = (Btu/h saved)/(capital investment to achieve this saving). Or, for this installation, S = 22,500,000/$100,000 = 225 Btu/h per $1 invested (65.9 W per $1 invested).

2. *Determine the return on investment*

Enter Fig. 20 at S = 225 Btu/h, and project horizontally to the right to the fuel-cost curve for $1.25 per 10^6 Btu. From the intersection project vertically downward to the ROI scale to read 225 percent.

Related Calculations: This procedure can be used for any type of energy-saving installation—be it new machinery, more effective insulation, alternative heat sources (such as solar, wood,

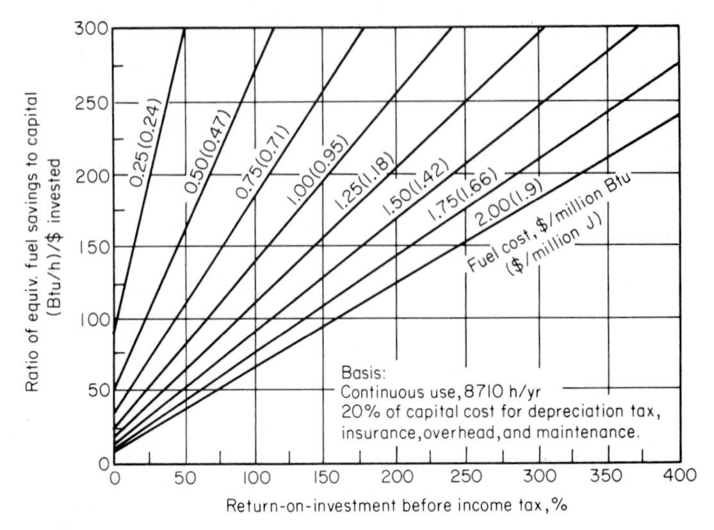

FIG. 20 ROI evaluation of energy-conservation projects. *(Chemical Engineering.)*

coal, gas, etc.), or other investments which save energy. Further, the ROI concept is valid for industrial, commercial, and residential installations where profit is a prime factor in business decisions. Here is how the ROI is put to use in business decisions:

(1) The ROI for alternative energy-saving schemes is computed, as described above. (2) These values are compared, and the highest value is chosen for further study. (3) By using the firm's target ROI for new projects, the new energy-saving project ROI is compared with that of the target. (4) If the energy-saving project ROI exceeds tha target ROI, the project is attractive from an investment standpoint. But if the energy-saving ROI is less than the firm's target ROI, the investment is not attractive from a business standpoint. Note, however, that in times of acute fuel shortages, an energy-saving project with a low ROI might still be attractive if it saves fuel. Further, each firm's management chooses the target ROI to be used in evaluating new projects of various kinds.

ROI is one of the most popular measures or indices for judging the business attractiveness of a proposed investment. When expressed as a percentage, ROI is the annual rate of return on the original investment made for an energy-saving project, new product, new structure, etc.

This procedure is the work of Jack Robertson, Senior Technical Specialist, Dow Chemical U.S.A., as reported in *Chemical Engineering* magazine.

ENERGY SAVINGS FROM REDUCED BOILER SCALE

A boiler generates 16,700 lb/h (2.1 kg/s) at 100 percent rating with an efficiency of 75 percent. If $\frac{1}{32}$ in (0.79 mm) of "normal" scale is allowed to form on the tubes, determine what savings can be made if 144,000-Btu/gal (40,133-MJ/m^3) fuel oil costs $1 per gallon ($1 per 3.8 L) and the boiler uses 16.74 million Btu/h (4.9 MW) operating 8000 h/year.

Calculation Procedure:

1. *Determine the annual energy usage*

Compute the annual energy usage from (million Btu/h) (hours of operation annually)/efficiency. For this boiler, annual energy usage = (16.74)(8000)/0.75 = 178,560 million Btu (188,380 kJ).

2. *Find the energy loss caused by scale on the tubes*

Enter Fig. 21 at the scale thickness, $\frac{1}{32}$ in (0.79 mm), and project vertically upward to the "normal" scale (salts of Ca and Mg) curve. At the left read the energy loss as 2 percent. Hence, the

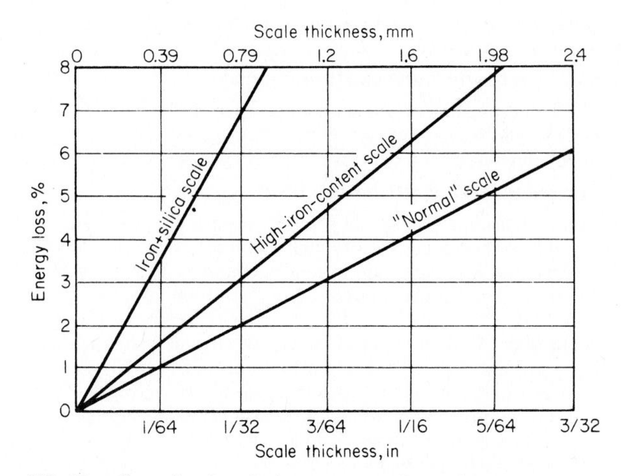

FIG. 21 Effects of scale on boiler operation. *(Chemical Engineering.)*

annual energy loss in heat units = (178,560 million Btu/year)(0.02) = 3571 million Btu/year (130.8 kW).

3. *Compute the annual savings if the scale is removed*

If the scale is removed, then the energy lost, computed in step 2, will be saved. Thus, the annual dollar savings after scale removal = (heat loss in energy units) (fuel price, \$/gal)/(fuel heating value, Btu/gal). Or, savings = $(3751 \times 10^6)(\$1.00)/144{,}000 = \$26{,}049$.

Related Calculations: This approach can be used with any type of boiler—watertube, firetube, etc. The data are also applicable to tubed water heaters which are directly fired.

Note that when the scale is high in iron and silica that the energy loss is much greater. Thus, with scale of the same thickness [1/32 in (0.79 mm)], the energy loss for scale high in iron and silica is 7 percent, from Fig. 21. Then the annual loss = 178,560(0.07) = 12,500 million Btu/year (3.63 MW). Removing the scale and preventing its reformation will save, assuming the same heating value and cost for the fuel oil, $(12{,}500 \times 10^6$ Btu/year$)(\$1.00)/144{,}000 = \$86{,}805$ per year.

While this calculation gives the energy savings from reduced boiler scale, the results also can be used to determine the amount that can be invested in a water-treatment system to prevent scale formation in a boiler, water heater, or other heat exchanger. Thus, the initial investment in treating equipment can at least equal the projected annual savings produced by the removal of scale.

This procedure is the work of Walter A. Hendrix and Guillermo H. Hoyos, Engineering Experiment Station, Georgia Institute of Technology, as reported in *Chemical Engineering* magazine.

GROUND AREA AND UNLOADING CAPACITY REQUIRED FOR COAL BURNING

An industrial plant is considering switching from oil to coal firing to reduce fuel costs. Determine the ground area required for 60 days' coal storage if the plant generates 100,000 lb/h (45,360 kg/h) of steam at a 60 percent winter load factor with a steam pressure of 150 lb/in^2 (gage) (1034 kPa), average boiler evaporation is 9.47 lb steam/lb coal (4.3 kg/kg), coal density = 50 lb/ft^3 (800 kg/m^3), boiler efficiency is 83 percent with an economizer, and the average storage pile height for the coal is 20 ft (6.096 m).

Calculation Procedure:

1. *Determine the storage area required for the coal*

The storage area, A ft^2, can be found from $A = 24WFN/EdH$, where H = steam generation rate, lb/h; F = load factor, expressed as a decimal; N = number of days storage required; E =

average boiler evaporation rate, lb/h; d = density of coal, lb/ft^3; H = height of coal pile allowed, ft. Substituting yields A = 24(100,000)(0.6)(60)/[(9.47)(50)(20)] = 9123 ft^2 (847 m^2).

2. ***Find the maximum hourly burning rate of the boiler***

The maximum hourly burning rate in tons per hour is given by $B = W/2000E$, where the symbols are as defined earlier. Substituting, we find B = 100,000/2000(9.47) = 5.28 tons/h (4.79 t/h). With 24-h use in any day, maximum daily use = 24 × 5.28 = 126.7 tons/day (115 t/day).

3. ***Find the required unloading rate for this plant***

As a general rule, the unloading rate should be about 9 times the maximum total plant burning rate. Higher labor and demurrage costs justify higher unloading rates and less manual supervision of coal handling. Find the unloading rate in tons per hour from $U = 9W/2000E$, where the symbols are as defined earlier. Substituting gives U = 9(100,000)/2000(8.47) = 47.5 tons/h (43.1 t/h).

Related Calculations: With the price of oil, gas, wood, and waste fuels rising to ever-higher levels, coal is being given serious consideration by industrial, central-station, commercial, and marine plants. Factors which must be included in any study of conversion to (or original use of) coal include coal delivery to the plant, storage before use, and delivery to the boiler.

For land installations, coal is usually received in railroad hopper-bottom cars in net capacities ranging between 50 and 100 tons with 50- and 70-ton (45.4- and 63.5-t) capacity cars being most common.

Because cars require time for spotting and moving on the railroad siding, coal is actually delivered to storage for only a portion of the unloading time. Thawing of frozen coal and car shaking also tend to reduce the actual delivery. True unloading rate may be as low as 50 percent of the continuous-flow capacity of the handling system. Hence, the design coal-handling rate of the conveyor system serving the unloading station should be twice the desired unloading rate. So, for the installation considered in this procedure, the conveyor system should be designed to handle 2(47.5) = 95 tons/h (86.2 t/h). This will ensure that at least six rail cars of 60-ton (54.4-t) average capacity will be emptied in an 8-h shift, or about 360 tons/day (326.7 t/day).

With a maximum daily usage of 126.7 tons/day (115 t/day), as computed in step 2 above, the normal handling of coal, from rail car delivery during the day shift, will accumulate about 3 days' peak use during an 8-h shift. If larger than normal shipments arrive, the conveyor system can be operated more than 8 h/day to reduce demurrage charges.

This procedure is the work of E. R. Harris, Department Head, G. F. Connell, and F. Dengiz, all of the Environmental and Energy Systems, Argonaut Realty Division, General Motors Corporation, as reported in *Combustion* magazine.

HEAT RECOVERY FROM BOILER BLOWDOWN SYSTEMS

Determine the heat lost per day from sewering the blowdown from a 600-lb/in^2 (gage) (4137-kPa) boiler generating 1 million lb/day (18,939.4 kg/h) of steam at 80 percent efficiency. Compare this loss to the saving from heat recovery if the feedwater has 20 cycles of concentration (that is, 5 percent blowdown), ambient makeup water temperature is 70°F (21°C), flash tank operating pressure is 10 lb/in^2 (gage) (69 kPa) with 28 percent of the blowdown flashed, blowdown heat exchanger effluent temperature is 120°F (49°C), fuel cost is \$2 per 10^6 Btu [\$2 per (9.5)^6 J], and the piping is arranged as shown in Fig. 22.

Calculation Procedure:

1. ***Compute the feedwater flow rate***

The feedwater flow rate, 10^6 lb/day = (steam generated, 10^6 lb/day)/(100 − blowdown percentage), or 10^6/(100 − 5) = 1.053 × 10^6 lb/day (0.48 × 10^6 kg/day).

2. ***Find the steam-production equivalent of the blowdown flow***

The steam-production equivalent of the blowdown = feedwater flow rate − steam flow rate = 1.053 − 1.0 = 53,000 lb/day (24,090 kg/day).

3. ***Compute the heat loss per lb of blowdown***

The heat loss per lb (kg) of blowdown = saturation temperature of boiler water − ambient temperature of makeup water. Or, heat loss = (488 − 70) = 418 Btu/lb (973.9 kJ/kg).

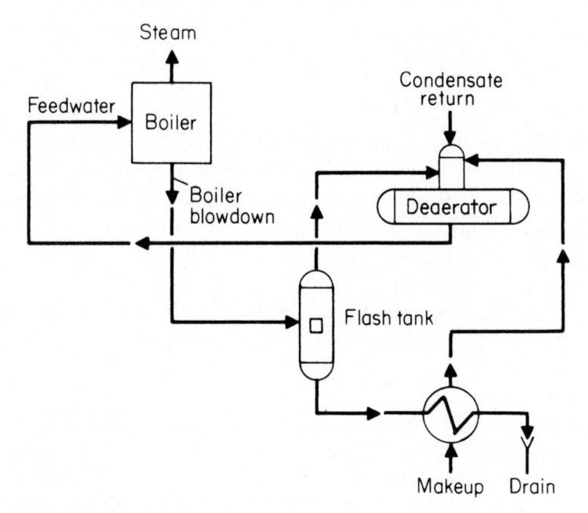

FIG. 22 Typical blowdown heat-recovery system. *(Combustion.)*

4. *Find the total heat loss from sewering*

When the blowdown is piped to a sewer (termed *sewering*), the heat in the blowdown stream is lost forever. With today's high cost of all fuels, the impact on plant economics can be significant. Thus, total heat loss from sewering = (heat loss per lb of blowdown) (blowdown flow rate, lb/day) = 418 Btu/lb (53,000 lb/day) = 22.2×10^6 Btu/day (23.4×10^6 J/day).

5. *Determine the fuel-cost equivalent of the blowdown*

The fuel-cost equivalent of the blowdown = (heat loss per day, 10^6 Btu)(fuel cost, \$ per 10^6 Btu)/(boiler efficiency, %), or (22.2)(2)/0.8 = \$55.50 per day.

6. *Find the blowdown flow to the heat exchanger*

With 28 percent of the blowdown flashed to steam, this means that 100 − 28 = 72 percent of the blowdown is available for use in the heat exchanger. Since the blowdown total flow rate is 53,000 lb/day (24,090 kg/day), the flow rate to the blowdown heat exchanger will be 0.72(53,000) = 38,160 lb/day (17,345 kg/day).

7. *Determine the daily heat loss to the sewer*

As Fig. 22 shows, the blowdown water which is not flashed, flows through the heat exchanger to heat the incoming makeup water and then is discharged to the sewer. It is the heat in this sewer discharge which is to be computed here.

With a heat-exchanger effluent temperature of 120°F (49°C) and a makeup water temperature of 70°F (21°C), the heat loss to the sewer is 120 − 70 = 50 Btu/lb (116.5 kJ/kg). And since the flow rate to the sewer is 38,160 lb/day (17,345 kg/day), the total heat loss to the sewer is 50(38,160) = 1.91×10^6 Btu/day (2.02×10^6 kJ/day).

8. *Compare the two systems in terms of heat recovered*

The heat recovered − heat loss by sewering = heat loss with recovery = 22.2×10^6 Btu/day − 1.91×10^6 Btu/day = 20.3×10^6 Btu/day (21.4×10^6 J/day).

9. *Determine the percentage of the blowdown heat recovered and dollar savings*

The percentage of heat recovered = (heat recovered, Btu/day)/(original loss, Btu/day) = (20.3/22.2)(100) = 91 percent. Since the cost of the lost heat was \$55.50 per day without any heat recovery, the dollar savings will be 91 percent of this, or 0.91(\$55.50) = \$50.51 per day, or \$18,434.33 per year with 365 days of operation. And as fuel costs rise, which they are almost certain to do in future years, the annual saving will increase. Of course, the cost of the blowdown heat-recovery equipment must be offset against this saving. In general, the savings warrant the added investment for the extra equipment.

Related Calculations: This procedure is valid for any type of steam-generating equipment for residential, commercial, industrial, central-station, or marine installations. (In the latter installation the "sewer" is the sea.) The typical range of blowdown heat recovery is in the 80 to 90 percent area. In view of the rapid rise in fuel prices, this range of heat recovery is significant. Hence, much wider use of blowdown heat recovery can be expected in all types of steam-generating plants.

To reduce scale buildup in boilers, low cycles of boiler water concentration are preferred. This means that high blowdown rates will be used. To prevent wasting expensive heat present in the blowdown, heat-recovery equipment such as that discussed above is used. In industrial plants (which are subject to many sources of condensate contamination), cycles of concentration are seldom allowed to exceed 50 (2 percent blowdown). In the above application, the cycles of concentration = 20, or 5 percent blowdown.

To prevent boiler scale buildup, good pretreatment of the makeup is recommended. Typical current selections for pretreatment equipment, by using the boiler operating pressure as the main criterion, are thus:

Boiler pressure, lb/in^2 (kPa)	Pretreatment equipment
0–600 (4137)	Sodium zeolite softening
600–900 (4137–6205)	Hot line/demineralizers
Above 900 (6205)	Demineralizers

This procedure is the work of A. A. Askew, Betz Laboratories, Inc., as reported in *Combustion* magazine.

AIR-COOLED HEAT EXCHANGER: PRELIMINARY SELECTION

Kerosene flowing at a rate of 250,000 lb/h (31.5 kg/s) is to be cooled from 160°F (71°C) to 125°F (51.6°C), for a total heat duty of 4.55 million Btu/h (1.33 MW). How large an air cooler (sometimes called a *dry* heat exchanger) is needed for this service if the design dry-bulb temperature of the air is 95°F (35°C)?

Calculation Procedure:

1. *Determine the temperature rise of the air during passage through the cooler*

From Table 5 estimate the overall heat-transfer coefficient for an air cooler handling kerosene at 55 $Btu/(h \cdot ft^2 \cdot °F)$ [312.3 $W/(m^2 \cdot K)$]. Then the air-temperature rise is $t_2 - t_1 = 0.005U[(T_2 + T_1)/2 - t_1]$, where t_1 = inlet air temperature, °F or °C; t_2 = outlet air temperature, °F or °C; U = overall heat-transfer coefficient, $Btu/(h \cdot ft^2 \cdot °F)$ [$W/(m^2 \cdot K)$]; T_1 = cooled fluid outlet temperature, °F or °C; T_2 = cooled fluid inlet temperature, °F or °C. Substituting yields $t_2 - t_1 = 0.005(55)[(160 + 125)/2 - 95] = 13.06$°F (7.2°C).

Next, from Fig. 23, the correction factor for a process-fluid temperature rise of 160 − 125 = 35°F (19.4°C) is 0.92. So the corrected temperature rise = $f(t_2 - t_1) = 0.92(13.06) = 12.02$°F (6.7°C). Therefore, $t_2 = 95 + 12.02 = 107.02$°F (41.6°C).

TABLE 5 Heat-Transfer Coefficients for Air-Cooled Heat Exchangers*

	Heat-transfer coefficient	
Liquid cooled	$Btu/(h \cdot ft^2 \cdot °F)$	$W/(m^2 \cdot K)$
Diesel oil	45–55	255.5–312.3
Kerosene	55–60	312.3–340.7
Heavy naphtha	60–65	340.7–369.1

**Chemical Engineering.*

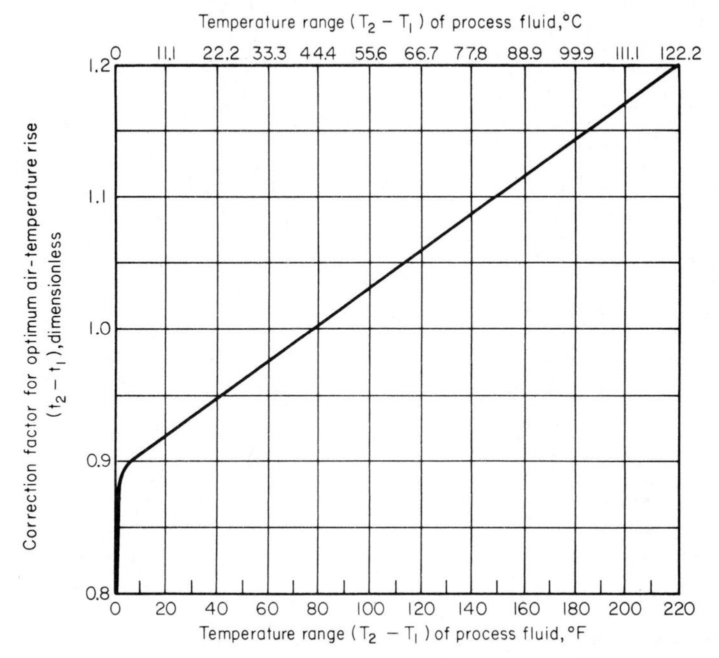

FIG. 23 Correction factors for estimated temperature rise. *(Chemical Engineering.)*

2. *Find the log mean temperature difference (LMTD) for the heat exchanger*

Use the relation LMTD = $(\Delta t_2 - \Delta t_1)$ ln $(\Delta t_2/\Delta t_1)$. Or, LMTD = [(160 − 107.02) − (125 − 95)] ln [(160 − 107.02)/(125 − 95)] = 40.41. This value of the LMTD must be corrected by using Fig. 24 for temperature efficiency P and a correlating factor R. Thus, $P = (t_2 - t_1)/(T_1 - t_1)$ = (107.02 − 95)/(160 − 95) = 0.185. Also, R = (160 − 125)/(107.02 − 95) = 2.91. Then, from Fig. 24, LMTD correction factor = 0.95, and the corrected LMTD = f(LMTD) = 0.95(40.41) = 38.39°F (21.3°C).

3. *Determine the hypothetical bare-tube area needed for the exchanger*

Use the relation $A = Q/U\Delta T$, where A = hypothetical bare-tube area required, ft^2 (m^2); Q = heat transferred, Btu/h (W); ΔT = effective temperature difference across the exchanger = corrected LMTD. Substituting gives A = 4,550,000/55(38.39) = 2154 ft^2 (193.9 m^2).

4. *Choose the cooler size and number of fans*

Enter Table 6 with the required bare-tube area, and choose a 12-ft (3.6-m) wide cooler with either four rows of 40-ft (12-m) long tubes with two fans, for a total bare surface of 2284 ft^2 (205.6 m^2), or five rows of 32-ft (9.6-m) long with two fans for 2288 ft^2 (205.5 m^2) of surface. From Fig. 25, the fan horsepower for the cooler would be 1.56(2284/100) = 35.63 hp (25.6 kW).

Related Calculations: Air coolers are widely used in industrial, commercial, and some residential applications because the fluid cooled is not exposed to the atmosphere, air is almost always available for cooling, and energy is saved because there is no evaporation loss of the fluid being cooled.

Typical uses in these applications include process-fluid cooling, engine jacket-water cooling, air-conditioning condenser-water cooling, vapor cooling, etc. Today there are about seven leading design manufacturers of air coolers in the United States.

The procedure given here depends on three key assumptions: (1) an overall heat-transfer coefficient is assumed, depending on the fluid cooled and its temperature range; (2) the air temperature rise $t_2 - t_1$ is calculated by an empirical formula; (3) bare tubes are assumed and fan horsepower (kW) is estimated on this basis to avoid the peculiarities of one fin type. By using the

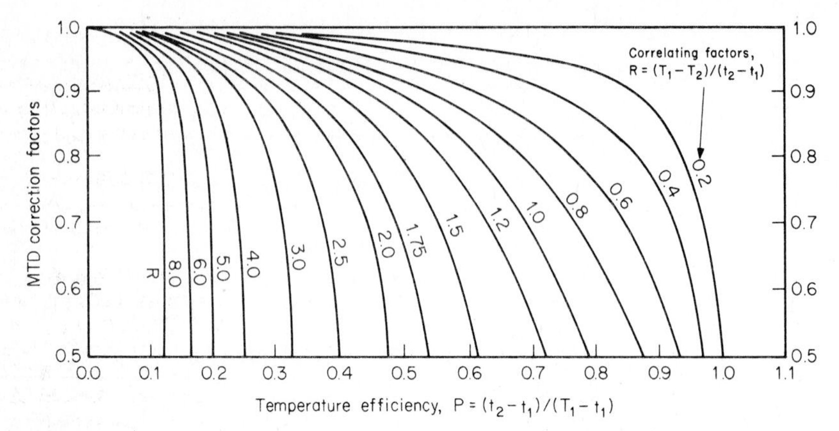

FIG. 24 MTD correction factors for one-pass crossflow with both shell side and tube side unmixed. T represents hot-fluid characteristics, and t represents cold-fluid characteristics. Subscripts 1 and 2 represent inlet and outlet, respectively. *(Chemical Engineering.)*

empirical formula given in step 1 of this procedure, the size air cooler obtained will be within 25 percent of optimum. This is adjusted for greater accuracy through use of the correction factor shown in Fig. 23.

Since no existing computer program is capable of considering all variables in optimizing air coolers, the procedure given here is useful as a first trial in calculating an optimum design. The flow pattern and correction factors used for this estimating procedure are those for one-pass crossflow with both tube fluid and air unmixed as they flow through the exchanger.

Where additional correction factors are needed for different flow patterns across the exchanger, the designer should consult the standards of the Tubular Exchanger Manufacturers Association (TEMA). Similar data will be found in reference books on heat exchange.

The procedure given here is the work of Robert Brown, General Manager, Happy Division, Therma Technology, Inc., as reported in *Chemical Engineering* magazine. Note that the procedure given is for a preliminary selection. The final selection will usually be made in conjunction with advice and guidance from the manufacturer of the air cooler.

FUEL SAVINGS PRODUCED BY DIRECT DIGITAL CONTROL OF THE POWER-GENERATION PROCESS

A 200-MW steam-turbine generating unit supplied steam at 1000°F (538°C) and 2400 lb/in^2 (gage) (16,548 kPa) has an existing variability of 20°F (6.7°C) in the steam-temperature control. It is desired to reduce the variability of the steam temperature and thus allow a closer approach

TABLE 6 Typical Air-Cooled Heat-Exchanger Cooling Area, ft^2 (m^2)*

Approximate cooler width		Tube length		Fans per unit	No. of 1-in (2.5-cm) tube rows in depth on 2⅜-in (6-cm) pitch	
ft	m	ft	m		4	5
12	3.66	32	9.8	2	1827 (169.7)	2288 (212.6)
		36	10.9	2	2056 (191.0)	2574 (239.1)
		40	12.2	2	2284 (212.2)	2861 (265.8)
14	4.27	14	4.3	1	931 (86.5)	1166 (108.3)
		16	4.9	1	1064 (98.8)	1333 (123.8)

**Chemical Engineering*.

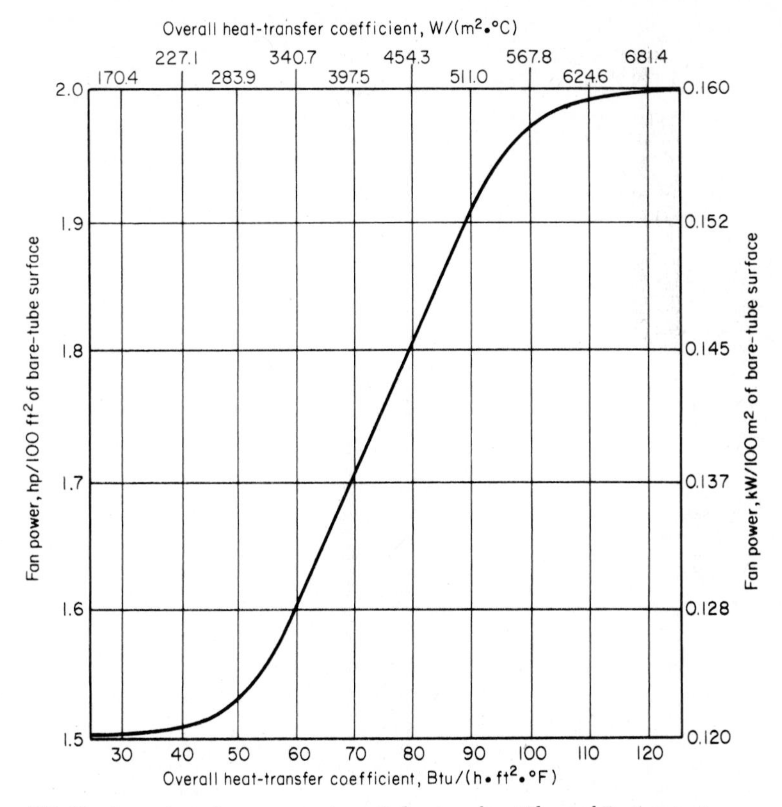

FIG. 25 Approximate fan power requirements for air coolers. *(Chemical Engineering.)*

to the turbine design warrantee limits of 1050°F (566°C). A digital-control system will allow a 30°F (16.7°C) higher operating temperature at the turbine throttle. What will be the effect of this more precise temperature control on the efficiency and fuel cost of this unit? Fuel cost is $2.50 per 10^6 Btu ($2.38 per 10^6 J), and the plant heat rate is 9061 Btu/kWh (9559 kJ/kWh), with a turbine backpressure of 1 inHg (2.5 cmHg).

Calculation Procedure:

1. *Sketch the unit flow diagram; write the overall efficiency equation*

Figure 26 shows the flow diagram for this unit with the input, output, and losses indicated. The overall efficiency e of the system is determined by dividing the power output H_w, Btu/h (W), by the fuel input F, Btu/h (W).

2. *Express the efficiency equation with the system losses shown*

The losses in a typical steam-turbine generating unit are the stack loss L_s, Btu/h (W); mechanical loss in turbine L_m, Btu/h (W); condenser loss L_c, Btu/h (W). The power output can now be expressed as $H_w = F - L_s - L_m - L_c$. Hence, $e = (F - L_s - L_m - L_c)/F$.

3. *Write the loss relations for the unit*

The boiler efficiency and turbine efficiency can each be assumed to be about 90 percent. This is a safe assumption for such an installation. Then $L_s = 0.1F$; $L_m = 0.1H_c$; $H_i = 0.9F$; $H_o = 0.9H_i$, by using the symbols shown in Fig. 26.

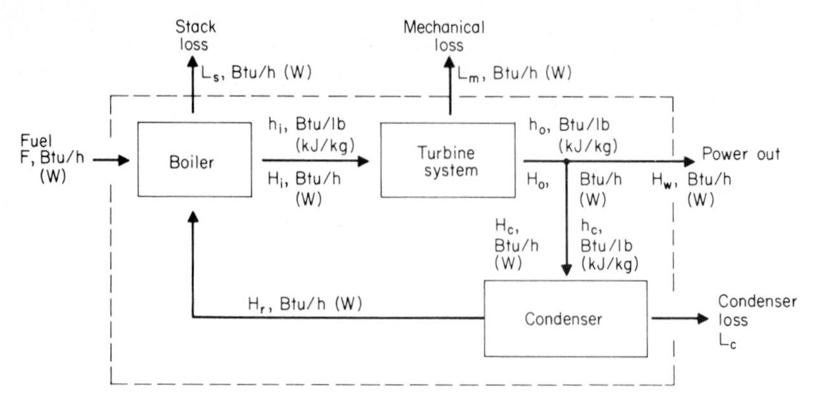

FIG. 26 Flow diagram showing input, output, and losses. *(Combustion.)*

4. *Write the plant efficiency equation at the higher temperature*

With the temperature at the turbine inlet increased by 30°F (16.7°C), the condenser inlet enthalpy h_i' will change to 1480.9 Btu/lb (3444.6 kJ/kg), based on steam-table values. Setting up a ratio between the condenser inlet enthalpy after the throttle-temperature increase and before yields $1480.9/1023 = 1.0135h_i$. This ratio can be used for the other values, if we allow the prime symbol to indicate the values at the higher temperatures. Or, $H_i' = 1.0135H_i$; $L_m' = 1.0135H_i$; $L_c' = 1.0058L_c$. Then $e' = (F' - L_s' - L_m' - L_c')/F'$.

5. *Compute the efficiency and heat-rate improvement*

The improvement in plant efficiency $\Delta e = (e' - e)/e = e'/e - 1$. Substituting values, we find $\Delta e = [0.8209 - (L_c/F)(1.0058)]/\{1.0135[0.81 - (L_c/F)] - 1\}$.

With a heat rate of 9061 Btu/kWh (9559 kJ/kWh), the overall efficiency $e = (3412$ Btu/kWh$/9061)100 = 37.65$ percent. Substituting this value of e in the general efficiency relation given in step 2 above shows that $L_c/F = 0.4335$. Then, substituting this value in the Δe equation above, we find $\Delta e \doteq 0.3849/0.3816 - 1 = 0.0086$, or 0.86 percent.

6. *Convert the efficiency improvement into annual fuel-cost savings*

The annual fuel cost C can be computed from $C = 3.412$(fuel cost, \$ per 10^6 Btu) (hours of operation per year) (plant MW capacity)/e. Or, $C = 3.412\ (2.5)\ (8760)(200)/0.3765 =$ \$39,693,386 per year. Then, with an efficiency improvement of 0.86 percent, the annual fuel-cost saving $S = 0.0086$(\$39,693,386) = \$341,362. In 10 years, with no increase in fuel costs (a highly unlikely condition), the fuel-cost savings with more precise steam-temperature control would be nearly \$3.5 million.

Related Calculations: Although this procedure is based on the use of digital-control systems, it is equally applicable for all forms of advanced control systems which can improve operator effectiveness and thereby save money in operating costs. Table 7 shows the basic reasons for using automatic control in both central-station and industrial power plants. With the ever-increasing energy costs forecast for the future, automatic control of power equipment will become of greater importance.

Table 8 shows the functions of direct digital controls (DDCs) for a variety of power-plant types. Performance improvements of up to 5:1 or more have been reported with DDC. With the expected life of today's plants at 40 years, the annual savings produced by DDC can have a significant impact on life-cycle costs. Table 9 shows the evolution of boiler controls—six phases of evolution. Central-station and large industrial plants today are in a distributed digital "revolution." Changes will occur in measurement, control, information, systems, and actuators. Clear benefits will be measured in terms of installed cost, ease of startup, reliability, control performance, and flexibility. This procedure is based on the work of M. A. Keys, Vice-President, Engineering, M. P. Lukas, Manager, Application Engineering, Bailey Controls Company, as reported in *Combustion* magazine.

TABLE 7 Basic Reasons for Using Automatic Control°

1. Increase in quantity or number of products (generation for fixed investment)
2. Improved product quality
3. Improved product uniformity (steam-temperature variability)
4. Savings in energy (improved efficiency or heat rate)
5. Raw-material savings (fixed savings)
6. Savings in plant equipment (more capacity from fixed investment)
7. Decrease in human drudgery (increased operator effectiveness)

° *Combusion* magazine.

TABLE 8 Functions of Direct Digital Controls°

1. Feedwater
2. Air flow and furnace draft
3. Fuel flow
4. RH and SH steam temperature
5. Primary air pressure and temperature
6. Minor loop control
   a. Cold-end metal temperature
   b. Cold-end metal temperature
   c. Turbine lube oil temperature
   d. Generator stator coolant to secondary coolant pressure differential
   c. Hydrogen temperature controls

° *Combustion* magazine.

TABLE 9 Evolution of Boiler Controls°

1. 1905–1920 Hand control with regulator assistance
2. 1920–1940 Analog boiler control systems acceptance
3. 1940–1950 Pneumatic direct connected analog systems
4. 1950–1960 Pneumatic transmitted analog systems
5. 1960–1970 Discrete component solid-state electric analog systems, burner control, and digital computers
6. 1970–1980 Integrated-circuit digital and analog systems

° *Combustion* magazine.

SMALL HYDRO POWER CONSIDERATIONS AND ANALYSIS

A city is considering a small hydro power installation to save fossil fuel. To obtain the savings, the following steps will be taken: refurbish an existing dam, install new turbines, operate the generating plant. Outline the considerations a designer must weigh before undertaking the actual construction of such a plant.

Calculation Procedure:

1. *Analyze the available head*

Most small hydro power sites today will have a head of less than 50 ft (15.2 m) between the high-water level and tail-water level, Fig. 27. The power-generating capacity will usually be 25 MW or less.

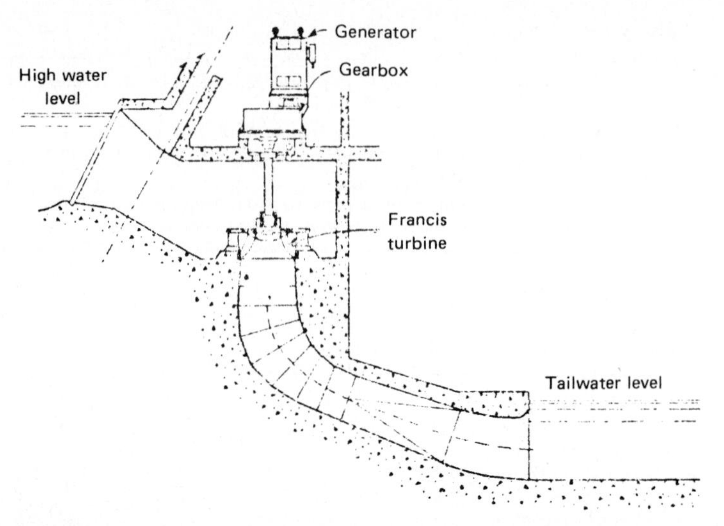

FIG. 27 Vertical Francis turbine in open pit was adapted to 8-m head in an existing Norwegian dam. *(Power.)*

2. *Relate absolute head to water flow rate*

Because heads across the turbine in small hydro installations are often low in magnitude, the tail-water level is important in assessing the possibilities of a given site. At high-water flows, tail-water levels are often high enough to reduce turbine output, Fig. 28*a*. At some sites, the available head at high flow is extremely low, Fig. 28*b*.

The actual power output from a hydro station is $P = HQwe/550$, where P = horsepower output; H = head across turbine, ft; Q = water flow rate, ft^3/s; w = weight of water, lb/ft^3; e = turbine efficiency. Substituting in this equation for the plant shown in Fig. 28*b*, for flow rates of 500 and 1500 m^3/s, we see that a tripling of the water flow rate increases the power output by only 38.7 percent, while the absolute head drops 53.8 percent (from 3.9 to 1.8 m). This is why the tail-water level is so important in small hydro installations.

Figure 28*c* shows how station costs can rise as head decreases. These costs were estimated by the Department of Energy (DOE) for a number of small hydro power installations. Figure 28*d* shows that station cost is more sensitive to head than to power capacity, according to DOE estimates. And the prohibitive costs for developing a completely new small hydro site mean that nearly all work will be at existing dams. Hence, any water exploitation for power must not encroach seriously on present customs, rights, and usages of the water. This holds for both upstream and downstream conditions.

3. *Outline machinery choice considerations*

Small-turbine manufacturers, heeding the new needs, are producing a good range of semistandard designs that will match any site needs in regard to head, capacity, and excavation restrictions.

The Francis turbine, Fig. 27, is a good example of such designs. A horizontal-shaft Francis turbine may be a better choice for some small projects because of lower civil-engineering costs and compatibility with standard generators.

Efficiency of small turbines is a big factor in station design. The problem of full-load versus part-load efficiency, Fig. 29, must be considered. If several turbines can fit the site needs, then good part-load efficiency is possible by load sharing.

Fitting new machinery to an existing site requires ingenuity. If enough of the old powerhouse is left, the same setup for number and type of turbines might be used. In other installations the powerhouse may be absent, badly deteriorated, or totally unsuitable. Then river-flow studies should be made to determine which of the new semistandard machines will best fit the conditions.

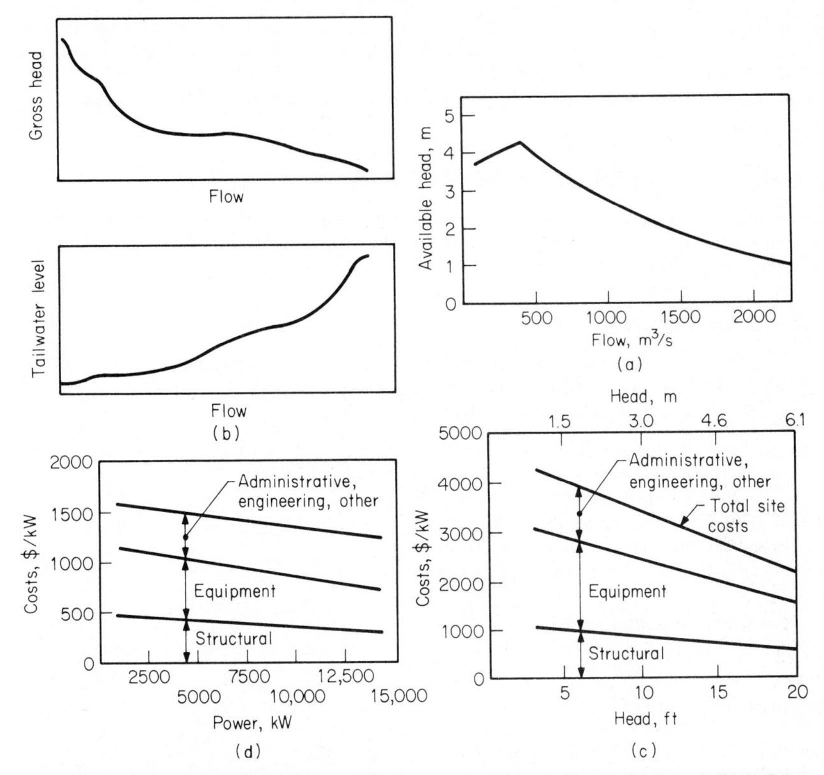

FIG. 28 (*a*) Rising tail-water level in small hydro projects can seriously curtail potential. (*b*) Anderson-Cottonwood dam head dwindles after a peak at low flow. (*c*) Low heads drive DOE estimates up. (*d*) Linear regression curves represent DOE estimates of costs of small sites. *(Power.)*

Personnel costs are extremely important in small hydro projects. Probably very few small hydro projects centered on redevelopment of old sites can carry the burden of workers in constant attendance. Hence, personnel costs should be given close attention.

Tube and bulb turbines, with horizontal or nearly horizontal shafts, are one way to solve the problem of fitting turbines into a site without heavy excavation or civil engineering works. Several standard and semistandard models are available.

In low head work, the turbine is usually low-speed, far below the speed of small generators. A speed-increasing gear box is therefore required. A simple helical-gear unit is satisfactory for vertical-shaft and horizontal-shaft turbines. Where a vertical turbine drives a horizontal generator, a right-angle box makes the turn in the power flow.

Governing and control equipment is not a serious problem for small hydro plants.

Related Calculations: Most small hydro projects are justified on the basis of continuing inflation which will make the savings they produce more valuable as time passes. Although this practice is questioned by some people, the recent history of inflation seems to justify the approach.

As fossil-fuel prices increase, small hydro

FIG. 29 Steep Francis-turbine efficiency falloff frequently makes multiple units advisable. *(Power.)*

installations will become more feasible. However, the considerations mentioned in this procedure should be given full weight before proceeding with the final design of any plant. The data in this procedure were drawn from an ASME meeting on the subject with information from papers, panels, and discussion summarized by William O'Keefe, Senior Editor, *Power* magazine, in an article in that publication.

SIZING FLASH TANKS TO CONSERVE ENERGY

Determine the dimensions required for a commercial flash tank if the flash tank pressure is 5 lb/in^2 (gage) (34.5 kPa) and 14,060 lb/h (1.77 kg/s) of flash steam is available. Would the flash tank be of the centrifugal or top-inlet type?

Calculation Procedure:

Two major types of flash tanks are in use today: top-inlet and centrifugal-inlet tanks, as shown in Fig. 30. Tank and overall height and outside diameter are also shown in Fig. 30.

1. *Determine the rating and type of flash tank required*

Refer to Table 10. Locate the 5-lb/in^2 (gage) (34.5-kPa) flash tank pressure column, and project downward to the minimum value that exceeds 14,060 lb/h (1.77 kg/s). Note that a no. 5 centrifugal flash tank with a maximum rating of 20,000 lb/h (2.5 kg/s) of flash steam is appropriate, and no standard top-inlet type has sufficient capacity at this pressure for this flow rate.

2. *Determine the dimensions of the tank*

In Table 10 locate tank no. 5, and read the dimensions horizontally to the right. Hence, the dimensions required for the tank are 60-in (152.4-cm) OD, 78-in (198.1-cm) tank height, 88-in (223.5-cm) overall height, inlet pipe size of 6 in (15.2 cm), steam outlet pipe of 8 in (20.3 cm), and a water outlet pipe of 6 in (15.2 cm).

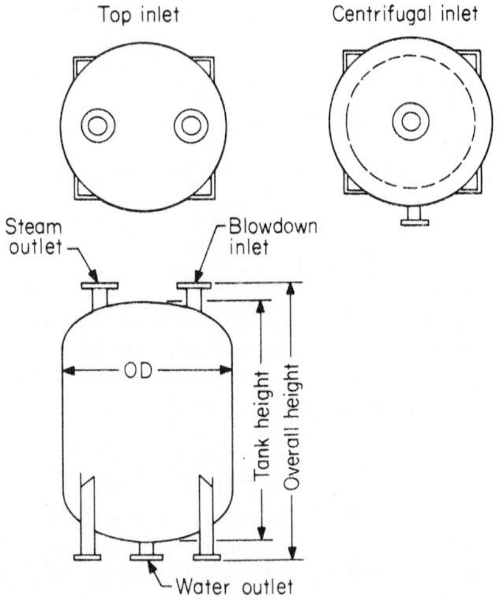

Related Calculations: Use this procedure for choosing a flash tank for a variety of applications—industrial power plants, central stations, marine steam plants, and nuclear stations. Flash tanks can conserve energy by recovering steam that might otherwise be wasted. This steam can be used for space heating, feedwater heating, industrial processes, etc. Condensate remaining after the flashing can be used as boiler feedwater because it is usually pure and contains valuable heat. Or the condensate may be used in an industrial process requiring pure water at an elevated temperature.

FIG. 30 Centrifugal and top-inlet flash-tank dimensions. *(Chemical Engineering.)*

Flashing steam can cause a violent eruption of the liquid from which the steam is formed. Hence, any flash tank must be large enough to act as a separator to remove entrained moisture from the steam. The dimensions given in Table 10 are for flash tanks of proven design. Hence, the values obtained from Table 10 are satisfactory for all normal design activities. The procedure given here is the work of T. R. MacMillan, as reported in *Chemical Engineering.*

DETERMINING WASTE-HEAT BOILER FUEL SAVINGS

An industrial plant has 3000 standard ft^3/min (1.42 m^3/day) of waste gas at 1500°F (816°C) available. How much steam can be generated by this waste gas if the waste-heat boiler has an efficiency of 85 percent, the specific heat of the gas is 0.0178 Btu/(standard ft$^3 \cdot$°F) (1.19 kJ/cm^2), the exit gas temperature is 400°F (204°C), and the enthalpy of vaporization of the steam to be

TABLE 10 Maximum Ratings for Centrifugal and Top-Inlet Flash Tanks, 1000 lb/h (1000 kg/s)[*]

Tank no.	Flash-tank pressure, lb/in^2 (gage) (kPa)					
	1 (6.9)	5 (34.5)	10 (69.0)	20 (138.0)	50 (345.0)	100 (690.0)
	Centrifugal flash tanks					
4	6.0 (0.76)	7.1 (0.89)	8.8 (1.11)	12.0 (1.51)	21.0 (2.64)	34.0 (4.28)
5	16.0 (2.01)	20.0 (2.52)	24.0 (3.02)	32.0 (4.03)	58.0 (7.30)	100.0 (12.59)
6	27.0 (3.40)	34.0 (4.28)	42.0 (5.29)	58.0 (7.30)	105.0 (13.22)	180.0 (22.66)
	Top-inlet flash tanks					
2	1.1 (0.14)	1.3 (0.16)	1.7 (0.21)	2.2 (0.28)	4.0 (0.50)	6.90 (0.87)
3	2.2 (0.28)	2.9 (0.37)	3.5 (0.44)	4.9 (0.62)	8.7 (1.10)	14.80 (1.86)
4	4.3 (9.54)	5.2 (0.65)	6.5 (0.82)	8.7 (1.10)	15.0 (1.89)	25.0 (3.15)

Dimensions of commercial flash tanks

Tank no.	Outside diameter		Tank height		Overall height		Inlet pipe		Outlet pipe			
									Steam		Water	
	in	cm	in	cm	in	cm	in	cm	in	cm	in	cm
	Centrifugal flash tanks											
4	48	121.9	67	170.2	77	195.6	4	10.2	6	15.2	4	10.2
5	60	152.4	78	198.1	88	223.5	6	15.2	8	20.3	6	15.2
6	72	182.9	89	226.1	99	251.5	8	20.3	10	25.4	6	15.2
	Top-inlet flash tanks											
2	24	60.9	56	142.2	65.5	166.4	3	7.6	3	7.6	1.5	3.8
3	36	91.4	62	157.5	71.5	181.6	4	10.2	4	10.2	2	5.1
4	48	121.9	67	170.2	76.5	194.3	6	15.2	6	15.2	4	10.2

[*] *Chemical Engineering* magazine.

generated is 970.3 Btu/lb (2256.9 kJ/kg)? What fuel savings will be obtained if the plant burns no. 6 fuel oil having a heating value of 140,000 Btu/gal (39,200 kJ/L) and a current cost of $1.00 per gallon ($1 per 3.785 L) and a future cost of $1.35 per gallon ($1.35 per 3.785 L)? The waste-heat boiler is expected to operate 24 h/day, 330 days/year. Efficiency of fuel boilers in this plant is 80 percent.

Calculation Procedure:

1. *Compute the steam production rate from the waste heat*

Use the relation $S = C_v V(T - t)60E/h_v$, where S = steam production rate, lb/h; C_v = specific heat of gas, Btu/(standard ft$^3 \cdot$°F); V = volumetric flow rate of waste gas, standard ft^3/min; T = waste-gas temperature at boiler exit, °F; E = waste-heat boiler efficiency, expressed as a decimal; h_v = heat of vaporization of the steam being generated by the waste gas, Btu/lb. Substituting gives $S = 0.0178(3000)(1500 - 400)60(0.85)/970.3 = 3087.7$ lb/h (1403.3 kg/h).

2. *Find the present and future fuel savings potential*

The cost equivalent C dollars per hour of the savings produced by using the waste-heat gas can be found from $C = Sh_v K/E_b$, where the symbols are as given earlier and K = fuel cost, \$ per Btu as fired (\$ per 1.055 kJ), E_b = efficiency of fuel-fired boilers in the plant. Substituting for the current fuel cost of \$1 per gallon, we find $C = 3087.4(970.3)(\$1/140{,}000)/0.8 = \$26.75$. Since the waste-heat boiler will operate 24 h/day, the daily savings will be 24(\$26.75) = \$642. With 330-days/year operation, the annual saving is (330 days)(\$642 per day) = \$211,860. This saving could be used to finance the investment in the waste-heat boiler.

Where the exit gas temperature from the waste-heat boiler will be different from 400°F (204.4°C), adjust the steam output and dollar savings by using the difference in the equation in step 1.

Related Calculations: This procedure can be used for finding the savings possible from recovering heat from a variety of gas streams such as diesel-engine and gas-turbine exhausts, process-gas streams, refinery equipment exhausts, etc. To apply the procedure, several factors must be known or assumed: waste-heat boiler steam pressure, feedwater temperature, final exit gas temperature, heating value of fuel being saved, and operating efficiency of the waste-heat and fuel-fired boilers in the plant. Note that the exit gas temperature must be higher than the saturation temperature of the steam generated in the waste-heat boiler for heat transmission between the waste gas and the water in the boiler to occur.

As a guide, the exit gas temperature should be 100°F (51.1°C) above the steam temperature in the waste-heat boiler. For economic reasons, the temperature difference should be at least 150°F (76.6°C). Otherwise, the amount of heat transfer area required in the waste-heat boiler will make the investment uneconomical.

This procedure is the work of George V. Vosseller, P. E., Toltz, King, Durvall, Anderson and Associates, Inc., as reported in *Chemical Engineering* magazine.

Nuclear Engineering

REFERENCES: Klema and West—*Public Regulation of Site Selection for Nuclear Power Plants*, Johns Hopkins University Press; Hagel—*Alternative Energy Strategies: Constraints and Opportunities*, Holt, Rinehart and Winston; Komanoff—*Power Plant Cost Escalation*, Van Nostrand Reinhold; Seeley—*Elements of Thermal Technology*, Marcel Dekker; Hunt—*Handbook of Energy Technology*, Van Nostrand Reinhold; Munn—*Environmental Impact Assessment*, Wiley; Canapathy—*Applied Heat Transfer: A Complete Handbook for Power and Process Engineers*, PennWell Books; El-Wakil—*Nuclear Power Engineering*, McGraw-Hill; Sachs—*Nuclear Theory*, Addison-Wesley; Hoegerton and Grass—*Reactor Handbook*, U.S. Atomic Energy Commission; Schwenk and Shannon—*Nuclear Power Engineering*, McGraw-Hill; Murphy—*Elements of Nuclear Engineering*, Wiley; Rockwell—*Reactor Shielding Design Manual*, Van Nostrand; Price—*Radiation Shielding*, Pergamon; Hollaender—*Radiation Biology*, McGraw-Hill; Glasstone and Edlund—*The Elements of Nuclear Reactor Theory*, Van Nostrand; Murray—*Nuclear Reactor Physics*, Prentice-Hall; Etherington—*Nuclear Engineering Handbook*, McGraw-Hill; Glasstone—*Principles of Nuclear Reactor Engineering*, Van

Nostrand; Bonilla—*Nuclear Engineering*, McGraw-Hill; Glasstone and Lovberg—*Controlled Thermonuclear Reactions*, Van Nostrand; Schultz—*Control of Nuclear Reactors and Power Plants*, McGraw-Hill; International Atomic Energy Agency—*Directory of Nuclear Reactors;* Henley—*Advances in Nuclear Science and Technology*, Academicol'denblat—*Calculation of Thermal Stresses in Nuclear Reactors*, Consultants Bureau; Greenspan—*Computing Methods in Reactor Physics*, Gordon and Breach; International Atomic Energy Agency—*Programming and Utilization of Research Reactors;* Marchuk—*Theory and Methods of Nuclear Reactor Calculations*, Consultants Bureau.

NUCLEAR POWER REACTOR SELECTION

Select a nuclear power reactor to generate 60,000 kW at a thermal efficiency of 35 percent or more. If the selected unit is a 10-ft (3.0-m) diameter reactor that uses a fluidized bed containing 20×10^6 fuel pellets each 0.375 in (9.5 mm) in diameter with a density of 700 lbm/ft^3 (11,213 kg/m^3) and the reactor fluid is pressurized water at 600°F (315.6°C), determine the bed pressure drop when fluidized. Also, compute the reactor fuel volume, the collapsed fuel bed height, and the density of the pressurized water.

Calculation Procedure:

1. *Select the type of reactor to use*

Table 1 summarizes the operating characteristics of six types of power reactors. Study shows that a pressurized-water reactor will provide the desired thermal efficiency. Further, this type of reactor is successfully used for large-scale power generation. Hence, a pressurized-water reactor will be the first tentative choice for this plant.

2. *Compute the reactor fuel volume*

Use the relation $v_f = nv_p$, where v_f = fuel volume, ft^3; n = number of fuel pellets in the reactor; v_p = volume of each pellet, ft^3. Substituting yields $v_f = 20 \times 10^6\pi(0.375)^3/[6(1728)] = 320$ ft^3 (9.1 m^3).

3. *Compute the fuel volume in the collapsed form*

With the fuel bed not fluidized, the porosity P with packed spheres is about 0.40. Then collapsed volume $v_c = v_f/(1 - P) = 320/0.60 = 534$ ft^3 (15.1 m^3).

4. *Compute the collapsed fuel-bed height*

Use the relation $h = v_c/A_r$, where h = collapsed height of fuel bed, ft; A_r = reactor fuel bed area, ft^2. So $h = 534/(\pi 10^2/4) = 6.78$ ft (2.1 m).

5. *Determine the density of the pressurized water*

Using the steam tables shows $d_w = 42.45$ lb/ft^3 (680.0 kg/m^3) at 600°F (315.6°C) for saturated liquid.

6. *Compute the pressure loss through the fluidized bed*

Use the relation $p = 2.9h[(1 - P)d_f + Pd_w]$, where p = pressure loss through fluidized fuel bed, lb/ft^2; d_f = fuel density, lbm/ft^3; other symbols as before. Substituting, we find $p = 2.9\,[(1 - 0.4)700 + 0.4 \times 42.45] = 1268$ lb/ft^2 or 8.79 lb/in^2 (60.6 kPa).

Related Calculations: This general procedure is valid for preliminary selection of the type of nuclear reactor to use for a given power application. Since reactors are expensive, a complete economic analysis must be made of the alternatives available before the final choice is made.

NUCLEAR POWER-PLANT CYCLE ANALYSIS

A nuclear power plant using two coolants, Na and NaK, is arranged as shown in Fig. 1. Sodium, the first coolant, enters the reactor at 600°F (315.6°C) and leaves at 1000°F (537.8°C); NaK, the second coolant, enters the intermediate heat exchanger at 550°F (287.8°C) and leaves at 950°F (510.0°C). Neglecting heat and pressure losses in the piping, plot the enthalpy-temperature diagram for the plant if steam leaves the boiler at 1200 lb/in^2 (8273 kPa). What are the Na and NaK flow rates with the cycle arrangement shown in Fig. 1, a reactor capacity of 400,000 kW of heat energy, and a 155,000-kW turbine output? Determine the plant thermal efficiency if the auxiliary-power needs = 12,000 kW.

TABLE 1 Nuclear-Power Reactor Characteristics

Reactor type	Typical thermal efficiency, %	Typical power density, thermal, kW/ft^3 (MW/m^3)	Typical reactor pressure, lb/in^2 (gage) (kPa)	Average heat flux, Btu/(h·ft^2) (MW/m^2)	Typical fuel enrichment, %	Reactor coolant
Pressurized-water	36	1,600 (56.5)	1,500 (10,341)	300,000 (945.6)	1.5–3.0	Light water
Boiling-water	22–30	800 (28.3)	1,000 (6,894)	100,000 (315.2)	1.5	Light water
Gas-cooled	30	200 (7.1)	600–1,000 (4,136–6,894)	. . . . . .	0.70–2.5	Carbon dioxide
Liquid-metal	33	300 (10.6)	100 (689.4)	. . . . . .	. . .	Sodium, bismuth, lead, etc.
Fast-breeder	32	20,000 (706.5)	100 (689.4)	650,000 (2,049)	. . .	Sodium
Fluid-fueled	30	400 (14.1)	1,000–2,000 (6,894–13,788)	Varies (varies)	Varies	Reactor fuel solution

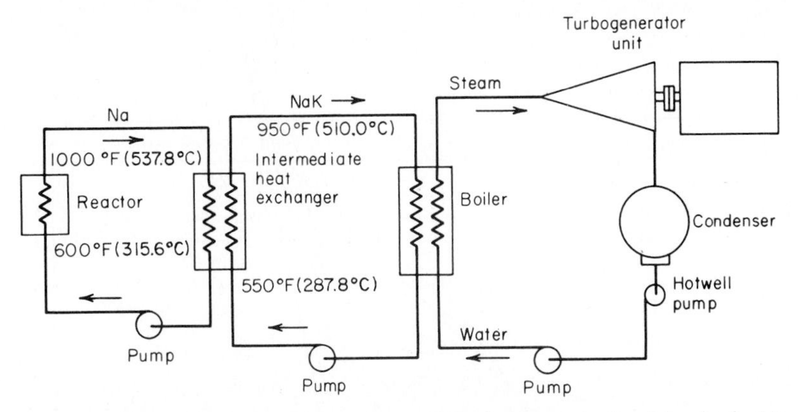

FIG. 1 Reactor plant with two-coolant system uses Na in the reactor circuit and transfers heat to the intermediate NaK circuit, which acts as a buffer against making the steam circuit radioactive.

Calculation Procedure:

1. *Determine the steam outlet and saturation temperature*

Figure 1 shows that NaK enters the boiler at 950°F (510.0°C). Draw a horizontal line on the enthalpy-temperature *(h-t)* diagram (Fig. 2), indicating the 950°F (510.0°C) NaK temperature entering the boiler. Also draw a horizontal line on the *h-t* diagram, Fig. 2, at 1000°F (537.8°C), indicating the Na temperature leaving the reactor.

The steam outlet temperature from the boiler will be less than 950°F (510.0°C) because transfer of heat between the NaK and the water and steam in the boiler provides the energy required to convert the water to steam. A temperature difference between the NaK and the steam is needed to produce the desired heat transfer.

Assume a 50°F (27.8°C) temperature difference between the boiler outlet steam and the NaK, which is a typical temperature difference for this type of cycle. With such a temperature difference the outlet steam temperature = 950 − 50 = 900°F (482.2°C). From the steam tables find the saturation temperature of steam at 1200 lb/in^2 (abs) (8273 kPa) as 567.2°F (297.3°C). Hence the steam will be superheated when it leaves the boiler.

2. *Compute the boiler evaporator coolant outlet temperature*

Incoming feedwater enters the boiler evaporator section where it is heated by the NaK before entering the boiler steam section. To provide heat transfer between the NaK leaving the evaporator section of the boiler and the incoming boiler feedwater, a temperature difference between the two fluids is necessary. Assume that the NaK coolant leaves the boiler evaporator section at a temperature 40°F (22.2°C) higher than the incoming feedwater. With the incoming feedwater at the saturation temperature, or 567.2°F (297.3°C), the NaK coolant outlet temperature from the boiler evaporator = 567.2 + 40 = 607.2, say 607°F (319.4°C).

3. *Plot the boiler coolant temperature path*

Locate the boiler outlet steam state on the *h-t* diagram, Fig. 2, on the 1200-lb/in^2 (abs) (8273-kPa) pressure curve and the 900°F (482.2°C) temperature horizontal. From this point, project vertically upward to the 950°F (510°C) NaK temperature horizontal to locate point 1, the temperature of the NaK entering the boiler, Fig. 2.

Next, locate the point 1*a* where the liquid enthalpy line of the *h-t* diagram, Fig. 2, intersects the 1200-lb/in^2 (abs) (8273-kPa) evaporation enthalpy line. From point 1*a*, project vertically upward to 607°F (319.4°C), point 2, the temperature of the NaK coolant leaving the boiler evaporator section.

Points 1 and 2 are the NaK *temperature path* in the boiler evaporator and steam-generating sections. Assuming that the NaK has a constant specific heat while flowing through the boiler evaporator and steam-generating sections (a completely valid assumption), draw a straight line

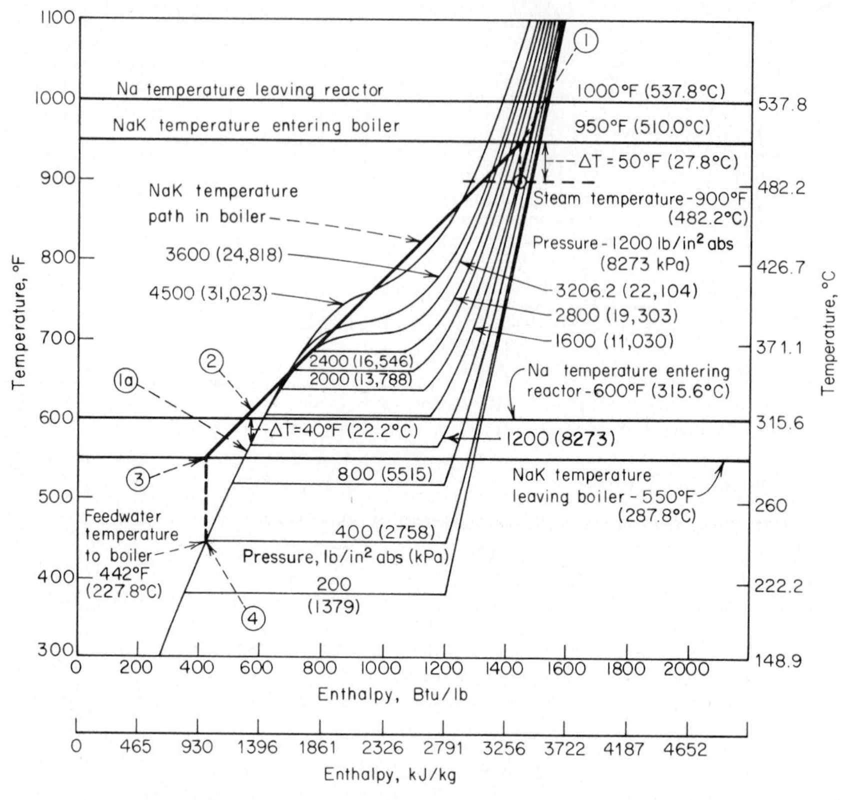

FIG. 2 Steam-water enthalpy-temperature diagram shows the relation between NaK circuit and steam circuit. Keeping the steam temperature high raises the thermal efficiency of the plant.

between points 1 and 2 and extend it to intersect the 550°F (287.8°C) temperature line at point 3. Note that point 3 represents the temperature of the NaK entering the intermediate heat exchanger.

4. *Determine the boiler feedwater inlet temperature*

Feedwater enters the boiler at a yet unknown temperature. During passage between the boiler inlet and the evaporator section inlet, the feedwater absorbs heat from the NaK coolant, leaving the evaporator at 607°F (319.4°C).

Draw a line vertically downward from point 3 until the liquid enthalpy curve is intersected, point 4. Point 4 represents the boiler feedwater inlet temperature, or 442°F (227.8°C), based on the valid assumption that the feedwater leaving the condenser hot well is in the saturated state.

5. *Compute the reactor coolant flow rate*

Sodium enters the reactor at 600°F (315.6°C) and leaves at 1000°F (537.8°C), Fig. 1. Thus, the temperature rise of the Na during passage through the reactor is 1000 − 600 = 400°F (222.2°C). Also, the average specific heat of Na is 0.306 Btu/(lb·°F) [1.28 kJ/(kg·°C], found from a tabulation of Na properties in an engineering handbook.

Compute the Na flow from $f = 3413\ kw/\Delta tc$, where f = Na flow rate, lb/h; kw = reactor heat rating, kW; Δt = Na temperature rise during passage through the reactor, °F; c = specific heat of the Na coolant, Btu/(lb·°F). Substituting gives us $f = 3413(400{,}000)/[400(0.306)] =$ 11,130,000 lb/h (1402.4 kg/s).

6. *Compute the boiler heating liquid flow rate*

Use the same relation as in step 5, substituting the temperature change and specific heat of NaK. Since the NaK enters the boiler at 950°F (510.0°C) and leaves at 550°F (287.8°C), its temperature change is 950 − 550 = 400°F (222.2°C). Also, the specific heat of NaK is 0.251 Btu/(lb·°F) [1.05 kJ/(kg·°C)], as found from NaK properties tabulated in an engineering handbook. So f = 3413(400,000)/[400(0.251)] = 13,600,000 lb/h (1713.6 kg/s).

7. *Compute the plant thermal efficiency*

The net station output kw = gross output of turbine, kW, minus the total plant auxiliary demand, kW = 155,000 − 12,000 = 143,000 kW. Then overall plant thermal efficiency = net station output, kW/reactor heat output, kW = 143,000/400,000 = 0.357, or 35.7 percent.

Related Calculations: This analysis is valid for a cycle in which the reactor coolant does not do work in the turbine. In general, designers prefer to avoid using the reactor coolant in the turbine. Although the thermodynamic aspects of a nuclear cycle are important, the cost of the plant must also be considered before a final choice of a cycle is made. The method presented is the work of Henry C. Schwenk and Robert H. Shannon, as reported in *Power* magazine.

REACTOR FUEL CONSUMPTION, ATOM BURNUP, AND NEUTRON FLUX

Determine the amount of fissionable material used in a 500-mW reactor having 3×10^{10} fissions per watt-second. The reactor core has a volume of 1360 ft^3 (38.5 m^3) and the fuel (99.3 percent U 238 plus 0.7 percent U 235) occupies 6 percent of the reactor volume. How much fissionable material is consumed if the plant operates 8760 h/year at an 80 percent load factor and the capture cross section/fission cross section ratio = 1.2? What are the maximum allowable atom burnup, the average fuel-cycle time, and the reactor neutron flux?

Calculation Procedure:

1. *Compute the reactor fission rate*

Use the relation $F_r = P_T C$, where F_r = reactor fission rate, fissions/(W·s); P_T = total reactor power, W; C = fissions (W·s). So $F_r = 500 \times 10^6(3 \times 10^{10}) = 1.5 \times 10^{19}$ fissions/s.

2. *Compute the total volume of the fuel*

Since the fuel occupies 6 percent of the reactor volume, the fuel volume $V_f = 0.06 \times 1360 =$ 81.6 ft^3 (2.3 m^3). Since reactor fuel quantities are often expressed in cubic centimeters, convert the fuel volume in cubic feet by multiplying by the conversion factor 2.832×10^4, or $V_{fc} = 2.832 \times 10^4(81.6) = 2.31 \times 10^6$ cm^3.

3. *Compute the U 235 nuclei in the reactor*

First determine the uranium nuclei per cm^3 N_U, using the relation N_U = [(uranium density, g/cm^3)/uranium atomic weight] (Avogadro's constant) = $(18.68/238.07)(6.023 \times 10^{23}) = 0.0472 \times 10^{24}$ nuclei/cm^3. In this relation the following constants are used: uranium density = 18.68 g/cm^3; uranium atomic weight = 238.07; Avogadro's constant = $N_m = 6.023 \times 10^{23}$ atoms/(g·atom).

With the uranium nuclei per cm^3 known, compute the U 235 nuclei in the reactor from $N_{U\,235} = 0.007 N_U V_{fc} = 0.007(0.0472 \times 10^{24})(2.31 \times 10^6) = 7.64 \times 10^{26}$ U 235 nuclei in the reactor.

4. *Compute the U 235 fissionable material consumed*

Use the relation $F_{U\,235} = F_r G_m / N_m$, where $F_{U\,235}$ = fissionable U 235 material consumed or burned up for power only, g/s; G_m = g/mol of the fissionable material; other symbols as before. Substituting gives $F_{U\,235} = (1.5 \times 10^{19})(235)/6.023 = 5.85 \times 10^{-3}$ g/s.

5. *Compute the annual consumption of fissionable material*

Use the relation $A_c = F_{U\,235}\,YL/1000$, where A_c = annual consumption of fissionable material, kg; Y = s/year; L = load factor; other symbols as before. Substituting reveals $A_c = 5.85 \times 10^{-3}(3600 \times 8760)(0.8)/1000 = 147.4$ kg/year.

6. *Compute the U 235 annual consumption*

The U 235 is consumed by fissioning for power and is also lost by absorption. The proportion of these two forms of consumption is expressed by α = U 235 total capture cross section/U 235 fission cross section. With $\alpha = 1.2$ for a typical reactor, the total annual U 235 consumption = 1.2(147.4) = 177 kg/year.

7. *Compute the maximum allowable atom burnup*

Both U 235 and U 238 are regarded as reactor fuel. The allowable percentage of burnup depends on the total integrated radiation dosage and radiation energy level, and the effect on fuel material dimensional stability, thermal conductivity, and reduction in effective multiplication factor. Assuming a maximum allowable burnup of 20 percent, which is a typical value, compute B_{ma} = (percentage of burnup)(fuel atoms per cm^3)(total cm^3 of fuel), where B_{ma} = maximum allowable atom burnup, atoms. Substituting gives us $B_{ma} = (0.002)(0.0472 \times 10^{24})(2.31 \times 10^6) = 2.18 \times 10^{26}$ atoms.

8. *Compute the average fuel-cycle time*

Use the relation $A_f = B_{ma}/F_r$, where A_f = average fuel-cycle time, s. Thus $A_f = 2.18 \times 10^{26}/(1.5 \times 10^{19}) = 1.45 \times 10^7$ s = 4040 h = 30 weeks, approximately.

9. *Compute the reactor neutron flux*

Use the reaction $N_f = P_T C/\sum f V_f$, where N_f = reactor neutron flux; $\sum f = N_{U\ 235} \times \sigma_{f\ 235}$, where $\sigma_{f\ 235}$ = total microscopic absorption cross section for U 235; other symbols as before. So $N_f = 500 \times 10^6 (3 \times 10^{10})/(0.00033 \times 10^{24})(549 \times 10^{-24})(2.31 \times 10^6) = 3.57 \times 10^{13}$. Note that values of $\sigma_{f\ 235}$ are obtained from nuclear data sources.

Related Calculations: Use this general method for any reactor designed to generate power. The method presented is the work of Henry C. Schwenk and Robert H. Shannon, as reported in *Power* magazine.

VALUE OF FISSIONABLE MATERIAL FOR POWER GENERATION

How many tons of coal are required to produce the heat equivalent of 1 lb (0.45 kg) of fissionable U 235? If heat is worth 40 cents per million Btu (37.9 cents per 10^6 kJ), what is 1 g of fissionable U 235 worth? One ton of coal contains 24×10^6 Btu (25.3×10^6 kJ).

Calculation Procedure:

1. *Compute the heat produced by 1 lb (0.45 kg) of fissionable material*

When all the nuclei in the atoms of 1 lb (0.45 kg) of fissionable U 235 fission, about 0.001 lb (0.45 g) of material converts to heat energy. Since by Einstein's mass-energy equation, 1 lbm = 11.3×10^9 kWh of energy, 1 lb (0.45 kg) of fissioning U 235 produces $0.001(3413)(11.3 \times 10^9) = 39.5 \times 10^9$ Btu/lb (91.9×10^9 kJ/kg). In this relation, the constant 3413 (3600.9) converts kW to Btu (kJ).

2. *Compute the heat equivalent of the fissionable material*

Use the relation equivalent tons of coal per pound of U 235 = heat released per pound of U 235, Btu/heat released by 1 ton of coal, Btu = $39.5 \times 10^9/(24 \times 10^6)$ = 1645 tons of coal per pound of U 235 (3290.0 t of coal per 1 kg U 235). Thus, it takes 1645 tons of coal to equal the potential heat produced by 1 lb (3290 t/kg) of U 235 in a nuclear reactor.

3. *Compute the monetary worth of the nuclear material*

Since heat is worth 40 cents per million Btu in this plant, the value of 1 lb (0.45 kg) of U 235 is $(39.5 \times 10^9)(0.4/10^6)$ = \$15,800, or about \$34.80 per gram of U 235.

Related Calculations: Use this general procedure for other fissionable materials used for fuels in nuclear plants. The method presented is the work of Henry C. Schwenk and Robert H. Shannon, as reported in *Power* magazine.

EFFECT OF NUCLEAR RADIATION ON HUMAN BEINGS

What is the total radiation dose in rems for a worker exposed to 0.3 rad of 1.0 MeV beta particles and 0.05 rad of 1.0 MeV neutrons each day? Is the total dose dangerous to this worker? Use National Bureau of Standards data (Tables 2 to 4) in the analysis.

Calculation Procedure:

1. *Compute the total radiation dose*

Use the relation, total dose, rem = $\sum$(dose, rad)(RBE), where rem = roentgen equivalent per man; rad = radiation absorbed dose; RBE = relative biological effectiveness. Table 2 lists the RBE values for various types of radiation. By substituting the appropriate values from Table 2, total dose = (0.3)(1.0) + 0.05(10.5) = 0.825 rem.

2. *Determine whether the dose is dangerous*

Table 3 lists the exposure tolerance of the human body. This listing shows that a dose of 1 rem/day is believed to cause debilitation within 3 to 6 months and death within 3 to 6 years. Since the daily dose to which this worker is exposed—0.825 rem—is close to the 1.0-rem danger level, the dose is excessive and dangerous.

Table 4 lists the recommended weekly maximum dosage for various types of radiation on different parts of the body. Study of this list also indicates that the radiation to which this worker is exposed is dangerous.

Related Calculations: The effects of radiation can be fatal to all living organisms. Hence, extreme care must be used in computing the dose received by anyone exposed to radiation. Since the allowable dose and the effects of various doses are under constant study, be certain to refer to the latest available data from the Nuclear Regulatory Commission before permitting exposure of any worker to radiation of any kind.

TABLE 2 Conversion: Rad to Rem*†

Radiation effects on humans: Definitions

One r (*r*oentgen) is the quantity of gamma or x-radiation that produces an energy absorption of 83 ergs/g of dry air.

One rep (*r*oentgen *e*quivalent *p*hysical) is the quantity of radiation that produces an energy absorption of 93 ergs/g of aqueous tissue.

One rad (*r*adiation *a*bsorbed *d*ose) is required to deposit 100 ergs/g in any material by any kind of radiation.

One rem (*r*oentgen *e*quivalent *m*an) is the unit of particulate radiation that produces tissue damage in humans.

The conversion factor from rad to rem is the RBE (relative biological effectiveness), i.e., dose in rem = dose in rad × RBE.

Type of radiation	RBE*	Type of radiation	RBE*
X-rays	1	Neutrons, 0.5 MeV	10.2
Gamma rays	1	Neutrons, 1.0 MeV	10.5
Beta particles, 1.0 MeV	1	Neutrons, 10 MeV	6.4
Beta particles, 0.1 MeV	1.08	Protons, 100 MeV	1–2
Neutrons, thermal	2.8	Protons, 1 MeV	8.5
Neutrons, 0.0001 MeV	2.2	Protons, 0.1 MeV	10
Neutrons, 0.005 MeV	2.4	Alpha particles, 5 MeV	15
Neutrons, 0.02 MeV	5	Alpha particles, 1 MeV	20

*Example for total dose: For a given exposure time, a dose of 0.2 rad of γ radiation plus 0.04 rad of thermal neutrons gives a total dose of (0.2 × *I*RBE) + (0.04 × 2.8 RBE) = 0.312 rem.

†Based on most detrimental chronic biological effects for continuous low-dose exposures.

TABLE 3 Exposure Tolerance Values for Humans*

0.001 rem/day	Natural background radiation
0.01 rem/day	Permissible dose range, 1957
0.1 rem/day	Permissible dose range, 1930 to 1950
1 rem/day	Debilitation 3 to 6 months; death 3 to 6 years (projected from animal data)
10 rem/day	Debilitation 3 to 6 weeks; death 3 to 6 years (projected from animal data)
100 rem—1 day 150 rem—1 week 300 rem—1 month	Survivable emergency exposure dose but permitting no further exposure for life
25 rem	Single emergency exposure
100 rem	Twenty-year career allowance
500 rem	Maximum permissible 20-year-career allowance

*Whole-body radiation doses.

ANALYSIS OF NUCLEAR POWER AND DESALTING PLANTS

Analyze the feasibility of building and operating nuclear-powered combined electric generating and water-desalting plants. Sketch the different types of cycles that might be used. Determine the cycle to use for a water production of 100×10^6 gal/day (4.4×10^3 L/s), electric power net output of 500 mW, and a desalting heat performance of 100.

Calculation Procedure:

1. *Draw the cycle diagrams*

Three cycles will be considered: the back-pressure, extraction, and multishaft cycles.

Figure 3 shows the back-pressure cycle in which the entire exhaust steam flow from the turbine is used to heat brine in the water-desalting system. For a given amount of water produced, this cycle generates large quantities of electric power.

In the extraction cycle (Fig. 4), the steam for brine heating is removed from the turbines at some midpoint during expansion. The exhaust steam goes to a standard condenser. This cycle can have a high product ratio (PR), that is, the ratio of the electric power to desalted water. If desired, large amounts of water can be produced when needed.

The multishaft cycle (Fig. 5) is fundamentally the same as Fig. 3, but it uses parallel condensing and noncondensing turbines. The electric output can vary over a wide range without

TABLE 4 Maximum Weekly Dosage*

Radiation	Skin: Total body	Skin: Appendages	Lens of eye	Gonads	Blood-forming organs	Intermediate tissue (0.07–5.0 cm depth)
X-rays or γ-rays < 3 MeV	0.45	1.5	0.45	0.3	0.4	0.4–0.45
Electrons or β	0.6	1.5	0.3	0.3	0.3	0.3–0.6
Protons	0.6	1.5	0.3	0.3	0.3	0.3–0.6
Fast neutrons	0.3–0.6	0.75–1.5	0.3	0.3	0.3	0.3–0.6
Thermal neutrons	0.5	1.2	0.3	0.1	0.17	0.17–0.5
Alpha particles	1.5	1.5	0.3	0.3	0.3	0.3–1.5
Heavy nuclei (O, N, C, locally generated)	1.5	1.5	0.3	0.3	. . .	0.3–1.6

*Rems per week.

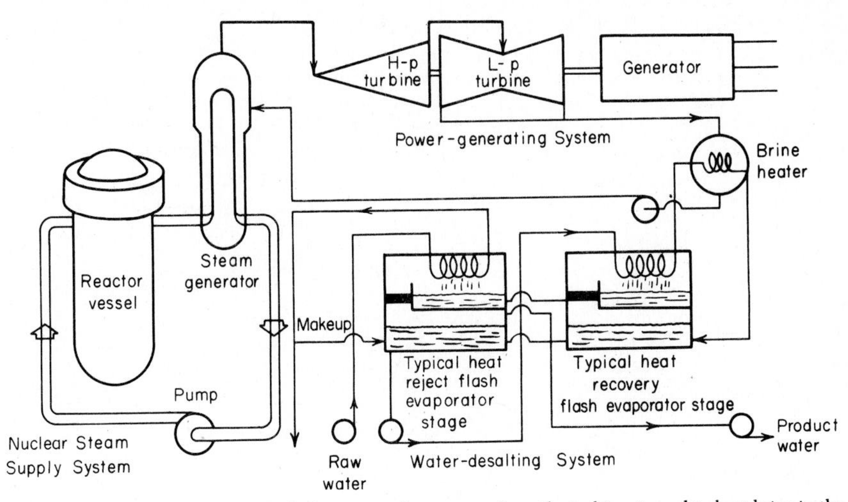

FIG. 3 Back-pressure cycle in which the entire exhaust steam from the turbines is used to heat brine in the water-desalting system.

changing the water-desalting production. Although many other cycles are possible, all are variations of the three basic arrangements described above.

2. *Choose the type of cycle and reactor size to use*

Figure 6 allows quick *estimates* of the type of cycle and reactor size. Any of the four plotted quantities can be determined from Fig. 6 when the other three are known.

Enter Fig. 6 at the bottom at the water production rate of 100×10^6 gal/day (4.4×10^3 L/s), and project vertically upward (1) to the desalting heat performance of 100. From the intersection with the appropriate curve, project horizontally to the left-hand scale of Fig. 6. Next project upward (2) parallel to the index scale. Then project vertically downward (3) from the net electric power output, 500 mW, on the top scale. From the intersection between lines 2 and 3, draw line 4 horizontally to the left-hand scale. At the intersection, read the reactor power as 2250 thermal mW.

The *type of cycle* is determined by the location of the point of intersection between lines 2 and 3. If lines 2 and 3 intersect to the *right* of the full back-pressure (FBP) line, the cycle used is

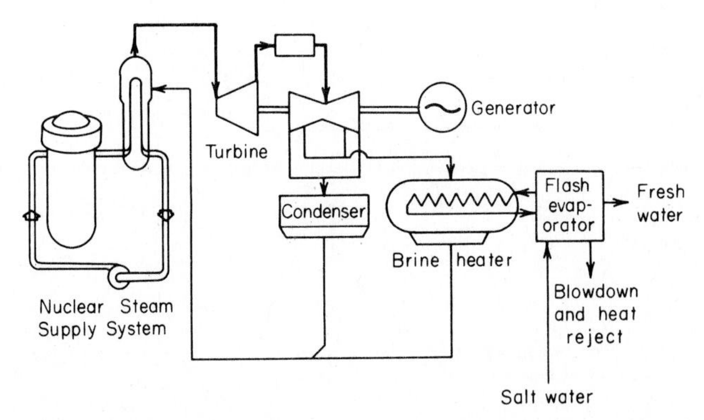

FIG. 4 Extraction cycle in which the steam for brine heating is removed from the turbines at some midpoint during expansion.

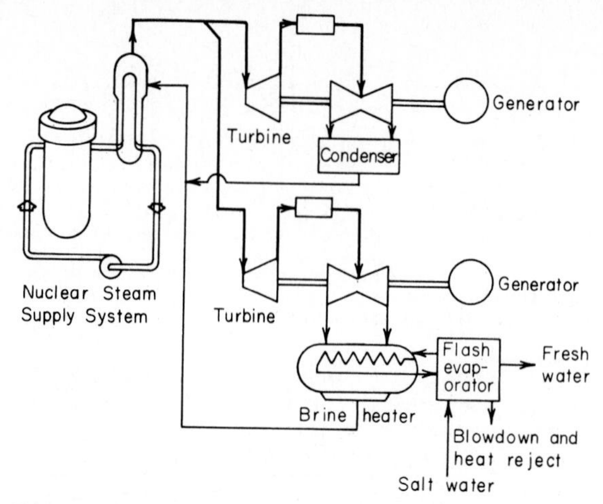

FIG. 5 Multishaft cycle is the same as the back-pressure cycle, but it uses parallel condensing and noncondensing turbines.

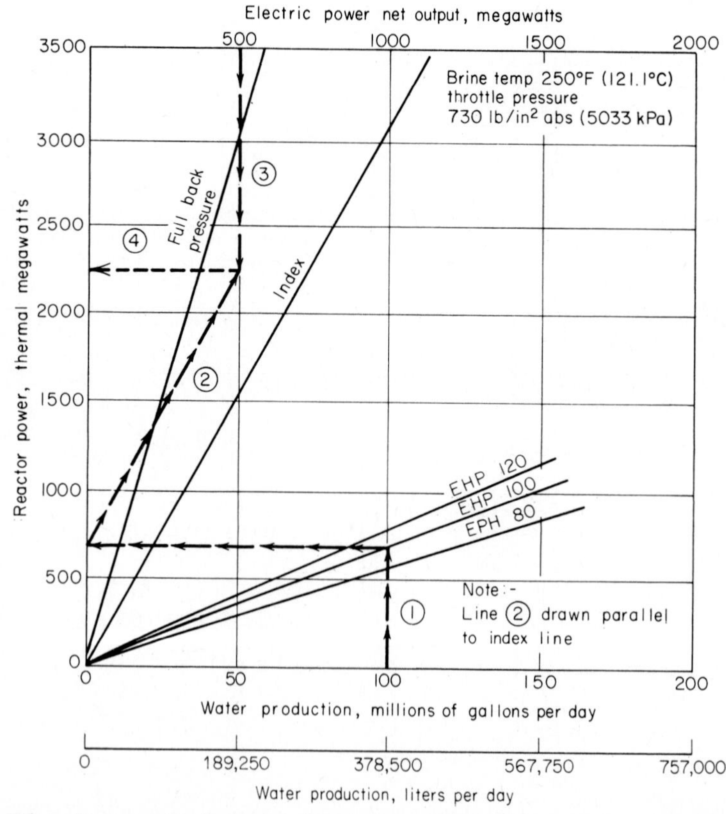

FIG. 6 Nomogram for plant ratings relates the four variables important in desalting when combined with power generation.

the extraction or multishaft type. When the intersection falls directly on the FBP line, a backpressure cycle is indicated. An intersection to the left of the FBP line indicates that some of the steam to the brine heater is bypassed around the turbine regardless of the cycle used. Since the intersection in Fig. 6 occurs to the right of the FBP line, either an extraction or multishaft type of cycle could be used. The final choice of a cycle would depend on the water output required.

Related Calculations: The data presented here were developed by W. H. Comtois, Westinghouse Electric Corp., and were reported in *Mechanical Engineering*. Studies made at Westinghouse show that:

1. The fixed-annual-charge rate exerts the greatest single influence on water cost, increasing the cost by about two-thirds for a factor of 2 increase in the rate. This effect is moderated somewhat for large plant sizes.
2. The plant load factor gives the expected result of decreasing product costs with increasing load factor. The effect is a 1 to 2 percent decrease (increase) for every percentage increase (decrease) in load factor in the range from 75 to 95 percent.
3. Plant design life is of little consequence in the range normally considered (30 to 40 years).
4. The range of maximum brine temperatures studied was 200 to 250°F (93.3 to 121.1°C). Without exception, the computed optimum brine temperature was 250°F (121.1°C).
5. The single-shaft cycles (backpressure or extraction) enjoy a small (5 to 10 percent) water cost advantage over the multishaft cycle.

INDEX

ABOUT THE EDITOR

Tyler G. Hicks, P.E., is a consulting engineer with International Engineering Associates. He has worked in both plant design and operation in a variety of industries, taught at several engineering schools, and lectured both in the United States and abroad on engineering topics. He is a member of ASME and IEEE and holds a bachelor's degree in mechanical engineering from Cooper Union School of Engineering. Mr. Hicks is the author of numerous engineering reference books on equipment and plant design and operation.